Lecture Notes in Networks and Systems

Volume 374

The series "Lecture Notes in Networks and Systems" publishes the latest developments in Networks and Systems—quickly, informally and with high quality. Original research reported in proceedings and post-proceedings represents the core of LNNS.

Volumes published in LNNS embrace all aspects and subfields of, as well as new challenges in, Networks and Systems.

The series contains proceedings and edited volumes in systems and networks, spanning the areas of Cyber-Physical Systems, Autonomous Systems, Sensor Networks, Control Systems, Energy Systems, Automotive Systems, Biological Systems, Vehicular Networking and Connected Vehicles, Aerospace Systems, Automation, Manufacturing, Smart Grids, Nonlinear Systems, Power Systems, Robotics, Social Systems, Economic Systems and other. Of particular value to both the contributors and the readership are the short publication timeframe and the world-wide distribution and exposure which enable both a wide and rapid dissemination of research output.

The series covers the theory, applications, and perspectives on the state of the art and future developments relevant to systems and networks, decision making, control, complex processes and related areas, as embedded in the fields of interdisciplinary and applied sciences, engineering, computer science, physics, economics, social, and life sciences, as well as the paradigms and methodologies behind them.

Indexed by SCOPUS, INSPEC, WTI Frankfurt eG, zbMATH, SCImago.

All books published in the series are submitted for consideration in Web of Science.

For proposals from Asia please contact Aninda Bose (aninda.bose@springer.com).

More information about this series at https://link.springer.com/bookseries/15179

Sotir S. Sotirov · Tania Pencheva ·
Janusz Kacprzyk · Krassimir T. Atanassov ·
Evdokia Sotirova · Galya Staneva
Editors

Contemporary Methods in Bioinformatics and Biomedicine and Their Applications

Springer

Editors
Sotir S. Sotirov
Faculty of Technical Sciences
University Prof. D-r Asen Zlatarov
Burgas, Bulgaria

Janusz Kacprzyk
Systems Research Institute
Polish Academy of Sciences
Warsaw, Poland

Evdokia Sotirova
Intelligent Systems Laboratory
University Prof. D-r Asen Zlatarov
Burgas, Bulgaria

Tania Pencheva
Institute of Biophysics and Biomedical Engineering
Bulgarian Academy of Sciences
Sofia, Bulgaria

Krassimir T. Atanassov
Institute of Biophysics and Biomedical Engineering
Bulgarian Academy of Sciences
Sofia, Bulgaria

Galya Staneva
Institute of Biophysics and Biomedical Engineering
Bulgarian Academy of Sciences
Sofia, Bulgaria

ISSN 2367-3370 ISSN 2367-3389 (electronic)
Lecture Notes in Networks and Systems
ISBN 978-3-030-96637-9 ISBN 978-3-030-96638-6 (eBook)
https://doi.org/10.1007/978-3-030-96638-6

This Springer imprint is published by the registered company Springer Nature Switzerland AG
The registered company address is: Gewerbestrasse 11, 6330 Cham, Switzerland

Preface

The present book collects selected and peer-reviewed papers from the first *International Symposium on Bioinformatics and Biomedicine* (*BioInfoMed'2020*) held in Burgas, Bulgaria, on 8–10 October 2020. More than 100 abstracts were submitted for the Symposium, yet only those which were topically relevant and scientifically sound were considered and accepted.

Nine invited lecturers, highly reputed scientists from different countries around the world—Australia, France, India, Mexico, Northern Macedonia, South Africa, the UK and the USA—took part online, while more than 50 participants, including a large number of young scientists, made 66 oral or poster presentations in presence. The undeniable success of *BioInfoMed'2020* was mainly due to the high-quality investigations in the fields of biomedicine, human medicine, health care, bioinformatics, artificial intelligence and mathematical modelling, which were submitted for presentation by the Bulgarian and the international scientific community.

Some of the presentations at the *BioInfoMed'2020* Symposium have been distinguished with diplomas, and the authors have been especially invited for full-length paper submissions. After a double-blind peer-review process, the current book comprises 46 selected papers as the most representative of the topics and the scientific level of the Symposium.

The book is organized in accordance with the topics of the Symposium, namely

- Decision Making in Bioinformatics, Biomedicine and Healthcare
- Mathematical Modelling in Biomedicine and Healthcare
- Artificial Intelligence in Biomedicine and Healthcare
- Intuitionistic Fuzzy Sets with Applications in Bioinformatics and Biomedicine
- Biomedical Approaches and Applications
- Biomedical Imaging and Image Processing
- Excitable Structures and Motor Activity

At the end of *BioInfoMed'2020,* all members of the programme and organizing committees of the first *International Symposium on Bioinformatics and Biomedicine* expressed a strong desire and determination to make the needed efforts for this successful event to be the starting point of a biennial forum for exchanging novel knowledge in the extremely fast developing topics in the fields of bioinformatics and biomedicine.

The Editors

Sotir Sotirov
Tania Pencheva
Janusz Kacprzyk
Krassimir Atanassov
Evdokia Sotirova
Galya Staneva

Contents

Decision Making in Bioinformatics, Biomedicine and Healthcare

Comparative Analysis of Decision-Making Models in National Healthcare Systems of EU Member-States: Change-Drivers' Identification

Desisslava Atanasova(✉) and Nataliya Venelinova

University of Ruse "Angel Kanchev", 8, Studentska Street, 7017 Ruse, Bulgaria
{datanasova,nvenelinova}@uni-ruse.bg

Abstract. The paper is aiming to outline the typical decision-making models, applicable at the healthcare systems of EU member-state, through comparing key features of clinical and non-clinical decisions, related to the decision-making processes at a national level. An assessment of the identified models in terms of their relevance to the dynamic context of Industry 4.0 and Industry 5.0. was performed, based on a comparative analysis of the predominant decision-making models of six national healthcare systems. The sample includes two old democracies, two countries from the enlargement in 2004 and two countries from the wave in 2007. The central issue is finding out how the applied decision-making models ensure the conformity between the national and the European healthcare priorities, and how these models refer to the ability of rapid and elastic adaptiveness of the national systems to the ethical, information and socio-economic needs of the stakeholders of the healthcare systems. The national models were also compared to a range of existing decision support systems (DSS) with the purpose to identify change-drivers and gaps.

Keywords: Healthcare policy · Decision-making models · Healthcare technology assessment · Comparative analysis

1 Health Care Systems vs. Health Systems: An Introduction to the Importance of the System Care for the Healthcare Decision-Making

Modern models of functioning, and in particular decision-making in national healthcare systems, are much less often subject of scientific research than decisions taken by the different systemic players (e.g. hospitals, health insurance funds, pre-hospital care, general practitioners, emergency centres, etc.). One of the reasons for such consequence is the continuing desire of the specialists in the healthcare sector to transform the nature of the traditional models, based on asymmetric communication between patients and medical staff into a more empathetic process of mutual information and shared responsibility in clinical and non-clinical decisions within this interaction. The expectation for a more efficient and result-oriented diagnostic, treatment, therapy and non-clinical activities is

S. S. Sotirov et al. (Eds.): BioInfoMed 2020, LNNS 374, pp. 3–15, 2022.
https://doi.org/10.1007/978-3-030-96638-6_1

not justified in terms of population quality of life and satisfaction of the care, because the efficiency of the healthcare system is not a simple function of the quality of decision taken by its units, not a cumulative sum of the workable decision applied at the different levels. In the meantime, governments in the EU, which are directly responsible for the implementation of common health policy, regularly report progress on improving one or more sub-systems - from the legislative framework enhancement, through patient inclusion raise in the decision-making, health and social care quality performance, and to the state-of-the-art digital systems introduced in the sector.

This inevitably raises the question of which system refers to improvement: to the healthcare system or the health system? Even fast overview of the scientific papers published in the last decade shows a lack of terminological precision in the usage of three different terms: "health care system", "healthcare system" and "health system". The governmental web-sources nay of some of the member-states does not differ the scope of the terms. Although such a clarification of the nature of the terms may seem superfluous, understanding the difference between their scope is key to the decision-making models that apply to the different functions of these systems. The Merriam-Webster Dictionary [1] online introduces additional confusion, defining health care as: "Efforts made to maintain or restore physical, mental, or emotional well-being especially by trained and licensed professionals—usually hyphenated when used attributively". This definition covers somewhat matters to healthcare but doesn't include the insurance component, nor the entire system. Usually "health care" and "healthcare" are used interchangeably, even though they mean and encompasses different things and respectively refers to different and often contradicting decisions to be taken. To lighten the difference between the usage of "health care" (as two words) and "healthcare" as one word, we introduce the following consideration: the healthcare is the system through which the population receive the health care they need [2, 3]. Based on this clarification a reconsideration of the four-level model of health care system of Reid PP, Comton WD, Grossman JH, et al. [4] is made (Fig. 1(a) and Fig. 1(b)).

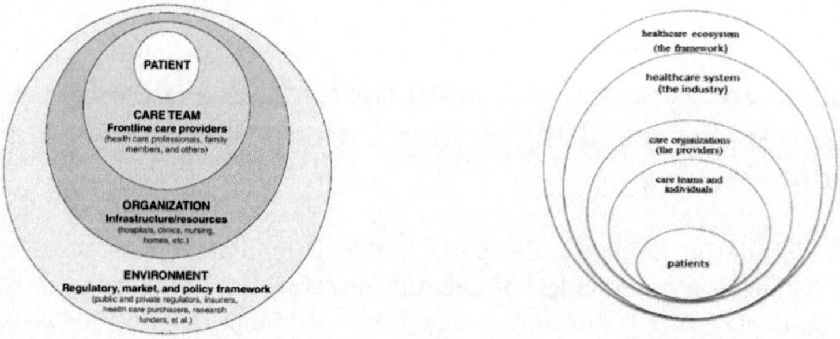

(a) (b)

Fig. 1. (a) Conceptual drawing of a four-level health care system [4]. (b) Reconsidered four-level healthcare system

In the initial conceptual drawing are incorporated the interactions between the patient and the health system (called "care team" [4]), the healthcare system referred to the "organization" and the policy framework ("environment"). Obviously, the model is patient-centered and relatively symmetric which is the ideal case of interactions between these four levels in every direction (from center to the periphery and back as well as bottom-up and up-bottom).

Despite the four-level health care system presupposes the enabling of "ongoing asynchronous communication between patients and care providers. In the future, with the advent of remote monitoring devices and wireless communication systems, information/communications systems have the potential to support continuous monitoring of a patient's health status at home, rapid diagnosis by clinicians, and timely, effective therapeutic interventions in the home by the patient or a family member, with guidance by health professionals." [4], the asynchronicity which refers to the asymmetry of the communication represents much more technical consequence of the decisions taken, split between the levels than the real dissonance in the system because the model stays still the patient's centered and presenting the idea of shared-decision-making. The model is featuring the decision-making between the sub-systems within the healthcare ecosystem but is equity-oriented in the interactions. Such idea almost covers the EU priorities for 2016–2020 for the public health policy [5] as well as the collaborative model of healthcare ecosystem of the future, proposed by IBM Institute of Business values [6] as well as Kostadinova's transforming model [7] but is the ideal case of dialogical-based active interaction of mutual understanding. In real terms the process is much more asymmetric regardless the innovative technologies introduced due to the fact that the ecosystem level is dominantly governed by the states with more or less decentralized decision-making depending on the healthcare system type introduced nationally.

2 Research Methodology

To confirm this consideration, a comparative analysis of the health systems of different socio-economic, political and technological developed countries in the European Union has been undertaken. The research framework centralized two main questions: what kind of criteria for comparing national health systems to be applied and what criteria for comparing the main decision-making model in these systems are common for all countries. This requires the development of a study roadmap incorporating all applied research method (in whole or partially) for the purpose of the comparative analysis (see Fig. 2).

Based on the map the approach was defined as representation of consistently applied research method incorporating comparative analysis with elements of gap analysis and system analysis. After a preliminary on-desk literature review of the open access available resources and the data from the last executed Health System Performance (HSP) of World Health Organization (WHO) [6], the assessment grid was designed, including two types of indicators to be measured by Likert or nominative scaling. The essence of this matrix approach is shown in the table below (Table 1):

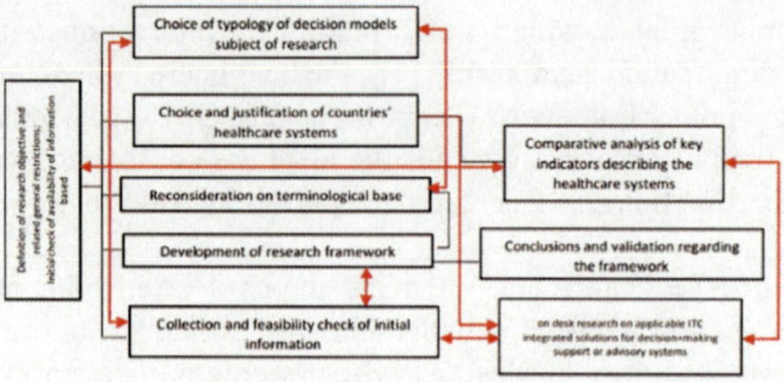

Fig. 2. Research road map

In terms of the presented approach the following limitations must be introduced:

- The approach has been designed for comparing national healthcare systems with aggregation of the assigned scores by using indexing method;
- The assessment is based on Eurostat data for financial, social and technological performance of the healthcare systems which is last updated through national reporting between 2016 and 2018 regarding the metadata [8, 9];
- The initial assessment is qualitative converted to quantitative data through performing of secondary scoring;
- The qualitative assessment is based on information of the relevant governmental websites [10–17, 19] of the responsible ministries and additional information from previously performed assessment of the countries' healthcare systems [20–23, 27–29];
- The approach can not be directly applied to lower governance level of the healthcare system.

3 Key Findings

The decision-making in old democracy healthcare systems: the cases of Germany and France.

The essentials of German healthcare system: the German population is covered through a system known as "gesetzliche krankenversicherung" (GKV), providing a standardized level of coverage for the whole territory of Germany. The general contribution rate in the social healthcare insurance (SHI) is 14.6% of salary, of which the employer pays the half [10, 11]. Each of the operating funds can also charge an additional premium. Private health insurance and a combination of the public and private models are featuring the case of German healthcare system. Persons with private health insurance (the so called PKV) may also be subsidized by their employer or pension funds [8–11] but the PKV is only opportunity available on German territory, limited by the condition that the resident should declare annual gross income at least equal to €63000.

The most common case of interaction within the healthcare system is based on medical-cost decision model [11, 24], shared between the provider and the patient is the most economical choice for health insurance – that means the most often choice of the patient is the government-regulated public health insurance system. In Germany it

Table 1. Assessment grid for comparative analysis of macrosystems of EU countries

Criteria	Measurement			
1. Organization and governance: maximum score assigned to fully decentralized healthcare system (score from 0 to 10)				
1.1. level of centralization	Fully centralized (FC) Score (0 to 1)	Rather centralized (RC) Score (2 to 4)	Rather decentralized (RD) Score (5 to 7)	Fully decentralized (FD) Score (8 to 10)
1.2. standardization of criteria applied to healthcare providers *maximum score (10) assigned to unified standardized system requirements introduced nationwide	Unified standard /requirements (UF) Score (8 to 10)	Rather unified (RU) Score (5 to 7)	Rather variety of fast changing standards/ requirements Score (2 to 4)	Fully confusing standards/ Requirements (FCS) Score (0 to 1)
2. Financing (source of funding, predominantly use for the system *Nominal score of 1 point assigned to all the cases due to lack of basic comparability of the social-economic development	Fully public (state budget) (FPs) Score = 1	Public from the municipality) (PM) Score = 1	Public-private partnerships (PPP) Score = 1	Fully private (FP) Score = 1
3.Introduction of technologies: maximum score assigned to fully digitally transformed healthcare system (score from 0 to 10) including e-health, DSS, integrated data management, remote access to health services				
3.1. DSS software for clinical-decisions spread nationwide	Entirely covering the sector Score (8 to 10)	Partially covering the sector Score (5 to 7)	Locally introduced for certain clinical needs Score (2 to 4)	Not introduced or only for certain clinical process Score (0 to 1)
3.2. Integrated data management	Nationwide introduced for clinical and non-clinical purposes (e-health) Score (8 to 10)	Partially introduced nationwide (for clinical or non-clinical purposes) Score (5 to 7)	Locally introduced only for keeping records on separate clinical or non-clinical decisions of providers Score (2 to 4)	Not introduced or introduced only for certain clinical process Score (2 to 4)
3. Patient involvement and empowerment	Entirely involved (shared decision-making) Score (8 to 10)	Rather involved (patient-oriented evidence-based decision-making) Score (5 to 7)	Rather non-involved (based on informed consent) Score (2 to 4)	System-centered decisions Score (0 to 1)
maximum score per country = 51 points				

Total score	Germany	France	Estonia	Hungary	Romania	Bulgaria
	39	36	49	30	22	21

is administered by more than 100 operating "Krankenkassen" or "Sickness funds" [10, 11], and price of the public services is based on gross income per capita. The operators (insurance funds) reimburse all charged cost for treatment and medicines directly.

The German system is highly decentralized and multilevel but there are three important levels acting nationally with inter-shared responsibility of governing the healthcare system: the federal level, represented by the Federal Assembly (Bundestag), the Federal Council (Bundesrat) and the Federal Ministry of Health (Bundesministerium für Gesundheit). They are the key actors in the healthcare system. The Federal Commissioner for Narcotics (Drogenbeauftragter der Bundesregierung) and the Federal Commissioner for the Concerns of Patients (Beauftragter der Bundesregierung für die Belange der Patientinnen und Patienten) have been assigned to the structure of the Ministry [25];

Private and public providers are free to regulate their relations themselves, within a statutory defined framework, known as "the principle of self-administration". We can consider that the German system is market-based and the state plays a subordinate role. Such decentralized and self-regulating system is fast adaptive to the changes in the EU healthcare ecosystem and the market uncertainty, showing its corporatist nature, which encompasses a hidden preventive mechanism in case of governance crises. In terms of clinical decision-making we also identified the following types of decision models characterizing the German healthcare system:

- The predominant diagnostic decision-making model, concerning the interaction "provider of health care services" – "patient" is patient-oriented but not patient-centred, due to the corporatist nature of the system where leading is result-based performance;
- The prevailing decision-making model on treatment is strongly evidence-based with an element of the transformation of typical patient consent to shared-decision model. It doesn't mandatory means that the established system is changing but the physicians and medical staff's behaviour faced the need to be much more collaborative and supportive in patients" decisions on treatment which is expected to be both relevant to the evidence but also cost affordable in respect with the full reimbursement opportunity for the insured patient; Although the highly decentralized systems offer much more affordable decision-making models, being more sensitive to the patients' needs in terms of cost-effectiveness of the treatment, this financial performance must be further developed in shared responsibility about the treatment quality and the treatment costs. Almost the same can be considered on the decision-model on therapy and medications which are also evidence-based and patient-oriented but not patient-centred.
- Considering that the German system is market-based it expected the non-clinical decision to be also market-oriented moreover when the state is playing a subordinating role in terms of financing and ownership of the assets and main facilities in healthcare. The decisions on the budget and those on the development of the human resources are also decentralized and based on annual performance indicators of the hospitals first (as organizational structure) and second – as individual staff annual performance.

The role of the state increases dramatically when it concerns the education of the specialists needed for the providers to ensure permanently extending the scope of care. The state intervention is also remarkable on the technological acquisition and upgrade and it can be easily proven with the decision of the Bundestag and Bundesrat for extended investment in innovations taken last year (the planned investment for technological improvements of the healthcare system was extended with 200 000 euro per year until 2024 regarding the Digital Healthcare Act, approved 7.11.2019 by the Bundesrat [10]). Such decision renders the control over the healthcare policy-making to the federal level and balances the decision-making responsibilities in a collaborative way where the state is investing in innovative means for the system which is expected to provide a better quality of the services. The above-mentioned Act will change the traditional decision-making model because foresees digitalization of the healthcare system by the development of different Apps. The last will be subject to approval after testing the safety, functionality and data protection. Under the Act, mandatory digital network for the health sector will be established. Patients are to be able to use digital offers such as the electronic patient record, nationwide and as soon as possible (compulsory for pharmacies by the end of 2020) and hospitals (by 1 January 2021) to get connected to the Telematics Infrastructure (TI). The statement of interchangeable controlling role, split between the state and the providers can be proven also the restriction introduced with the above-mentioned Act e.g. "Physicians who still refuse to be connected will be subject to an increased fee deduction of 2.5%, starting 1 March 2020 (up to now this deduction is 1%)" [10]. In the same timethe Act impose the collaborative decision-making model through the introduction of the concept of equal participation and access to services, where the health insurance funds will be required to offer to the insured members opportunities to promote their digital health skills. The skills formation and upgrade fully depend on the effective and collaborative interaction between physicians/medical staff and the patient.

The Essentials of French Healthcare System: The second healthcare system subject of our assessment is the opposite of the German. In order to ensure the comparability but to illustrate different system organization, we chose the French healthcare system. By applying the same criterial set (see Table 1) we achieved similar result: the total final score of Germany is 39 points and those of France -36, despite the fact that the French healthcare system is centralized and the annual contributions of the compulsory coverage provided to all residents by non-competitive statutory health insurance funds. It is regulated and determined by the Parliament and government-funded agencies covering more than 75% of the health expenditures. The Ministry of Social Affairs and Health (Ministere des Solidarites et de la Sante) administrates public healthcare in France, with primary and secondary care services delivered by the various different healthcare providers. France offers a high level of preventative healthcare, with available services including addiction prevention, regular medical check-ups, and the promotion of physical activity and healthy eating [12] and since 2016 the healthcare system for foreigners, known as Protection Universelle Maladie (PUMA), allows access to state healthcare after three months of residence. The hospitals are mostly public owned by state intermediates and the private insurance which is voluntary is complementary and voluntary, representing a type of co-payment for certain diagnostic activities and dental care which are minimally reimbursed by the SHI. Despite the complementary character of the private insurance providers and services covered by their packages, after the last healthcare reform (as of 2016) the insured employees (with SHI) can also benefit from employer, who is sponsoring almost the half of the voluntary health insurance (VHI).

The precise extend to which the costs are covered depends on separate contracts but the logic of VHI is to cover the difference between the SHI reimbursement and the official fixed cost (prices and fees) in the national fixed cost list, approved on annual basis. The healthcare system is centralized at two main levels the national and the regional, but as it can be considered is much more patient-centered than the German system. This is also and the predominant orientation of decision-making processes which are more collaborative and including advisory and negotiating interactions between the key players, but also between the patients and the key players.

In terms of decision-making models, we can consider that the French system is dominantly market-based decision-making process but framed by the cost-sharing between the actors. The distribution of the responsibilities between the leading agencies within the system also proves this consideration:

- The Agency for Information on Hospital Care is responsible for data management systematically collected from all hospitals for hospital planning and financing purposes;
- The National Agency for the Quality Assessment of Health and Social Care Organizations is responsible for prevention and evaluates providers and services;
- The National Health Authority is responsible for the main health technology assessment structure, accrediting health care organizations, and organizing the certification procedures of the physicians [12];

Regarding the clinical decisions the biggest difference between the French and the German system is dealing with the level of patient-centrism, for both systems the decisions on treatment, diagnosis, therapy and medication are evidence-based but in the French system, they are more patient-centered and collaborative. It doesn't mean directly that it is the better system and we can't consider it moreover the purpose of this system examination is not to assess the patients' sense and notion for the system, we can only conclude that the decision models introduced in France are more patients-collaborative. This might be considered both an advantage in social perspective but can also lead to certain weaknesses e.g. how the information from a care visit after-hours by available physicians on a voluntary base, to be transferred to the patient's GP. Partially this problem has been solved with the legalization of the teleconsultations in 2018 which are the logical substituting after-hours care, but the telemedicine is still doesn't have the needed extend and coverage to provoke essential reconsidering of the outpatient care.

The decision-making in new democratic countries: the cases of the healthcare systems of Estonia and Hungary.

The Essentials of Estonian Healthcare System: a remarkable case of patient-centered decision-making in healthcare refers to the highly centralized system of Estonia where we can find strictly separated functions, spread between the main actors – the Estonian Ministry of Social Affairs (EMSA) and its agencies, the Estonian Health Insurance Fund (EHIF), and the independent providers of healthcare services, operating under the corporate (private entities) legislation. The EHIF regulates and governs the national mandatory health insurance scheme and execute quality control through a range of quality assurance activities. In the same time most of the hospitals are owned by the local governance level or intentionally established foundations or limited companies owned by the state, municipalities or public agencies subordinated to EMSA. Only limited number of hospitals remain privately-owned [17]. The provision of care and related services is 99% digitalized and follows transparent procedures subject of governmental and public control. Despite the very first impression from the system is as for a rule-stressing[1] and restrictive system the second sight seems much friendlier after analyzing the features of completely different environment determining the healthcare ecosystem – the so called "E-Estonia".

- The Electronic Health Record (e-Health Record) is a nationwide system integrating data from Estonia's different healthcare providers to create a common record for each patient (e-Patient portal) [17, 18].
- The system integrity and coordination are ensured by blockchain solution;
- Each patient acquires an electronic ID-card – the system functionality allows patient to review doctor's visits and current prescriptions, and check which doctors have had access to his/her files [18]. The feedback is active and ensuring the patients' control;
- e-Ambulance is the quickest response that can be provided and the applied solution in e- Estonia can detect the patient's call and refer it to the nearest positioned ambulance (within 30 s);
- e-Prescription is a centralized paperless system for issuing and handling medical prescriptions [18]. The physicians prescribe medicine using the system electronically, through an online form. Almost 99% of all prescriptions in the country are issued electronically. This frees up time for patients and doctors and reduces the administrative burden on hospitals.

To summarize the typology of decision-making models introduced with the digital governance of the healthcare system in Estonia we can consider that the highly centralized environment and system do not exclude the shared-decision-model with the highest known performance of patient-centrism functionally incorporated in the system design. The clinical decisions on diagnosis, treatment, therapy and medication are both evidence-based and patient-centered. The non-clinical decision is also patient-oriented, transparent and medical-cost efficiency oriented. The decisions concerning the adaptiveness of the system to the changing environment are shared with the patients and their feedback and functionally improving the further development of the system through the chosen blockchain solution.

The Essentials of Hungarian Healthcare System: completely opposite is the case of Hungary where the traditional conservative model of healthcare governance can be met. The system is centralized administrated by the Ministry of Human Capacities (Emberi Erőforrások Minisztériuma) [6, 14]. Medical care in Hungary is, as a gene-ral rule, free of charge but the National Institute of Health Insurance Fund Management (NEAK) covers part or all of the cost for diagnosis, treatment and medicine prescribed, only when

[1] e.g. *Patient can be referred to a specialist only if his/her family doctor decides that the health concern requires the involvement of a specialist. The family doctor issued a referral letter for appointment to specialized one; the patient can choose the hospital and the specialists* [18].

they are included in the reimbursement list [22]. Home-based treatments are intended to reduce the number of patients who have to be hospitalized. Managed mostly by local governments, with funds from the National Health Insurance Fund (NHI) in Hungary are operating 104 public hospitals and more than 100 specialized clinics [22]. As it is described the Hungarian system is a representation of the old traditional models of governance and centralized decision making with minimized collaboration between the interacting stakeholders, low effectiveness and system-driven not patient-oriented decision-making models. In terms of contemporary trends in healthcare management, such kind of system is highly unwieldy both for the patients and for the medical staff, which creates an inevitable reflection on the quality, equity, transparency and the whole functionality of the system.

The decision-making in transitioning healthcare systems: the cases of Romania and Bulgaria.

The Essentials of Romanian Healthcare System: The Ministry of Health is primarily responsible for healthcare in Romania. It is both responsible for the regulatory framework and policies as well as the management of the healthcare system. In the last 25 years the Ministry was headed by a total of 28 Ministers - a fact illustrating the unstable political and healthcare ecosystem environment which affect its development. The healthcare system is divided in actors at the national and district level and the district level of 41 counties (judete) and the municipality of Bucharest (total of 42 administrative units). The system is strongly centralized at the national level. The district level formally decentralized because it is mainly responsible for the execution of decisions taken at the national level. Key actors at the national level are the Ministry of Health, the National Health Insurance House (NHIH) and the professional organizations.

Decisions about health by the various governmental institutions are taken without respecting the health needs of the population, due to the lack of unified centralized integrated information system. The existing centralized electronic information system named "Sistemul Informatic Unic Integrat" (SIUI) that connects healthcare providers to the NHIH is vulnerable and often inaccessible for days. The health outcomes are limited by the constraints faced by the whole country – lack of enough financing, lack of coordination and doubtable transparency in the decision-making. The lack of sound information makes the decision-making formal and unjustified in the same type SIUI acquire and store large scale of data which are not relevant to the system performance nor pre-processed for raising the usability and the efficiency of the system. The development of the healthcare system is only possible with substantial change in the quality and rationalism of the decision-making – much more related with the patient needs than the political expectations.

The Essentials of Bulgarian Healthcare System: similar to the Romanian is the Bulgarian healthcare system which *is* also highly centralized and based on compulsory (SHI) and voluntary health insurance (VHI): covering diagnostic, treatment and rehabilitation services as well as full or partial reimbursement of medications only for those included in the national reimbursement list. The Bulgarian system is state-centered both in terms of policy-making and as typology of decision-making, regardless the existence of regional level represented by district structures of National Health Insurance Fund and the regional structure of the Ministry of health – the so-called Regional Health Inspectorates (RHI). The Bulgarian MoH is empowered for decisions, concerning the provision and funding of public health services; the planning and ensuring HR for health system, development of medical science, data management about population health status; development and improvement of public health facilities [26]. The health care providers are both private and public. Formally the National (regional) Health Insurance Fund NHIF (RHIF) is the centralized body with representation of government, employers, syndicates, patients' organizations aiming to guarantee the equal access to health services of insured population (through signing the national framework contract on annual basis for medical and dental care, defining the rights and the obligations of all key players as well as the implementation of monitoring and control mechanism. This is deeply contradicting with the possible paths illustrating the formal and informal interactions between the patient and the structures involved.

4 Conclusion

The presented analysis and assessment of the healthcare decision-making models in EU member-states outline the differences, the limitations and the opportunities created through the different systems ins regard of possible further developments. The assessment grid can be applied as a feasibility study tool for identifying gaps and areas of interventions in process of planning improvements of the national healthcare systems or related decision-making models, rule and procedures or entire framework. The analysis is helping avoiding the discrepancies between the patients' needs and the healthcare system functioning, as well as it is finding out the change drivers for further positive developments. On the ground of the present analysis the following general considerations are being proved:

- The digital transformation of the system changes the decision-making model from informed consent to patient-centered or from patient-centered to shared decision-making, regardless of how centralized or decentralized is the system;
- The change of the financial decision-making model from medicine-cost effective to patient-cost-benefitting and from market-based to shared-responsibility oriented increases the efficiency of the whole healthcare system regardless of its type;
- The patient-centrism is not confronting with the evidence-based decision-making typical for clinical decisions. The integration of the two models strengthens the patient-centrism and establishes the sustainable collaborative ground for further system upgrades.

Comparing the analytical findings with the nation statistics for the healthcare expenditures on annual basis or the demographic picture of the state population, estimation and scenarios for the exosystemic influences that might affect the healthcare system can be prepared.

References

1. https://www.merriam-webster.com/dictionary/health%20care. Accessed 14 Oct 2020
2. https://isilanguagesolutions.com/2019/10/15/health-care-vs-healthcare/. Accessed 14 Oct 2020
3. https://www.arcadia.io/final-word-healthcare-vs-health-care/. Accessed 14 Oct 2020
4. Reid, P.P., Compton, W.D., Grossman, J.H., et al. (eds.): National Academy of Engineering (US) and Institute of Medicine (US) Committee on Engineering and the Health Care System;, editors. Building a Better Delivery System: A New Engineering/Health Care Partnership. National Academies Press, Washington (DC) (2005). 2, A Framework for a Systems Approach to Health Care Delivery. https://www.ncbi.nlm.nih.gov/books/NBK22878/. Accessed 12 Oct 2020
5. https://ec.europa.eu/health/policies/overview_en
6. Papanicolas, I., Smith, P. (eds.): WHO. Health System Performance. An Agenda for policy, information and research. Open University Press, NY. ISBN-10: 0-33-524726-1
7. Kostadinova, I.: Digital models for transformation of corporate social responsibility education. In: Conference Proceedings of 29th Annual Conference of the European Association for Education in Electrical and Information Engineering – EAEEIE 2019. IEEE Digital Xplore (2019). ISBN 978-1-7281-3222-8. https://doi.org/10.1109/EAEEIE46886.2019.9000436
8. https://ec.europa.eu/eurostat/statisticsexplained/index.php?title=Health_in_the_European_Union_%E2%80%93_facts_and_figures. Accessed 13 Oct 2020
9. http://ec.europa.eu/health/dyna/echi/datatool/index.cfm?indlist=77a. Accessed 20 Dec 2020
10. https://www.bundesamtsozialesicherung.de/en/federal-office-for-social-security/about-the-federal-office-for-social-security/. Accessed 12 Oct 2020
11. https://medium.com/@myGermanExpert/everything-you-need-to-know-about-health-insurance-in-germany-d01f90630548. Accessed 5 Oct 2020
12. https://solidarites-sante.gouv.fr/. Accessed 10 Oct 2020
13. https://eurohealthnet.eu/members-and-partners/romania-ministry-health. Accessed 10 Oct 2020
14. https://www.kormany.hu/en/ministry-of-human-resources. Accessed 5 Oct 2020
15. http://www.ms.ro/. Accessed 5 Oct 2020
16. https://www.sm.ee/en/minister-health-and-labour-jevgeni-ossinovski. Accessed 5 Oct 2020
17. https://www.haigekassa.ee/en/people/health-care-services/estonian-health-care-system. Accessed 5 Oct 2020
18. https://e-estonia.com/. Accessed 5 Oct 2020
19. Busse, R., Blümel, M.: Germany: health system review. Health Syst. Transit. **16**(2), 1–296 (2014)
20. https://www.euro.who.int/__data/assets/pdf_file/0005/383054/HiT-Bulgaria-2018-web.pdf
21. https://martamuriscot.com/2020/07/12/trends-within-the-healthcare-industry-i/. Accessed 10 Oct 2020
22. https://eurohealthnet.eu/sites/eurohealthnet.eu/files/publications/Health_Equity_in_the_EU_Factsheet_October_2019_A4_online%20version.pdf. Accessed 10 Oct 2020
23. Scholl, I., Zill, J.M., Härter, M., Dirmaier, J.: An integrative model of patient-centeredness – a systematic review and concept analysis. PLoS One **9**(9), e107828 (2014)
24. Meltzer, D.: Accounting for future costs in medical cost-effectiveness analysis. J. Health Econ. **16**(1), 33–64 (1997)
25. https://www.bundesgesundheitsministerium.de/fileadmin/Dateien/3_Downloads/G/Gesundheitssystem/German_Health_System.pdf. Accessed 10 Oct 2020
26. Dimova A, et al.: Bulgaria: health system review. Health Syst. Transit. **20**(4), 20–167

27. Lameire, N., Joffe, P., Wiedemann, M.: Healthcare systems—an international review: an overview. Nephrol. Dialysis Transpl. **14**(suppl_6), 3–9 (1999). https://doi.org/10.1093/ndt/14.suppl_6.3
28. lo Storto, C., Goncharuk, A.G.: Efficiency vs effectiveness: a benchmarking study on european healthcare systems. Econ. Sociol. **10**(3), 102–115 (2017). https://doi.org/10.14254/2071-789X.2017/10-3/8
29. Konrad, T.R., Link, C., Shackelton, R.J., et al.: It's about time: physicians' perceptions of time constraints in primary care medical practice in three national healthcare systems. Med. Care **48**(2), 95–100 (2010). https://doi.org/10.1097/MLR.0b013e3181c12e6a

A Comparative Machine Learning Modelling Approach for Patients' Mortality Prediction in Hospital Intensive Care Unit

Mahmoud Aldraimli[1](✉), Nodira Nazyrova[2], Abdumalik Djumanov[2], Ikboljon Sobirov[2], and Thierry J. Chaussalet[1]

[1] School of Computer Science and Engineering, University of Westminster, London W1W 6UW, UK
m.aldraimli@my.westminster.ac.uk

[2] School of Technology, Westminster International University in Tashkent, Tashkent 100047, Uzbekistan
nnazyrova@wiut.uz

Abstract. Mortality prediction in a hospital Intensive Care Unit (ICU) is a challenge that must be addressed with high precision. Machine Learning (ML) is a powerful tool in predictive modelling but subject to the problem of class imbalance. In this study, we tackle class imbalance with combining new features, data re-sampling, ensemble learning and an appropriate selection of evaluation metrics in a clinical setting. We built and evaluated 126 ML models to predict mortality in 48546 ICU admissions extracted from the Medical Information Mart for Intensive Care III (MIMIC-III) repository.

In our study design, six mortality prediction datasets are extracted; five of which are legacy dataset sets while the remainder is our new constructed dataset. For our combined data models, when testing on isolated data, our selection of features enhanced the prediction performances beyond those for the traditional legacy sets used in research. The legacy datasets are the Simplified Acute Physiology Score (SAPS II), the Sequential Organ Failure Assessment score (SOFA), the Glasgow Coma Scale (GCS), Elixhauser Comorbidity Index (ECI) and Demographics & Disease Groups (DDG). Our approach has a considerable impact on the classification; it resulted in an improvement in the mortality status prediction. For evaluation, we implement a comparative multi-stage evaluation filter for a binary classification to compare all models. The best models are identified. The Area Under Receiver Operator Characteristic curves of the tested models range from 0.57 to 0.94. These encouraging results can guide further development of models to allow for more reliable ICU mortality predictions.

Keywords: Classification · Imbalanced data · MIMIC III · Tomek links · SMOTE · Ensembles · ICU mortality prediction

1 Introduction

Real-world data are often imbalanced and lack a uniform distribution across classes. Classification of imbalanced datasets is one of the challenges across several industrial

S. S. Sotirov et al. (Eds.): BioInfoMed 2020, LNNS 374, pp. 16–31, 2022.
https://doi.org/10.1007/978-3-030-96638-6_2

and research domains [1]. There are multiple approaches to tackle class imbalance [2], simplest approaches include, for example, Data Enrichment. Others more sophisticated methods include various sampling techniques [3], cost-sensitive learning [4, 5] and feature selection; or more complex strategies including meta-learning [6], combining classifiers [1], and algorithmic modifications [7].

When sampling methods are applied, questions over their suitability are often raised [8]. For example: is the new re-sampled dataset representative of the population in relation to the response variable? Is it acceptable to artificially generate synthetic data of class subjects when training Machine Learning (ML) classification models? It has been argued that by using sampling methods, the original class ratio is lost during the training process and that this affects the accuracy metrics [9]. Similarly, training ML models with synthetic data may also compromise accuracy measures by deceiving the process of cross-validations sampling [10].

In this paper, we compare the classification performance of nine ML algorithms Naïve Bayes, Logistic Regression, Support Vector Machine, K-Nearest Neighbours, Extreme Gradient Boosting, Adaptive Boost and Random Forest in (default, balanced and weighted settings). The models are built with enhanced training sets using SMOTE [3] and SMOTE/TOMEK hybrid oversampling techniques [3] and the original highly imbalanced data to predict patients' mortality in an intensive care unit (ICU) in a binary-class classification problem; and to suggest the most suitable models that would compete to meet a clinical domain experts' success criteria. The data imbalance characteristic causing the transition in classifier training performance was monitored numerically via the Mean Decrease Impurity (MDI) Filter [11].

The paper is structured as follows: In Sect. 2, the domain problem and all the methods and approaches used in this study are presented. Then in Sect. 3, the experiment results are introduced, while Sect. 4 is reserved for discussion and conclusion.

2 Materials and Methods

2.1 The Domain Problem

Predicting the mortality of patients admitted to the ICU is crucial for assessing the severity of illness and adjusting the treatments and health care policies [12]. Many severity scores have been proposed, but validation studies have concluded that they are not adequately suitable [13]. The scores from the SAPS II dataset were subsequently developed using statistical modelling approaches [14]. Until the current times, the SAPS II mortality prediction statistical models remain the most widely used in clinical practice. However, since first being published, they have been modified several times to improve their predictive performance [15].

2.2 Study Design

A binary classification modelling is performed to predict patients' mortality in ICU. Five Legacy datasets were extracted from the Medical Information Mart for Intensive Care III (MIMIC III) repository. Our study is a cross-sectional assessment of 48546 adults admitted to ICU between 2001 and 2012. The Data collection procedures comply with the Health Insurance Portability and Accountability (HIPAA) regulations. Our Application number is 8638306.

All the severity scores were calculated for the same cohort of patients in PostgreSQL. The extracted legacy datasets were the SOFA, SAPS II, GCS, ECI and DDG, each set with its associated variables. These datasets together account for 29 different diseases that are used in the analysis. A sixth new dataset is constructed by combining all features with non-zero Gini importance from legacy sets.

The Full extracted MIMIC III cohort had $n = 61532$; $n = 12986$ subjects were excluded from the analysis; these were new-born admission type, children below the age of 18 years and erroneous data (length of stay < 0). The remaining cohort $n = 48546$ were used in the analysis. We followed the Pareto principle [16] and performed a randomised data split with a ratio of 80% training observations to 20% test. The training-test split was performed before data processing. Table 1 shows the list of variables extracted for each dataset and included in the analysis.

In total, there are 18 training sets and 6 test sets to train and evaluate nine learners; each training set represents one of the six extracted mortality prediction datasets but in three different settings in addition to the test set:

The Imbalanced State contains 80% of the original cohort with variables to form the SOFA, SAPS II, GCS, ECI, DDG and Combined training sets, where the discharged patients are a majority class, and the deceased patients are the minority.

The SMOTE Re-sampling State, where the imbalanced training datasets are re-sampled to balance the distribution of instances between both class labels. In order to obtain such balance, the Synthetic Minority Oversampling Techniques (SMOTE) is applied.

The Hybrid Re-sampling State where the imbalanced training datasets (SOFA, SAPS II, GCS, ECI, DDG and Combined sets) were re-sampled to balance the distribution of instances between both class labels. To obtain such balance, the SMOTE oversampling and Tomek links removal techniques were applied.

The Test Dataset is sampled without replacement with stratification from the original cohort. It contains 20% of the subjects and variables to form the SOFA, SAPS II, GCS, ECI, DDG and Combined test sets, where the discharged patients are a majority class, and the deceased patients are the minority. All models are tested with the same test instances with the build model's corresponding input features.

2.3 Methodology

Binary-class classification ML models are built to predict mortality for patients admitted to ICU based on a new selection of score-transformed features from multiple ICU mortality legacy datasets. The newly formed dataset with combined scored features

Table 1. A list of variables included per dataset used in the comparative modelling

Dataset	Input features	Features count
DDG	Age, Gender, Race (Asian, Black/African American, Hispanic/Latino, Other/Unknown, White), Marital Status (Divorced, Life Partner, Married, Separated, Single, Unknown, Widowed), Religion (Religious, Unobtainable, Not Specified), Admission Type (Elective, Emergency, Urgent), Insurance Type (Government, Medicare, Medicaid, Private, Self-Pay), Length of Stay in ICU, 17 ICD 9 groups of diseases (blood, circulatory, congenital, digestive, endocrine, genitourinary, ill-defined, infectious, injury, mental, muscular, neoplasms, nervous, pregnancy, prenatal, respiratory, skin)	43
SAPS II	Age score, heart rate, systolic blood pressure, temperature score, Glasgow coma scale, PaO2/FiO2 score, urine output, blood urea nitrogen, sodium, potassium, bicarbonate, bilirubin, white blood cell, chronic disease, type of admission, SAPS II score	16
SOFA	Respiratory score, cardiovascular score, hepatic score, coagulation score, renal score, neurological systems score, overall sofa score	7
GCS	GCS eyes, GCS Verbal, GCS Motor, overall GCS score	4
ECI	Congestive heart failure, cardiac arrhythmias, valvular disease, pulmonary circulation disorders, peripheral vascular, hypertension uncomplicated, hypertension complicated, paralysis, other neurological disorders, chronic pulmonary disease, diabetes uncomplicated, diabetes complicated, hypothyroidism, renal failure, liver disease, peptic ulcer disease, AIDS/HIV, lymphoma, metastatic cancer, solid tumor without metastasis, rheumatoid arthritis, coagulopathy, obesity, weight loss, fluid and electrolyte disorders, blood loss anaemia, deficiency anaemia, alcohol abuse, drug abuse, psychosis, depression	30
Combined	Age, Gender, Race (Asian, Black/African American, Hispanic/Latino, Other/Unknown, White), Marital Status (Divorced, Life Partner, Married, Separated, Single, Unknown, Widowed), Religion (Religious, Unobtainable, Not Specified), Admission Type (Elective, Emergency, Urgent), Insurance Type (Government, Medicare, Medicaid, Private, Self-Pay), Length of Stay in ICU, 17 ICD 9 groups of diseases (blood, circulatory, congenital, digestive, endocrine, genitourinary, ill-defined, infectious, injury, mental, muscular, neoplasms, nervous, pregnancy, prenatal, respiratory, skin), Age score, heart rate, systolic blood pressure, temperature score, Glasgow coma scale, PaO2/FiO2 score, urine output, blood urea nitrogen, sodium, potassium, bicarbonate, bilirubin, white blood cell, chronic disease, type of admission, Respiratory score, cardiovascular score, hepatic score, coagulation score, renal score, neurological systems score, overall sofa score, GCS eyes, GCS Verbal, GCS Motor, overall GCS score, Congestive heart failure, cardiac arrhythmias, valvular disease, pulmonary circulation disorders, peripheral vascular, hypertension uncomplicated, hypertension complicated, paralysis, other neurological disorders, chronic pulmonary disease, diabetes uncomplicated, diabetes complicated, hypothyroidism, renal failure, liver disease, peptic ulcer disease, AIDS/HIV, lymphoma, metastatic cancer, solid tumor without metastasis, rheumatoid arthritis, coagulopathy, obesity, weight loss, fluid and electrolyte disorders, blood loss anemia, deficiency anemia, alcohol abuse, drug abuse, psychosis, depression	95

represents baseline patient's measurements, geographic and demographic variables and health-related conditions. A subset of 38836 subjects was randomly selected from the original 48546 adults and used nine ML algorithms using 10-fold cross-validation in three different scenarios. The models were tested on the remaining 9710 cases.

Figure 1 shows the methodology. Multiple legacy imbalanced mortality scores datasets and one newly formed (combined dataset) each with different predictor variables and the same subjects are modified with state-of-the-art oversampling techniques and used for modelling using the six ML algorithms. In addition, three ensemble learners were used to model the unsampled imbalanced training datasets. The AUC of the models and classes' accuracy rates were compared after the testing phase. Mean Decrease Impurity Filter was used for Feature Selection and feature importance levels were monitored with Gini Purity Scores.

2.4 Data Pre-processing

Multiple necessary steps are performed to transform the data in a shape that enhances modelling performance and reduces any potential bias when modelling. Such steps are data imputation, scaling, encoding, feature selection and data re-sampling.

Mean Imputation. The use of this single imputation technique here is deemed appropriate since the missing observations within each variable accounted for less than 5% [17].

Feature Scaling. All input variables were scaled to a common magnitude using a robust standardisation approach [18, 19].

Data Encoding. Diseases and procedures in the MIMIC-III dataset are encoded using the International Classification of Diseases version 9 (ICD9) codes. There are 15693 distinct diagnoses for 61532 admissions. To avoid high dimensionality and very sparse representation for patients, the diseases are categorised to capture group-level diseases using the first three letters of the original length codes. We have used the International Statistical Classification of Diseases and the classification of related health problems to group the diseases into 17 chapters [20].

Data Re-sampling. Medical data is often imbalanced; this results in an uneven distribution of samples between the class labels. In the extracted dataset, there are 42658 discharged (negative) cases and 5888 deceased (positive) instances. To tackle the class imbalance problem, two synthetic data generation techniques are used to oversample the minority class; those are Synthetic Minority Oversampling Technique (SMOTE) [21], and Hybrid SMOTE-Tomek Links oversampling [22]. The K-parameter is set to $(k = \sqrt[2]{\textit{number of input features}})$.

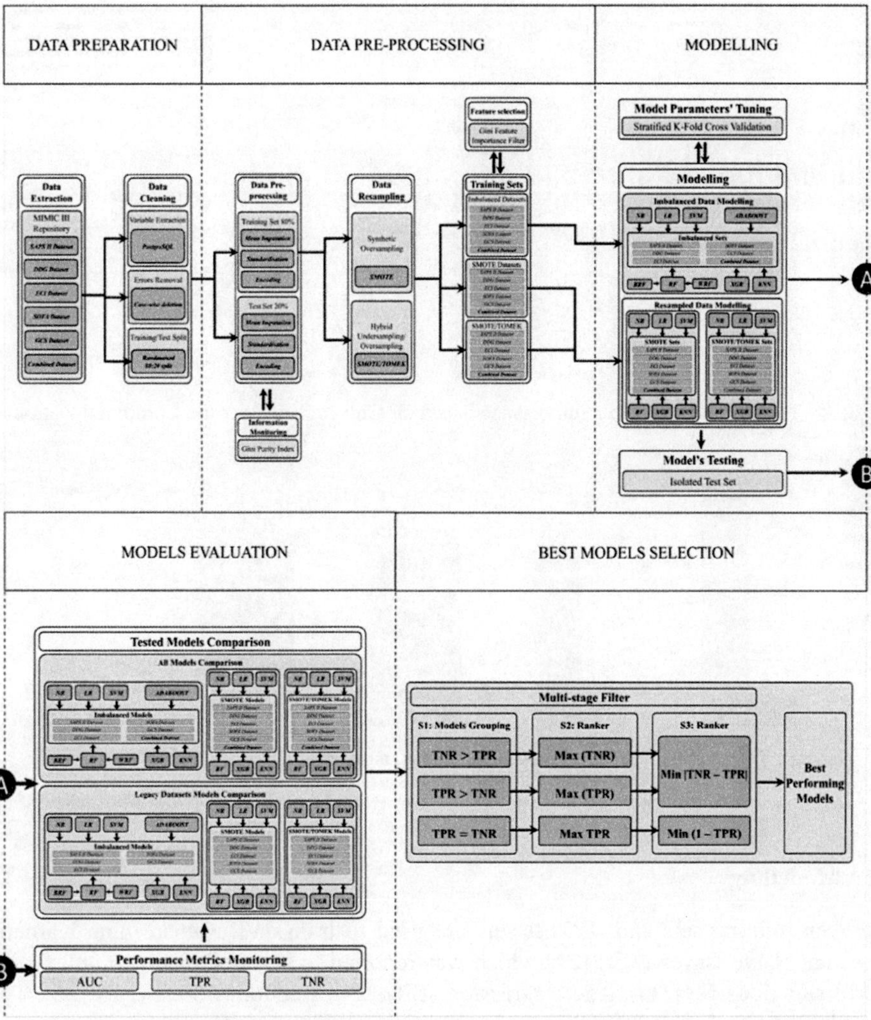

Fig. 1. The methodology adopted in this work, showing the different steps followed.

Feature Selection. Features were selected for modelling based on importance measured with Gini Impurity filter [23] only for the new combined dataset. The combined dataset includes demographic and disease-related data, along with the combination of all severity scores. Features with zero importance were excluded from the analysis. Figure 2 shows the features with their Gini index. In total, there are 95 features used for mortality classification all in the new combined dataset extracted from the five legacy datasets. The combined dataset, the legacy SAPS II, SOFA, DDG, GCS, and ECI datasets proceeded to the modelling phase in three different settings. These are the imbalanced state and two re-sampled states; the SMOTE and the SMOTE-Tomek. Figure 3 shows the distribution of instances among all three training dataset settings as well as the isolated test set.

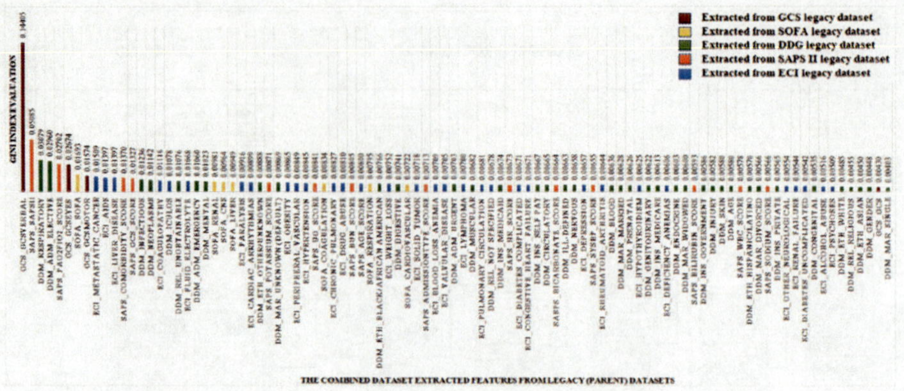

Fig. 2. The extracted features importance based on Gini Impurity for the Combined Dataset

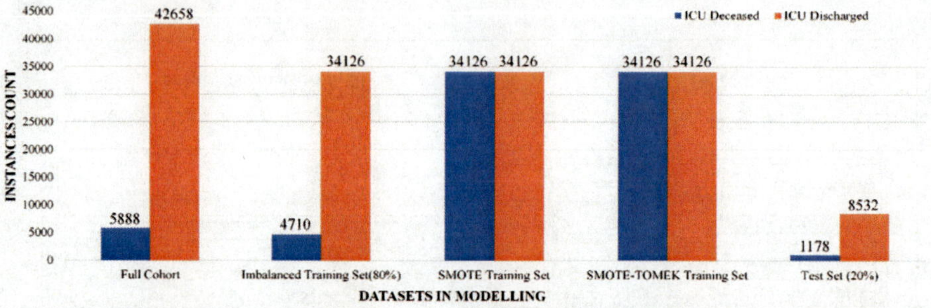

Fig. 3. The distribution of records across the imbalanced and re-sampled datasets settings

2.5 Modelling

Eighteen training sets and six test sets are used to train and evaluate nine learners, those are Naïve Bayes (NB) [24] which was reported to be effective in complex real-world situations [24], Logistic Regression (LR) [25], Random Forest (RF) [26], The default Random Forest learner is applied, and its performance was compared to two of its versions; *Balanced RF (BRF)* [27] and *Weighted RF (WRF)* [28], K- Nearest Neighbour (KNN) [29]. The selection of voting neighbourhood size (K value) is set to $k = \sqrt[2]{number\ of\ input\ features})$ in our analysis with the Euclidean Distance linear search distance function [29], Support Vector Machine (SVM) with a poly kernel function [30], Adaptive Boosting (AdaBoost) AKA Easy Ensemble learner implementation in python SciKit learn ensemble learner [31] and Extreme Gradient Boosting (XGB) [32]. All the nine techniques are implemented in Python SciKit Learn (with default parameters) [33]. 126 models were built and then tested with the same isolated test set for evaluation.

2.6 Models Evaluation and Selection

Medical experts often seek a trade-off between the cost of misclassification and its impact. Misclassifying a subject from the mortality group has an extreme impact of a loss of life while incorrectly predicting a patient from the negative group will incur additional costs from further tests and care [34].

There is not a balanced metric to evaluate the performance of machine learning models [35]; hence, additional metrics need to be examined. In this study, the True Positive Rate (also known as sensitivity or recall, 'TPR') [36], the True Negative Rate (also known as specificity or', TNR') [37] and the Area Under Receiver Characteristic Curve (AUC) [38] are examined. Practically, a high TPR and TNR values indicate that both precisions are high, meaning fewer subjects are misdiagnosed. Since the impact of misclassification could mean loss of life, the AUC is chosen to measure the trade-off between correctly predicted deceased subjects (true positives) and discharged patients who are incorrectly classified as deceased (false positives incurring cost).

Once the test performances are obtained, the metrics are passed through our developed multi-stage performance evaluation filter to select the best performing models based on a high AUC with a balanced TPR and TNR. Figure 4 shows the mechanism of the models' selection filter. At stage 1 (S1), the models are grouped into three comparative groups TPR and TNR. At S2, where the TPR exceeds the TNR, the model with the maximum TPR AND minimum |TPR-TNR| is selected. In the case where the TNR exceeds TPR, the model with the maximum TNR AND minimum |TPR-TNR| is selected. If both TNR and TPR are the same, then the model with the lowest 1-TPR is selected. At S3, the selected models can then be ranked by the highest corresponding AUC score.

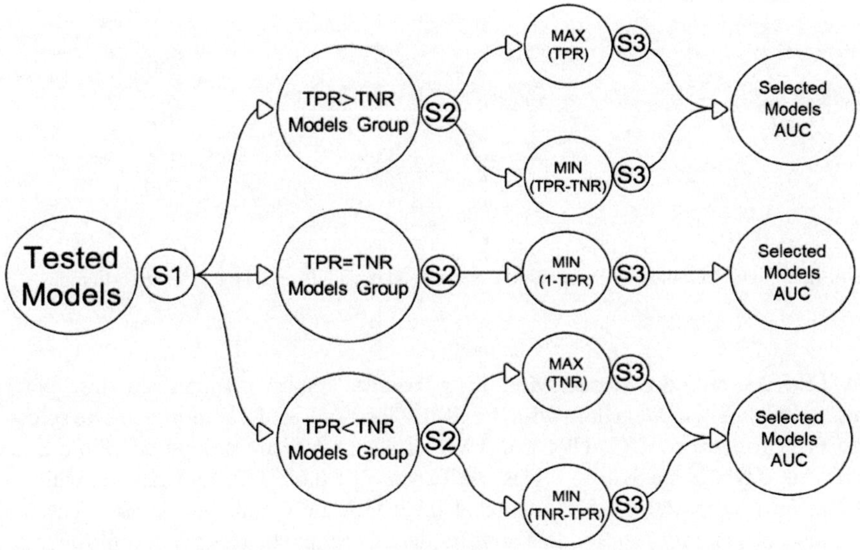

Fig. 4. A multi-stage models evaluation filter for best models selection

3 Results

3.1 Models Test Results

Imbalanced Datasets Modelling Results. Table 2 shows the test performance of fifty-four models built with the imbalanced datasets setting. The selected performance metrics: AUC, TPR and TNR of all models are computed. Table 2 also shows that SVM achieved the lowest AUC score of 0.57 with ECI dataset while the XGB and LR achieved the highest AUC score of 0.94 with the Combined dataset. The table also shows that thirty-seven models (highlighted in orange) had a greater class accuracy difference than 10% between the TPR and TNR; thirty of which has an AUC score of a minimum of 0.70. One of those models reported 99.8% difference between TNR and TPR in the case of LR trained with the ECI dataset with an AUC score of 0.70.

It is also observed that seventeen models (highlighted in blue) reported a minimum AUC score of 0.70 while achieving a difference below 10% between TPR and TNR; these models offer a more balanced classification between the deceased and the discharged patient classes sixteen of which are models build with ensemble learners (RF, BRF, XGB, AdaBoost). Figure 5-(A) shows that the class imbalance problem could result in misleading large AUC scores. Both Fig. 5-(B) and Fig. 5-(C) show the advantage offered by using ensemble techniques in tackling the impact of higher classification accuracy bias towards the majority group in severely imbalanced datasets. Therefore, ensemble models maintain a high AUC score and TNR, while reducing incorrect predictions in the deceased group (the minority class).

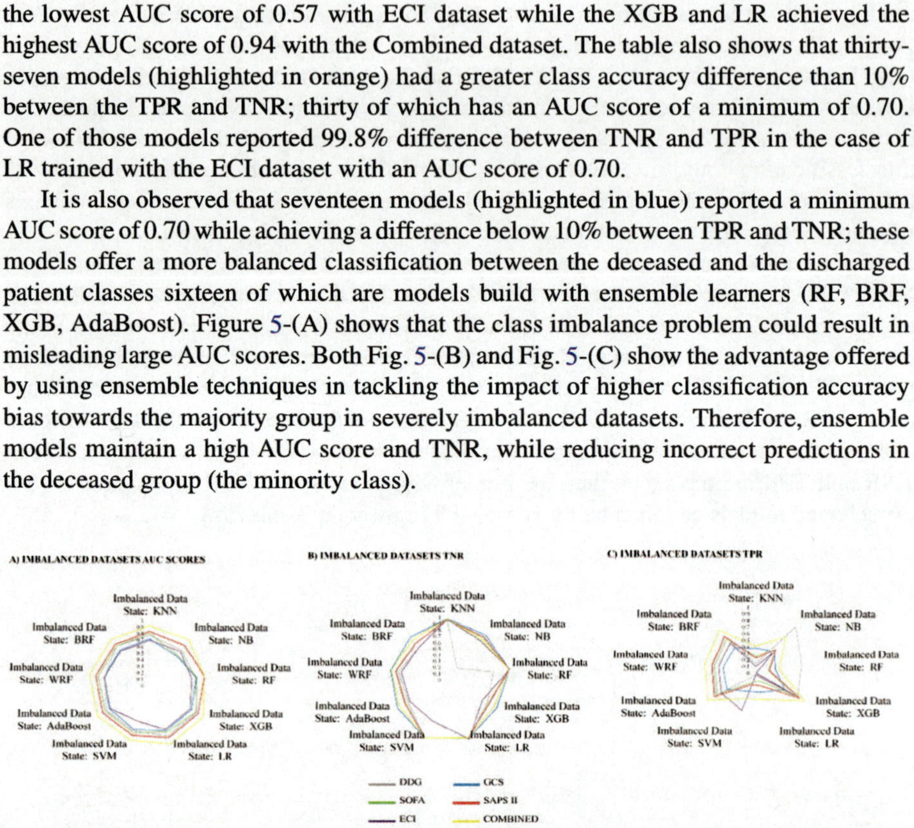

Fig. 5. The imbalanced datasets ML models AUC, TNR and TPR test performances

SMOTE Re-sampled Datasets Modelling Results. Table 3 shows the test performance of thirty-six models built with the SMOTE re-sampled training sets. The selected performance metrics AUC, TPR and TNR of all models are computed. Table 3 also shows that KNN achieved the lowest AUC score of 0.60 with ECI dataset while the XGB achieved the highest AUC score of 0.93 with the Combined dataset. The table also shows that twenty-four models (highlighted in orange) had a greater difference than 10% between the TPR and TNR; seventeen of which have an AUC score of a minimum

Table 2. The test performance of nine ML models trained with six imbalanced datasets

Dataset	Metrics	Tested Machine Learning Imbalanced Models								
		KN N	NB	RF	XGB	LR	SV M	AdaBoos t	WR F	BR F
DDG	AU C	0.77	0.7 3	0.7 0	0.81	0.81	0.73	0.76	0.70	0.80
	TPR	0.14	0.9 2	0.7 0	0.72	0.05	0.04	0.69	0.66	0.74
	TNR	0.97	0.2 4	0.7 4	0.73	0.99	0.99	0.71	0.77	0.69
GCS	AU C	0.72	0.6 1	0.7 1	0.72	0.72	0.71	0.71	0.71	0.72
	TPR	0.38	0.4 3	0.3	0.5	0.31	0.3	0.50	0.50	0.50
	TNR	0.98	0.9 4	0.9 9	0.9	0.99	0.99	0.90	0.90	0.90
SOFA	AU C	0.80	0.6 3	0.7 2	0.80	0.75	0.76	0.78	0.80	0.80
	TPR	0.25	0.4 7	0.2 0	0.70	0.13	0.16	0.66	0.63	0.68
	TNR	0.97	0.8 8	0.9 9	0.75	0.99	0.99	0.75	0.81	0.77
SAPS II	AU C	0.83	0.8 1	0.8 4	0.84	0.84	0.8	0.83	0.84	0.83
	TPR	0.23	0.4 7	0.2 1	0.76	0.21	0.19	0.72	0.67	0.73
	TNR	0.97	0.9 0	0.9 9	0.76	0.98	0.99	0.76	0.82	0.77
ECI	AU C	0.70	0.7 0	0.6 4	0.70	0.7	0.57	0.70	0.71	0.70
	TPR	0.10	0.4 0	0.0 1	0.67	0.02	0.61	0.60	0.57	0.60
	TNR	0.97	0.8 2	1.0 0	0.64	1.00	0.69	0.68	0.73	0.69
Combine d	AU C	0.92	0.9 2	0.8 4	0.94	0.94	0.88	0.92	0.93	0.92
	TPR	0.41	0.7 2	0.3 6	0.85	0.43	0.42	0.83	0.77	0.85
	TNR	0.98	0.8 5	0.9 9	0.86	0.98	0.99	0.85	0.89	0.84

0.70. All models built with SMOTE re-sampled ECI training set reported the poorest AUC scores in test. It is also observed that eleven models out of twelve (highlighted in blue) reported a minimum AUC score of 0.74 while achieving a difference below 10% between TPR and TNR. These models offer a more balanced classification between the deceased and the discharged patients' classes. Table 3 also shows that the models built with SMOTE re-sampled Combined training set dominated the highest AUC scores in test.

Table 3. The test performance of six ML models trained with six SMOTE re-sampled datasets

Dataset	Metrics	SMOTE Re-sampling State					
		KNN	NB	RF	XGB	LR	SVM
DDG	AUC	0.74	0.76	0.74	0.81	0.81	0.75
	TPR	0.71	0.83	0.30	0.76	0.67	0.56
	TNR	0.67	0.32	0.93	0.70	0.67	0.80
GCS	AUC	0.71	0.71	0.66	0.71	0.71	0.72
	TPR	0.33	0.43	0.42	0.42	0.50	0.50
	TNR	0.98	0.94	0.96	0.96	0.90	0.90
SOFA	AUC	0.76	0.76	0.73	0.78	0.78	0.76
	TPR	0.45	0.62	0.43	0.22	0.66	0.67
	TNR	0.89	0.80	0.93	0.98	0.74	0.76
SAPS II	AUC	0.71	0.81	0.75	0.84	0.84	0.80
	TPR	0.55	0.58	0.37	0.77	0.70	0.58
	TNR	0.84	0.84	0.95	0.74	0.75	0.83
ECI	AUC	0.60	0.68	0.64	0.67	0.67	0.69
	TPR	0.32	0.75	0.31	0.82	0.61	0.18
	TNR	0.86	0.35	0.89	0.37	0.67	0.95
Combined	AUC	0.89	0.91	0.92	0.93	0.91	0.90
	TPR	0.89	0.75	0.76	0.84	0.82	0.75
	TNR	0.70	0.79	0.90	0.84	0.84	0.79

It is expected that the models' TPR's benefit from applying the SMOTE re-sampling technique to the original imbalanced datasets. However, Fig. 6 shows that six models' TPR's deteriorated as a result of applying SMOTE; these are the KNN-GCS, NB-DDG, RF-DDG, XGB-GCS, XGB-SOFA and SVM-ECI models.

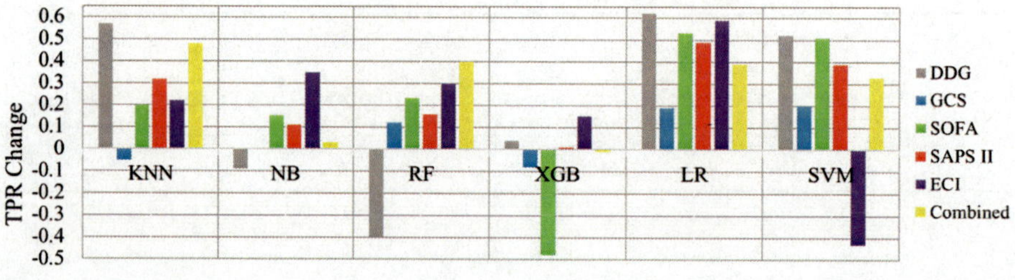

Fig. 6. The TPR change as a result of re-sampling the imbalanced training sets with SMOTE

SMOTE-Tomek Hybrid Re-sampled Datasets Modelling Results. Table 4 shows the test performance of thirty-six models built with the SMOTE-Tomek Hybrid re-sampled training sets. The selected performance metrics AUC, TPR and TNR of all models are computed. Table 4 also shows that RF and SVM models built with the re-sampled ECI set achieved the lowest AUC score of 0.64 while the XGB model built with the re-sampled Combined dataset achieved the highest AUC score of 0.93. The table also shows that thirty-one models (highlighted in orange) had a greater class accuracy difference than 10% between the TPR and TNR; twenty-five of which have an AUC score of a minimum 0.70. All models built with SMOTE re-sampled ECI training set reported the poorest AUC scores in test except for KNN model. It is also observed that four out of five models (highlighted in blue) reported a minimum AUC score of 0.89 while achieving a difference below 10% between TPR and TNR; These models offer a more balanced classification between the deceased and the discharged patient classes. Table 4 also shows that the models built with the Hybrid re-sampled Combined training set dominated the highest AUC scores in test.

It is known that the models' TPR's benefit from applying SMOTE-Tomek re-sampling to the original imbalanced datasets. Nevertheless, Fig. 7 shows that six models' TPR's deteriorated as a result of applying SMOTE-Tomek technique; these are the KNN-GCS, NB-DDG, RF-DDG, XGB-SOFA, XGB-Combined and SVM-ECI models. This TPR deterioration behaviour is similar to that observed in SMOTE models in Fig. 6 except for XGB-GCS and XGB-Combined.

Table 4. The test performance of six ML models trained with six Hybrid re-sampled datasets

Dataset	Metrics	HYBRID Re-sampling State					
		KNN	NB	RF	XGB	LR	SVM
DDG	AUC	0.68	0.78	0.74	0.81	0.81	0.81
	TPR	0.63	0.85	0.38	0.79	0.58	0.59
	TNR	0.73	0.31	0.91	0.65	0.75	0.80
GCS	AUC	0.71	0.70	0.71	0.71	0.71	0.71
	TPR	0.35	0.43	0.47	0.50	0.43	0.50
	TNR	0.97	0.94	0.93	0.9	0.94	0.90
SOFA	AUC	0.8	0.77	0.75	0.78	0.78	0.76
	TPR	0.50	0.59	0.54	0.44	0.58	0.60
	TNR	0.86	0.81	0.88	0.90	0.82	0.83
SAPS II	AUC	0.79	0.81	0.74	0.84	0.84	0.80
	TPR	0.56	0.56	0.45	0.80	0.63	0.57
	TNR	0.83	0.86	0.93	0.68	0.81	0.85
ECI	AUC	0.70	0.69	0.64	0.67	0.67	0.64
	TPR	0.28	0.82	0.38	0.93	0.51	0.42
	TNR	0.89	0.30	0.85	0.21	0.76	0.82
Combined	AUC	0.89	0.92	0.92	0.93	0.84	0.91
	TPR	0.85	0.73	0.51	0.84	0.79	0.75
	TNR	0.77	0.82	0.97	0.84	0.87	0.79

Feature Importance Evaluation Results. The consideration of monitoring information with the Gini index for each feature in the training and test sets in Fig. 8 shows a minimal difference between the test features and the training features information. This shows that the information contained in the test set is representative of that in the training set; hence, the test results are not subject to large information bias.

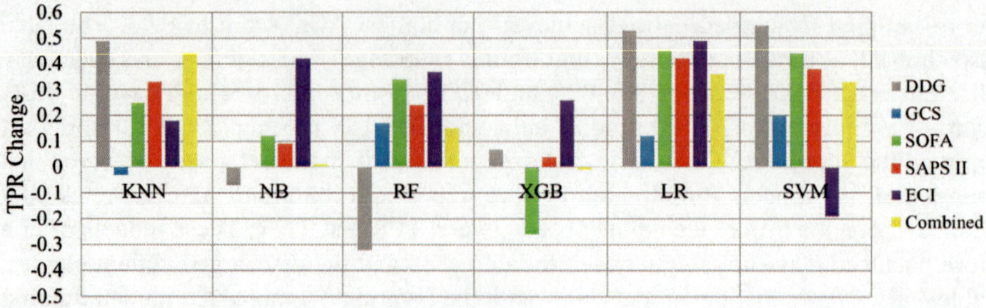

Fig.7. The TPR change as a result of re-sampling the imbalanced training sets with SMOTE-Tomek hybrid technique

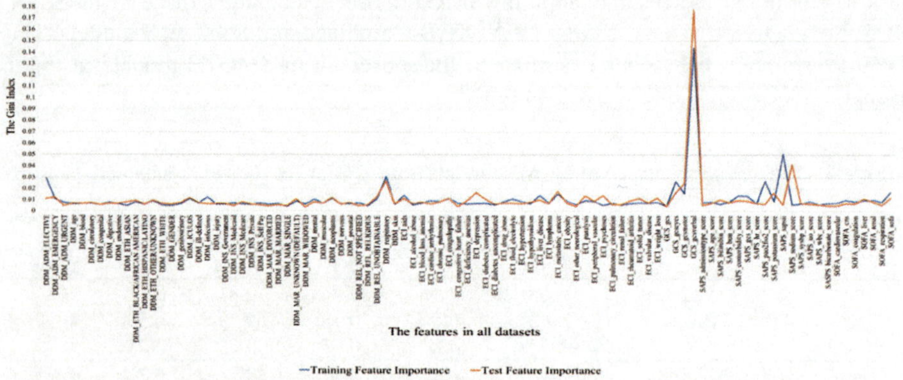

Fig. 8. Information monitoring with Gini Index per variable in each dataset

3.2 Models Selection

All the 126 tested models underwent a selection process with a multi-stage filter to select the models which satisfy the above success criteria. Four models were identified, and those are XGB model built with the Combined imbalanced dataset (AUC = 0.94, TPR = 0.85, TNR = 0.86), XGB model built with the SMOTE Combined datasets (AUC = 0.93, TPR = 0.84, TNR = 0.84), XGB built with the SMOTE-Tomek Combined dataset (AUC = 0.93, TPR = 0.84, TNR = 0.84) and BRF built with the imbalanced dataset (AUC = 0.92, TPR = 0.85, TNR = 0.84).

4 Discussion

The prediction of ICU patient mortality is a critical task; we tackled it with a formation of a dataset with a combined set of features extracted from legacy datasets dedicated to predicting patient mortality. Our prediction results outperformed those based on the legacy datasets. In this paper, we acknowledged that imbalanced classes have a significant impact on the performance of standard machine learning algorithms. Classification performance in the training phase is severely impacted by class separability. Training the standard ML algorithms with highly imbalanced overlapping classes without any adjustment to the training set or the learning algorithm results in an accuracy bias towards the majority class. In this study, two methods (Ensembles learning and Re-sampling techniques) were applied to adjust the class imbalance in the classification training phase at the algorithmic and dataset levels. It remains unclear as to whether other remedies for imbalanced data classifications, such as Cost-Sensitive meta-learning with explicit cost matrix could result in better performances [39, 40]. The advantages of sampling techniques evaluated here, however, include simplicity and transportability. Nevertheless, they are limited by the amount of information manipulation as a result of their application resulting in biased prediction towards the minority class. The excessive use of such techniques could result in overfitting of the models.

In this study, traditional ML algorithms were sensitive to higher purity levels and tended to produce better performance results with re-sampling techniques. Nevertheless, ensemble learners and specifically XGB dominated the highest ranking in TPR, TNR and AUC scores for both imbalanced and re-sampled data states for the combined dataset.

The MIMIC III dataset used in this study shows that with our newly combined set of features and the use of ensemble learners, together with the desired choice of performance metrics and slight manipulation of information levels produces a mortality prediction solution which could be tested further with patients' screening alongside other currently available models. Among all 126 models presented in this study, only four satisfied the success criteria for this specific domain problem (XGB model built with the Combined imbalanced dataset, XGB model built with the SMOTE model Combined datasets, XGB model with the SMOTE-Tomek Combined dataset and BRF model built with the imbalanced dataset). This domain problem is the first to use the suggested combinations of features to predict mortality of patients in ICU. We empirically compared our dataset performance to the performance of the same algorithms on the legacy datasets. The results of this study point out that there is a significant improvement were made in ICU mortality predictions. Future work will take into account the predictions from this current paper and compare them to the actual predictions of mortality in the other cohorts in other research publications where data is available.

References

1. Ali, A., Shamsuddin, S., Ralescu, A.: Classification with class imbalance problem: a review. Int. J. Adv. Soft Comput. Appl. **5**(3), 176–204 (2015)
2. Bekker, M., Alitouche, T.: Imbalanced data learning approaches review. Int. J. Data Min. Knowl. Manag. Process (IJDKP) **3**(4), 15 (2013)
3. Rahman, M., Davis, D.N.: Addressing the class imbalance problem in medical datasets. Int. J. Mach. Learn. Comput. **3**(2), 224 (2013)
4. Ling, C., Sheng, V.: Cost-sensitive learning and the class imbalance problem. Springer (2008)
5. Thai-Nghe, N., Gantner, Z., Schmidt-Thieme, L.: Cost-sensitive learning methods for imbalanced data. In: The 2010 International Joint Conference on Neural Networks (IJCNN) (2010)
6. Ling, C., Sheng, V.: Class imbalance problem. In: Encyclopedia of Machine Learning, p. 171 (2011)
7. Krawczyk, B.: Learning from imbalanced data: open challenges and future directions. Progr. Artif. Intell. **5**(4), 221–232 (2016). https://doi.org/10.1007/s13748-016-0094-0

8. Lango, M.: Tackling the problem of class imbalance in multi-class sentiment classification: an experimental study. Found. Comput. Decis. Sci. **44**(2), 151–178 (2019)
9. Johnson, J.M., Khoshgoftaar, T.M.: Survey on deep learning with class imbalance. J. Big Data **6**(1), 1–54 (2019). https://doi.org/10.1186/s40537-019-0192-5
10. Rankin, D., Black, M., Bond, R., Wallace, J., Mulvenna, M., Epelde, G.: Reliability of supervised machine learning using synthetic data in health care: model to preserve privacy for data sharing. JMIR Med. Inform. **8**(7), e18910 (2020)
11. Behnamian, A., Millard, K., Banks, S., White, L., Richardson, M., Pasher, J.: A systematic approach for variable selection with random forests: achieving stable variable importance values. IEEE Geosci. Remote Sens. Lett. **14**(11), 1988–1992 (2017)
12. Awad, A., Bader-El-Den, M., McNicholas, J., Briggs, J.: Early hospital mortality prediction of intensive care unit patients using an ensemble learning approach. Int. J. Med. Informatics **108**, 185–195 (2017)
13. Pirracchio, R., Petersen, M., Carone, M., Rigon, M., Chevret, S., van der Laan, M.: Mortality prediction in intensive care units with the Super ICU Learner Algorithm (SICULA): a population-based study. Lancet Respir. Med. **3**(1), 42–52 (2015)
14. Pirracchio, R.: Mortality prediction in the ICU based on MIMIC-II results from the Super ICU Learner Algorithm (SICULA) project. In: Secondary Analysis of Electronic Health Records, pp. 295–313. Springer, Cham (2016). https://doi.org/10.1007/978-3-319-43742-2_20
15. Le Gall, J.R., Lemeshow, S., Saulnier, F.: A new Simplified Acute Physiology Score (SAPS II) based on a European/North American Multicenter study. JAMA: J. Am. Med. Assoc. **270**, 2957–2963 (1993)
16. Von Rosing, M., Scheer, A., Zachman, J., Jones, D., Womack, J., von Scheel, H.: Phase 3: process concept evolution. In: The Complete Business Process Handbook, pp. 37–77 (2015)
17. Gelman, A., Hill, J.: Missing-data imputation. In: Data Analysis Using Regression and Multilevel/Hierarchical Models, pp. 529–544 (2006)
18. Suarez-Alvarez, M., Pham, D., Prostov, M., Prostov, Y.: Statistical approach to normalization of feature vectors and clustering of mixed datasets. Proc. R. Soc. A: Math. Phys. Eng. Sci. **468**(2145), 2630–2651 (2012)
19. Iglewicz, B.: Robust scale estimators and confidence intervals for location. In: Hoaglin, D.C., Mosteller, M., Tukey, J. W. (eds.) Understanding Robust and Exploratory Data Analysis. Wiley, New York (1983)
20. Medicode (Firm). ICD-9-CM: International Classification of Diseases, 9th Revision, Clinical Modification. Medicode, Salt Lake City (1996)
21. Karabulut, E.M., Ibrikci, T.: Effective automated prediction of vertebral column pathologies based on logistic model tree with SMOTE preprocessing. J. Med. Syst. **38**(5), 1–9 (2014). https://doi.org/10.1007/s10916-014-0050-0
22. Zeng, M., Zou, B., Wei, F., Liu, X. Wang, L.: Effective prediction of three common diseases by combining SMOTE with Tomek links technique for imbalanced medical data, pp. 225–228 (2016)
23. Jiang, L., Zhang, B., Ni, Q., Sun, X., Dong, P.: Prediction of SNP sequences via Gini impurity based gradient boosting method. IEEE Access **7**, 12647–12657 (2019)
24. Berrar, D.: Bayes' theorem and naive bayes classifier. Encycl. Bioinform. Comput. Biol. 403–412 (2018)
25. Wright, R.E.: Logistic regression. In: Grimm, L.G., Yarnold, P.R. (eds.) Reading and Understanding Multivariate Statistics, pp. 217–244. American Psychological Association (1995)
26. Lee, T., Ullah, A., Wang, R.: Bootstrap aggregating and random forest. In: Fuleky, P. (ed.) Macroeconomic Forecasting in the Era of Big Data. Advanced Studies in Theoretical and Applied Econometrics, vol. 52, pp. 389–429. Springer, Cham (2019). https://doi.org/10.1007/978-3-030-31150-6_13

27. Liaw, A., Wiener, M.: Classification and regression by random forest. R News **2**, 18–22 (2002)
28. Winham, S., Freimuth, R., Biernacka, J.: A weighted random forests approach to improve predictive performance. Stat. Anal. Data Min.: ASA Data Sci. J. **6**(6), 496–505 (2013)
29. Cover, T., Hart, P.: Nearest neighbor pattern classification. IEEE Trans. Inf. Theor. **13**(1), 21–27 (2006)
30. Hearst, M., Dumais, S., Osuna, E., Platt, J., Scholkopf, B.: Support vector machines. IEEE Intell. Syst. Their Appl. **13**(4), 18–28 (1998)
31. Dietterich, T.G.: Ensemble methods in machine learning. In: Kittler, J., Roli, F. (eds.) MCS 2000. LNCS, vol. 1857, pp. 1–15. Springer, Heidelberg (2000). https://doi.org/10.1007/3-540-45014-9_1
32. Chen, T., Guestrin, C.: XGBoost. In: Proceedings of the 22nd ACM SIGKDD International Conference on Knowledge Discovery and Data Mining, pp. 785–794 (2016)
33. Hao, J., Ho, T.: Machine learning made easy: a review of scikit-learn package in Python programming language. J. Educ. Behav. Stat. **44**(3), 348–361 (2019)
34. Morey, R., Fine, D., Loree, S., Retzlaff-Roberts, D., Tsubakitani, S.: The trade-off between hospital cost and quality of care. Med. Care **30**(8), 677–698 (1992)
35. Luque, A., Carrasco, A., Martín, A., de las Heras, A.: The impact of class imbalance in classification performance metrics based on the binary confusion matrix. Pattern Recogn. **91**, 216–231 (2019)
36. Wang, H., Zheng, H.: True positive rate. In: Encyclopedia of Systems Biology, pp. 2302–2303 (2013)
37. Ting, K.M.: True negative rate. In: Sammut, C., Webb, G.I. (ed.) Encyclopedia of Machine Learning. Springer, Boston (2011)
38. Melo, F.: Area under the ROC curve. In: Encyclopedia of Systems Biology, pp. 38–39 (2013)
39. Kulluk, S., Özbakır, L., Tapkan, P., Baykasoğlu, A.: Cost-sensitive meta-learning classifiers: MEPAR-miner and DIFACONN-miner. Knowl.-Based Syst. **98**, 148–161 (2016)
40. Shilbayeh, S.: Cost sensitive meta learning. Ph.D. School of Computing, Science and Engineering University of Salford Manchester, UK (2015)

Unified Approach for Designing a Database for the Needs of the Anesthesiologists in Bulgaria

Rumen Rusev[1](✉), Ivo Rusev[1], and Vladin Petrov[2]

[1] University of Ruse "Angel Kanchev", 8, Studentska Street, 7017 Ruse, Bulgaria
rumen.rusev@uni-ruse.bg

[2] University Hospital "Kaneff" Ruse, 2, Nezavisimost, 7002 Ruse, Bulgaria
vladin.petrov@hospitalruse.org

Abstract. This paper addresses the need for building an information system and a database which can improve the functioning of the anesthesiology departments in Bulgaria. It also provides means for the analysis of the collected anesthesiology data.

An overview of already existing similar developments worldwide has been made. Based on the conducted analysis and the experience of the authors in anesthesia and software development, is proposed a set of data on the different medical conditions, the parameters of anesthesia and the information on the postoperative condition of patients. It is proposed a structure of a database and an architecture of an information system which will allow to work with the collected information.

The information system project is designed to operate on two different levels-hospital level and a national level. The separation of data is done based on the possible different usages of the data- storing of a hospital level archive, analyzing national data anesthesiologic data with the purpose to improve the results in that area and reducing the complication while applying anesthesia including the post-operative period. A multitier client-server architecture will be used during the design of the system. The back-end provides a REST API that would also allow external system integration. Measures for information protection are proposed, which is especially critical in the analysis of a national database with the possibility of use by all registered medical specialists and organizations. In the national database the patient-sensitive information (e.g. names and identifiers) will be stored using one way encryption. The purpose of the proposed information system and database is to improve the quality of work of anesthesiologists and to use the collected data for future research. At the end of this paper is given a short overview of possible usage of IoT devices to automatically collect data during post-operative conditions.

Keywords: Medical database · Information systems in medicine · Anesthesiology

S. S. Sotirov et al. (Eds.): BioInfoMed 2020, LNNS 374, pp. 32–38, 2022.
https://doi.org/10.1007/978-3-030-96638-6_3

1 Introduction

1.1 Overview of the Information Management in the Anesthesiologic Activity in the Medical Practice

Anesthesia is an important part of a surgery which enables patients to undergo an operation or a surgery without feeling distress and pain. A huge number of anesthesias are performed each year where in such procedures different parameters are involved, i.e. already existing patient medical conditions, anesthesia parameter and others.

Currently most of the anesthesiologists keep track of the historical data with the goal to analyze and improve it.

Research in the anesthesiology field often includes finding relations and medical algorithms in patient's data [1, 2]. This is especially important when looking for demographic aspects because a large set of data is needed [3]. At the moment most of the anesthesiology data is collected manually and does not contain large enough statistical sample for accurate analysis. An information system will significantly improve this process by providing an easy to use interface for storing data from many different places. Worldwide there are already similar systems that aim to improve medical activities using computer technology, particularly in anesthesiology [4]. Solutions for information systems for the needs of anesthesiology at the national level are also known. An excellent example of this is the Danish Anesthesia Database system [5]. However, the use of information technology in this field in Bulgaria is rarely done and in the Bulgarian hospitals collects data on the anesthesia on department level. Other specialists whether from the same hospital or other healthcare facilities cannot access the gathered data. In the republic of Bulgaria no data is collected on a national level. The collection of such information can allow further analysis which can reduce patient complications and improve the anesthesiologists work.

Analysis of this data could significantly improve anesthesia and the overall patient experience so it is important to store digitally the collected data so that it can be used in the medical practice.

1.2 Justification for the Need of a National Unified System for the Needs of the Anesthesiologists

Large databases have traditionally been the basis for statistical analysis. Today they are also used in artificial intelligence systems. Specifically, in the field of anesthesiology, we focus on, but are not limited to, several possible applications of the national anesthesiology information system.

First - automatic determination of the functional state of the patient (ASA status). There are reports that this is possible [6] and allows avoiding the subjective assessment of the anesthesiologist or reinsurance against risks and accidents by determining a higher ASA class.

Second - prognosis of risks and adverse events [7], which allows for the prevention of adverse events or the use of different than planned anesthesia techniques.

Third - assistance in choosing the appropriate anesthesia [8].

Fourth - planning and management of the operation of operating rooms [9] by analyzing data on the type and duration of the upcoming operation, the accompanying diseases of the patient, the type of planned anesthesia.

Fifth - analysis of data related to research and studies in the field of anesthesiology.

Decisions made after analysis of large databases cannot replace the clinical experience, skills, and judgment of anesthesiologists, but they can support and facilitate their work.

1.3 Unified System for the Needs of Anesthesiologists in Bulgaria

The proposed system is a unified national information system that allows the collection of patient's anesthesia data first on a hospital level. This will also allow to control and evaluate the anesthesiology departments activities.

In order to further improve the anesthesiology, work the solution will work also on a national level. Each healthcare facility will be able to collect its own data and the users will be able to access other hospitals' data.

This will enable the anesthesiologists to use large volumes of data on a national level. The specialists can further use this statistical information through the information system for similar cases and track patient's condition in a more long-term period.

1.4 Identifying Key Anesthesiology Data that Will Be Stored

Key anesthesia data that needs to be persisted in the information system was identified using the experience of the authors. For each patient needs to be collected personal data. This includes:

- general patient data (name, age and contact data),
- pre-existing medical conditions,
- lab results,
- pre-surgical therapy and
- type of surgical procedure that is being performed

This information can be stored locally in the anesthesia system or extracted from the hospital administration system.

The database will contain information about the performed anesthesia - type, medications used, values of different vital signs during the operation and etc. In order to monitor the condition of the patients in the postoperative period, it is planned to store data from the equipment that monitors the condition of the patients. It is possible to use the widely available IoT (Internet of Thins) technologies to collect patient status data [10] for a longer period after the end of a hospital stay. Postoperative condition data will be useful for assessing the relations between anesthesia and the patient's recovery process.

2 System Architecture and Design

The proposed information system will be web application that will be accessible through the Internet (Fig. 1) via Graphical User Interface (GUI) or an Application Programming Layer (API). Each hospital or a healthcare facility will be able through the Internet to access the information system and add data to the national anesthesia system. The different hospitals would be able to use the system either via the graphical user interface or in case they want to integrate their internal information systems the API can be used.

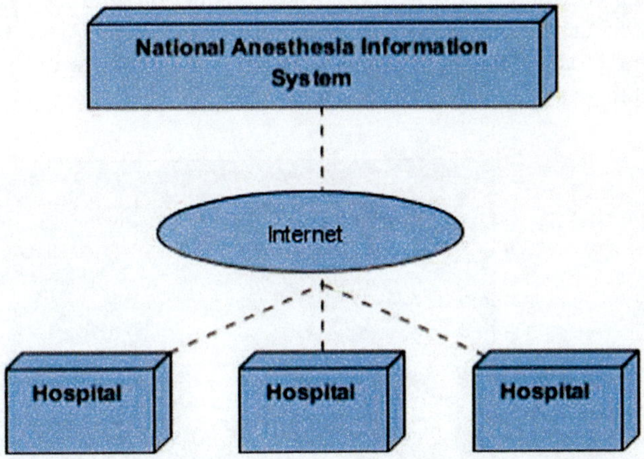

Fig. 1. A high-level overview of the national anesthesiology information system

2.1 System Roles

The system will have different roles that will be able to execute different set of actions. Those roles help for the separation of information and controlling access to sensitive information (like patient's data).

Anesthesiologists. This role would be assigned to specialists who work at hospitals and healthcare facilities and are directly involved with patient's anesthesiology data. They would be able to enter patient data, anesthesia data and to follow up on the patient's condition. This role is reserved mainly for medical doctors who will enter their own patient's data.

Head-of-Department. The head of department is a role will be able to execute everything that the anesthesiologist role can, plus will be able to control and assess the anesthesia for all patients in the anesthesiology department.

Head-of-Administration. This role is reserved for the head of administration of the hospital and healthcare facility. It will be able to request statistical data but will not have direct access to patient's data.

External Users. Access will be allowed for people that are not working in the current health care facility. For example, those are medical staff from other hospitals. The external users will be able to access anesthesia and patient's data with explicit approval from the treating physician of the patient, which can be requested through the system.

2.2 Use Case View

Here are explained the critical paths of the application for the main use cases. The shown in (Fig. 2) use case displays the Anesthesiologist interaction with the system. As seen in the diagram the anesthesiologist can manage his own patient's data. He can create, edit and update the data. The delete in the diagram is a soft delete that does not erase information but moves it into archived database table.

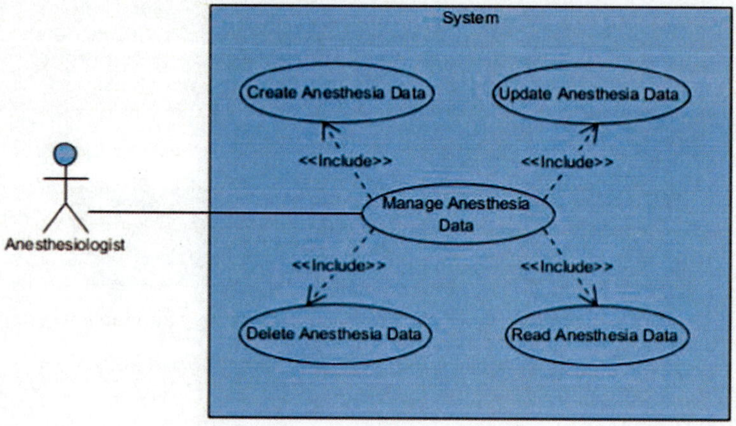

Fig. 2. Anesthesiologist use case view

2.3 Used Technologies

The most common modern approaches for building information systems were used in the design of the system [11]. The chosen architecture allows easy extension of the system's functionalities, integration between the local hospital systems and allow to access a large amount of collected information for doctors and researchers. Modern technologies such as artificial intelligence and machine learning can be used for automated analysis of the collected data [12].

Back-end (Third Level)
The software architecture of the back-end will be written in Java with the Springboot framework, which provides an embedded Tomcat web server. This framework also provides a lot of plug-ins which can decrease the development time and provide easy maintenance. It also reduced the boilerplate code and the number of configurations that

need to be made in a typical web-based system. Each instance of the application will be deployed in a docker container. Docker provides isolation and will increase the speed in which new instances of the application are created. Additionally Openshift or Azure can be used to manage the application instances.

The load balancer included in the diagram (Fig. 3) will route the request to the different docker nodes. This is done in order to be able to scale the application better horizontally. The nodes represent a single instance of the application and can be increased in case of high load. The database where the data is stored will be a separate component.

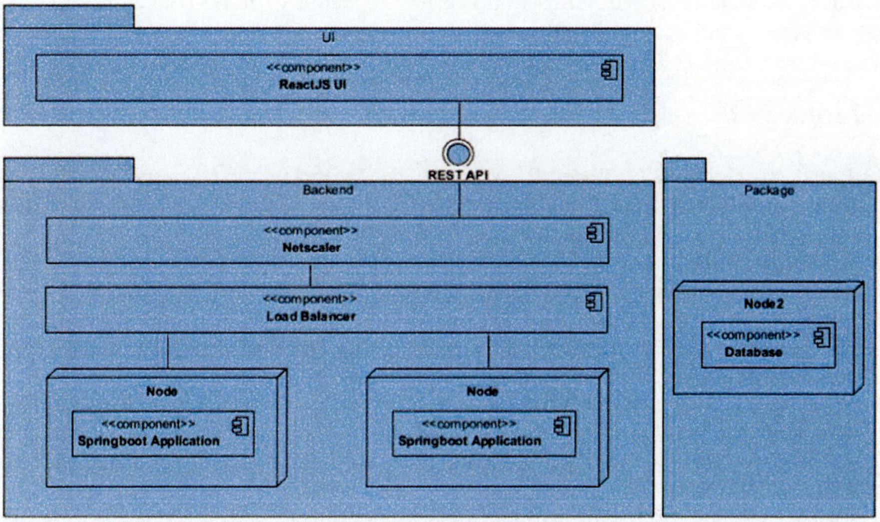

Fig. 3. Component diagram of the architecture.

Front-end (Third Level). The UI will connect to securely the back end by calling its endpoint using an OAuth security. The back end will expose an API for the front-end which will be behind a Netscaler component. The Netscaler can be configured to protect the system from DOS attack and can also provide logs. Behind the Netscaler there will be a load balancer which will take care to which node to send the requests.

The front-end will be written in a Javascript and HTML using the framework react.js. This is done in order to ensure the support of maximum number of mobile devices.

3 Conclusion and Future Work

After an analysis of the current state of the computer information collection of the data needed to aid the activity of the anesthesiologists in Bulgaria, it is proposed to build a modern information system. A design of the system is proposed, taking into account the experience of anesthesiologists and the specifics of medical activity in Bulgaria. It is possible to work both at the local level (department, hospital) and to summarize the information at the national level. The future work is in the direction of:

- refining the scope of the stored data and integration with the most commonly used in Bulgaria hospital information systems;
- the creation of an hardware infrastructure for an information system for anesthesiology at the national level
- Implementation of the National Information System for the Anesthesiologists and its integration in the healthcare facilities in Bulgaria.

The development of the proposed system will improve the quality of the work of the anesthesiologists in Bulgaria and will aid its control. The collection of a big amount of data on a national level will allow to do further research with the goal of finding best practices and reducing complication in the post-operative period.

References

1. Petrov, Vl., Georgiev, G., Radev, R., Arabadzhieva, D., Radev, Vl.: Comparison of hemodynamic effects of continuous spinal and combined spinal-epidural anesthesia for lower limb revascularization. Anaesthesiol. Intensive Care, **1**, 9–13 (2016)
2. Arabadzhieva, D., Radev, Vl., Radev, R., Petrov, Vl.: BIS – based algorithm for assessing adequacy of anesthesia in obese patients. Anaesthesiolo. Intensive Care **5**, 13–16 (2014)
3. Radev, V., Radev, R., Arabadzhieva, D., Petrov, V., Kolev, N.: Original article. Demographic changes in the population of bulgaria as challenges for anaesthesiologists. J. Biomed. Clin. Res. **9**(2), 130–133 (2016). https://doi.org/10.1515/jbcr-2016-0019
4. Stonemetz, J., Ruskin, K.: Anesthesia Informatics. Springer, London (2008). https://doi.org/10.1007/978-0-387-76418-4
5. Kristian, A., Rosenstock, C., Lundstrøm, L., The Danish anaesthesia database. Clin. Epidemiol. J. (2016)
6. Zhang, L., Fabbri, D., Lasko, T.A., Ehrenfeld, J.M., Wanderer, J.P.: A system for automated determination of perioperative patient acuity. J. Med. Syst. **42**(7), 1–11 (2018). https://doi.org/10.1007/s10916-018-0977-7
7. Lin, C.-S., Chiu, J.-S., Hsieh, M.-H., et al.: Predicting hypotensive episodes during spinal anesthesia with the application of artificial neural networks. Comput. Methods Programs Biomed. **92**, 193–197 (2008). https://doi.org/10.1016/j.cmpb.2008.06.013
8. Hancerliogullari, G., Hancerliogullari, K.O., Koksalmis, E.: The use of multi-criteria decision making models in evaluating anesthesia method options in circumcision surgery. BMC Med. Inform. Decis. Mak. **17**, 14 (2017). https://doi.org/10.1186/s12911-017-0409-
9. Combes, C., Meskens, N., Rivat, C., Vandamme, J.-P.: Using a KDD process to forecast the duration of surgery. Int. J. Prod. Econ. **112**, 279–293 (2008). https://doi.org/10.1016/j.ijpe.2006.12.068
10. Rusev R., Rusev, I.: Study of the possible usage of Internet of Things devices in the sphere of healthcare. In: Proceedings University of Ruse "Angel Kanchev", vol. 58, book 6.5 ISSN 1311-3321
11. Garcia-Molina, H., Ullman, J.W.: Database Systems - The Complete Book, 2nd edn. Pearson Prentice Hall, London (2009)
12. Berman, J.J.: Confidentiality issues for medical data miners. In: Artificial Intelligence in Medicine, vol. 26 (2002)

Computer-Based Rehabilitation of Cognitive Impairments in Patients with Multiple Sclerosis

Valentina Ignatova[1](✉) and Lyudmila Todorova[2]

[1] Clinic of Neurology, Multiprofile Hospital for Active Treatment – National Heart Hospital, 1309 Sofia, Bulgaria
valyaig@abv.bg

[2] Institute of Biophysics and Biomedical Engineering, Bulgarian Academy of Sciences, 1113 Sofia, Bulgaria

Abstract. Multiple sclerosis (MS) is an inflammatory disease of the central nervous system, which leads to demyelinating lesions in brain and spinal cord and further neurodegeneration. It results in manifestation of motor, cognitive and neuropsychological symptoms, which can occur simultaneously or independently of each other. Approximately 2/3 of people with MS suffer from cognitive impairment which lead to deteriorated quality of life.

In unison with the global world tendencies, for the first time in Bulgaria a com-puter system Rila Cogni Soft for assessment and rehabilitation of cognitive im-pairment has been developed. The system is composed of 3 modules: 1) electronic record of demographic and clinical data; 2) module for assessment of the cognitive status and 3) module for rehabilitation of the registered cognitive impairment. The rehabilitation training programs are developed on the basis of established international standards. Based on the affected cognitive domains and degree (severity) of damage, the therapist administers individual rehabilitation programs.

Using the system, we create our own norm (based on results from 70 healthy persons) for each of the diagnostic tests. A statistically significant difference ($p < 0.05$) in the level of cognition was found between healthy controls and individuals with MS ($n = 23$). InterCriteria Analysis was used to study the relationships between the degree of cognitive impairment and clinical features. Improved cognitive functions were registered after 3-month period of rehabilitation ($p > 0.05$).

Keywords: Multiple sclerosis · Cognitive impairment · Cognitive rehabilitation · InterCriteria analysis

1 Introduction

1.1 A Subsection Sample

Multiple sclerosis (MS) is an inflammatory disease of the central nervous system, which leads to demyelination and subsequent neurodegeneration. The wide distribution of lesions in brain and spinal cord in patients with MS results in manifestation of

S. S. Sotirov et al. (Eds.): BioInfoMed 2020, LNNS 374, pp. 39–49, 2022.
https://doi.org/10.1007/978-3-030-96638-6_4

motor, cognitive and neuropsychological symptoms, which can occur simultaneously or independently of each other.

Neurorehabilitation is based on two main approaches –restorative and compensatory (strategic). The restorative approach depends on the brain's ability for cortical reorganization after injury and leads to alleviation of the patient's deficit in processing and interpreting the information. The compensatory approach for training is based on strategy to adapt the person to the environment against his/her personal background. As can be expected in such a disease with pronounced cerebral pathology, a significant number [1–3] of MS patients are cognitively compromised in both the earlier and later stages of the disease. Cognitive impairments affect up to 70% of people with MS and negatively affect employment, social functioning, quality of life and daily activities [4].

Naturally, need to overcome cognitive deficits in these patients is increasingly the focus of clinical practice and global research. General cognitive symptoms include deficits in attention, efficiency and speed of data processing and long-term memory, which are the basis for today's efforts for rehabilitation and treatment of neuropsychiatric disorders in individuals with MS [5].

Although the rehabilitation of cognitive impairments has been applied relatively recently, based on a comparative analysis of studies on the problem, it is necessary to update the approach to these patients. While in the beginning the emphasis was on memory and learning new information, now the efforts are focused on improving the executive functions, attention and speed of information processing [6].

The potential of individuals to improve their own cognitive abilities through the systematic exercise of thought strategies with increasing difficulty is intriguing and is only now beginning to be used more fully. Complex mental activity has been shown to cause improvements in cognitive and brain function, but it is not clear to what extent the brain is capable of such plasticity [1]. Cognitive plasticity is thought to be due to a long-term mismatch between the body's functional reserves and environmental requirements and depends on the brain's ability to anatomically adapt to reactive changes in behavioral flexibility.

An extensive up-to-date review of research on the topic [6] found conflicting results regarding the effectiveness of different cognitive rehabilitation techniques, which means that no definite conclusions can be drawn about their effect on cognition, mood, quality of life, fatigue and self-efficacy. The lack of convincing evidence in these studies may be due to the heterogeneity of rehabilitation approaches, methodological weaknesses, and the small sample size that characterize most studies. COGNIsoft-I is a software, designed for rehabilitation of more severe and less selective cognitive impairment due to head trauma, stroke and multiple sclerosis (https://www.cognisoft.info/cognisoft-i.html).

There is currently no developed standard for cognitive rehabilitation in patients with MS, which is natural given the individual course of the disease both in terms of physical disability and cognitive aspect. Techniques with different cognitive architecture have been developed: 6-week home-based program for training of memory, attention, planning and oral performances [2, 3]; BrainStim computer training program [7], which trains spatial orientation, visual memory and the working memory and adapts the level of difficulty; Computer training program CogniFit Personal [8] with 15 learning tasks;

RoProCog-SEP program [9] for memory training, executive functions, associative memory, visual-spatial abilities, attention; MAPSS-MS [10] which builds cognitive compensatory strategies followed by computer-assisted home-based cognitive rehabilitation; RehaCom system [11] for training the attention, memory, executive functions, visual-motor abilities at different levels; Specific training based on the Attention Processing Training program (APT) [12].

The aim of the study was presentation of innovative computer system for individual targeted rehabilitation of specific impaired cognitive domains in persons with MS based on actual international trends.

2 Rehabilitation Module Rila Cogni Soft

The computer based cognitive rehabilitation, presented in the paper, was developed and introduced for the first time in Bulgaria. The system is implemented in a cloud infrastructure and will be provided as an IT service for neurological wards and clinics.

The system, named Rila Cogni Soft, consisting of 3 modules. The first module contains patient's demographic and clinical data, inserted by the therapist into the system. The second module is designed for cognitive assessment [13]. The third module allows rehabilitation of the registered impairment in the specific cognitive domains: working memory, delayed recall, attention, visuo-spatial orientation skills, emotional recognition, executive functions. Based on the affected cognitive domains and degree (severity) of damage, the therapist selects appropriate rehabilitation programs, taking into account the level of difficulty and the frequency of execution.

The module for rehabilitation of cognitive disorders in people with MS includes the following training programs:

2.1 Training Program for Rehabilitation of Speed of Information Processing

The training program for rehabilitation of speed of information processing is based on Symbol Digit Modality Test (SDMT) [14]. The program starts with presentation of a key field, which shows a correspondence between a symbol/stylized image and a number. The patient has time for a short training (30 s), followed by the actual execution of the program. During the rehabilitation the patient is required to fill the empty cells with digits corresponding to the appropriate symbols above (see Fig. 1).

The training program includes 3 levels of complexity - with different number of fields to fill (45, 60, 90 respectively). In order to be more attractive and to avoid habituation, various combinations of symbols and stylized images have been proposed.

2.2 Training Program for Rehabilitation of Visual-Spatial Memory and Orientation

The training program for rehabilitation of visual-spatial memory and orientation is a short program, which also train episodic memory. The game follows the principle of Brief Visual-Memory Test (BVMT) [15]. The program starts with presentation of combination of 3-dimensional geometric figures. Their number varies according to its level

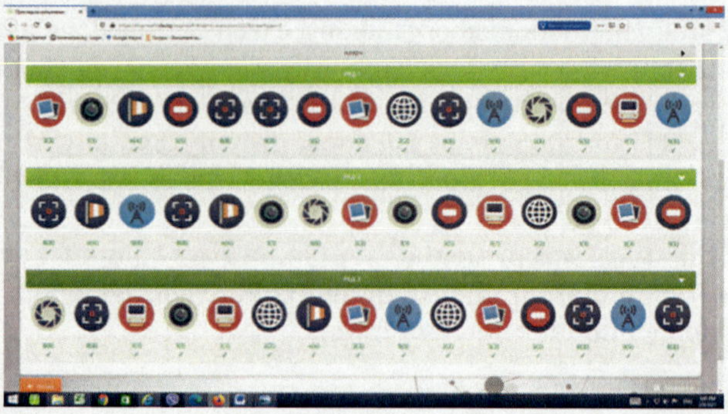

Fig. 1. Sample screen from a successfully completed (green ticks) training program for rehabilitation of speed of information processing

of difficulty (6–12). The figures present for 10 s. A subsequent screen displays an empty grid which should be filled through choice from a set of doubled number of images, half of which are the correct ones. The patient is required to memorize the figures and their location and to reproduce the configuration (see Fig. 2). The success rate depends on the choice of the correct figure as well as its correct positioning. Three performance attempts are provided and the best result is reported.

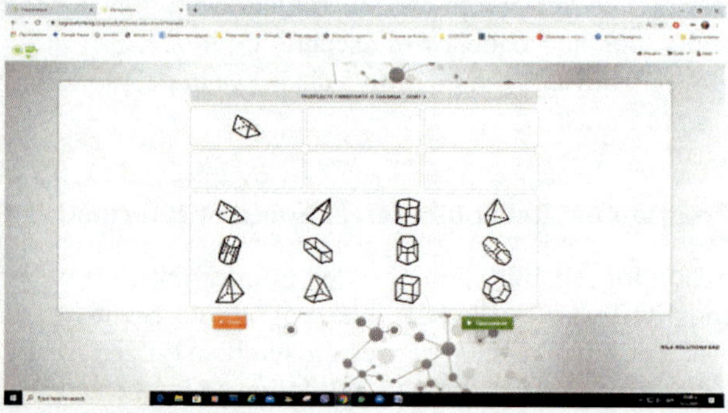

Fig. 2. Sample screen from process of execution of training program for rehabilitation of visual-spatial memory and orientation.

2.3 Training Program for Rehabilitation of Verbal Learning

The training program for rehabilitation of verbal learning is based on presentation of list of auditory stimuli and aims to improve the episodic verbal learning. The game consists of 5 attempts in which a series of 16 words is pronounced "behind the screen" (see Fig. 3). After each listening, the system automatically displays two consecutive screens, which contain 32 fields-buttons with words, among which are the spoken 16 words. The person must use the mouse to select the words he thinks he has heard. He can change his choice of words until he decides to move forward by pressing the "Continue" button. In case of difficulties, the option for recognition is facilitated – the words on buttons are accompanied by explanatory image.

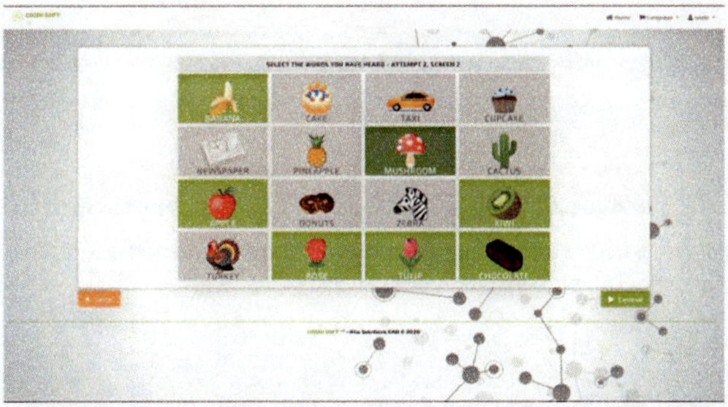

Fig. 3. Sample screen from training program for rehabilitation of verbal learning (the chosen answers from the patient are marked in green).

The computer implementation of the offered Training program for rehabilitation of verbal learning includes different categories of images: national flags; animals; plants; vegetables; fruits; beverages; food; tools; cartoon characters; sports equipment; vehicles; numbers; musical instruments; decoration.

2.4 Picture Memory Training Program

The Picture memory training program is a computer adaptation of the Picture memory test [16], which is based on the idea that human memory is extremely sensitive to the symbolic modality of presenting event information. The patient should remember set of pictures and recognize them from list of pictures from grid. The recognition of sequence/position of image are not required (see Fig. 4). Three levels of difficulty are proposed. Complicating the levels is achieved by increasing the number of images and reducing both the time for exposure and time for execution.

2.5 Training Program for Rehabilitation of Executive Functions

Two variants of rehabilitation games for improving the executive functions are developed. Both games are (A and B) based on the idea to remember and to arrange set of images with specific order. The rehabilitation starts with basic level for the appropriate game and the task is complicated after successful completion of the current level. In Game A, the person is required to recognize the originally set colored symbols from a set twice as large and to position them correctly (see Fig. 5).

In game B, the person is required to memorize a configuration of two-color checkers and restore it to an empty grid (see Fig. 6).

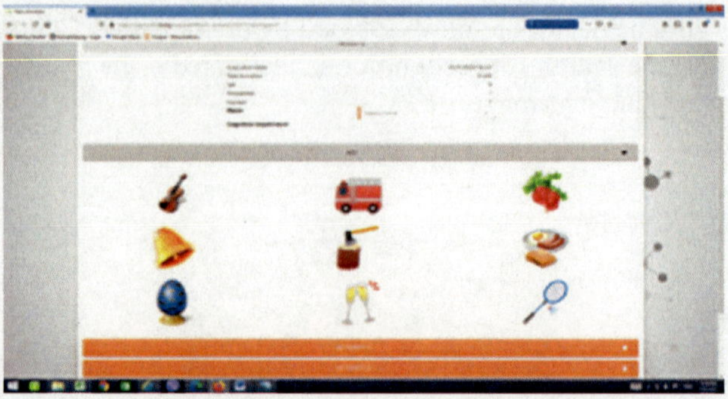

Fig. 4. Sample screen with results from completion of picture memory training program. Basic configuration for the specific performance.

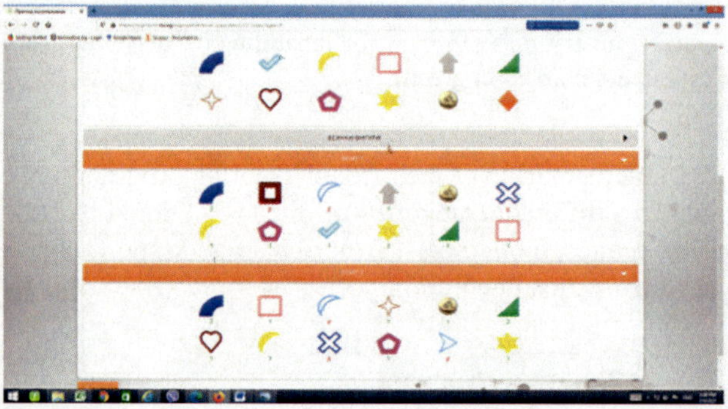

Fig. 5. Sample screen with results from completion of training program for rehabilitation of executive functions (game A). The basic configuration, followed by results from 2 attempts, is visible. The correct responses are marked with green tick and the wrong ones- with red crosses.

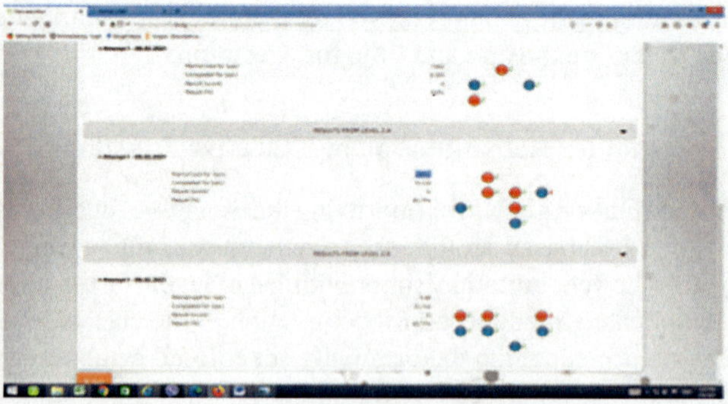

Fig. 6. Sample screen with grids of different levels of complexity from completion of training program for rehabilitation of executive functions (game B).

2.6 Training Program for Rehabilitation of Emotional Recognition

Computer adaptation of Visual empathy test is used for composition of Visual empathy training program of the platform Rila Cogni Soft for cognitive rehabilitation of patients with MS. When starting the test, a facial expression (only the area around the eyes) of a person (man or woman) and 2 or 4 words corresponding to 2/4 possible emotional states are displayed on the screen (see Fig. 7). Only one of the words corresponds exactly to the visualized expression, i.e. - the correct answer.

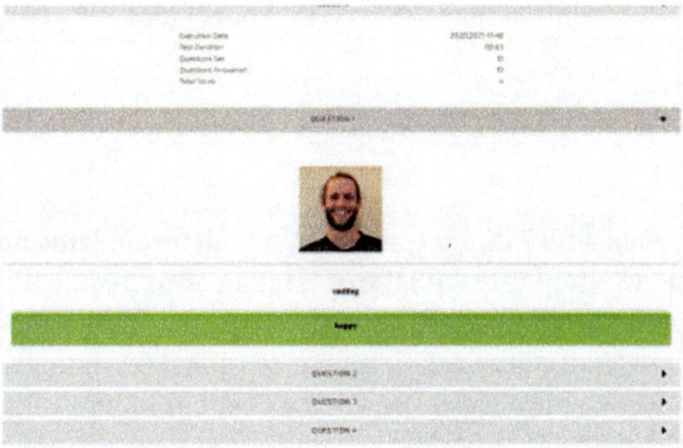

Fig. 7. The sample screen illustrates choice of emotional state (green field) during completion of training program for rehabilitation of emotional recognition.

Animation version of the program is also developed. The emotional state should be defined through analysis of body language (see Fig. 8).

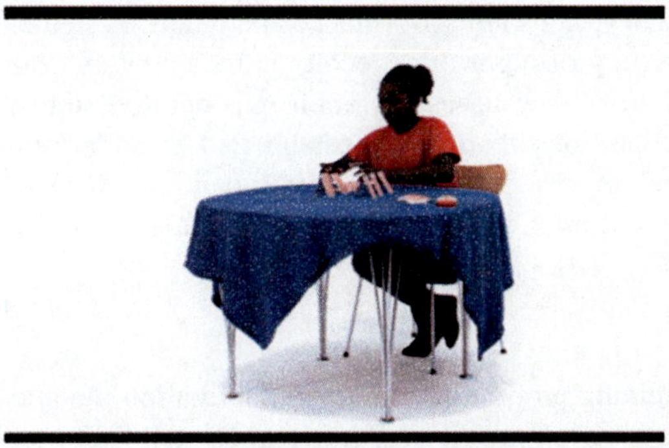

Fig. 8. The sample screen from clip presenting emotional state which should be recognized.

The aim is to rehabilitate the non-verbal communication, which is one of the main channels through which people transmit socially and psychologically significant information.

3 Results

Based on the results of 70 healthy individuals from the tests performed with the cognitive assessment module of RilaCognisoft, a range of normative values was developed for the three main cognitive domains that are affected in MS - Table 1.

Table 1. Cognitive function. Comparison of results between MS patients and healthy controls.

	n	Speed of information processing	Visuo-spatial orientation	Verbal learning and referred memory
Healthy controls	70	55.8 ± 3.62	13.5 ± 5.15	11.26 ± 0.32
Patients with MS	23	37.52 ± 17.02	9.57 ± 2.63	12.91 ± 3.78

Twenty three patients with MS were also tested with through the presented diagnostic battery. The results were compared with the obtained norms and on this basis the levels of cognitive impairment in the respective domains were determined. The cognitive decline registered in the patients had statistical significance ($p < 0.05$) compared to the cognitive status of the studied healthy individuals. In view of the finding in the particular patient, the therapist has developed an individual rehabilitation strategy, which includes various rehabilitation programs with an appropriate level of complexity. Rehabilitation sessions were conducted with an average duration of 15 min and an average frequency of once every 5 days for 3 months. At the end of the rehabilitation period, a control test of cognition was performed. A positive effect on cognitive skills was reported ($p > 0.05$).

InterCriteria Analysis (ICA) [17] is a recently developed approach used to measure the ordinal relationship between two or more observed parameters. The method uses the apparatus of the indexed matrices [18] and Intuitionistic Fuzzy Sets [19]. ICA is successfully applied in various fields of science and practice - in the field of economics, e-learning, the efficiency of algorithms, ecology, medicine and others [20–22]. After applying the ICA-approach for disease duration and cognitive status, no stable tendency for deterioration of cognition over time was established. Similar results were obtained for the degree of physical impairment and cognitive impairment. Due to the relatively small number of patients monitored, these dependencies remain to be further investigated.

4 Discussion

The choice of the training programs from the rehabilitation module Rila Cogni Soft is justified for the following reasons: it is effective; it is intended for international use; the results are observable for the patient and easy for statistical processing and interpretation. On the other hand, the rehabilitation provides options for many and varied combinations. In this way a habituation is excluded, the interest of the patients is provoked and kept. Neurorehabilitation is based on 2 main approaches:

1. Restorative/functional training approach/- alleviation of the patient's deficit in the processing and interpretation of information

2. Compensatory/strategic/approach to training (eg changing the patient's environment.

The selection of the above listed rehabilitation programs in the RilaCognisoft system provides complete neurorehabilitation with both approaches.

Following the current trends for cognitive rehabilitation in people with MS of the listed rehabilitation programs in the RilaCognisoft, the system provides complete neurorehabilitation with both approaches.

Following the current trends for cognitive rehabilitation in people with MS [23], we have relied not only on training of working memory and attention, but also on the speed of information processing, visual-spatial orientation, verbal learning, executive functions and socio-cognitive skills. Stimulation of the cognitive functioning would help the adaptation of the homologous to the damaged cortical regions, the expansion of the associative local areas and outside the region activation of the cognitive networks. It is not yet possible to clearly differentiate adaptive from non-adaptive neuroplasticity, which would allow the development of specific cognitive rehabilitation programs for all MS phenotypes and would encourage the validation of the most effective cognitive rehabilitation interventions for people with MS [24].

According to Mike, A et al. [25], both a disconnection mechanism due to lesions in the white matter of the brain and cortical thinning of specific brain areas can cause cognitive deficits in MS, impairing the processing of emotions and mental state by facial expression. As a result, daily activities are disrupted and difficulties arise in the social life of patients [25].

5 Conclusion

The present study shows that computer cognitive rehabilitation leads to adaptive cortical reorganization. Systematic performance of cognitive exercises is the basis for maintaining and even improving the cognitive brain reserve. The presented rehabilitation module Rila CogniSoft is a step towards optimizing the rehabilitation potential of persons with MS.

Further studies of cognitive rehabilitation interventions are highly recommended as hopeful therapeutic approach for cognitive improvement considering limited drug treatment yet. More research are needed for expanding the knowledge about the effectiveness of long-term interventions, the duration, frequency and intensity of training sessions to achieve maximum effect in terms of maintaining and improving cognitive reserve in people with MS.

Acknowledgement. This research is supported by the Bulgarian National Science Fund under Grant Ref. No. KP-06-N22/1/2018 "Theoretical research and applications of InterCriteria Analysis".

References

1. Chapman, S.B., et al.: Neural mechanisms of brain plasticity with complex cognitive training in healthy seniors. Cerebr. Cortex **25**(2), 396–405 (2013)

2. Brenk, A., Laun, K., Haase, C.G.: Short-term cognitive training improves mental efficiency and mood in patients with multiple sclerosis. Eur. Neurol. **60**(6), 304–309 (2008)
3. Baller, G.: Cognitive Training: 2-Week Course to Improve Cognitive Performance, Parts IIII. Bad Honnef, Hippocampus (2005)
4. Doskas, T., et al.: Neurocognitive impairment and social cognition in multiple sclerosis. Int. J. Neurosci. 1–26 (2021)
5. Chiaravalloti, N.D., DeLuca, J.: Cognitive impairment in multiple sclerosis. Lancet Neurol. **7**(12), 1139–1151 (2008)
6. Mitolo, M., Venneri, A., Wilkinson, I.D., Sharrack, B.: Cognitive rehabilitation in multiple sclerosis: a systematic review. J. Neurol. Sci. **354**(1–2), 1–9 (2015)
7. Penner, I.K., Kobel, M., Opwis, K.: BrainStim – a recently developed tool to train different aspects of working memory. In: Proceedings of the INS/GNP Conference, pp.17–18 (2006)
8. Shatil, E., Metzer, A., Horvitz, O., Miller, A.: Home-based personalized cognitive training in MS patients: a study of adherence and cognitive performance. Neuro Rehabil. **26**(2), 143–153 (2010)
9. Brissart, H., Leroy, M., Morele, E., Baumann, C., Spitz, E., Debouverie, M.: Cognitive rehabilitation in multiple sclerosis. Neurocase **19**(6), 553–565 (2013)
10. Stuifbergen, A.K., Becker, H., Perez, F., Morison, J., Kullberg, V., Todd, A.: A random-ized controlled trial of a cognitive rehabilitation intervention for persons with multiple sclerosis. Clin. Rehabil. **26**(10), 882–893 (2012)
11. Parisi, L., Rocca, M.A., Valsasina, P., Panicari, L., Mattioli, F., Filippi, M.: Cognitive rehabilitation correlates with the functional connectivity of the anterior cingulate cortex in patients with multiple sclerosis. Brain Imaging Behav. **8**(3), 387–393 (2014). https://doi.org/10.1007/s11682-012-9160-9
12. Amato, M.P., et al.: Computer-assisted rehabilitation of attention in patients with multiple sclerosis: results of a randomized, double-blind trial. Multiple Scler. J. **20**(1), 91–98 (2014)
13. Todorova, L., Ignatova, V., Surchev, J.: Computerized neuropsychological test battery CogniSoft for assessment of cognition in patients with multiple sclerosis. Int. J. Bioautom. **24**(4), 371–380 (2020)
14. Smith A. Symbol Digit Modalities Test: Manual. Western Psychological Services, Los Angeles (1982)
15. Benedict, R.H., Schretlen, D., Groninger, L., Dobraski, M., Shpritz, B.: Revision of the brief visuospatial memory test: studies of normal performance, reliability, and validity. Psychol. Assess. **8**(2), 145 (1996)
16. Yuille, J.C.: Imagery, memory and cognition (PLE: Memory): Essays in honor of Allan Paivio. Psychology Press (2014)
17. Atanassov, K., Mavrov, D., Atanassova, V.: InterCriteria Decision Making: A New Approach for Multicriteria Decision Making Based on Index Matrices and Intuitionistic Fuzzy Sets. Issues Intuition. Fuzzy Sets General. Nets **11**, 1–8 (2014)
18. Atanassov, K.: Index Matrices: Towards an Augmented Matrix Calculus. Studies in Computational Intelligence, vol. 573. Springer, Heidelberg (2014). https://doi.org/10.1007/978-3-319-10945-9
19. Atanassov, K.: On Intuitionistic Fuzzy Sets Theory. Springer, Berlin (2012). https://doi.org/10.1007/978-3-642-29127-2
20. Stratiev, D.S., et al.: Investigation of relationships between bulk properties and fraction properties of crude oils by application of the intercriteria analysis. Petrol. Sci. Technol. **34**(13), 1113–1120 (2016)
21. Fidanova, S., Roeva, O., Gepner, P., Paprzycki, M.: InterCriteria analysis of ACO start strategies. In: 2016 Federated Conference on Computer Science and Information Systems, pp. 547–550. FedCSIS (2016)

22. Roeva, O., Fidanova, S.: Comparison of different metaheuristic algorithms based on InterCriteria analysis. J. Comput. Appl. Math. **340**, 615–628 (2018)
23. Goverover, Y., Chiaravalloti, N.D., O'Brien, A.R., DeLuca, J.: Evidenced-based cognitive rehabilitation for persons with multiple sclerosis: an updated review of the literature from 2007 to 2016. Arch. Phys. Med. Rehabil. **99**(2), 390–407 (2018)
24. Chiaravalloti, N.D., Genova, H.M., DeLuca, J.: Cognitive rehabilitation in multiple sclerosis: the role of plasticity. Front. Neurol. **6**, 67 (2015)
25. Mike, A., Strammer, E., Aradi, M., Orsi, G., Perlaki, G., Hajnal, A., et al.: Disconnection mechanism and regional cortical atrophy contribute to impaired processing of facial expressions and theory of mind in multiple sclerosis: a structural MRI study. PLoS One **8**, e82422 (2013)

Point of Care Lung Diagnostic in ICU – Our Experience with Bedside Lung Ultrasound

Filip Naumovski[1,2(✉)] and Biljana Kuzmanovska[1,2]

[1] Department of Anesthesia, Reanimation and Intensive Care, University "St. Cyril and Methodius", Skopje, North Macedonia

[2] Department of Anesthesia, Reanimation and Intensive Care, University Clinical Center Mother Theresa, Skopje, North Macedonia

Abstract. Lung ultrasound as a Point of Care (POC) diagnostic technique might be a helpful tool in clinical decision making in critically ill patients. The aim of this study is to present 6 months experience of using lung ultrasound as a point of care tool in ICU. We performed a retrospective study of clinical records in patients with respiratory failure hospitalized in the ICU. The examination was made according to the BLUE Protocol (1). LUS was used in 48 out of 180 patients (26.6%). Pathological findings were noted in 42 patients (87,5%). Pleural effusion was detected in 10 patients (23.8%). Signs of interstital syndrome were detected in 21 patients (50%). Two patients (4.7%) had pneumothorax. Lung consolidation was detected in 18 patients (42.8%); 8 patients (44.5%) with pneumonia and 10 patients (55,5%) with atelectasis. According to the results of the LUS in 19 patients (45.3%) some sort of invasive procedure was undertaken. In 7 patients (36,8%) thoracic drainage was performed and bronchoaspiration in 4 patients (21%). In 12 patients (63.1%), mechanical ventilation was initiated. POCLUS is a helpful tool for rapid assessment of underlying pathological substrate in critically ill patients with acute respiratory failure having important role in clinical decision making.

Keywords: Point of care · Lung ultrasound · Critically ill

1 Introduction

Acute respiratory failure is one of the most frequent causes for admission of critically ill patients in the Intensive Care Unite (ICU). No matter etiology, due to critical illness, acute respiratory failure (ARF) could be developed lately in the ICU, also. According to the dynamic nature of critical illness, rapid and bedside assessment of the failing lungs in a critically ill patient is more than essential. It is well known that the goal of Point of Care methods usage is providing immediate diagnosis and fast treatment decision when it is used near to the site of patient wherever medical care is needed, especially in critical illness and emergencies [1]. Therefore, we considered that Point of Care (POC) Lung Ultrasound (LUS) might be a helpful tool while providing immediate diagnosis and care in patients with acute respiratory failure.

S. S. Sotirov et al. (Eds.): BioInfoMed 2020, LNNS 374, pp. 50–55, 2022.
https://doi.org/10.1007/978-3-030-96638-6_5

2 Aim of Study

The aim of this study is to present our 6 months experience of using LUS as a point of care diagnostic tool in a tertiary University Clinic ICU while treating critically ill patients who exhibit signs and symptoms of acute respiratory failure.

3 Material and Methods

We performed a retrospective study of the clinical records of patients experiencing respiratory failure hospitalized in the ICU of Clinical Center "Mother Teresa" in Skopje, North Macedonia for the period from January 1st 2020, till June 30th 2020. We have defined ARF as a clinical condition in patients with SpO_2 lower than 90%; patients having PaO_2 lower than 60 mmHg and/or with $PaCO_2$ higher than 45 mmHg as well as those patients with PaO_2/FiO_2 ratio lower than 300. All patients included in the study were examined according to the BLUE Protocol [2]. While using the well-known BLUE Protocol for LUS assessment, 12 separated lung areas, 6 in the left and another 6 segments in the right lung should be examined. Each of the examined areas right upper anterior and right upper posterior, right lower anterior and right lower posterior, left upper anterior and left upper posterior as well as left lower anterior and left lower posterior segments should be examined due to LUS assessment in a timeframe no longer than 3 min as described within BLUE Protocol. Our primary goal was to detect any presence of pathological finding in the lungs making distinction from the patients with normal findings, described in the literature as a presence of A-lines and intact pleural sliding in all areas examined. Under the term "pathological findings", we considered disappearing A-lines and abolished lung-sliding, appearance of a more than 3 B-lines, detecting lung consolidations with static or dynamic bronchograms and detecting pleural effusions. We have examined 48 critically ill patients exhibiting signs and symptoms of ARF. Patients were examined in semirecumbent position or in a supine position if they were already intubated. Because our hospital was not designated as a regional COVID Center, in this study we did not include patients experiencing COVID Pneumonia. LUS exam was made with Ultrasound Machine Mindray TP2200 using the curvilinear probe.

4 Results

In our study we have used LUS assessment for ARF in 48 out of 180 critically ill patients (26.6%) where 27 (56,25%) were male and 21 (45,75%) were female. According to the results, pathological findings were detected in 42 patients (87,5%), while the other 6 patients (12,5%) have had normal findings. Pleural effusion was detected in 10 patients (23.8%). More than 3 B-lines in a examined rib interspace with present pleural sliding as a sign of interstitial syndrome were detected in 21 patients (50%). Presence of A-lines but absent pleural sliding suggesting pneumothorax was detected in 2 patients (4.7%). In 18 patients (42.8%) presence of tissue like structure suggesting lung consolidation was found; where in 8 patients (44.5%) the consolidation was accompanied with presence of dynamic bronchograms suggesting pneumonia and in 10 patients (55,5%) atelectasis was found where consolidations were accompanied with static bronchograms. In 9 out of 42 patients (21,4%) we have detected overlapping of two or more findings at the time of examination. According to the results of the LUS in 19 patients (45.3%) some sort of invasive procedure was undertaken. In 7 patients (36,8%) thoracic drainage was established. Bronchoaspiration was indicated in 4 patients (21%). In 12 patients (63.1%), mechanical ventilation was initiated (Tables 1 and 2).

Table 1. Total number of examined patients and patients with normal and pathological findings

Patients examined with LUS	Number of patients
Total number of examined patients	48 (100%)
Normal LUS findings	6 (12,5%)
Pathological findings	42 (87,5%)

Table 2. Types of pathological findings detected by LUS

Findings	Description of LUS result	Patients
Pathological findings	Interstital syndrome (Presence of ≥3 B-lines in one or more intercostal spaces)	21 (50%)
Pathological findings	Pneumothorax (Presence of A-lines, but abolished or absent pleural slidinw with "Barcode" sign in M-mode)	2 (4.7%)
Pathological findings	Pleural effusions (anechoic areas)	10 (23,8%)
Pathological findings	Consolidation with bronchogram	18 (42,8%)

5 Discussion

Critically ill patients experiencing signs and symptoms of acute respiratory failure demand emergent and bedside diagnosis of the cause why lungs are failing. According to the review of literature made by Gentle S. et al. usage of LUS in diagnosis of the most frequent causes of ARF as Pneumonia, Pulmonary edema, Pleural effusions and Pneumothorax is superior when compared to chest radiograph and is comparable to the usage of chest CT [3] also is faster, easy to use, repeatable and without radiation exposure [4]. We have used LUS as a tool supposed to help in diagnosing the cause or underlying condition that leads to respiratory failure in our patients in a six months period. Practicing bedside LUS has given us the answer why lungs are failing in 87.5% of the examined patients. Lichtenstein D and Meziere G. have reported similar results where LUS has provided immediate diagnosis of acute respiratory failure in 90.5% of critically ill patients [2].

Pleural effusion was detected in 23.8% of patients while half of them had small bilateral effusions and the other half were unilateral needing thoracic drainage. Pleural effusions were observed as dark anechoic areas in the dependent regions surrounding the lungs and eventually compressing the lung that in some cases leaded to aeration impairment. Diagnostic accuracy of bedside lung ultrasound for detecting pleural effusions is 93% [5–7] (Fig. 1).

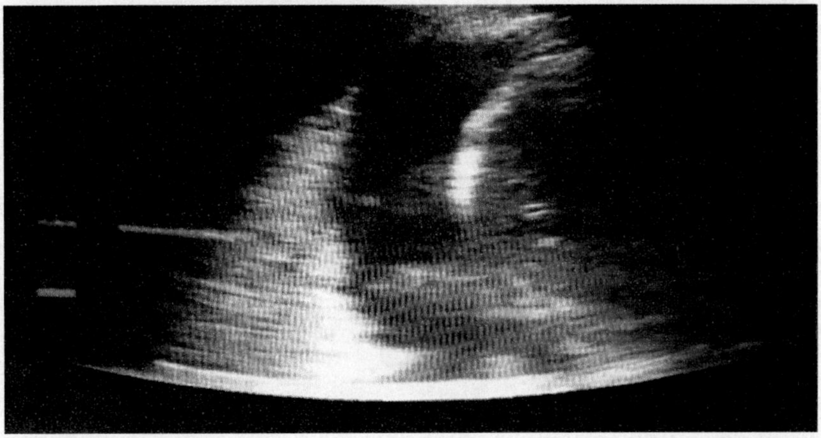

Fig. 1. Pleural effusion surrounding and compressing already atelectatic lung

More than 3 B-lines arousing from the pleural line, hyperechoic by their nature, well defined, laser like and moving left and right during breathing with intact plural sliding were observed in 50% of the patients implying presence of pulmonary edema. This finding in the literature is described as interstitial syndrome which could be detected by the bedside LUS with the accuracy of 93–95% [2, 6]. Detecting of more than 3 B-lines implies erasing the normal A-lines as the result of the accumulation of extravascular lung water no matter primary cause of the edema (Fig. 2).

Lung sliding where visceral and parietal pleura slide one against another with detectable A-lines under the pleural line are normal findings. Abolished or absent pleural sliding in the anterior parts of the examined areas when the patient is in a supine position with present A-lines is typical for pneumothorax [2, 8, 9] which was found in 2 patients (4.7%). Presence of a "Barcode sign" in M-mode Lung ultrasonography has a confirmatory value when diagnosing pneumothorax [3]. Diagnosing pneumothorax with LUS is superior to supine chest radiography where ultrasound sensitivity is 92–100% and the specificity is 91–100% [10].

We observed tissue like structure of the lung or/and the "shred sign" in 18 (42,5%) of the patients with pathological findings, suggesting presence of lung consolidation. All consolidations were observed in the lower anterior and posterior segments and in 5 patients were accompanied with surrounding pleural effusion. When tissue like structure of the lung is present it can be compared with the structure of the liver/spleen and it is called translobar consolidation [3, 5] but more frequently smaller lung consolidations are observed as subpleural areas with irregular border from the surrounding normal lung tissue named as "shred sign" [3, 11]. In 44.5% of the patients with consolidation, we detected presence of dynamic bronchograms that are pathological findings typical

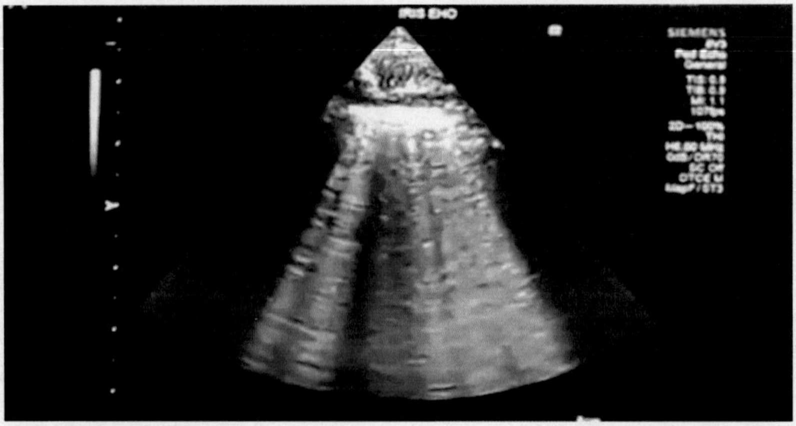

Fig. 2. Thickened pleural line and multiple well defined, hyperechoic B-lines suggesting interstitial syndrome

for pneumonia. Dynamic bronchograms are pathological findings of air-filled bronchi moving for more than 1mm during inspiration and have a sensitivity of 94% when differentiating pneumonia from atelectasis [12]. Static bronchograms were found in 55,5% of the patients with pathological findings suggesting atelectasis. LUS has 97% accuracy when it comes to diagnosing of alveolar consolidation [6] (Fig. 3).

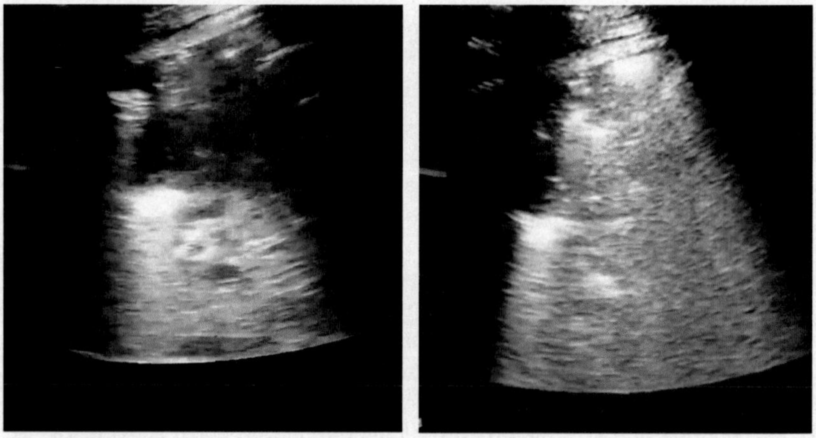

Fig. 3. Lung consolidation with presence of hyperechoic bronchogram

6 Conclusions

Based on the results from the study we can conclude that Point of Care LUS is a helpful tool for rapid assessment of underlying pathological substrate in critically ill patients with acute respiratory failure and have important role in clinical decision making.

References

1. Kost, G.J., Ferguson, W.J., Kost, L.E.: Principles of point of care culture, the spatial care path™, and enabling community and global resilience. e-J. IFCC **25**, 134–53 (2014)
2. Lichtenstein, D.A., Mezière, G.A.: Relevance of lung ultrasound in the diagnosis of acute respiratory failure: the BLUE protocol. Chest **134**(1), 117–125 (2008). [published correction appears in Chest **144**(2), 721 (2013)]
3. Shrestha, G.S., Weeratunga, D., Baker, K.: Point-of-care lung ultrasound in critically ill patients. Rev. Recent Clin. Trials **13**(1), 15–26 (2018). https://doi.org/10.2174/1574887112666170911125750. PMID: 28901850
4. Rudas, M., Orde, S., Nalos, M.: Bedside lung ultrasound in the care of the critically ill. Crit. Care Resusc. **19**(4), 327–336 (2017). PMID: 29202259
5. Lichtenstein, D.: Lung ultrasound in the critically ill. Ann. Intensive Care **4**, 1 (2014)
6. Lichtenstein, D., Goldstein, I., Mourgeon, E., Cluzel, P., Grenier, P., Rouby, J.J.: Comparative diagnostic performances of auscultation, chest radiography, and lung ultrasonography in acute respiratory distress syndrome. Anesthesiology **100**(1), 9–15 (2004). https://doi.org/10.1097/00000542-200401000-00006. PMID: 14695718
7. Inglis, A.J., et al.: Bedside lung ultrasound, mobile radiography and physical examination: a comparative analysis of diagnostic tools in the critically ill. Crit. Care Resusc. **18**(2), 124 (2016). PMID: 27242110
8. Lichtenstein, D.A., et al.: Ultrasound diagnosis of occult pneumothorax. Crit. Care Med. **33**(6), 1231–1238 (2005). https://doi.org/10.1097/01.ccm.0000164542.86954.b4. PMID: 15942336
9. Chan, S.S.: Emergency bedside ultrasound to detect pneumothorax. Acad. Emerg. Med. **10**(1), 91–94 (2003). https://doi.org/10.1111/j.1553-2712.2003.tb01984.x. PMID: 12511323
10. Zechner, P.M., et al.: Arbeitsgruppe des Moduls 5 in Anästhesie Fokussierte Sonographie der DGAI. Lungensonographie in der Akut - und Intensivmedizin [Lung ultrasound in acute and critical care medicine]. Anaesthesist **61**(7), 608–617 (2012). https://doi.org/10.1007/s00101-012-2046-9. PMID: 22772347
11. Lichtenstein, D.A., Lascols, N., Mezière, G., Gepner, A.: Ultrasound diagnosis of alveolar consolidation in the critically ill. Intensive Care Med. **30**(2), 276–281 (2004)
12. Lichtenstein, D., Meziere, G., Seitz, J.: The dynamic air bronchogram. An ultrasound sign of alveolar consolidation ruling out atelectasis. Chest **135**, 1421–1425 (2009)

Application of Digital Health and Telemedicine to Deal with COVID-19

Vladimir Gonchev(✉) and Valentin Vasilev

Faculty of Public Health and Health Care, Prof. Assen Zlatarov University, 1 Prof. Yakimov str., 8010 Burgas, Bulgaria
gonchev@btu.bg, vailvasilev@mail.bg

Abstract. Artificial intelligence, digital health, telemedicine, portable sensors, virtual reality - these new technologies will completely change the way patients and doctors perceive healthcare. Digital health already has a huge impact on the way doctors work, it is very important what tasks it allows them to take on. There are many repetitive and monotonous tasks that most medical professionals do not like to do. Digital health solutions can make them better, faster and cheaper. These tasks usually do not require any creativity or empathy. The process of digital technologies entering healthcare is more complex than simply saying that artificial intelligence (AI) or robotics will take over medical work and activities.

We live in unprecedented times. The new strain of coronavirus is spreading at an alarming rate, so much so that the World Health Organization (WHO) declared a pandemic on March 11. However, this strain of COVID-19, which is thought to have originated either by transfer from a non-human host or evolved into a pathogenic strain in the human population, is very real and causes suffering at every level of society. In particular, its virulent nature puts the elderly and the immunocompromised at higher risk.

The development of the pandemic contributes to the widespread use of telemedicine. Prior to the pandemic, only 1 in 10 American patients used telemedicine services, according to a study by J.D. Power. But with COVID-19 the use of telehealth applications is increasing to 158% in the United States [3, 5]. The coronavirus pandemic has caused an explosive increase in the need for digital services, including those related to health and medicine. Many medical organizations (public and private) are starting to use telemedicine platforms and have announced the opening of free remote consultations between patients and doctors on COVID-19.

Keywords: Artificial intelligence · Digital health · Telemedicine · COVID-19

1 Introduction

In this paper we will look at some digital health and telemedicine applications and activities that have proven effective in dealing with COVID-19.

For the purposes of our study, a review was made in the databases of Google, Google Scholar, Medscape, Medline, LinkedIn, Twitter by keywords: telemedicine, digital health, artificial intelligence, COVID 19.

S. S. Sotirov et al. (Eds.): BioInfoMed 2020, LNNS 374, pp. 56–59, 2022.
https://doi.org/10.1007/978-3-030-96638-6_6

From the study we found that artificial intelligence, telemedicine, portable sensors will be of great importance for the development of a number of medical specialties. These are general medicine, cardiology, pediatrics, radiology, ophthalmology, oncology [7].

Also, new technologies are relevant in the development of the use of interactive maps to determine the territory of the virus, the use of mobile phones to track the contacts of those infected with Covid 19 and compliance with quarantine, chatbot conversations.

The use of telemedicine technologies in dealing with coronavirus infection does not require legislative changes. Remote consultations on such issues must be carried out in accordance with strictly regulated scenarios, prepared in accordance with orders, instructions and methodological documents of the Ministry of Health. There can be no remote diagnosis or appointment of treatment here. It is about identifying risk factors and typical health conditions (cough, fever, etc.) in the process of taking a complete history and providing recommendations about the route or the necessary behavior.

The same is true with tele monitoring, which relies on the monitoring of the dynamics of the general condition and the typical symptoms and is applied entirely within the framework of the current legislation (including with regard to personnel, information systems, medical devices, etc.). Remote monitoring of people's health in self-isolation also includes an element of mandatory emergency response, which is provided through interaction with the emergency medical service.

Basic limitations.

A much more complex and controversial issue is remote diagnosis and prescribing treatment for non-coronavirus problems. Patient-physician consultation using telemedicine technology has three main limitations:

- lack of information (medical documents are not in sufficient quantity, in each clinical case)
- lack of communication (there is no constant interaction between the patient and the attending physician, most often only the primary and only contact takes place);
- lack of clinical influence (there is no way to control the process of therapy over time).

These restrictions significantly affect the quality of care provided through telemedicine technologies. This must be taken into account when developing normative and methodological documents [2, 8, 9, 11].

Based on systematic global experience, the leading reasons for conducting remote consultations for adults include:

- acute respiratory diseases (both viral and infectious), including sinusitis, nasopharyngitis, otitis; this also includes bronchitis/bronchiolitis, influenza;
- skin rashes (acne, herpes, eczema, urticarial, etc.);
- diseases of the lower urinary tract.

Our experience is within the framework of a two-year project in telemedicine, digital health and smart medicine, medical examinations and tests are performed, which include the following activities and procedures - examination of lungs, heart, ears, ECG, blood pressure, pulse, blood oxygenation, body temperature, body mass index, ultrasound of abdominal organs, prostate, breasts, thyroid gland.

We started in April 2020 with the above activities.

During this period, examinations of more than 600 people were performed. The ratio of men to women is 55% women 45% men.

Examinations revealed the following diseases.

Cardiovascular disease was found in 45% of the examinees.

In 25% with problems and diseases of the kidneys and prostate.

In 10% with problems and diseases of the thyroid gland.

In 10% with problems and diseases of the lungs.

40% have more than one disease.

Moderate to severe obesity was found in 35%

More than 30% of the patients had an additional consultation with medical specialists.

Through the use of information and communication technologies is provided equal access for health consultation with various specialists of chronically ill and quarantined people with the aim of avoiding contact and travel, which activity is directly related to the new epidemic situation caused by COVID 19.

2 Conclusion

The last ten years have seen the rapid development of digital health and telemedicine. This gives a new impetus to the development of medicine and medical science. But different medical specialties are affected to different degrees, some develop faster than others. The new epidemic situation caused by COVID-19 is already further catalyzing the faster application of telemedicine, digital health in most medical specialties and activities.

This review shows that the appearance of COVID-19 in 2020 and the need for effective coping has necessitated the rapid and widespread introduction of telemedicine and digital health applications into medical practice.

This ongoing global crisis has forced healthcare institutions and regulators to turn to alternative ways of providing healthcare, while limiting exposure to the virus. Telemedicine provides the ideal solution to these problems by restricting the movement of patients in hospitals, allocating hospital capacity for important cases, while limiting the spread of the disease.

This leads us to think that in the near future more and more telemedicine applications will find a place in medical practice for dealing with pandemics, as well as in other medical fields and specialties.

The last ten years have seen the rapid development of digital health and telemedicine. This gives a new impetus to the development of medicine and medical science. But different medical specialties are affected to different degrees, some develop faster than others. The new epidemic situation caused by COVID-19 will further catalyze the faster application of telemedicine, digital health in most medical specialties and activities.

References

1. Dvorin, E.L., Rothberg, M.B., Rood, M.N., Martinez, K.A.: Corticosteroid use for acute respiratory tract infections in direct to consumer telemedicine. Am. J. Med. S0002–9343(20)30161-3 (2020). https://doi.org/10.1016/j.amjmed.2020.02.014
2. Evangelista, L., Steinhubl, S.R., Topol, E.J.: Digital health care for older adults. Lancet **393**(10180), 1493 (2019)
3. Foster, C.B., Martinez, K.A., Sabella, C., Weaver, G.P., Rothberg, M.B.: Patient satisfaction and antibiotic prescribing for respiratory infections by telemedicine. Pediatrics **144**(3), e20190844 (2019). https://doi.org/10.1542/peds.201
4. Gonchev, V.: Digital health and some medical specialties with the greatest potential for development in the future, Management and education, Prof. Dr Assen Zlatarov University (2020)
5. Mesko, B., Dhunnoo, P.R.: Digital Health and the fight against COVID-19 pandemic 2020. In: The Medical Futurist Handbook (2020)
6. Meskó, B.: Social Media in Clinical Practice, pp. 139–141. Springer, London, UK (2013)
7. Meskó, B.: The medical futurist institute: a vision about the technological future of healthcare. Patterns **2**(4), 100234 (2021)
8. Mesko, B.: Future directions of digital health. In: Rivas, H., Wac, K. (eds.) Digital Health. Health Informatics, pp. 339–363. Springer, Cham (2018). https://doi.org/10.1007/978-3-319-61446-5_22
9. Ray, K.N., Shi, Z., Gidengil, C.A., Poon, S.J., Uscher-Pines, L., Mehrotra, A.: Antibiotic prescribing during pediatric direct-to-consumer telemedicine. Pediatrics **143**(5), e20182491 (2019). https://doi.org/10.1542/peds.2018-2491
10. Ray, K.N., Shi, Z., Poon, S.J., Uscher-Pines, L., Mehrotra, A.: Use of commercial direct-to-consumer telemedicine by children. Acad Pediatr. **19**(6), 665–669 (2019). https://doi.org/10.1016/j.acap.2018.11.016
11. Topol, E.: The Creative Destruction of Medicine: How the Digital Revolution Will Create Better Health Care: E, p. 303. Basic Books, Topol, New York, NY (2012). ISBN-13: 9780465025503

Virtual Assisted Technologies as a Helping Tool for Therapists in Assessment of Anxiety. Outcomes of a Pilot Trial with Chatbot Assistance

Milen Todorov(✉), Gergana Avramova-Todorova, Krasimira Dimitrova, and Valentin Irmov

University "Prof. Dr. Asen Zlatarov", Burgas, Bulgaria
mtodorov@btu.bg

Abstract. As a result of many factors, it is suggested, that mental health issues associated with anxiety, depression and other disorders have been increased lately. Developing appropriate support strategies, especially for highly prevalent problems, such as mood disorders and more specifically for anxiety disorders is a critical public health challenge. Generally, a special topic in the field of healthcare, turns to be accessibility of specialists immediately when needed.

Online interventions have the potential to address this gap for a variety of disorders and mental health issues. Implementation of special support by the means of digital communication provides easy accessibility and anonymity, therefore making it suitable option for clients with variety of psychological issues.

Present study puts into discussion and analysis the effective role of so-called 'virtual assistants' (also known as 'chatbots'). The focus is set on technical side of development of electronic questionnaire for anxiety based on standardized test - Anxiety Symptoms Questionnaire (ASQ). The chatbot is used to collect data for the need of psychometric properties of ASQ with a sample of students in Bulgarian university "Prof. Dr. Assen Zlatarov". The obtained results are processed by making use of intuitionistic fuzzy set (IFS) techniques.

Keywords: Virtual assisted diagnosis · Digital healthcare · Anxiety · Mental disorders

1 Introduction

University students are a unique group of people experiencing a critical transition period as they progress from adolescence into adulthood; accordingly, the university years can be among the most stressful times in a person's life [1]. In this crucial stage, many university students experience frustration due to the pressure of competition for good grades and the need to find a job. As a result, anxiety is a common and major problem that negatively affects academic performance and health [2–5]. Because the stress or difficulties that university students experience (e.g. competition for good grades, financial difficulties, finding jobs) may be different from those of other populations such

S. S. Sotirov et al. (Eds.): BioInfoMed 2020, LNNS 374, pp. 60–66, 2022.
https://doi.org/10.1007/978-3-030-96638-6_7

as adolescents, older adults, and clinical patients, this necessitates reliable, valid, and relatively brief screening tools for anxiety disorders among university students [6–8]. Such tools would facilitate early detection and proper, timely intervention, not only in primary care institutions but also in mental-health settings.

At present, given the worldwide crisis (as a result of many factors including Covid19), there is an urgent need to support mental health stability of university students. In order to overcome the problems associated with mental disorders such as anxiety the first step is to assess the level of this disorder for the students in a given university structure. Human–computer interaction is currently an area of singular importance, and a key to understanding it is an appreciation of the fact that interactive interfaces mediate the redistribution of cognitive tasks between humans and machines [9]. In fact, such interaction is a perfect example of the implementation of state-of-the-art consumer-oriented digital technologies (e.g. artificial intelligence (AI)). It simulates human behavior based on formal models, and it is interesting subject for research of patterns of human interaction as well as issues related to assigning social roles to others, finding patterns of successful and unsuccessful interactions, and establishing social relationships and bonds. Over the past few years, virtual help agents also known as chatbots (software program that interacts with users using natural language) have taken on surprisingly sensitive jobs in modern society offering help, support, and companionship [10]. As they gather mood data and processes any texts and emojis that a person might enter, the chatbot traces the branches of a decision tree to offer personal responses.

The current study aimed to explore the magnitude of anxiety among students from University "Prof. Dr. Assen Zlatarov" by making use of recent digital technology. The study also aimed to identify the most contributing factors for anxiety.

2 Virtual Assisted Technologies for Self Assesements

The promise of AI in healthcare is to provide a set of tools to augment and extend the effectiveness of the therapists in practically all fields [11]. The introduction of data-rich technologies to the clinic will require specialists to interpret and operationalize information from many (bio)medical sources. Thus, it is expected that decisions - particularly those made in urgent situations - could be more efficient, more effective, and personalized when they are supported by reliable digital technology. While AI is needed to support the processes leading to an ultimate medical decision, there are other possibilities that will bridge the gaps in healthcare [12].

Over the past few years, virtual help agents - also known as 'chatbots' (software program that interacts with users using natural language) - have taken on surprisingly sensitive jobs in modern society by offering help, support, and companionship [13]. As chatbots gather mood data and any texts or emojis that a patient might enter, the chatbot traces the branches of a decision tree to offer personal responses. In general, the chatbot asks questions like as "What is your energy like today?", "How are you feeling?", or "What's going on in your world right now?" Those prompts are modelled on today's most popular form of talk therapy, cognitive behavioral therapy (CBT), which instructs individuals in how to recast their negative thoughts in a more objective light. Users are encouraged to talk about their emotional responses to life events, and then stop to identify the psychological traps that cause their stress, anxiety, and/or depression.

By analyzing the dialogues that occurred during therapy sessions, a specific order of question could be constructed to identify the type of emotional imbalance. Then, it is possible to offer a suitable practice that will influence the needed emotion. If the question sequence is known, it can be technically structured in a simple If-Then-Else logic. Currently there are numerous open source and paid decisions for construction of such logic as interaction with chatbots.

3 Methodology for Assessment of Anxiety

In the current study the Anxiety Symptoms Questionnaire (ASQ) [14] is used for evaluation of anxiety of university students. The ASQ is a brief and practical self-report assessment for evaluating the severity of anxiety symptoms over the past week. The ASQ contains 17 items, selected by a panel of experts in anxiety disorders, that measure the frequency and intensity of anxiety symptoms on a 0–10 Likert-type scale, for example, "How intense or bothersome have the symptom(s) been in the past week and how frequently have you experienced the symptom(s) in the past week?". The ASQ was designed to be a global measure of anxiety (ie, measuring more than just physical symptoms), and to assess a range of symptoms central to anxiety, including nervousness, worrying, irritability, trouble relaxing, insomnia, lack of energy, difficulty concentrating, somatic symptoms and impairment in functioning due to anxiety.

4 Data Processing by Intuitionistic Fuzzy Sets

The obtained results are processed by making use of Intuitionistic Fuzzy Set (IFS) techniques. IFSs represent an extension of the concept of fuzzy sets, as defined by Zadeh, exhibiting function $\mu_A(x)$ defining the membership of an element x to the set A, evaluated in the [0; 1]-interval [15, 16]. The difference between fuzzy sets and IFSs is in the presence of a second function $\nu_A(x)$ defining the non-membership of the element x to the set A, where $\mu_A(x) \in [0; 1]$, $\nu_A(x) \in [0; 1]$, under the condition of $\mu_A(x) + \nu_A(x) \in [0; 1]$. The theory of IFS contains many developed techniques not only for registering but also quantification of the so inherent to the real processes and objects imprecision and uncertainty. Based on the received answers, the intuitionistic fuzzy assessments $\mu(x)$ and $\nu(x)$ for determine the degree of anxiety and degree of calmness respectively of each interviewed will be given. The evaluations are formed on the basis of a set of intuitionistic fuzzy estimations $\langle \mu(x),\ \nu(x) \rangle$ of real numbers from the set $[0, 1] \times [0, 1]$, and $\mu(x) + \nu(x) \leq 1$.

The degree of uncertainty $\pi(x) = 1 - \mu(x) - \nu(x)$ represents such cases wherein there is no information for the status of the current interviewed.

5 Methodology and Experimental Design of Chatbot for Assessment of Anxiety

Depending on type of technology that is used, the chatbot could interact by text messaging, voice, or combination of these modalities. In current work, the interaction process is based on text messages and visual elements. Because it is based on logical flow of information a structured type is used including menus, buttons, choices, and forms that need to be filled. To evaluate the anxiety, a sequence of questions based on Anxiety Symptoms Questionnaire [14] is used. The part of questions flow is schematically presented in Fig. 1.

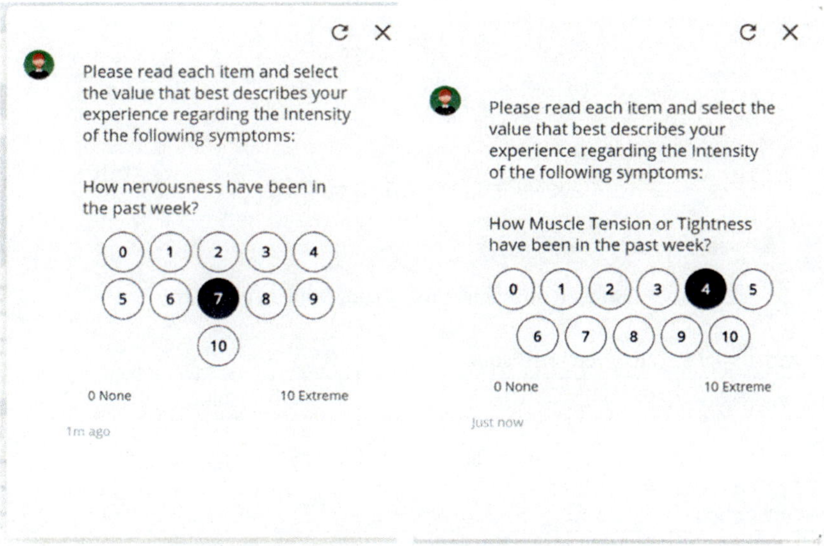

Fig. 1. Representation of dialog with the chatbot.

Following the question sequence, the user should give answers to seventeen questions. As a result of answers analysis based on intuitionistic fuzzy set technique potential problems with anxiety could be identified.

6 Evaluation of Student's Anxiety

The sample consisted of 280 students attending University "Prof. Dr. Asen Zlatarov" – Burgas, Bulgaria (Table 1). All participants completed electronic questionnaire based on ASQ items for intensity of symptoms related to anxiety. The theory of intuitionistic fuzzy sets extends above concepts by comparing belonging and not belonging real numbers in the interval [0, 1] and the sum of these numbers must also belongs to the interval [0, 1].

Let the number of students interviewed be n. The questionnaire contains m questions, each of them can be assessed with $l_j, j \in [1, m]$ and $l_j \in [0, 10]$. In current study $n = 280, m = 17$.

Let the number of questions, for which the i-th student did not provide an answer are $a_{.i}$.

The degree of anxiety $\mu_i(x)$, the degree of calmness $\nu_i(x)$ and the degree of uncertainty $\pi_i(x)$ can be evaluated as a follow:

$$c_j = \frac{l_j}{m} \tag{1}$$

$$\mu_i(x) = \frac{\sum_{j=1}^{m} c_j}{m} \tag{2}$$

$$\pi_i(x) = \frac{a_i}{m} \tag{3}$$

$$\nu_i(x) = 1 - \mu_i(x) - \pi_i(x) \tag{4}$$

Table 1. Representation of dialog with the chatbot.

Characteristics of the analyzed population	Value
Total participants	280
Male	91
Female	189
University degree	
Bachelor	208
Master	72

The obtained results are graphically presented in Fig. 2 and could be interpreted as follows:

1) Upon the global increasing of anxiety based on various reasons, the overall assessment of anxiety of students is relatively low. Based on the scale of ASQ test, the calculated value of 0.27 is categorized as "Mild" anxiety. It can be concluded that there is no significant stress factors that deserves a special attention for further psychological counselling assistance.
2) The main contributing factors for anxiety appear to be feeling restless, keyed Up, or On Edge (0.37), anxiety, nervousness, worrying, irritability and fatigue or lack of energy are in the range of 0.34–0.36.

Based on the fact that there is no significant differences in values of the anxiety items the logical further step of the study is to perform analysis on internal relations between items. It is expected that this analysis would provide more details for student's anxiety.

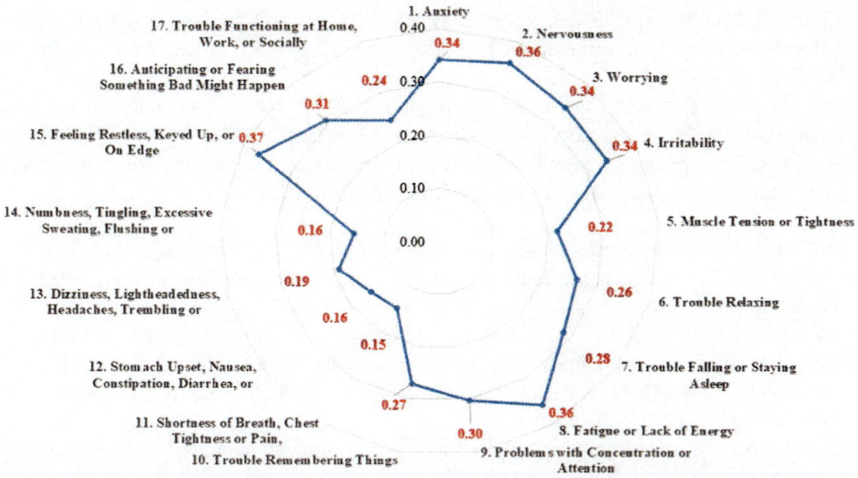

Fig. 2. Quantitative evaluation of the seventeen items related to anxiety.

7 Conclusions

Students at higher education institutions face many kinds of challenges that affect negatively their psychology feeling, causing them anxiety. Its effects vary among students in how to deal with it, overcome its symptoms, and find solutions. The study found that the most important factor causing anxiety in the current experiment is related to items - feeling restless, keyed Up, or On Edge. If the focus is set on anxiety in university environment and more specifically on communication between students it could be expected that current finding could be related to emotional symptoms of anxiety. In this aspect it would be of scientific interest to expand the current investigation in relation to explore the connection between anxiety and emotional awareness – more specifically, the state of one's self control.

Acknowledgements. This paper is supported by the National Scientific Program "Information and Communication Technologies for a Single Digital Market in Science, Education and Security (ICTinSES)", financed by the Ministry of Education and Science.

References

1. Buchanan, J.: Prevention of depression in the college student population: a review of the literature. Arch. Psychiatr. Nurs. **26**(1), 21–42 (2012)
2. Bayram, N., Bilgel, N.: The prevalence and socio-demographic correlations of depression, anxiety and stress among a group of university students. Soc. Psychiatr. Epidemiol. **43**, 667–672 (2008)
3. Dyson, R., Renk, K.: Freshmen adaptation to university life: depressive symptoms, stress, and coping. J. Clin. Psychol. **62**(10), 1231–1244 (2006)
4. Brook, C.A., Willoughby, T.: The social ties that bind: Social anxiety and academic achievement across the university years. J. Youth Adolesc. **44**(5), 1139–1152 (2015)

5. Erzen, E.: The effect of anxiety on student achievement. In: Karadag, E. (ed.) The Factors Effecting Student Achievement, pp. 75–94. Springer, Cham (2017). https://doi.org/10.1007/978-3-319-56083-0_5
6. ul Haq, M.A., Dar, I.S., Aslam, M., Mahmood, Q.K.: Psychometric study of depression, anxiety and stress among university students. J. Public Health **26**(2), 211–217 (2018)
7. Andrews, B., Wilding, J.M.: The relation of depression and anxiety to life-stress and achievement in students. Br. J. Psychol. **95**(4), 509–521 (2004)
8. Crawford, J.R., Henry, J.D.: The Depression Anxiety Stress Scales (DASS): normative data and latent structure in a large non-clinical sample. Br. J. Clin. Psychol. **42**(2), 111–131 (2003)
9. Fischer, G.: User modeling in human-computer interaction. User Model. User-Adap. Inter. **11**, 65–86 (2001). https://doi.org/10.1023/A:1011145532042
10. Rawassizadeh, R., Sen, T., Kim, S.J., et al.: Manifestation of virtual assistants and robots into daily life: vision and challenges. CCF Trans. Pervasive Comp. Interact. **1**, 163–174 (2019)
11. Davenport, T., Kalakota, R.: The potential for artificial intelligence in healthcare. Future Healthc. J. **6**(2), 94–98 (2019)
12. Graham, S., et al.: Artificial intelligence for mental health and mental illnesses: an overview. Curr. Psychiatry Rep. **21**(11), 1–18 (2019)
13. Adamopoulou, E., Moussiades, L.: An overview of chatbot technology. In: Maglogiannis, I., Iliadis, L., Pimenidis, E. (eds.) IFIP International Conference on Artificial Intelligence Applications and Innovations, pp. 373–383. Springer, Cham (2020). https://doi.org/10.1007/978-3-030-49186-4_31
14. Baker, A., et al.: Anxiety Symptoms Questionnaire (ASQ): development and validation. Gen. Psychiatry **32**(6), 1–11 (2019)
15. Atanassov, K.: Intuitionistic Fuzzy Sets. Springer, Heidelberg (1999). ISBN 978-3-7908-1870-3
16. Atanassov, K.T.: On Intuitionistic Fuzzy Sets Theory. Springer, Heidelberg (2012). ISBN 978-3-642-29127-2

InterCriteria Analysis Approach for Decision-Making in Virtual Screening: Comparative Study of Various Scoring Functions

Dessislava Jereva[1], Maria Angelova[1], Ivanka Tsakovska[1], Petko Alov[1], Ilza Pajeva[1], Maria Miteva[2], and Tania Pencheva[1](✉)

[1] Institute of Biophysics and Biomedical Engineering, Bulgarian Academy of Sciences, Acad. G. Bonchev Bl. 105, 1113 Sofia, Bulgaria
{dessislava.jereva,maria.angelova,itsakovska,pajeva, tania.pencheva}@biomed.bas.bg, petko.alov@biophys.bas.bg

[2] Inserm U1268 MCTR, CNRS UMR 8038 CiTCoM Université de Paris, Faculté de Pharmacie de Paris, 4 av de l'Observatoire, CEDEX 06, 75270 Paris, France
maria.mitev@inserm.fr

Abstract. The goal of this study is to evaluate the ability of the recently developed InterCriteria Analysis (ICrA) approach to support decision making in computer-aided drug design projects. ICrA is a method for multicriteria analysis based on two fundamental mathematical concepts – index matrices and intuitionistic fuzzy sets. The approach is designed to distinguish possible relations in the behavior of pairs of criteria when multiple objects are considered, meanwhile allowing to account for the uncertainty in information processing. In this study, ICrA has been applied to examine diverse scoring functions, implemented in AMMOS, FRED, and X-Score software platforms for docking performance. Two protein targets have been considered in this study – estrogen receptor and neuraminidase, selected for dissimilar physicochemical properties and topology of their binding sites. The ability of the scoring functions to evaluate the interactions between protein targets and receptor-based focused libraries, based on the ChemBridge diversity set of drug-like molecules, has been examined, and the relationships between the scoring functions have been analyzed, applying ICrA.

1 Introduction

The well-known fact that drug design is a process requiring lots of time and expenses motivates the intense development and use of approaches aiming at its optimization such as computer-aided (*in silico*) ones. These methods, themselves, are dependent on the available structural information and are generally classified as ligand- and structure-based. Molecular docking is the most commonly used technique among the latter ones. Docking is a modelling technique

S. S. Sotirov et al. (Eds.): BioInfoMed 2020, LNNS 374, pp. 67–78, 2022.
https://doi.org/10.1007/978-3-030-96638-6_8

that places the small molecules (ligands) in different orientations and conformations (known as docking poses) within the binding site of the protein and offers optimal binding modes based on the calculated binding energies between the ligand and the protein. Thus, a key element in the docking algorithms is the scoring function, whose purpose is to calculate quickly and accurately the energy of interaction between the protein and the ligand in their complexes and to allow for selection of appropriate ligand poses in the protein binding pocket. The scoring functions play an important role also in post-docking and re-scoring procedures as subsequent steps in the docking and virtual screening protocols.

Despite the large number of comparative studies on various scoring functions, the selection of an appropriate scoring function for a particular dataset of ligands binding to a specific receptor is not straightforward. Facing this problem, ICrA [1], recently developed as a new multicriteria decision making approach, has prerequisites to be a suitable tool. Over time, ICrA has gained an increasing scientific interest, both for theoretical development and for applications in different fields. ICrA has been proved as a promising decision-making technique for solving numerous real world problems in important areas, such as economy [6], ecology [7,8], medicine and biomedicine [9,10], artificial intelligence [11,12], metaheuristics [13,14], universities ranking [15], etc. In [16,17], for the first time ICrA has been applied to explore the performance of various docking scoring functions available in molecular modelling software packages for the purposes of *in silico* studies of bioactive compounds.

In this investigation, the capacity of ICrA as a supporting tool in the selection of a reliable docking scoring function is explored. The study is focused on the software platforms AMMOS [32,33], FRED (OpenEye Scientific Software, [19]) and X-Score [20,21] widely used in docking and virtual screening applications. Receptor-based focused libraries from ChemBridge diversity set of drug-like molecules have been subjected to re-scoring in two target proteins, estrogen receptor and neuraminidase. The performance of the scoring functions available in the software platforms has been explored based on the calculated binding energies. The obtained results have been subjected to ICrA and subsequently analyzed.

2 Materials and Methods

2.1 Target Proteins

Two protein targets have been selected for the purposes of this study – on the one hand, they are targets of high importance for human health, and, on the other hand, they are representatives of receptors whose binding pockets manifest dissimilar physicochemical properties and topology.

The first target in the study is the estrogen receptor alpha (ERα). The protein belongs to the group of the transcription factors and it is mainly expressed in tissues of the female reproductive system and breast, but also in the cardiovascular system, brain, bone, liver, adipose tissue, colon, skin, prostate, testes, salivary glands etc. [35]. Several diseases are associated with changes in its levels and function, such as diverse types of cancer (breast, ovaries, prostate, colon), but also metabolic disorders, cardiovascular diseases, neurodegeneration, inflammation, and osteoporosis. Selective ERα modulators are applied in cases of ERα-positive breast cancer, among them tamoxifen, raloxifene, and fulvestrant [34].

The other target, neuraminidase (NA), is a major surface tetrameric glycoprotein present in influenza A and B viruses [37]. Its main function is to facilitate the release of the virions from the infected cell. The NA's inhibitors aim to block NA activity and to prevent in this way the spread of the virions outside of the infected cell. Such inhibitors, authorized globally or in certain countries, are oseltamivir, zanamivir, peramivir, and laninamivir [36].

The crystal structures of ERα with 4-hydroxytamoxifen (OHT) (PDB ID 3ert, resolution 1.90 Å) and of NA with 4-(n-acetylamino)-3- [n-(2-ethylbutanoylamino)]benzoic acid (FDI) (PDB ID 1b9s, resolution 2.50 Å), were selected from Protein Data Bank (PDB) [22]. The selection was done according to criteria for a high resolution and a completeness of the protein structure. The water molecules and co-crystallized ligands were removed from the binding sites. Hydrogens were assigned using the program InsightII [23].

Both receptors possess binding pockets with different physicochemical properties and topology: the binding pocket of ERα is non-polar and closed (Fig. 1, left), while the binding pocket of NA is polar and open (Fig. 1, right).

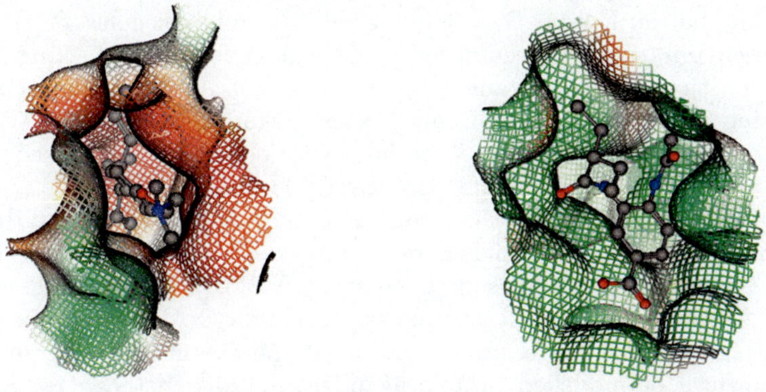

Fig. 1. Binding sites of the ERα (left) and NA (right) with bound ligands. The hydrophilic (polar) and lipophilic (non-polar) regions are shown respectively in green and orange. The ligands atoms are displayed in ball and stick rendering (adapted from [27]).

2.2 Dataset Preparation

The ChemBridge diversity set (CDS) [24] was used for the dataset collection of small molecules [27]. CDS contained 37,970 molecules after ADME/Tox filtering of drug-like molecules using the program Filter 1.0.2 [19]. Ten known active compounds for each protein target (with binding affinities from micromolar to nanomolar range) were also added to the 37,970 decoys collection. The resulting chemical library was transformed in a single 3D conformer library and saved in mol2 format using the program OMEGA 2.0 [19]. The multi-conformer states were then generated by the free program Multiconf-DOCK [25,26] with up to 50 conformers per molecule. Following the protocol described in detail in [27], two receptor-based focused libraries have been designed, respectively for ERα and NA.

2.3 Docking-Scoring Protocol

Aiming at scoring functions comparison, the same docking-scoring protocol has been applied in order to evaluate their performance on an equal basis. The docking-scoring protocol is described in detail in [27] and consists of the following steps. Firstly, the prepared receptor-based focused libraries were subjected to the shape complementarity filtering by MS-DOCK [25] using rigid body docking of multiple conformers with DOCK 6.0 [28,29]. Considering the binding site properties of ERα and NA, the receptor-based libraries were constructed by taking the 12,000 top scored molecules (about 30% of the initial library) after MS-DOCK for ERα while 22,000 molecules (about 50% of the initial library) were taken for NA, taking into account its large open binding pocket. Six actives for ERα and for NA passed the shape-based filter.

Next, to better handle the ligand flexibility, all molecules of the generated focused libraries were re-docked using a flexible ligand docking mode in DOCK 6.0 with optimized parameters, defined in [25]. All docked molecules were scored using the standard DOCK energy score involving soft van der Waals and distance-dependent electrostatic potentials. After clustering the docked poses up to 10 best scored poses per ligand were saved.

Prior to re-scoring with the other scoring functions, the multiple pre-generated protein-ligand complexes were subjected to optimization via energy minimization using the open source program AMMOS [31–33]. The software platform AMMOS (Automatic Molecular Mechanics Optimization for *in silico* Screening) provides tools for minimization of the energies of a large number of experimental or modelled protein-ligand complexes that can be generated through a user-selected docking program or via virtual screening web servers. AMMOS is based on the open-source program AMMP (All-featured Molecular Mechanics Program, embedded in the platform) [30,31] and utilizes the physics-based force field *sp4* of AMMP. AMMOS allows for optimization of protein-ligand interactions via five levels of flexibility of the protein; the ligand is always kept flexible:

- *Case 1* – the entire protein is flexible;
- *Case 2* – the protein sidechains are only flexible;
- *Case 3* – all protein atoms within a certain radius around the ligands are flexible (the radius is of a user-choice);

- *Case 4* – the protein sidechains within a certain radius around the ligands are flexible (the radius is of a user-choice);
- *Case 5* – the entire protein is rigid.

In the final step of the docking-scoring protocol all minimized ligand conformers were scored by the AMMOS protein-ligand interaction energy.

2.4 Re-scoring

The multiple AMMOS minimized docked poses were re-scored by using the program FRED [19]. The scoring functions available in FRED have been applied as follows[27]:

- *Shapegauss* scoring function that represents all atoms as smooth Gaussian functions.
- *Chemgauss* and *Chemgauss2* scoring functions that combine the *Shapegauss* with additional potentials for chemically matching positions for hydrogen bonds and aromatic interactions.
- *PLP* (Piecewise Linear Potential) scoring function that incorporates steric and hydrogen-bonding terms for all pairwise ligand and protein heavy atoms.
- *Chemscore* scoring function that involves terms for lipophilic and hydrogen bonds interactions, as well as penalties for clashes and frozen rotatable bonds due to ligand binding and possibility for optimization of the position of hydrogens involved in hydrogen bonding.
- *Screenscore* scoring function that is a sum of terms for steric interactions, hydrogen bonds, metal atom interactions, aromatic interactions and penalties for clashes and the number of rotatable ligand bonds.
- *ZapBind* scoring function that reflects electrostatic binding energy.

In addition, X-Score [21] developed for binding affinity estimation [20] was applied for re-scoring of the AMMOS minimized multiple docked poses. The program implements four scoring functions, namely *HSScore*, *HPScore*, *HMScore* and their average one. These scoring functions include terms for van der Waals interactions, hydrogen bonds, hydrophobic effects, a torsional entropy penalty and a regression constant. They differ in the approach to estimate hydrophobic effects.

3 InterCriteria Analysis

3.1 Theoretical Outline

Intercriteria analysis approach developed in 2014 by Atanassov et al. [1] brings together two fundamental mathematical concepts – the intuitionistic fuzzy sets [2,3] and the index matrices [4]. Here, for completeness, a brief outline of the theoretical background of ICrA is going to be presented.

An index matrix (IM) is usually denoted as $[O, C, e_{O,C}]$. In terms of ICrA, $O = O_1, \ldots, O_m$ are the evaluated objects, while $C = C_1, \ldots, C_n$ render the evaluation criteria applied to considered objects. The index matrix elements $e_{O,C}$ (for every i, j, where $1 \le i \le m$, $1 \le j \le n$) are the evaluations of the i-th object against the k-th criterion. The IM elements might be real numbers of a certain type or ordered pairs of real numbers. In general, all the values $e_{O,C}$, assigned to the objects by the criteria, are assumed to be known and usually to be real numbers. Thus, a two-dimensional IM might be presented in the following form:

$$A = \begin{array}{c|ccccc} & C_1 & \dots & C_j & \dots & C_n \\ \hline O_1 & e_{O_1,C_1} & \dots & e_{O_1,C_j} & \dots & e_{O_1,C_n} \\ \vdots & \vdots & \ddots & \vdots & \ddots & \vdots \\ O_i & e_{O_i,C_1} & \dots & e_{O_i,C_j} & \dots & e_{O_i,C_n} \\ \vdots & \vdots & \ddots & \vdots & \ddots & \vdots \\ O_m & e_{O_m,C_1} & \dots & e_{O_m,C_j} & \dots & e_{O_m,C_n} \end{array}$$

ICrA begins with an IM of dimensions m and n and ends with an IM of dimensions n and n, meanwhile performing a pairwise comparison between every two different criteria along all evaluated objects. Following these computations, two intuitionistic fuzzy (IF) counters $S^{\mu}_{k,l}$ and $S^{\nu}_{k,l}$ are formed and incremented following all mathematical justifications presented in details in [1].

For every k, l ($1 \le k \le l \le m$ and $m \ge 2$), two values are defined by normalizing the above counters $S^{\mu}_{k,l}$ and $S^{\nu}_{k,l}$ as follows:

- $\mu_{C_k,C_l} = 2\frac{S^{\mu}_{k,l}}{m(m-1)}$, which, in terms of ICrA, is known as *degree of agreement*, and
- $\nu_{C_k,C_l} = 2\frac{S^{\nu}_{k,l}}{m(m-1)}$, known as *degree of disagreement.*

In ICrA, *degrees of agreement/disagreement* between two criteria should be calculated just once, considering that $\mu_{C_k,C_l} = \mu_{C_l,C_k}$ and $\nu_{C_k,C_l} = \nu_{C_l,C_k}$. It is obvious, that both μ_{C_k,C_l} and ν_{C_k,C_l} are numbers in the interval [0, 1], and their sum is a number in this interval as well. The number, that complements the sum of μ_{C_k,C_l} and ν_{C_k,C_l} to 1, is known as a *degree of uncertainty.* In practice, just two of these three values are enough to describe the *degrees of agreement, disagreement* and *uncertainty* observed between two criteria.

The collected IF pairs $\langle \mu_{C_k,C_l}, \nu_{C_k,C_l} \rangle$, which play the role of IF evaluation of the relations between any two criteria C_k and C_l, form the resulting IM of the ICrA:

$$\begin{array}{c|ccccc} & C_1 & \dots & C_k & \dots & C_n \\ \hline C_1 & \langle 1,0 \rangle & \dots & \langle \mu_{C_1,C_k}, \nu_{C_1,C_k} \rangle & \dots & \langle \mu_{C_1,C_n}, \nu_{C_1,C_n} \rangle \\ \vdots & \vdots & \ddots & \vdots & \ddots & \vdots \\ C_k & \langle \mu_{C_k,C_1}, \nu_{C_k,C_1} \rangle & \dots & \langle 1,0 \rangle & \dots & \langle \mu_{C_k,C_n}, \nu_{C_k,C_n} \rangle \\ \vdots & \vdots & \ddots & \vdots & \ddots & \vdots \\ C_n & \langle \mu_{C_n,C_1}, \nu_{C_n,C_1} \rangle & \dots & \langle \mu_{C_n,C_k}, \nu_{C_n,C_k} \rangle & \dots & \langle 1,0 \rangle \end{array}.$$

The closer the values $\langle \mu_{C_k,C_l}, \nu_{C_k,C_l} \rangle$ are to $\langle 1, 0 \rangle$, the more similar is the behaviour of the C_k and C_l.

Following the concept of *consonance* and *dissonance* [5], if $0 \leq \beta \leq \alpha \leq 1$ are given, such that $\alpha + \beta \leq 1$, the criteria C_k and C_l are in:

- positive consonance, if $\mu_{C_k,C_l} > \alpha$ and $\nu_{C_k,C_l} < \beta$;
- negative consonance, if $\mu_{C_k,C_l} < \beta$ and $\nu_{C_k,C_l} > \alpha$;
- dissonance, otherwise.

3.2 ICrAData Software Package

ICrAData software package has been developed [18] for the implementation of the ICrA approach. ICrAData is a free for use software package written in Java (v1.7) and C++ (v2.3). The latest version ICrAData v2.3 can be downloaded from http://intercriteria.net/wp-content/uploads/2020/09/ICrAData23repack.zip (Last access September 28, 2020).

The user interface of ICrAData consists of three main parts: the left panel displays the input data, the central panel – the result of ICrA in a coloured table view, while the rightmost panel shows the graphical interpretation of the results.

For better visualization of the results, the table cells are coloured according to the following rules, depending on the user defined α and β thresholds:

- green colour for pairs in positive consonance;
- red colour for pairs in negative consonance;
- otherwise, violet colour for pairs in dissonance.

The default values used by the software ICrAData are $\alpha = 0.75$ and $\beta = 0.25$, respectively.

4 Results and Discussion

ICrA has been applied to evaluate the performance of the scoring functions, implemented in the software platforms AMMOS, FRED, and X-Score, for the two selected protein targets ERα and NA. In terms of ICrA, scoring functions play the role of criteria, while the ligands from the receptor-based focused libraries of ERα and NA act for objects. ICrA implementation allows for a comparative assessment of the performance of altogether twelve scoring functions, attempting to reveal intercriteria relations between them. Only the results for *Case 3* and *Case 5* among the five available cases of protein flexibility in AMMOS are presented here as more applicable for the purposes of *in silico* drug design: *Case 5* corresponds to the rigid docking – a protocol, available in many commercial, free-for-use and open source molecular modelling packages, while *Case 3*, with flexible atoms within a certain radius around the ligand, simulates more closely what happens in the protein binding site in reality.

Figure 2 presents the results from ICrA applied to evaluate the scoring functions available in the software platforms AMMOS, FRED, and X-Score toward ERα after minimization performed by *Case 3* in AMMOS.

Fig. 2. ICrA applied to ERα after minimization performed by *Case 3* in AMMOS. The following abbreviations are used: C* – the case of protein flexibility, in which the minimization by AMMOS has been performed (C3 codes for *Case 3*); Inp – for results after docking by DOCK, evaluated by AMMOS scoring function; AMMOS – for AMMOS scoring function in chosen case of flexibility (C*); Chmgs – for *Chemgauss* scoring function in FRED; Chmgs2 – for *Chemgauss2* scoring function in FRED; Chmsc – for *Chemscore* scoring function in FRED; *PLP* – for *Piecewise Linear Potential* scoring function in FRED; Scrsc – for *Screenscore* scoring function in FRED; Shpgs – for *Shapegauss* scoring function in FRED; ZpBnd – for *ZapBind* scoring function in FRED; HMS – for *HMScore* scoring function in X-Score; HPS – for *HPScore* scoring function in X-Score; HSS – for *HSScore* scoring function in X-Score; *Avrg* – for the *average* scoring function in X-Score.

In terms of ICrA implementation, the following results can be discussed:

- Positive consonance (calculations coloured in green), i.e. similar results, is observed between most of the scoring functions in FRED, which means their use is interchangeable.
- The X-Score scoring functions themselves are in positive consonance as well, which also suggests their interchangeability.
- There is neither positive, nor negative consonance between the scoring functions in AMMOS, FRED, and X-Score. The only exception is the relationship between the *Shapegauss* scoring function in FRED and *HPScore* scoring function in X-Score (calculations in red), suggesting they are in negative consonance, but with a value very close to dissonance. This observation implies that examined scoring functions could be combined in consensus docking studies as they give different information.
- For two of the scoring functions – *Chemscore* (5-th column) and *ZapBind* (9-th column) of FRED group of scoring functions, the obtained results fall in dissonance (calculations coloured in magenta), so the suggestion is to use those two scoring functions in addition to other functions, because they could reveal different results and outline different potential drug candidates.

Similar results have been obtained when applying ICrA to the data after minimization performed by *Case 5* in AMMOS for ERα, with very few differences that do not affect the aforementioned observations.

Figure 3 presents the results from ICrA applied to evaluate the scoring functions available in the software platforms AMMOS, FRED, and X-Score toward NA after minimization performed by *Case 5* in AMMOS (the same abbreviations as in Fig. 2 are used, only C5 codes for *Case 5* in AMMOS).

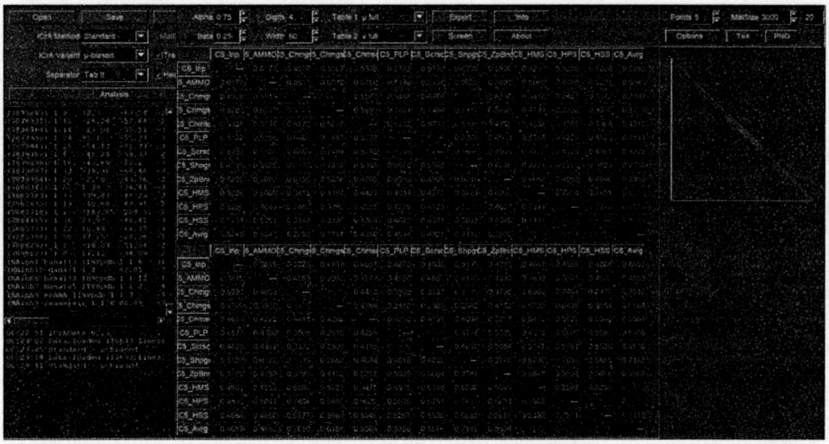

Fig. 3. ICrA applied to NA after minimization performed by *Case 5* in AMMOS.

What can be outlined from the obtained results is that almost all energies calculated by the scoring functions in the considered software platforms fall into dissonance (calculations coloured in magenta), i.e. each scoring function calculates the binding affinity of the ligand in its manner and outlines different potential drug candidates.

In analogy to the case of ERα, very similar results have been obtained when applying ICrA to the data after minimization performed by *Case 3* in AMMOS for NA, again with very few differences that do not affect the aforementioned observations.

Although based on a limited number of scoring functions and protein targets, the above analysis confirms once again that prioritizing scoring functions for a binding pocket with particular characteristics is a multifactor and complex task. As a further step, ICrA might be purposefully applied to the actives only for validation of the correlation results.

5 Conclusion

In this study, the ICrA approach has been applied to comparatively evaluate the performance of altogether twelve scoring functions, available in three software platforms, i.e. AMMOS, FRED, and X-Score. Two protein targets have been selected for this study – estrogen receptor alpha, with non-polar and closed binding site, and influenza virus B neuraminidase, with polar and open binding site. Re-scoring has been performed over the multiple poses, obtained after the minimization by two different cases of protein flexibility, available in AMMOS software platform. The relationships between the scoring functions have been evaluated, applying ICrA. No significant consonance has been outlined between the scoring functions, implemented in the three software platforms, suggesting that they can be combined in consensus docking studies. In addition, the scoring functions within each of the software packages, FRED and X-Score, fall into positive consonance, i.e. demonstrate similar results, and for this reason their use could be interchangeable. This study confirms again that for a particular protein/binding site more than one scoring functions should be used to reveal possible drug candidates with elevated probability to become drug leads.

Acknowledgements. The work is supported by the National Science Fund of Bulgaria, grant DN 17/6 "A New Approach, Based on an Intercriteria Data Analysis, to Support Decision Making in *in silico* Studies of Complex Biomolecular Systems". The authors thank the OpenEye Free Academic Licensing Program for providing a free academic license for molecular modelling and cheminformatics software.

References

1. Atanassov, K., Mavrov, D., Atanassova, V.: Intercriteria decision making: a new approach for multicriteria decision making, based on index matrices and intuitionistic fuzzy sets. Issues IFSs GNs **11**, 1–8 (2014)
2. Atanassov, K.: Intuitionistic fuzzy sets, VII ITKR session, Sofia, 20–23 June 1983. Reprinted: Int. J. Bioautom. **20**(S1), S1–S6 (2016)
3. Atanassov, K.: On Intuitionistic Fuzzy Sets Theory. Springer, Cham (2012). https://doi.org/10.1007/978-3-642-29127-2
4. Atanassov, K.: Index Matrices: Towards an Augmented Matrix Calculus. Springer, Berlin (2014). https://doi.org/10.1007/978-3-319-10945-9
5. Atanassov, K., Atanassova, V., Gluhchev, G.: InterCriteria analysis: ideas and problems. Notes IFSs **21**(1), 81–88 (2015)
6. Atanassova, V., Doukovska, L., Atanassov, K., Mavrov, D.: InterCriteria decision making approach to EU member states competitiveness analysis. In: Proceedings of the International Symposium on Business Modeling and Software Design - BMSD 2014, pp. 289–294 (2014)
7. Georgieva, V., Angelova, N., Roeva, O., Pencheva, T.: Intercriteria analysis of wastewater treatment quality. J. Int. Sci. Publ.: Ecol. Saf. **10**, 365–376 (2016)
8. Ilkova, T., Petrov, M.: Using intercriteria analysis for assessment of the pollution indexes of the Struma River. Adv. Intell. Syst. Comput. **401**, 351–364 (2016)
9. Todinova, S., et al.: Blood plasma thermograms dataset analysis by means of intercriteria and correlation analyses for the case of colorectal cancer. Int. J. Bioautom. **20**(1), 115–124 (2016)
10. Zaharieva, B., Doukovska, L., Ribagin, S., Radeva, I.: Intercriteria analysis of data obtained from patients with Behterev's disease. Int. J. Bioautom. **24**(1), 5–14 (2020)

11. Sotirov, S., et al.: A hybrid approach for modular neural network design using intercriteria analysis and intuitionistic fuzzy logic. Complexity **2018**, Art. ID 3927951 (2018)
12. Sotirov, S., et al.: Application of the intuitionistic fuzzy intercriteria analysis method with triples to a neural network preprocessing procedure. Comput. Intell. Neurosci. **2017**, Art. ID 2157852 (2017)
13. Pencheva, T., Roeva, O., Angelova, M.: Investigation of genetic algorithm performance based on different algorithms for InterCriteria relations calculation. In: Lirkov, I., Margenov, S. (eds.) LSSC 2017. LNCS, vol. 10665, pp. 390–398. Springer, Cham (2018). https://doi.org/10.1007/978-3-319-73441-5_42
14. Roeva, O., Fidanova, S.: Comparison of different metaheuristic algorithms based on intercriteria analysis. J. Comput. Appl. Math. **340**, 615–628 (2018)
15. Krawczak, M., Bureva, V., Sotirova, E., Szmidt, E.: Application of the intercriteria decision making method to universities ranking. Adv. Intell. Syst. Comput. **401**, 365–372 (2016)
16. Jereva, D., Pencheva, T., Tsakovska, I., Alov, P., Pajeva, I.: Exploring applicability of the InterCriteria analysis to evaluate the performance of MOE and GOLD scoring functions. In: Georgiev, I., Kostadinov, H., Lilkova, E. (eds.) BGSIAM 2018. SCI, vol. 961, pp. 198–208. Springer, Cham (2021). (in press). https://doi.org/10.1007/978-3-030-71616-5_18
17. Tsakovska, I., et al.: InterCriteria analysis implementation for exploration of the performance of various docking scoring functions. In: Dimov, I., Fidanova, S. (eds.) HPC 2019. SCI, vol. 902, pp. 88–98. Springer, Cham (2021). https://doi.org/10.1007/978-3-030-55347-0_8
18. Ikonomov, N., Vassilev, P., Roeva, O.: ICrAData - software for intercriteria analysis. Int. J. Bioautom. **22**(1), 1–10 (2018)
19. OpenEye Scientific Software. http://www.eyesopen.com
20. Wang, R., Lai, L., Wang, S.: Further development and validation of empirical scoring functions for structure-based binding affinity prediction. J. Comput. Aided Mol. Des. **16**, 11–26 (2002)
21. X-Score. http://sw16.im.med.umich.edu/software/xtool
22. Berman, H.M., et al.: The protein data bank. Nucl. Acids Res. **28**, 235–242 (2000)
23. InsightII. http://www.accelrys.com/
24. ChemBridge Corporation. http://chembridge.com/chembridge
25. Sauton, N., Lagorce, D., Villoutreix, B.O., Miteva, M.A.: MS-DOCK: accurate multiple conformation generator and rigid docking protocol for multi-step virtual ligand screening. BMC Bioinform. **9**, 184 (2008)
26. Multiconf-DOCK. http://dock.compbio.ucsf.edu/Contributed_Code/multiconfdock.htm
27. Pencheva, T., Soumana, O., Pajeva, I., Miteva, M.: Post-docking virtual screening of diverse binding pockets: comparative study using DOCK, AMMOS, X-Score and FRED scoring functions. Eur. J. Med. Chem. **45**(6), 2622–2628 (2010)
28. Moustakas, D.T., et al.: Development and validation of a modular, extensible docking program: DOCK 5. J. Comput. Aided Mol. Des. **20**, 601–619 (2006)
29. DOCK6. http://dock.compbio.ucsf.edu/DOCK_6/
30. AMMP. http://www.cs.gsu.edu/cscrwh/ammp/ammp.html
31. Pencheva, T., Lagorce, D., Pajeva, I., Villoutreix, B.O., Miteva, M.A.: AMMOS: automated molecular mechanics optimization tool for *in silico* screening. BMC Bioinform. **9**, 438 (2008)
32. Labbe, C., et al.: AMMOS2: a web server for protein-ligand-water complexes refinement via molecular mechanics. Nucl. Acids Res. **45**(W1), W350–W355 (2017)

33. AMMOS2. http://drugmod.rpbs.univ-paris-diderot.fr/ammosHome.php
34. Jia, M., Dahlman-Wright, K., Gustafsson, J.: Estrogen receptor alpha and beta in health and disease. Best Pract. Res. Clin. Endocrinol. Metab. **29**(4), 557–68 (2015)
35. Eyster, K.M.: The estrogen receptors: an overview from different perspectives. Methods Mol. Biol. **1366**, 1–10 (2016)
36. Bassetti, M., Castaldo, N., Carnelutti, A.: Neuraminidase inhibitors as a strategy for influenza treatment: pros, cons and future perspectives. Exp. Opin. Pharmacother. **20**(14), 1711–18 (2019)
37. Jagadesh, A., Salam, A.A.A., Mudgal, P.P., Arunkumar, G.: Influenza virus neuraminidase (NA): a target for antivirals and vaccines. Adv. Virol. **161**(8), 2087–2094 (2016). https://doi.org/10.1007/s00705-016-2907-7

Bayesian Hierarchical Modelling for Antimicrobial Resistance

Maya Zhelyazkova[1], Roumyana Yordanova[2,3], Iliyan Mihaylov[1], Stefan Kirov[4], Stefan Tsonev[5], David Danko[6], and Dimitar Vassilev[1(✉)]

[1] Faculty of Mathematics and Informatics, Sofia University "St. Kliment Ohridski", 5 James Bourchier Blvd., 1164 Sofia, Bulgaria
{zhelyazkova,dimitar.vassilev}@fmi.uni-sofia.bg
[2] Hokkaido University, Sapporo, Japan
[3] Bulgarian Academy of Sciences, Institute of Mathematics and Informatics, Acad. Georgi Bonchev Str., Block 8, Sofia, Bulgaria
[4] Bristol-Myers Squibb, 311 Pennington Rocky Hill Rd, Pennington, NJ, USA
[5] AgroBioInstitute, 8 Dragan Tsankov Blvd., 1164 Sofia, Bulgaria
[6] Weill Cornell Medical College, New York, NY, USA
dcd3001@med.cornell.edu

Abstract. Antimicrobial resistance poses a serious health problem. Computational and statistical approaches have been developed for bacterial antimicrobial resistance discovery and classification. Next generation sequencing technologies allows us to analyze complex metagenomes and the presence of resistomes in them. In this work we apply Bayesian hierarchical spatial model to estimate the relative risk of antimicrobial resistance related taxa by using the available spatial information of the samples.

Keywords: Bayesian hierarchical spatial inference · Spatial correlations · Antimicrobial resistance · Metagenomics

1 Background

Antimicrobial resistance is the property of microbes to evolve and become resistant to the effects of various antimicrobials, such as antibiotics. Diseases caused by resistent microbes pose medical challenges and associated with increased mortality. Many classes of microorganisms can develop resistance among which are bacteria, fungi and viruses. Both mutations and horizontal gene transfer can lead to Antimicrobial resistance [1]. The increased use of antibiotics, including self medication leads to new drug resistant pathogens. Resistant bacteria have also been found in human food chains which is another significant public health concern [2]. In addition resistant bacteria can be disseminated across international borders by travelers, animals, agricultural products which creates a global problem affecting both developed and developing countries [3].

S. S. Sotirov et al. (Eds.): BioInfoMed 2020, LNNS 374, pp. 79–87, 2022.
https://doi.org/10.1007/978-3-030-96638-6_9

Next generation sequencing technologies allow us to study complex metagenomes and the resistant genes that occur in them. A number of computational and statistical methods [4–6] have been developed to analyze antimicrobial resistance in metagenomics data.

The International Metagenomics and Metadesign of Subways and Urban Biomes (MetaSUB) Consortium [7] is a multidisciplinary initiative which collects and sequences large number of samples and use advanced methods to analyze and visualize the metagenomics profiles. MetaSUB has a large collection of data samples (more than 4000) across more than 60 cities. The sequence data are collected from the subway systems around the world. Additional collections of metadata, such as spatial information of the samples (longitude and latitude) and features related to environmental and climate conditions are also available. CAMDA2020 is a data analysis contest in collaboration with MetaSUB which provides the participants with sequence data for the samples.

In this work we analyze CAMADA2020 (http://camda.info/) dataset by using a Bayesian hierarchical model to derive estimates of the relative risk of antimicrobial related taxa. The estimations of the model are visualized on geographical maps to highlight regions with elevated antimicrobial resistance.

2 Data

CAMDA2020 dataset is sequence data for the samples taken on June 21st in 2016 and 2017. For each city there are about 50 samples spread across different areas of the subway system. Each sample was sequenced with an average of 6M 125bp paired-end reads using Illumina NGS sequencers. Participants have access to more than 1000 files with pair reads from 23 cities across different continents. Meta data for each sample and for each city include spatial information, weather data (temperature, pressure, precipitation, humidity) and demographics (population, population density, type of settlement).

We preprocess the data using Trimmomatic v0.39 [8] with the following parameters: minimum length of 50 and the average quality in a window size of 3 is no less than 20. The antimicrobial resistance database from NCBI [9] consists of sequences of genes related to antimicrobial resistant microorganisms. Each of the genes is associated with a specific microbial species (OTUs), where OTU stands for operational taxonomic unit. The corresponding fasta format sequences are then used in Kaiju metagenomics classifier [10]. The total number of species possessing antimicrobial resistance taxa (genes) in the study was 443. We further transformed the raw counts to RPKM (reads/kilobase/million mapped reads). Since the data are taken on different days (a year apart) when we examined them we observe a strong batch effect and therefore decided to analyze the data on each day separately.

3 Bayesian Inference

Spatial correlation among the samples in MetaSub based on their longitude and latitude information were observed in previous studies [11,12]. This violates the assumption that the residuals in ordinary regression are independent, which causes the estimates to be biased. To account for the spatial correlation structure we use a Bayesian hierarchical model which is often applied in ecology, education and epidemiology. These models include a set of spatially autocorrelated random effects which have a conditional autoregressive prior (CAR). The general form of the model as described in [13] is

$$
\begin{aligned}
Y_k|\mu_k &\sim f(y_k|\mu_k, \nu^2), k = 1, 2, ...K \\
g(\mu_k) &= X^T\beta + O_k + \psi_k \\
\beta &\sim N(\mu_\beta, \Sigma_\beta) \\
\nu^2 &\sim \text{Inverse-Gamma}(a, b)
\end{aligned}
$$

Here Y_k are the set of responses while O_k are the offsets. In our case the responses are observed OTU counts (O) for the antimicrobial resistant taxa while O_k are the expected counts (E). X is the matrix of covariates while β are the regression parameters. The spatial structure is defined by ψ_k which includes a set of random effects ϕ of the form $\phi \sim N(0, \tau^2 Q(W)^{-1})$. The matrix W is $K \times K$ neighboring matrix between the areas in the cities or countries which may be discrete or continuous. In the discrete case areas sharing common boundary are denoted by 1, otherwise 0. The diagonal elements of W are also zeros. If Y_k follows Poisson distribution and $ln(\mu_k) = X^T\beta + O_k + \psi_k$ the model becomes Besag-York-Mollie (BYM) [13,14]. Since our responses are counts and follow Poisson distribution we choose the BYM model. The input in the model is the so called Standardized Incidence Ratio (SIR), defined as Observed O/Expected E, which is considered a crude estimator of relative risk. The posterior estimates of the model are estimates of the relative risk. If the relative risk estimates are higher than 1, the counts of antimicrobial resistant taxa are elevated in the samples, compared to their expected values. The expected antimicrobial resistance can be chosen as the median abundance, proportional to the population of the area or other relevant quantity. Sometimes, the risk is elevated by chance due to small expected values because the disease is rare or the population in the area is small. This is the reason why we choose Bayesian hierarchical models which decrease the probability of having elevated risk by chance. We fit the model with the function S.CARbym from R package CARBayes [13] with and without covariates (e.g. surface material of the samples). We use Deviance Information Criteria (DIC) [15] and Geweke z-scores [16] to examine model fit and model convergence.

4 Results

The relative abundance of antimicrobial related species is a small portion of the metagenomics annotated genome (0 to 0.33% of the total reads). The Reads Per Kilobase per Million (RPKM) for antimicrobial related species for each city for CSD16 and CSD17 sampling days are plotted in Fig. 1 and Fig. 2. Ilorin, Offa and Lisbon show high abundance for CSD16 day while Fairbanks and Zurich have high variability among the samples. Doha, Denver and Stockholm have the lowest median antimicrobial resistance abundance.

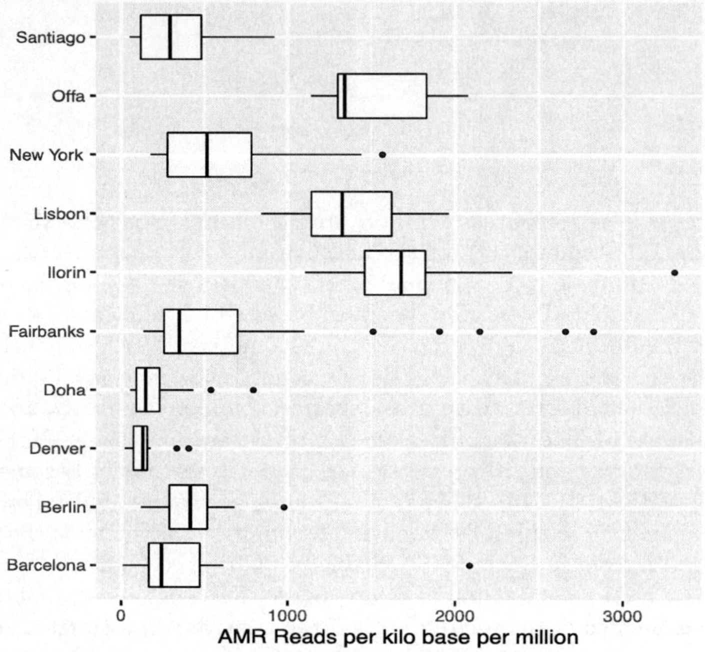

Fig. 1. Reads per kilobase per million (RPKM) for antimicrobial related species for each city for CSD16

To measure the spatial correlation we use Moran's I test [17]. Big metropolitan areas with many samples like New York, Hong Kong and Seoul show strong spatial correlations for many of the top abundant antimicrobial resistant taxa. Next we apply the spatial convolution model described in the Methods section to each of the taxa for which the spatial correlation measured by Moran's I test is significant (p-value < 0.01). The results are posterior estimates of the Standardized Incidence Ratio. We then create maps using BatchGeo [18] based on the latitude and longitude information from MetaSub and overlay the relative risk estimates. In Fig. 3 we show an example of two maps from London with

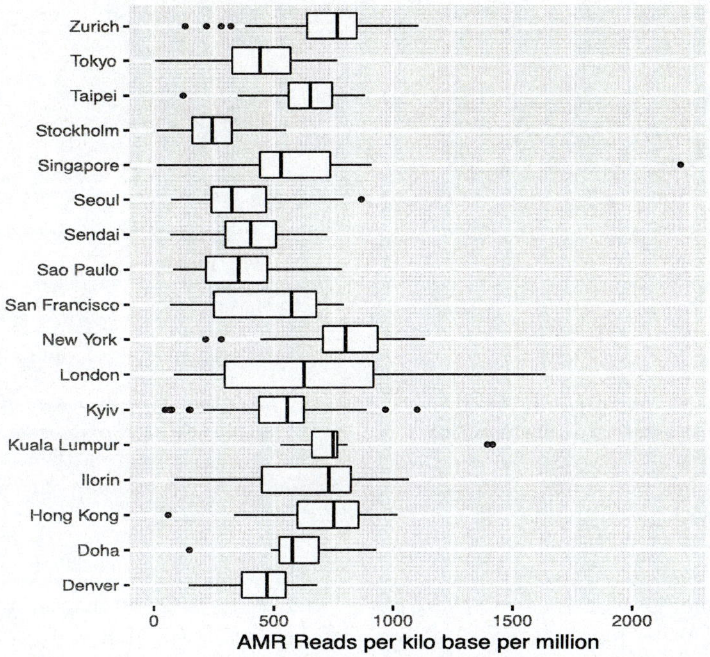

Fig. 2. The Reads Per Kilobase per Million (RPKM) for antimicrobial related species for each city for CSD17

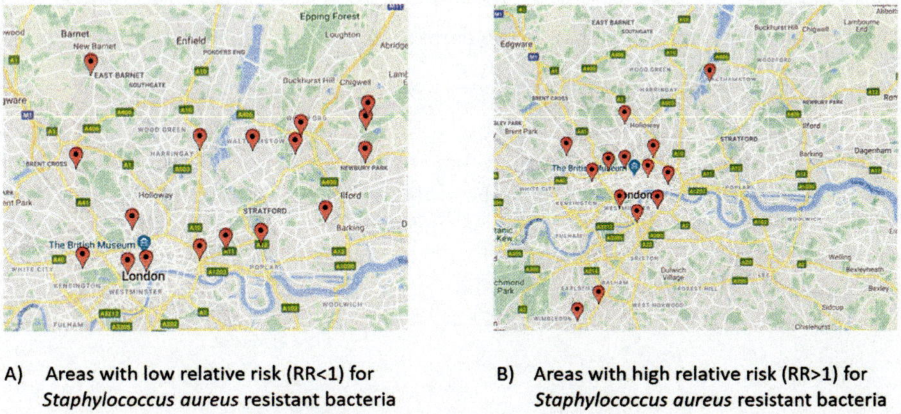

Fig. 3. An interactive map of the sampling locations in London with estimated low (A) and high (B) relative risk of *Staphylococcus aureus* AMR.

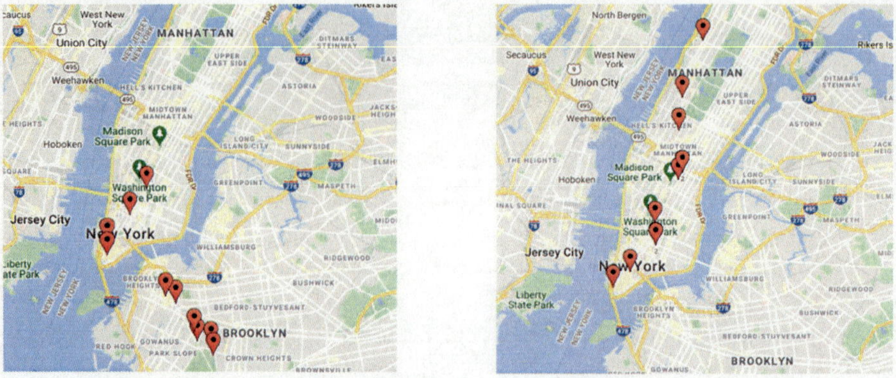

A) Areas with low relative risk (RR<1) for *Salmonella enterica* resistant bacteria

B) Areas with high relative risk (RR>1) for *Salmonella enterica* resistant bacteria

Fig. 4. An interactive map of the sampling locations in New York City with estimated low (A) and high (B) relative risk of *Salmonella enterica* AMR.

overlaid model estimates for *Staphylococcus aureus* resistant bacteria. On the left map (A) we have areas with lower posterior estimates (RR < 1) and on the right (B) we have areas with higher posterior estimates (RR > 1). Samples from subway stations in areas further from the center of the city (A) have lower risk while subway stations in areas in the center and close to touristic sites (B) have higher risk for *Staphylococcus aureus* resistant bacteria. Similarly, we estimate the relative risk in New York City, Fig. 4, where heavily used subway areas in Manhattan show higher relative risk compared to the less populated areas in Brooklyn.

Figures 5 and 6 show the relative risk estimates from the model for each city for the top abundant antimicrobial resistant taxa for CSD16 and CSD17 sampling days respectively. Cities like Ilorin and Offa (very close proximity and sharing similar profiles) and Lisbon show high relative risk for resistant bacteria compared to other cities from CSD16 sampling day. New York, Zurich and Ilorin also show relative high risk compared to other cities from CSD17 sampling day for the top abundant resistant taxa.

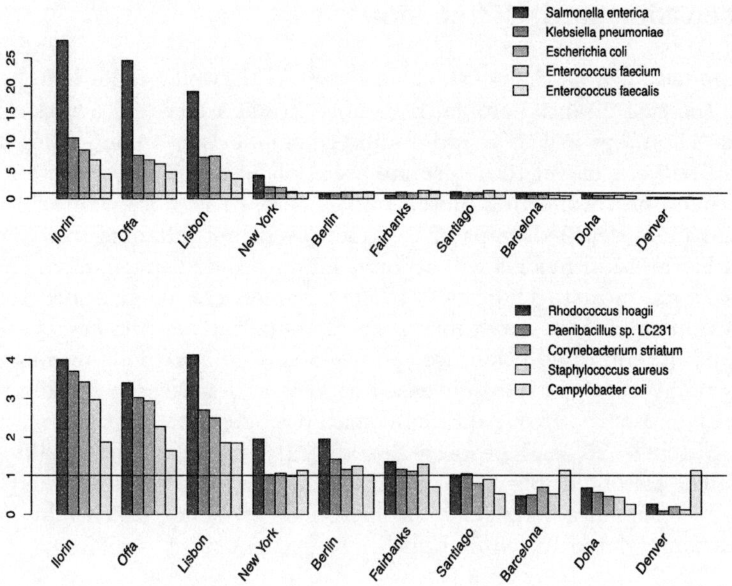

Fig. 5. Barplot of relative risk estimates of top ten abundant antimicrobial resistant species using CSD16 samples. The expected relative risk is the median across the samples for each taxon.

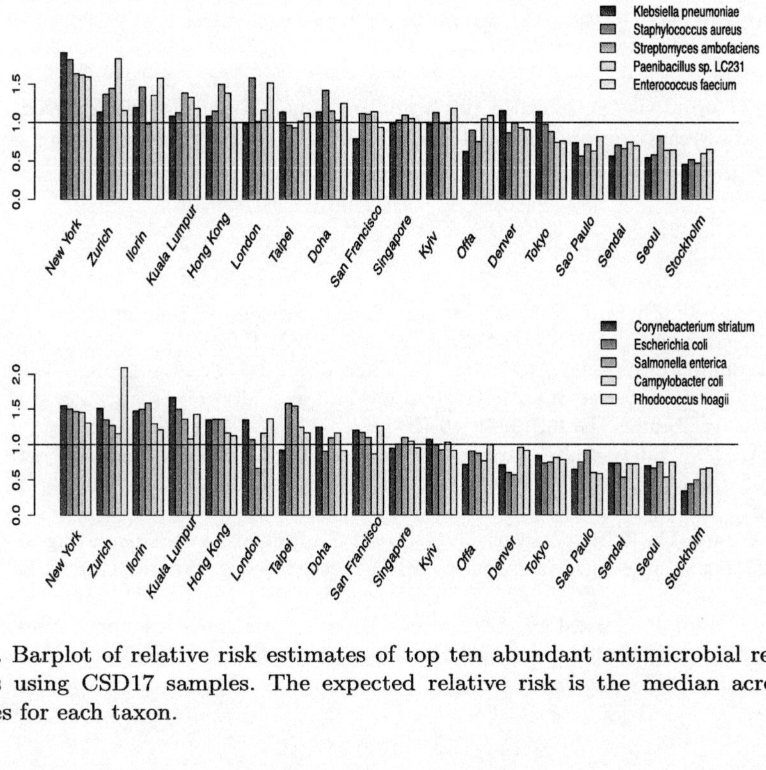

Fig. 6. Barplot of relative risk estimates of top ten abundant antimicrobial resistant species using CSD17 samples. The expected relative risk is the median across the samples for each taxon.

5 Discussion and Conclusion

Spatial models such as the convolution Besag-York-Mollie model are helpful in assessing the risk of antimicrobial resistance inside a city and across cities and countries. The maps in Fig. 3 and 4 and the comparison among cities in Fig. 5 and 6 can be very useful to determine areas of higher relative risk for specific resistant microbes. Collecting more samples across the globe will help to build better and more detailed maps of relative risk of antimicrobial resistant taxa. Bayesian hierarchical models will account for such overdispersion which is commonly observed in metagenomics data. The models can be extended to include more covariates: climate conditions such as temperature, pressure, population, density and others. In addition, we can use a zero-inflated Poisson model (mixture of a point mass distribution based at zero and a Poisson distribution) for counts with excessive zeros. Additional spatial models such as Leroux et al. [19], Lee and Mitchell [20] and Lee and Sarran [21] which consider locally smooth methods can also be applied. In conclusion we propose the estimated relative risk as a way to characterize the antimicrobial resistance which may be useful in applications related to public health.

Acknowledgements. This work was supported by project BG05M2OP001-1.001-0004 (UNITE) for the coverage of conference fee and the National Scientific Program Information and Communication Technologies for a Single Digital Market in Science, Education and Security (ICTinSES) for logistics. In addition, this work was partially supported by the financial funds allocated to the Sofia University St. Kl. Ohridski, grant No. 80-10-116/2020.

Note. The opinions expressed in this work are personal and do not represent in any way Bristol Myers-Squibb. No Bristol Myers-Squibb resources were used to generate results or prepare this paper.

References

1. Thomas, C., Nielsen, K.: Mechanisms of, and barriers to, horizontal gene transfer between bacteria. Nat. Rev. Microbiol. **3**, 711–721 (2005)
2. Bennani, H., Mateus, A., Mays, N., Eastmure, E., Stärk, K.D.C., Häsler, B.: Overview of evidence of antimicrobial use and antimicrobial resistance in the food chain. Antibiotics (Basel) **9**(2), 49 (2020)
3. Okeke, I.N., Edelman, R.: Dissemination of antibiotic-resistant bacteria across geographic 444 borders. Clin. Infect. Dis. **33**, 364–369 (2001). https://doi.org/10.1086/321877
4. Lal Gupta, C., Kumar Tiwari, R., Cytryn, E.: Platforms for elucidating antibiotic resistance in single genomes and complex metagenomes. Environ. Int. **138**, 105667 (2020)
5. Van Camp, P.J., Haslam, D.B., Porollo, A.: Bioinformatics approaches to the understanding of molecular mechanisms in antimicrobial resistance. Int. J. Mol. Sci. **21**(4), 1363 (2020)

6. Hüls, A., Frömke, C., Ickstadt, K., et al.: Antibiotic resistances in livestock: a comparative approach to identify an appropriate regression model for count data. Front Vet Sci. **4**, 71 (2017)
7. MetaSUB International Consortium: The metagenomics and metadesign of the subways and urban biomes (MetaSUB) international consortium inaugural meeting report. Microbiome **4**(1), 24 (2016)
8. Bolger, A.M., Lohse, M., Usadel, B.: Trimmomatic: a flexible trimmer for illumina sequence data. Bioinformatics **30**(15), 2114–2120 (2014)
9. National Database of Antibiotic Resistant Organisms (NDARO). https://www.ncbi.nlm.nih.gov/pathogens/antimicrobial-resistance/
10. Menzel, P., et al.: Fast and sensitive taxonomic classification for metagenomics with Kaiju. Nat. Commun. **7**, 11257 (2016)
11. Danko, D., et al.: Global genetic cartography of urban metagenomes and antimicrobial resistance. bioRxiv, accepted in Science (2019)
12. Feargal, J.R.: Application of machine learning techniques for creating urban microbial fingerprints. Biol. Direct **14**(1), 13 (2019)
13. Lee, D.: CARBayes: an R package for Bayesian spatial modeling with conditional autoregressive priors. J. Stat. Softw. **55**(13), 1–24 (2013)
14. Besag, J., York, J., Mollie, A.: Bayesian image restoration with two applications in spatial statistics. Ann. Inst. Stat. Math. **43**, 1–59 (1991)
15. Lawson, A.B.: Bayesian Disease Mapping: Hierarchical Modeling in Spatial Epidemiology, 3rd edn. Chapman and Hall/CRC (2018)
16. Geweke, J.: Evaluating the accuracy of sampling-based approaches to calculating posterior moments. In: Bernado, J.M., Berger, J.O., Dawid, A.P., Smith, A.F.M. (eds.) Bayesian Statistics, vol. 4. Clarendon Press, Oxford (1992)
17. Gittleman, J.L., Kot, M.: Adaptation: statistics and a null model for estimating phylogenetic effects. Syst. Zool. **39**, 227–241 (1990)
18. BatchGeo. https://www.batchgeo.com/
19. Leroux, B., Lei, X., Breslow, N.: Estimation of disease rates in small areas: a new mixed model for spatial dependence. In: Halloran, M., Berry, D. (eds.) Statistical Models in Epidemiology, the Environment and Clinical Trials, vol. 116, pp. 179–191. Springer, New York (2000). https://doi.org/10.1007/978-1-4612-1284-3_4
20. Lee, D., Mitchell, R.: Boundary detection in disease mapping studies. Biostatistics **13**, 415–426 (2012)
21. Lee, D., Sarran, C.: Controlling for unmeasured confounding and spatial misalignment in long-term air pollution and health studies. Environmetrics **26**, 477–487 (2015)

Mathematical Modelling in Biomedicine and Healthcare

Generalized Net Model of Telerehabilitation Program for Patients with Socially Significant Diseases

Simeon Ribagin[1,2](✉), Antoaneta Grozeva[2], and Gergana Popova[2]

[1] Department of Bioinformatics and Mathematical Modelling, Institute of Biophysics and Biomedical Engineering, Bulgarian Academy of Sciences, Sofia, Bulgaria
sim_ribagin@mail.bg

[2] Department of Health and Social Care, Medical College, University "Prof. D-R Asen Zlatarov", Burgas, Bulgaria

Abstract. The development of telecommunication technologies and artificial intelligence has led to a significant change in the healthcare trends. A strong example of innovative healthcare provision is the so-called telerehabilitation (TR), which is part of the rapidly evolving field of telemedicine. In the present paper we propose a reduced GN-model representing a real time patient/therapist interaction via video-conferencing platform. The model allows different treatment behavior depending on the disease. The model describes a telerehabilitation protocol and algorithm for choosing the best rehabilitation treatment based on the patient dysfunctions and the possible outcomes.

Keywords: Telerehabilitation · Generalized nets · GN-model

1 Introduction

A large number of patients with chronic and subacute diseases may need physical and cognitive rehabilitation treatment at home. In general, the reduced length of hospital stays implies that patients return home sicker and with incapacities. Most of the patients will suffer from neuromuscular weakness, reduced functional capacity, exercise tolerance, health related quality of life and social function. Systematic rehabilitation programs in home can positively influence the course of physical problems associated with patient disease. Nowadays the interactive approaches of in-home rehabilitation like teleconsultation, telemonitoring, telehomecare etc., are becoming more and more popular. Numerous applications are utilizing commercially available monitoring devices and the internet to enable home-based disease management and monitoring [1]. Telerehabilitation (TR), also known as e-rehabilitation, is a modern form of rehabilitation service that is delivered over the interactive devices. TR is a part of the rapidly evolving field of telemedicine [2–4]. The typical telemedicine encounter involves a short, intensive session between a clinician and a patient. The TR process, in contrast, requires the continuous and frequent monitoring of the patient's functionality to test the delivered therapy

S. S. Sotirov et al. (Eds.): BioInfoMed 2020, LNNS 374, pp. 91–99, 2022.
https://doi.org/10.1007/978-3-030-96638-6_10

and/or adapt it to the patient's progress [5]. It can address a wide spectrum of health conditions, from minor health issues to emergency medical conditions. Many patients with socially significant diseases suffering from disabilities, or other impairments can benefit greatly from TR, as it can reduce the costs and complications related to travel compared to traditional inpatient or person-to-person rehabilitation. It can be especially beneficial to people in rural or remote areas. One the main advantages of TR is that it reduces the number of trips to specialized clinics, reducing costs and improving the availability of the therapies [6]. Some studies on chronic patients [7] highlight the benefits of fewer journeys and flexible hours for therapies. Its effectiveness depends upon the specifics of the application, including the type and severity of disability, purpose of assessment, intervention goals, specifics of the technology, and infrastructure of both the local and remote sites [8]. Various studies have shown that TR can have similar effects to those of traditional rehabilitation. In general, TR can be divided in two major types of delivery. The first mode is remote and/or web-based rehabilitation and the second mode of telerehabilitation involves real-time interaction via video-conferencing or similar communication [9]. There are a number of different types of video-conferencing systems. This approach enabled the physiotherapist to watch participants performing the exercises and provide real-time feedback and modification, as required. The use of video-conferencing allows verbal and visual interactions and providing consultations, diagnostic evaluations and also therapeutic interventions. Most of the components of a standard rehabilitation assessment can be conducted via video-conference, thanks to the development of new technologies such as wearable devices, mobile equipment and devices that include peripherals, smartphone apps that are used to collect patient measures and others. They are able to monitor the physical activity, during the prescribed exercises or other physical activity targets and the collected data can be further processed to provide more effective health interventions. At the beginning and end of each video-conference telerehabilitation session, the therapist interacts directly with the patient and during task performance, the patient is able to see and hear the therapist. On the other hand, the cost of video equipment, limited network connectivity from home, complexity and cost of hardware, patient adherence, technical and health literacy, complicate the use of video-conferencing approach. In this manner an appropriate protocol with requirements for the use of such therapy is needed. In the present paper, an application of the apparatus of generalized nets [10–12] to telerehabilitation services and the advantages of using such model, is discussed. We propose a reduced GN-model representing a real time patient/therapist interaction via video-conferencing platform.

2 The Generalized Net Model

The GN-model (Fig. 1) has 6 transitions and 22 places with the following meaning:

- Transition Z_1 represents the patient device,
- Transition Z_2 represents the therapist device,
- Transition Z_3 represents the video-conference platform,
- Transition Z_4 represents the patient assessment components and monitoring
- Transition Z_5 represents the exercise program,

- Transition Z_6 represents the patient performance and the patient outcomes

The net contains six types of tokens: α, β, μ, η, γ and κ. Some of the model transitions contain a so called "special place" where a token stays and collect information about the specific parts of the screening process and serves as a database storage, which it represents as follows:

- In place l_3, token $\boldsymbol{\alpha}$ stays permanently and collects the information about the patient personal device,
- In place l_6, token $\boldsymbol{\beta}$ stays permanently and collects information about the therapist personal device,
- In place l_9, token $\boldsymbol{\mu}$ stays permanently and collects information about the videoconference platform,
- In place l_{12}, token $\boldsymbol{\eta}$, stays permanently and collects information about the personal data of the patient (general medical history, subjective symptoms and objective measures),
- In place l_{16}, token $\boldsymbol{\gamma}$ stays permanently and collects information about the exercise program (exercise description, intensity, duration, type, length and frequency) and recordings of the exercise session.
- In place l_{19}, token $\boldsymbol{\kappa}$ stays permanently and collects information about the patient performance and the patient outcomes.

At the time of duration of the GN-functioning, some of these tokens can split, generating new tokens, that will transfer in the net obtaining respective characteristics, and also in some moments they will unite with some of tokens: α, β, μ, η, γ and κ.

Token α enters the net with additional characteristics "*patient (user)*" in place l_1.

Token β enters the net with additional characteristics "*therapist*" in place l_4.

The transition Z_1 of the GN-model has the following form:

$$Z_1 = <\{l_1, l_3\}, \{l_3, l_3\}, r_1>,$$

where:

$$r_1 = \begin{array}{c|cc} & l_2 & l_3 \\ \hline l_1 & false & true \\ l_3 & W_{3,2} & true \end{array}$$

and $W_{3,2}$ = "*the patient(user) has access to a personal mobile/stationary device and reliable internet connection*".

On the next time-moment, token α splits to two tokens – the same token α and token α_1. When the predicate $W_{3,2}$ is true, token α_1 enters place l_2 and there it obtains a characteristic:

"*established videoconference connection with the therapist,*".

The transition Z_2 of the GN-model has the following form:

$$Z_2 = <\{l_4, l_6\}, \{l_5, l_6\}, r_2>,$$

where:

$$r_2 = \begin{array}{c|cc} & l_5 & l_6 \\ \hline l_4 & false & true \\ l_6 & W_{6,5} & true \end{array}$$

and $W_{6,5}$ = *"the therapist has access to a personal mobile/stationary device and reliable internet connection"*.

Token β splits to two tokens – the same token β and token β_1. When the predicate $W_{5,6}$ is true, token β_1 enters place l_5 and there it obtains a characteristic:

"established videoconference connection with the patient,".

The transition Z_3 has the following form:

$$Z_3 = \langle\{l_2, l_5, l_9\}, \{l_7, l_8, l_9\}, r_3\rangle,$$

where:

$$r_3 = \begin{array}{c|ccc} & l_7 & l_8 & l_9 \\ \hline l_2 & false & false & true \\ l_5 & false & false & true \\ l_9 & W_{9,7} & W_{9,8} & true \end{array}$$

and, $W_{9,7}$ = *"both the patient and the therapist have a real-time video and audio communication"*;

$W_{9,8}$ = "$\neg W_{9,7}$";

The tokens from the three input places of transition Z_3 enter place l_9 and unite with token μ with the above mentioned characteristic. On the next time-moment, token μ splits to three tokens – the same token μ and tokens μ_1 and μ_2. When the predicate $W_{9,7}$ is true, token μ_1 enters place l_7 and there it obtains a characteristic:

"start the video-conference telerehabilitation session with the current patient".

When the predicate $W_{9,8}$ is true, token μ_2 enters place l_8 and there it obtains a characteristic:

"restart the broadband connection".

The transition Z_4 has the following form:

$$Z_4 = \langle\{l_7, l_{12}, l_{13}, l_{20}\}, \{l_{10}, l_{11}, l_{12}, l_{13}\}, r_4\rangle,$$

where:

$$r_4 = \begin{array}{c|cccc} & l_{10} & l_{11} & l_{12} & l_{13} \\ \hline l_7 & false & false & W_{7,12} & W_{7,13} \\ l_{12} & W_{12,10} & W_{12,11} & true & true \\ l_{13} & false & false & W_{13,12} & true \\ l_{20} & false & false & true & false \end{array}$$

and,

$W_{7,12}$ = *"the patient has personal health e-record (name, age, diagnosis, platform ID, personal exercise program, etc.)"*;

$W_{7,13}$ = "$\neg W_{7,12}$";

$W_{12,10}$ = *"there is a current objective data from the measuring devises and a current data from the patient subjective complains"*;

$W_{12,11}$ = "$\neg W_{12,10}$";

When the predicate $W_{7,12}$ is true, token μ_1 enters place l_{12} and unite with token η with the above mentioned characteristic. On the next time-moment, token η splits to three tokens – the same token η and tokens η_1 and η_2. When the predicate $W_{12,10}$ is true, token η_1 enters place l_{10} and there it obtains a characteristic:

"*objective and subjective measures*: < *BP, HR,* SpO_2, *temperature, pain, tiredness, adherence, etc.* >".

When the predicate $W_{12,11}$ is true, token η_2 enters place l_{11} and there it obtains a characteristic:

"*reset the measuring devices and perform another patient symptoms evaluation*".

When the predicate $W_{7,13}$ is true, token μ_1 enters place l_{13} and there it obtains a characteristic:

"*create a new patient personal health e-record*".

$$Z_5 = \langle\{l_{10}, l_{17}, l_{20}\}, \{l_{14}, l_{15}, l_{16}, l_{17}\}, r_4\rangle,$$

where:

$$r_5 = \begin{array}{c|cccc} & l_{14} & l_{15} & l_{16} & l_{17} \\ \hline l_{10} & false & false & W_{10,16} & W_{10,17} \\ l_{17} & W_{17,14} & W_{17,15} & false & true \\ l_{20} & false & false & false & true \end{array}$$

and,

$W_{10,\,17}$ = "*objective and subjective measures of the patient (the functional status) are normal and allow for exercise*";

$W_{10,\,16}$ = "$\neg W_{10,\,17}$".

$W_{17,\,14}$ = "*the patient is classified according to its specific condition and functional status and in the database of the platform, there are predefined exercise programs for the current patient*";

$W_{17,\,15}$ = "$\neg W_{17,\,14}$".

When the predicate $\mathrm{W}_{10,\,17}$ is true, token η_1 enters place l_{17} and unites with token γ with the above mentioned characteristic. On the next time-moment, token γ splits to two tokens – the same token γ, token γ_1 and token γ_2.

When the predicate $\mathrm{W}_{10,\,16}$ is true, token η_1 enters place l_{16} and there it obtains a characteristic:

"*skip the exercise session and give patient consultation (identify problems and make recommendations)*".

When the predicate $\mathrm{W}_{10,\,14}$ is true, token γ_1 enters place l_{14} and there it obtains a characteristic:

"*select from the patient personal health e-record the description, intensity, duration, type, length and frequency of the exercise program (specific exercises for the current patient) presented via pictures or videos*".

When the predicate $\mathrm{W}_{16,\,15}$ is true, token $\boldsymbol{\gamma}_2$ enters place l_{15} and there it obtains a characteristic: description, intensity, duration, type, length and frequency.

"*classify patient and select a set of exercises, or define new ones specific for the current patient needs and upload them for the next session*".

$$Z_6 = \langle \{l_4, l_{22}\}, \{l_{18}, l_{19}, l_{20}, l_{21}, l_{22}\}, r_6 \rangle,$$

where:

$$r_6 = \begin{array}{c|ccccc} & l_{18} & l_{19} & l_{20} & l_{21} & l_{22} \\ \hline l_{14} & false & false & false & false & true \\ l_{22} & W_{22,18} & W_{22,19} & W_{22,20} & true & true \end{array}$$

and,

$W_{22,\,18}$ = "*the patient has performed the prescribed exercises for the current session under the supervision of the therapist, without any objective and subjective complains*";

$W_{22,\,19}$ = "*the patient has performed the prescribed exercises for the current session under the supervision of the therapist, with some objective and subjective complains*".

$W_{22,\,20}$ = "*during the performance of the exercises for the current session, under the supervision of the therapist, the patient feels physical, psychological and functional discomfort*".

The tokens from the two input places of transition Z_6 enter place l_{22} and unite with token κ with the above mentioned characteristic. On the next time-moment, token κ splits to five tokens – the same token κ and tokens $\kappa_1, \kappa_2, \kappa_3$ and κ_4. When the predicate $W_{22,\,18}$ is true, token κ_1 enters place l_{18} and there it obtains a characteristic:

"*end the current session, export the objective measures data, evaluate the patient performance and functional status after the exercises, report the patient outcomes and therapy goals*"

When the predicate $W_{22,\,19}$ is true, token κ_2 enters place l_{19} and there it obtains a characteristic:

"*end the current session, export the objective measures data, evaluate the patient performance and functional status after the exercises and consider changes in the exercise program*"

When the predicate $W_{22,\,20}$ is true, token κ_3 enters place l_{20} and there it obtains a characteristic:

"*stop the current session and make changes in intensity, duration, type, length and frequency of the exercise program*".

Token κ_4 enters place l_{21} and there it obtains a characteristic:

"*extend the information of the patient personal health e-record (number of exercises, objective and subjective data during the exercising, patient outcomes and therapy goals)*".

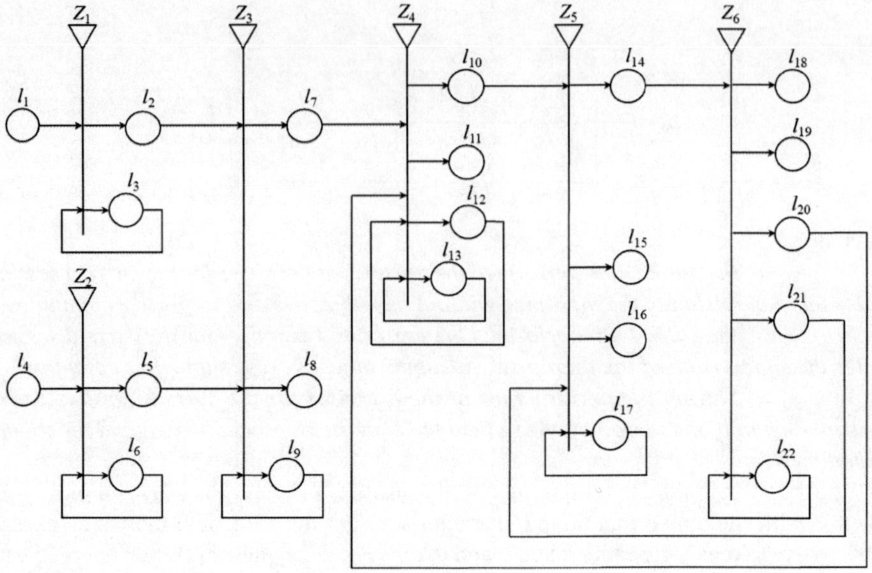

Fig. 1. Generalized Net model of Telerehabilitation program for patients with socially significant diseases.

3 Conclusions

The model describes a telerehabilitation protocol and algorithm for choosing the best rehabilitation treatment based on the patient dysfunctions and the possible outcomes. The model can be used for simulation of different situations, related to the condition of the patient and the treatment plan in order to observe, and analyze the possible results generated when specific information of the model's parameters is introduced. The results can be used as a powerful tool in the decision making process in order to optimize the telerehabilitation programs for patients with socially significant diseases. Moreover, the model can be complicated and detailed, which will significantly improve the accuracy of the assessment of the patients and the reliability of the proposed algorithm. Some of the structures of the proposed model represents database storages and all this data can be collected which in turn helps healthcare providers get a better understanding of the telerehabilitation approach, functional outcomes and optimize the processes for each individual.

Acknowledgements. The authors are grateful for the support provided by the University "Prof. Dr. Asen Zlatarov"- Burgas, Scientific research sector "Scientific and artistic activity" approved by № НИХ - 442/2020 "Exploring the possibilities of telemedicine, telerehabilitation and telepharmacy to improve patient care and medical education".

References

1. Demiris, G.: Electronic home healthcare: concepts and challenges. Int. J. Electron. Healthc. **1**(1), 4–16 (2004). https://doi.org/10.1504/IJEH.2004.004655
2. Peretti, A., et al.: Telerehabilitation: review of the state-of-the-art and areas of application. JMIR Rehabil. Assist Technol. **4**(2), e7 (2017). 22
3. Cottrell, M.A., Galea, O.A., O'Leary, S.P., Hill, A.J., Russell, T.G.: Real-time telerehabilitation for the treatment of musculoskeletal conditions is effective and comparable to standard practice: a systematic review and meta-analysis. Clin. Rehabil. **31**(5), 625–638 (2017). https://doi.org/10.1177/0269215516645148
4. Niknejad, N., et al.: Understanding telerehabilitation technology to evaluate stakeholders' adoption of telerehabilitation services: a systematic literature review and directions for further research. Arch. Phys. Med. Rehabil. **102**(7), 1390–1403 (2021)
5. Parmanto, B., Saptono, A.: Telerehabilitation: state-of-the-art from an informatics perspective Int. J. Telerehabil. **1**(1), 73–84 (2009). https://doi.org/10.5195/ijt.2009.6015, PMID:25945164, PMCID:PMC4296781
6. Navarro, E., et al.: Adaptive, multisensorial, physiological and social: the next generation of telerehabilitation systems. Front. Neuroinform. **12**, 43 (2018). https://doi.org/10.3389/fninf.2018.00043
7. Cranen, K., et al.: An exploration of chronic pain patients' perceptions of home telerehabilitation services. Health Expect. **15**, 339–350 (2012). https://doi.org/10.1111/j.1369-7625.2011.00668.x
8. Wakeford, L., et al.: Telerehabilitation position paper. Comm. Pract. Am. J. Occup. Ther. **59**(6), 656–60 (2005)
9. Pat, G.C.: The wicked problem of telerehabilitation: considerations for planning the way forward. AIMS Med. Sci. **5**(4), 357–369 (2018). https://doi.org/10.3934/medsci.2018.4.357
10. Atanassov, K.: Generalized Nets. World Scientific, Singapore (1999)
11. Atanassov, K.: On Generalized Nets Theory. Prof. M. Drinov Academic Publishing House, Sofia (2007)
12. Zoteva, D., Krawczak, M.: Generalized nets as a tool for the modelling of data mining processes. A survey. Issues Intuit. Fuzzy Sets Generalized Nets **13**, 1–60 (2017)

A Generalized Net Model with Intuitionistic Fuzzy Assessments of the Process of Cardiopulmonary Resuscitation

Hristo Bozov[1], Greta Bozova[2], Evdokia Sotirova[1](✉), and Anthony Shannon[3]

[1] Faculty of Public Health and Health Care, Prof. Assen Zlatarov University, 1 Prof. Yakimov str., Burgas, Bulgaria
esotirova@btu.bg

[2] Department of Nephrology, Military Medical Academy, 3 St. George Sofiiski, Sofia, Bulgaria
greti@abv.bg

[3] Resident Fellow, Warrane College, The University of New South Wales, Sydney, Kensington NSW, Australia
t.shannon@warrane.unsw.edu.au

Abstract. In the paper a Generalized Net (GN, [5, 6]) model of the Cardiopulmonary Resuscitation (CPR) process is presented. The CPR is a collection of diagnostic and therapeutic actions performed to restore suspended basic vital functions (respiration, cardiac activity, functions of the central nervous system) during cardiac arrest. The developed GN model is a set of transitions that present the sequence of actions of the CPR process with its three interconnected stages:

- basic life support - refers to maintaining an airway and supporting breathing and the circulation;
- advanced life support - restoring the spontaneous circulation, stabilization the condition of the patient, and starting the intensive treatment;
- prolonged life support - intensive treatment of the post-resuscitation syndrome.

From the input places of the GN model enter tokens that present the initial status of the person who needs from CPR. The new and enriched knowledge about the process of cardiopulmonary resuscitation of the output of the model is obtained.

In the research, intuitionistic fuzzy assessments $\langle \mu, \nu \rangle$ which estimate the execution time of the CPR process are suggested.

Keywords: Generalized nets · Cardiopulmonary resuscitation · Intuitionistic fuzzy sets

1 Introduction

The vitality of the central nervous system determines life expectancy [9]. What significantly distinguishes humans from other animal species is the presence of higher nervous activity (second signaling system), which anatomically correlates with the brain. It, unlike other organs of the human body, is very sensitive to lack of oxygen. In fact, it can

S. S. Sotirov et al. (Eds.): BioInfoMed 2020, LNNS 374, pp. 100–112, 2022.
https://doi.org/10.1007/978-3-030-96638-6_11

last only a few minutes without oxygenation. Cessation of efficient cardiac activity and inflow of oxygenated (oxygen-saturated) blood to the brain for more than 5 min leads to irreversible neurohistological changes [14]. The [15] is focused on the first crucial minutes of a cardiac arrest.

Acute, sudden cardiac arrest (cardiac arrest) is followed by a condition called clinical death. Blood flow to vital organs (heart, brain) is stopped. This is a reversible condition. The difference with biological death is that it has no recovery. Clinical death is an accidental death lasting 3–5 min (maximum 6 min). The period of clinical death is determined by the time during which the cerebral cortex can endure without irreversible degenerative changes after blood flow to it. The causes of clinical death are most often associated with cardiac pathology (82.5%). Non-cardiac causes (9%) are related to accidents - trauma, intoxication, asphyxia, drowning, and 8.5% with various diseases. Pathophysiological manifestations are dominated by the development of acute hypoxia (lack of sufficient oxygen in the tissues) and metabolic acidosis (acidic pH) ([9, 12, 14]). In [11] the etiology in the patients resuscitated from cardiac arrest is presented. The problems in early identification of patients and out-of-hospital cardiac arrest incidence are discussed in [13] and [17].

The diagnosis is based on the following symptoms: loss of consciousness, lack of heart function and pulse of a large artery, lack of breathing, dilation of the pupils, areflexia without light reaction, pale skin and/or cyanosis (dark blue to purple).

Depending on the electrocardiographic (ECG) image in cardiac arrest, there are two forms of clinical death - asystole (no ECG activity) and ventricular fibrillation (ECG shows electrical fibrillation (twitching), which does not lead to effective systole (contraction of the heart muscle).

Treatment consists of performing emergency cardiopulmonary (cardiac-pulmonary) resuscitation, which aims to restore basic vital functions. Without the timely application of CPR, any terminal condition ends in biological death [16].

Within the last 20 years a lot of GN models, related to description of biomedical processes in the living organism, and models of the ways of functioning of the different systems and organs in the human body were constructed [8]. The modelling of the biomedical processes and systems with a uniform instrument gives the possibility for their comparing, testing, analysis and control.

In the current research, a GN model of the Cardiopulmonary Resuscitation (CPR) process is presented. The CPR is a collection of diagnostic and therapeutic actions performed to restore suspended basic vital functions (respiration, cardiac activity, functions of the central nervous system) during cardiac arrest ([9, 18]).

Assessments which estimate the success of the CPR process are suggested. They are formed on the basis of a set of intuitionistic fuzzy estimations $\langle \mu, \nu \rangle$ of real numbers from the set $[0, 1] \times [0, 1]$, related to the respective patients. These intuitionistic fuzzy estimations reflect the degree of each patient's good condition μ, or bad condition ν, for each CPR procedure, and for them is valid that $\mu + \nu \leq 1$. The degree of uncertainty $\pi = 1 - \mu - \nu$ represents the cases in which the CPR procedure has not been completed or additional information is needed. Within the paper the ordered pairs were defined in the sense of intuitionistic fuzzy sets [7].

2 Generalized Net Model of the Process of Cardiopulmonary Resuscitation

The developed GN-model present the sequence of actions of the CPR process (Fig. 1). For the construction of the model the algorithms form [1, 2, 4, 9], It contains 12 transitions and 47 places, and five types of the tokens:

- α-tokens represent the persons, that perform CPR and their activities,
- β-tokens represent the special equipment, apparatus, drugs and fluids, etc.,
- χ-tokens represent the patients,
- δ-tokens represent the activities over the patients,
- ε-tokens represent the criteria for evaluation the CPR process.

For brevity, we shall use the notation α-, β-, χ-, δ- and ε- tokens instead of α_j-, β_k-χ_l -, δ_i-, and ε_s - tokens, where *j*, *k*, *l*, *i*, *s*, are numerations of the respective tokens.

Initially, the tokens α, β, χ and ε stay in places l_3, l_6, l_9, $l_{e,2}$. They will be in their own places during the whole time during which the GN functions. While they may split into two or more tokens, the original token will remain in its own place the whole time. The original tokens have the following initial and current characteristics:

- token α: "*initial or current status of the persons, that perform CPR and their activities*" (in place l_3),
- token β: "*initial or current status of the special equipment, apparatus, drugs and fluids, etc.*" (in place l_6),
- token χ: "*initial or current status of the patients*" (in place l_9).
- token ε: "*criteria and initial or current status of CPR process*" (in place $l_{e,2}$).

From places l_1, l_4 and l_7 enter tokens with characteristics, respectively, *"new person, that can perform CPR"* in place l_1, *"new special equipment/apparatus/drugs or fluids"* in place l_4, and *"new patient"* in place l_7.

From place $l_{e,1}$ enter ε-tokens with characteristics *"new criteria for evaluation the CPR process"*.

All tokens that enter transitions Z_1, Z_2 and Z_3 will unite with the corresponding original token. All information generated by the respective original token will be put as an initial characteristic of a token, generated by them.

The transitions of the GN-model represent the following functions:

Z_1 - activities of the persons, that perform CPR,
Z_2 - activities with the special equipment, apparatus, etc. for CPR,
Z_3 - activities of the persons, needed from CPR,
Z_4 - measures performed in CPR,
Z_5 - providing free airways,

Z_6 - activities for prolonged life support,
Z_7 - assessment of the patient's condition,
Z_8 - performing respiratory resuscitation,
Z_9 - performing of the artificial circulation (heart massage),
Z_{10} - performing of the assisted ventilation,
Z_{11} - activities for pharmacotherapy,
Z_{12} - activities for electrical cardioversion and defibrillation.

The transitions have the following forms:

$$Z_1 = \langle \{l_1, l_3, l_{13}\}, \{l_2, l_3\}, \begin{array}{c|cc} & l_2 & l_3 \\ \hline l_1 & false & true \\ l_3 & W_{3,2} & W_{3,3} \\ l_{13} & false & true \end{array} \rangle,$$

where:

- $W_{3,2}$ = "The person/persons must perform CPR",
- $W_{3,3} = \neg W_{3,2}$.

The α-tokens do not obtain new characteristic in place l_3 and they obtain the characteristic.

"persons, that perform CPR; patient needed from CPR;
needed special equipment, apparatus, etc."

in place l_2.

$$Z_2 = \langle \{l_4, l_6, l_{14}\}, \{l_5, l_6\}, \begin{array}{c|cc} & l_5 & l_6 \\ \hline l_4 & false & true \\ l_6 & W_{6,5} & W_{6,6} \\ l_{14} & false & true \end{array} \rangle,$$

where:

- $W_{6,5}$ = "The special equipment, apparatus, etc. is included in current characteristic of α-token",
- $W_{6,6} = \neg W_{6,5}$.

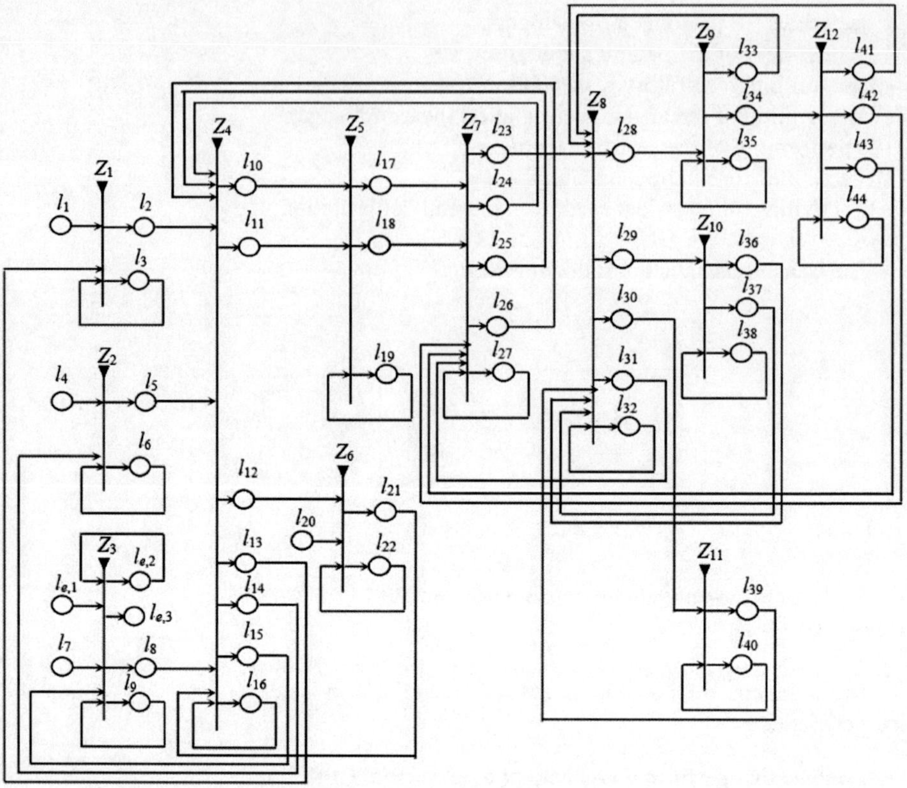

Fig. 1. Generalized net model of the process of cardiopulmonary resuscitation.

The β-tokens do not obtain new characteristic in place l_6 and they obtain the characteristic.

"special equipment, apparatus, drugs and fluids, etc."

in place l_5.

$Z_3 = \langle \{l_7, l_9, l_{15}, l_{e,1}, l_{e,2}\}, \{l_8, l_9, l_{e,2}, l_{e,3}\},$

	l_8	l_9	$l_{e,2}$	$l_{e,3}$
l_7	*false*	*true*	*false*	*false*
l_9	$W_{9,8}$	$W_{9,9}$	$W_{9,e}$	*false*
l_{15}	*false*	*true*	*false*	*false*
$l_{e,1}$	*false*	*false*	*true*	*false*
$l_{e,2}$	*false*	*false*	*true*	$W_{e,3}$

$\rangle,$

where:

- $W_{9,8}$ = "There is a person needed from CPR",
- $W_{9,9} = \neg W_{9,8}$,
- $W_{9,e}$ = "There is a CPR for evaluation",
- $W_{e,3}$ = "The current evaluations for CPR are calculated".

The χ-tokens do not obtain new characteristic in place l_9 and they obtain the characteristic.

"patient needed from CPR; current time-moment".

in place l_8.

The ε-tokens do not obtain new characteristic in place $l_{e,2}$ and they obtain the characteristic.

"current status of CPR procedures; current time-moment t; $\langle \mu(t), \nu(t) \rangle$"

in place $l_{e,3}$.

Here, and later t is the moment in the time when the values of the numbers $\mu(t)$ and $\upsilon(t)$ are determined. The values of the $\mu(t)$ and $\upsilon(t)$ are calculated according to the respective number of tokens in places l_{24} and l_{25}. The degree of uncertainty is determined by the number of tokens in place l_{26}. In this case the evaluations of the performed CPR procedures can be calculated according formula:

$$\langle \mu(t), \nu(t) \rangle = \left\langle \frac{a}{c}, \frac{b}{c} \right\rangle \quad (1)$$

where:

a is the number of patients with restored self-breathing and heart rhythm in the current time-moment t,

b is the number of dead patients in the current time-moment t,

c is the number of patients with performed CPR procedures in the current time-moment t.

Therefore, the degree of uncertainty here is determined by the number of patients needed from additional procedures.

Initially, when no information has been derived, all estimations take on initial values of $\langle 0, 0 \rangle$. When $k \geq 0$, the current $(k + 1)^{th}$ estimation is calculated on the basis of the previous estimations according to the recursive formula:

$$\langle \mu_{k+1}, \upsilon_{k+1} \rangle = \left\langle \frac{\mu_k k + \mu(t)}{k+1}, \frac{\upsilon_k k + \nu(t)}{k+1} \right\rangle \quad (2)$$

where $\langle \mu_k, \upsilon_k \rangle$ is the previous estimation, and $\langle \mu(t), \upsilon(t) \rangle$ is the estimation of the latest CPR, for $\mu(t), \nu(t) \in [0, 1]$ and $\mu(t), \nu(t) \leq 1$. Thus the model forms the final estimation of the CPR procedure.

$$Z_4 = \langle \{l_2, l_5, l_8, l_{16}, l_{21}, l_{24}, l_{25}, l_{26}\}, \{ l_{10}, l_{11}, l_{12}, l_{13}, l_{14}, l_{15}, l_{16}\},$$

	l_{10}	l_{11}	l_{12}	l_{13}	l_{14}	l_{15}	l_{16}
l_2	*false*	*false*	*false*	*false*	*false*	*false*	*true*
l_5	*false*	*false*	*false*	*false*	*false*	*false*	*true*
l_8	*false*	*false*	*false*	*false*	*false*	*false*	*true*
l_{16}	$W_{16,10}$	$W_{16,11}$	$W_{16,12}$	$W_{16,13}$	$W_{16,14}$	$W_{16,15}$	*false*
l_{21}	*false*	*false*	*false*	*false*	*false*	*false*	*true*
l_{24}	*false*	*false*	*false*	*false*	*false*	*false*	*true*
l_{25}	*false*	*false*	*false*	*false*	*false*	*false*	*true*
l_{26}	*false*	*false*	*false*	*false*	*false*	*false*	*true*

$\rangle$,

where:

- $W_{16,10}$ = "There is a patient for basic life support activities",
- $W_{16,11}$ = "There is a patient for advanced life support activities",
- $W_{16,12}$ = "There is a patient for prolonged life support activities",
- $W_{16,13}$ = "There is an information for current status of a persons, that perform CPR",
- $W_{16,14}$ = "There is an information for current status of a special equipment, apparatus, drugs and fluids, etc.",
- $W_{16,15}$ = "There is an information for current status of a patient".

The δ-tokens do not obtain new characteristic in place l_{16} and they obtain the characteristic.

"patient for basic life support activities; current time-moment" in place l_{10},
and "patient for advanced life support activities; current time-moment" in place l_{11}.

The χ-tokens obtain characteristics:

"patient for prolonged life support activities; current time-moment" in place l_{12}.
and "information for current status of a patient" in place l_{15}.

The α-tokens obtain characteristic.

"information for current status of a persons, that perform CPR" in place l_{13}.

The β-tokens obtain characteristic.

"information for current status of a special equipment,
apparatus, drugs and fluids, etc." in place l_{14}.

$$Z_5 = \langle \{l_{10}, l_{11}, l_{19}\}, \{l_{17}, l_{18}, l_{19}\}, \begin{array}{c|ccc} & l_{17} & l_{18} & l_{19} \\ \hline l_{10} & false & true & true \\ l_{11} & false & false & true \\ l_{19} & W_{19,17} & W_{19,18} & W_{19,19} \end{array} \rangle,$$

where:

- $W_{19,17} = W_{19,18} =$ "Free airways are provided",
- $W_{19,19} = \neg W_{9,18}$.

The δ-tokens do not obtain new characteristic in place l_{19} and they obtain the characteristic respectively:

"patient; current time-moment; main activities of the CPR, carried out without special equipment" in place l_{17},

and

"patient; current time-moment; main activities of the CPR, carried out with special equipment" in place l_{18}.

The ε-tokens enter the net from place l_{20} with characteristic "criteria for evaluation the patients".

$$Z_6 = \langle \{l_{12}, l_{20}\}, \{l_{21}, l_{22}\}, \begin{array}{c|cc} & l_{21} & l_{22} \\ \hline l_{12} & false & true \\ l_{20} & W_{22,21} & true \end{array} \rangle,$$

where.

- $W_{22,21} =$ "There are results from the intensive treatment of the post-resuscitation syndrome".

The δ-tokens do not obtain new characteristic in place l_{22} and they obtain the characteristic respectively:

"patient; indicators after the implementation of long-term therapy and care aimed at supporting and restoring the normal function of other organs and systems period for treatment" in place l_{21}.

$$Z_7 = \langle \{l_{17}, l_{18}, l_{27}, l_{31}, l_{42}, l_{43}\}, \{l_{23}, l_{24}, l_{25}, l_{26}, l_{27}\},
\begin{array}{c|ccccc}
 & l_{23} & l_{24} & l_{25} & l_{26} & l_{27} \\ \hline
l_{17} & false & false & false & false & true \\
l_{18} & false & false & false & false & true \\
l_{27} & W_{27,23} & W_{27,24} & W_{27,25} & W_{27,26} & W_{27,27} \\
l_{31} & false & false & false & false & true \\
l_{42} & false & false & false & false & true \\
l_{43} & false & false & false & false & true
\end{array} \rangle,$$

where:

- $W_{27,23}$ = "There is a patient for performing respiratory resuscitation",
- $W_{27,24}$ = "There is a patient with restored self-breathing and heart rhythm",
- $W_{27,25}$ = "There is a dead patient",
- $W_{27,26}$ = "There is a patient for additional procedures",
- $W_{27,27}$ = "The assessment of the patient's condition is not finished".

The δ-tokens do not obtain new characteristic in place l_{27} and they obtain the characteristic respectively:

"patient; current time-moment; activities of the CPR" in place l_{23},
"patient; current time-moment; restored self-breathing
and heart rhythm" in place l_{24},
"dead patient; current time-moment" in place l_{25},

and

"patient; current time-moment; query for additional procedures" in place l_{26}.

$$Z_8 = \langle \{l_{23}, l_{32}, l_{33}, l_{36}, l_{37}, l_{39}, l_{41}\}, \{l_{28}, l_{29}, l_{30}, l_{31}, l_{32}\},
\begin{array}{c|ccccc}
 & l_{28} & l_{29} & l_{30} & l_{31} & l_{32} \\ \hline
l_{23} & false & false & false & false & true \\
l_{32} & W_{32,28} & W_{32,29} & W_{32,30} & W_{32,31} & W_{32,32} \\
l_{33} & false & false & false & false & true \\
l_{36} & false & false & false & false & true \\
l_{37} & false & false & false & false & true \\
l_{39} & false & false & false & false & true \\
l_{41} & false & false & false & false & true
\end{array} \rangle,$$

where:

- $W_{32,28}$ = "There is a patient for compressions",
- $W_{32,29}$ = "There is a patient for breaths",
- $W_{32,30}$ = "There is a patient for pharmacotherapy",
- $W_{32,31}$ = "There is a patient for assessment of its current status",
- $W_{32,32}$ = "The assessment of the patient's respiratory resuscitation is not finished".

The δ-tokens do not obtain new characteristic in place l_{32} and they obtain the characteristic.

"patient for compressions; status; current time-moment" in place l_{28},
"patient for rescue breaths; status; current time-moment" in place l_{29},
"patient for pharmacotherapy; status; current time-moment" in place l_{30},
"the status of a patient, has to be assessed" in place l_{31}.

$$Z_9 = \langle \{l_{28}, l_{35}\}, \{l_{33}, l_{34}, l_{35}\}, \begin{array}{c|ccc} & l_{33} & l_{34} & l_{35} \\ \hline l_{28} & false & false & true \\ l_{35} & W_{35,33} & W_{35,34} & W_{35,35} \end{array} \rangle,$$

where:

- $W_{35,33}$ = "The compressions of the patient are finished",
- $W_{35,34}$ = "The electrocardiography and fibrillation treatment must be applied",
- $W_{35,35}$ = "The artificial circulation (chest compression)".

The δ-tokens do not obtain new characteristic in place l_{35} and they obtain the characteristic.

"patient, current status; completed compressions;
current time-moment" in place l_{33},
and "patient, current status; necessity from electrocardiography and fibrillation treatment current time-moment" in place l_{34}.

$$Z_{10} = \langle \{l_{29}, l_{38}\}, \{l_{36}, l_{37}, l_{38}\}, \begin{array}{c|ccc} & l_{36} & l_{37} & l_{38} \\ \hline l_{29} & false & false & true \\ l_{38} & W_{38,36} & W_{38,37} & W_{38,38} \end{array} \rangle,$$

where:

- $W_{38,36}$ = "The assisted ventilation of the patient without special equipment is finished",
- $W_{38,37}$ = "The assisted ventilation of the patient with special equipment is finished",
- $W_{38,38} = \neg W_{38,36}$ or $\neg W_{38,37}$.

The δ-tokens do not obtain new characteristic in place l_{38} and they obtain the characteristic.

"patient, current status; completed assisted ventilation without special equipment; current time-moment" in place l_{36},
and "patient, current status; completed assisted ventilation with special equipment; current time-moment" in place l_{37}.

$$Z_{11} = \langle \{l_{30}, l_{40}\}, \{l_{39}, l_{40}\}, \begin{array}{c|cc} & l_{39} & l_{40} \\ \hline l_{30} & false & true \\ l_{40} & W_{40,39} & W_{40,40} \end{array} \rangle,$$

where:

- $W_{40,39}$ = "The pharmacotherapy of the patient is finished",
- $W_{40,40} = \neg W_{40,39}$.

The δ-tokens do not obtain new characteristic in place l_{40} and they obtain the characteristic.
"patient, current status" in place l_{39}.

$$Z_{12} = \langle \{l_{34}, l_{44}\}, \{l_{41}, l_{42}, l_{43}, l_{44}\}, \begin{array}{c|cccc} & l_{41} & l_{42} & l_{43} & l_{44} \\ \hline l_{34} & false & false & false & true \\ l_{44} & W_{44,41} & W_{44,42} & W_{44,43} & W_{44,44} \end{array} \rangle,$$

where:

- $W_{44,41}$ = "There is a patient with asystole",
- $W_{44,42}$ = "The pacemaker of a patient is placed",
- $W_{44,43}$ = "There is a patient with effective systole",
- $W_{44,44}$ = "There is a patient with arrhythmias".

The δ-tokens do not obtain new characteristic in place l_{44} and they obtain the characteristic.

"patient in asystole, current status; current time-moment" in place l_{41},
"patient with pacemaker, current status; current time-moment" in place l_{42},
and "patient with effective systole, current status;
current time-moment" in place l_{43}.

3 Conclusions

The constructed here GN model can be changed adaptively by using hierarchical operators. It can be used for investigating the behavior of the CPR algorithm. Another application of the model can be prediction of some activities in the CPR and suggesting possible preventive measures, especially in the phase of the prolonged life support. The collecting the information about CPR parameters - fluids, drugs, therapy, etc., as well as qualitative observations can be used for: preparing the recommendations for improving the algorithm; collecting medical information for scientific investigations; training of students ([3, 10]).

Acknowledgements. The authors are grateful for the support provided by the Bulgarian Ministry of Education and Science under the National Research Program "Information and Communication Technologies for a Digital Single Market in Science, Education and Security" approved by DCM # 577/17.08.2018".

References

1. American Heart Association, Highlights of the 2020 American Heart Association Guidelines for CPR and ECC. https://cpr.heart.org/-/media/cpr-files/cpr-guidelines-files/highlights/hghlghts_2020_ecc_guidelines_english.pdf. Accessed 30 Oct 2020
2. American Heart Association, 2020 American Heart Association Guidelines for CPR and ECC. Algorithms. https://cpr.heart.org/en/resuscitation-science/cpr-and-ecc-guidelines/algorithms. Accessed 30 Oct 2020
3. American Heart Association, 2020 American Heart Association Guidelines for CPR and ECC. Part 6: Resuscitation Education Science. https://cpr.heart.org/en/resuscitation-science/cpr-and-ecc-guidelines/resuscitation-education-science. Accessed 30 Oct 2020
4. ACLS Training Center, Algorithms for Basic Life Support (2020). https://www.acls.net/blsalg. Accessed 20 Dec 2020
5. Atanassov, K.: On Generalized Nets Theory. Prof. M. Drinov Academic Publ. House, Sofia, Bulgaria (2007)
6. Atanassov, K.: Generalized Nets. World Scientific, Singapore (1991)
7. Atanassov, K.: On Intuitionistic Fuzzy Sets Theory. Springer, Berlin (2012)
8. Atanassov, K., Chakarov, V., Shannon, A., Sorsich, J.: Generalized Net Models of the Human Body. Prof. M. Drinov Publishing House, Sofia, Bulgaria (2008)
9. Bozov, H.: Guide to Marine Medicine. Military Medical Academy, Sofia, Bulgaria (2018). (in Bulgarian)

10. Cheng, A., et al.: Resuscitation education science: educational strategies to improve outcomes from cardiac arrest: a scientific statement from the American heart association. Circulation **138**(6), 82–122 (2018)
11. Chen, N., et al.: Arrest etiology among patients resuscitated from cardiac arrest. Resuscitation **130**, 33–40 (2018)
12. Eisenberg, M., Horwood, B., Cummins, R., Reynolds-Haertle, R., Hearne, T.: Cardiac arrest and resuscitation: a tale of 29 cities. Ann. Emerg. Med. **19**(2), 179–186. (1990)
13. Jentzer, J., Clements, C., Murphy, J., Wright, R.: Recent developments in the management of patients resuscitated from cardiac arrest. J. Crit. Care **39**, 97–107 (2017)
14. Lenjani, B., et al.: Cardiac arrest–cardiopulmonary resuscitation. J. Acute Dis. **3**(1), 31–35. Elsevier (2014)
15. Marsch, S., Müller, C., Marquardt, K., Conrad, G., Tschan, F., Hunziker, P.: Human factors affect the quality of cardiopulmonary resuscitation in simulated cardiac arrests. Resuscitation **60**(1), 51–56. Elsevier (2004)
16. Mathiesen, W., Bjorshol, C., Braut, G., Soreide, E.: Reactions and coping strategies in lay rescuers who have provided CPR to out-of-hospital cardiac arrest victims: a qualitative study. BMJ Open **6**(5), e010671 (2016). https://doi.org/10.1136/bmjopen-2015-010671
17. Nichol, G., Thomas, E., Callaway, C., et al.: Regional variation in out-of-hospital cardiac arrest incidence and outcome. JAMA **300**(12), 1423–1431 (2008)
18. Varon, J., Sternbach, G.: Cardiopulmonary resuscitation: lessons from the past. J. Emerg. Med. **9**(6), 503–507 (1991)

A Generalized Net Model of the Pattern of Behavior in Patients with eGFR < 20 mL/min (CKD Stage IV-V)

Martin Lubich[1], Anthony Shannon[2], Chavdar Slavov[3], Tania Pencheva[4], Simeon Ribagin[4,5(✉)], and Krassimir Atanassov[4,5]

[1] University Hospital "Sofiamed", 16 G. M. Dimitrov Blvd., 1797 Sofia, Bulgaria
martinlub11@abv.bg
[2] Warrane College, The University of New South Wales, Kensington, NSW 2033, Australia
tshannon38@gmail.com, t.shannon@warrane.unsw.edu.au
[3] University Hospital "Tsaritsa Yoanna - ISUl" Medical University, 8 Bialo More Str., 1527 Sofia, Bulgaria
ch.k.slavov@gmail.com
[4] Institute of Biophysics and Biomedical Engineering, Bulgarian Academy of Sciences, Acad. G. Bonchev Str., Bl. 105, 1113 Sofia, Bulgaria
tania.pencheva@biomed.bas.bg, simribagin@gmail.com, krat@bas.bg
[5] Prof. D-r Asen Zlatarov University, 8010 Bourgas, Bulgaria

Abstract. In the present study a mathematical model of behavior in patients with eGFR < 20 mL/min. (Chronic kidney disease (CKD) stage IV-V) based on generalized nets (GNs) is proposed. The developed reduced GN-model allows for optimization of clinical behavior in patients with CKD stage IV-V.

Keywords: Kidney disease · CKD stage IV-V · Generalized nets · Generalized net model

1 Introduction

Chronic kidney disease (CKD) is a long term condition caused by damage to both kidneys. The Global Burden of Disease Study 2015 estimated that worldwide deaths due to CKD increased by 31.7% in 10 years, from 938,000 in 2005 to 1.2 million deaths in 2015 [1]. There is no single cause and the damage is usually irreversible and can lead to ill health [2]. Population aging and increasing prevalence of diabetes and hypertension contributed mainly to the increased burden of CKD In some cases dialysis or transplantation may become necessary. The condition has been defined as decreased kidney function and/or kidney damage persistent for at least 3 months. Kidney dysfunction is indicated by a estimated glomerular filtration rate (eGFR) of less than $60\,\text{mL/min}/1.73\,\text{m}^2$, while kidney damage most frequently is manifested as increased urinary albumin excretion. Within this framework, CKD has been categorized into five stages:

– Stage I, with normal or high eGFR (eGFR > 90 mL/min)

S. S. Sotirov et al. (Eds.): BioInfoMed 2020, LNNS 374, pp. 113–120, 2022.
https://doi.org/10.1007/978-3-030-96638-6_12

- Stage II, Mild CKD (eGFR = 60–89 mL/min)
- Stage III, Moderate CKD (eGFR = 30–59 mL/min)
- Stage IV, Severe CKD (eGFR = 15–29 mL/min)
- Stage V, End Stage CKD (eGFR $<$ 15 mL/min)

In the first 3 stages of the disease, there are no specific symptoms, but kidney function can slowly decline. In stage IV the kidney function is very low, and treatment for kidney failure may be needed soon. During the stage V, kidneys can no longer keep up with removing waste products and extra water. This is called kidney failure. Although there is no cure, treatment options are available. Preclinical studies have identified several approaches that reduce fibrosis in experimental models, including targeting cytokines, transcription factors, developmental and signalling pathways and epigenetic modulators, particularly microRNAs [3]. The transition from CKD to the final stage of the course of kidney disease is a process that can take place over a period of time and is not strictly determined in duration [4]. Timely referral in stage IV (or earlier if CKD is progressing rapidly) allows for adequate physical and psychological preparation, including optimization of vascular access and work-up for potential kidney transplantation [5]. Comprehensive multi-disciplinary care by a nephrology team that includes nurses, social workers and dieticians is recommended. Such care can help to slow the renal disease progression, optimize management of CKD complications prior to initiation of dialysis and provide a higher likelihood of permanent vascular access placement (preferably an AV fistula) prior to beginning dialysis [8–11]. During this period of time it is appropriate to monitor the patient by specialists in various fields – nephrologist, nutritionist, vascular surgeon, urologist, social worker, clinical psychologist, etc. [6]. The goal for this multidisciplinary team should be:

1. Monitoring of the main parameters in advanced CKD.
2. Aggressive diet therapy.
3. Monitoring of complications in CKD stage IV-V – anemic syndrome, mineral-bone disease, malnutrition, etc.
4. Preparation for kidney transplantation, the possibility of preemptive transplantation is the method of choice in the treatment of these patients.
5. Selection of the appropriate method of dialysis treatment: peritoneal dialysis and hemodialysis. When choosing a hemodialysis method of treatment to make an appropriate vascular access.
6. Giving the necessary pre-dialysis information to the patient.
7. Determining the exact moment for starting dialysis procedures, in case of previously performed vascular access or access for peritoneal dialysis [7].

In the present study a multimodal mathematical model of behavior in patients with eGFR $<$ 20 mL/min (CKD stage IV-V) based on generalized nets (GNs) [12–15] is proposed.

2 The Generalized Net Model

The GN model describes the process of diagnostic and behaviour of patients with eGFR < 20 mL/min (CKD stage IV-V). It is presented in Fig. 1. The model contains 9 transitions, 23 places and 6 types of tokens:

- π – patients,
- ν – a nephrologist,
- δ – a dietologist,
- μ – a medical care specialist,
- κ – a clinical psychologist,
- σ – a vascular surgeon.

The π-tokens represent the patients with eGFR < 20 mL/min (CKD stage IV-V).

Tokens $\pi_1, \pi_2, \ldots$ enter sequentially place l_1. For brevity, we will denote them by π (without their current index). Each token π enters l_1 with an initial characteristic

"there is the typical biochemical constellation".

In the present GN-model, the π, ν, δ, μ, κ and σ tokens with the above mentioned characteristics stay permanently in places $l_2, l_3, l_6, l_8, l_{15}$. In future, in a more detailed model, in which the functioning of hospital unit with a lot of patients who will be cured in parallel, these tokens will receive new specific characteristics.

$$Z_1 = \langle \{l_1, l_2\}, \{l_2, l_3\}, \begin{array}{c|cc} & l_2 & l_3 \\ \hline l_1 & true & false \\ l_2 & false & true \end{array} \rangle.$$

Entering place l_2, token π obtains the characteristic

"perform a nephrological observation of the patient"

and in place l_3 it obtains the characteristic

"results of the nephrological observation".

$$Z_2 = \langle \{l_3, l_4\}, \{l_4, l_5\}, \begin{array}{c|cc} & l_4 & l_5 \\ \hline l_3 & true & false \\ l_4 & false & true \end{array} \rangle.$$

Entering place l_4, token π obtains the characteristic

"perform a dietological observation of the patient"

and in place l_5 it obtains the characteristic

"results of the dietological observation; concrete patient's diet".

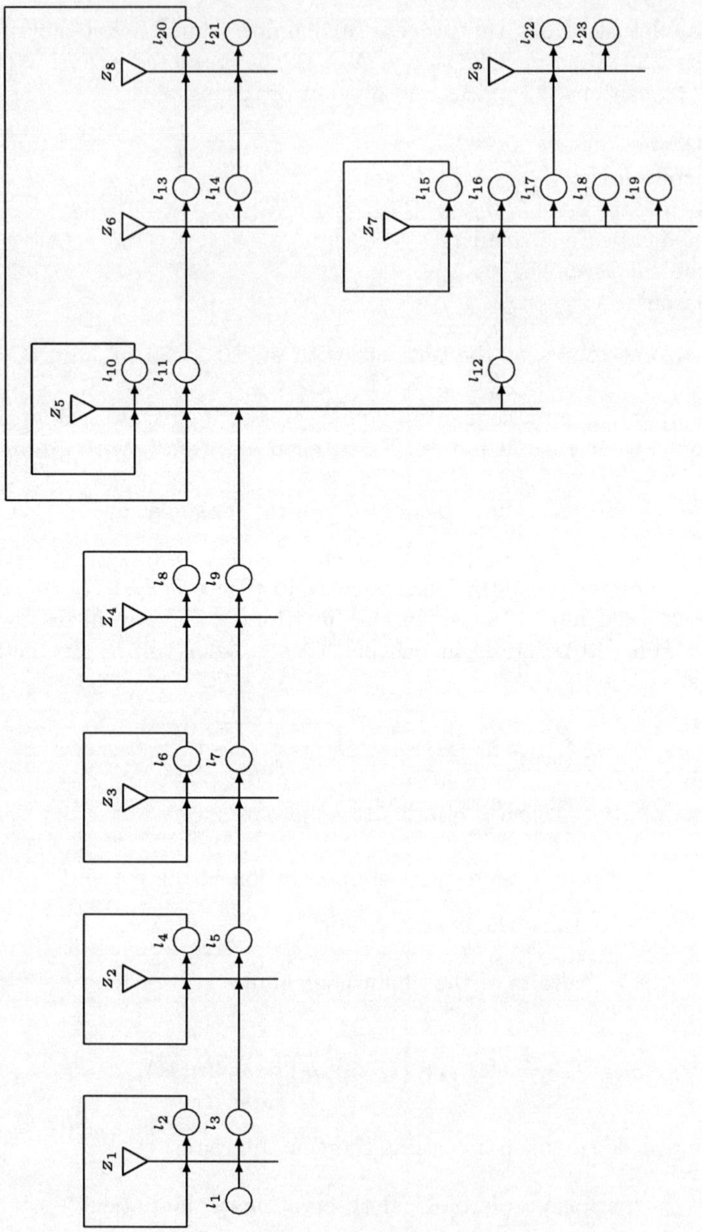

Fig. 1. GN model of the pattern of behavior in patients with eGFR $< 20\,\text{mL/min}$ (CKD stage IV-V)

$$Z_3 = \langle \{l_5, l_6\}, \{l_6, l_7\}, \begin{array}{c|cc} & l_6 & l_7 \\ \hline l_5 & true & false \\ l_6 & false & true \end{array} \rangle.$$

Entering place l_6, token π obtains the characteristic

"provide a consultation with a medical care specialist"

and in place l_7 it obtains the characteristic

"results of the consultation with the medical care specialist".

$$Z_4 = \langle \{l_7, l_8\}, \{l_8, l_9\}, \begin{array}{c|cc} & l_8 & l_9 \\ \hline l_7 & true & false \\ l_8 & false & true \end{array} \rangle.$$

Entering place l_8, token π obtains the characteristic

"perform a consultation with a vascular surgeon"

and in place l_9 it obtains the characteristic

"results of the consultation with the vascular surgeon".

$$Z_5 = \langle \{l_9, l_{10}, l_{20}\}, \{l_{10}, l_{11}, l_{12}\},$$

$$\begin{array}{c|ccc} & l_{10} & l_{11} & l_{12} \\ \hline l_9 & true & false & false \\ l_{10} & W_{10,10} & W_{10,11} & W_{11,12} \\ l_{20} & true & false & false \end{array} \rangle,$$

where
$W_{10,10} = \neg W_{10,11}$ and $\neg W_{10,12}$,
$W_{10,11} =$ "the patient is ready for renal transplantation",
$W_{10,12} =$ "the patient is ready for hemodialysis",
and $\neg P$ is the negation of the predicate P.
π-tokens obtain the characteristic

"treatment of anemic syndrome, Ca/P-disorders and others"

in place l_{10},

"patient is directed for a renal transplantation"

in place l_{11} and

"patient is directed for a hemodialysis"

in place l_{12}.

$$Z_6 = \langle \{l_{11}\}, \{l_{13}, l_{14}\}, \begin{array}{c|cc} & l_{13} & l_{14} \\ \hline l_{11} & W_{11,13} & W_{11,14} \end{array} \rangle,$$

where
$W_{11,13}$ = "there is a kidney from a life donor",
$W_{11,14}$ = "there is a kidney from a donation after brain death".
Entering place l_{13}, the π-token obtains the characteristic

"expected duration of functioning of the renal transplant from a life donor"

and in place l_{14}, it obtains the characteristic

"expected duration of functioning of the renal transplant from a brain death donor".

$$Z_7 = \langle \{l_{12}, l_{15}\}, \{l_{15}, l_{16}, l_{17}, l_{18}, l_{19}\}, \begin{array}{c|ccccc} & l_{15} & l_{16} & l_{17} & l_{18} & l_{19} \\ \hline l_{12} & true & false & false & false & false \\ l_{15} & false & W_{15,16} & W_{15,17} & W_{15,18} & W_{15,18} \end{array} \rangle,$$

where
$W_{15,16}$ = "there is a possibility for AV-fistula",
$W_{15,17}$ = "there is a possibility for AV-graft",
$W_{15,18}$ = "there is a possibility for permcath catheter",
$W_{15,19}$ = "there is a necessity for abdominal surgeon".
Entering place l_{15}, token π unites with token σ that stays permanently in place l_{15} and token π obtains the characteristic

"solution of the vascular surgeon for the type of vascular access".

In place l_{16}, the π-token obtains the characteristic

"blood volume in the AV-fistula of the forearm",

in place l_{17} it obtains the characteristic

"there is a solution for the place of the AV-graft",

in place l_{18} it obtains the characteristic

"a permcath is put in vena JI",

and in place l_{19} it obtains the characteristic

"the patient is directed to an abdominal surgeon".

$$Z_8 = \langle \{l_{13}, l_{14}\}, \{l_{20}, l_{21}\}, \begin{array}{c|cc} & l_{20} & l_{21} \\ \hline l_{13} & W_{13,20} & W_{13,21} \\ l_{14} & W_{14,20} & W_{14,21} \end{array} \rangle,$$

where

$W_{13,21} = W_{14,21} =$ "the renal transplantation is successful",

$W_{13,20} = W_{14,20} = \neg W_{13,21}$.

Entering place l_{20}, the π-token obtains the characteristic

"patient is directed for next procedures"

and in place l_{21} it obtains the characteristic

"already the patient does not have chronic kidney disease,

provide follow up recommendations".

$$Z_9 = \langle \{l_{17}\}, \{l_{22}, l_{23}\}, \begin{array}{c|cc} & l_{22} & l_{23} \\ \hline l_{17} & W_{17,22} & W_{17,23} \end{array} \rangle,$$

where

$W_{17,22} =$ "the AV-graft must be placed in the forearm",

$W_{17,23} =$ "the AV-graft must be placed in the hip".

Entering place l_{22}, the π-token obtains the characteristic

"blood volume of hemodialysis procedure with a AV-graft in a forearm"

and in place l_{23}, it obtains the characteristic

"blood volume of hemodialysis procedure with a AV-graft in a hip".

3 Conclusion

With this presentation we try, to explain the different stages of diagnosis, treatment and follow-up of patients with CKD IV-V stage using the method of generalized nets. The proposed model would be useful in the training of students, doctors, and also specialists in the field of nephrology. The construction of this and similar networks can serve for practical training of physicians in preparing patients for the transition from conservative treatment of CKD to some of the renal replacement methods of treatment. It is also suitable for assessing the level of theoretical readiness of students and specialists.

Acknowledgement. The present research has been supported by the Bulgarian National Science Fund under Grant Ref. No. DN-02/10 "New Instruments for Knowledge Discovery from Data and Their Modelling".

References

1. GBD 2015 Mortality and Causes of Death Collaborators. Global, regional, and national life expectancy, all-cause mortality, and cause-specific mortality for 249 causes of death, 1980–2015: a systematic analysis for the Global Burden of Disease Study. Lancet **388**, 1459–1544 (2016)

2. Xie, Y., Bowe, B., Mokdad, A.H., et al.: Analysis of the Global Burden of Disease Study highlights the global, regional, and national trends of chronic kidney disease epidemiology from 1990 to 2016. Kidney Int. **94**, 567–581 (2018)
3. Ruiz-Ortega, M., et al.: Targeting the progression of chronic kidney disease. Nat. Rev. Nephrol. **16**, 1–20 (2020)
4. Ramos, R., Molina, M.: New models of integrated health care management in nephrology. Nefrologia **33**(3), 301–307 (2013)
5. Smart, N.A., Dieberg, G., Ladhani, M., Titus, T.: Early referral to specialist nephrology services for preventing the progression to end-stage kidney disease. Cochrane Database Syst. Rev. **6**, Cd007333 (2014)
6. Goldstein, M., Yassa, T., Dacouris, N., McFarlane, P.: Multidisciplinary predialysis care and morbidity and mortality of patients on dialysis. Am. J. Kidney Dis. **44**(4), 706–174, PMID 15384022 (2004)
7. Elhoseny, M., Shankar, K., Uthayakumar, J.: Intelligent diagnostic prediction and classification system for chronic kidney disease. Sci. Rep. **9**(1), 1–14 (2019)
8. Papazov, V., Yonova, D.: Types of central venous catheters for hemodialysis - advantages and disadvantages. Mr. Nephrol. **1**, 27–29 (2006)
9. Papazov, V., Yonova, D., Terziev, D., Lubih, M.: A rare complication of a permanent venous catheter implantation to a patient on regular hemodialysis. Hippokratia J. **10**(1), 41–43 (2006)
10. Papazov, V., Terziev, D., Lubich, M., Yonova, D.: Complicated implantation of a permanent catheter for hemodialysis. In: V National Conference Dialysis in the 21st Century, Pamporovo, 13–14 May (2005). (in Bulgarian)
11. Papazov, V.: Use of angiographic methods in dialysis practice. School of Nephrology, Pamporovo, from 13 to 15 May 2014
12. Alexieva, J., Choy, E., Koycheva, E.: Review and bibliography on generalized nets theory and applications. In: Choy, E., Krawczak, M., Shannon, A., Szmidt, E. (eds.) A Survey of Generalized Nets, Raffles KvB Monograph, no. 10, pp. 207–301 (2007)
13. Atanassov, K.: Generalized Nets. World Scientific, Singapore (1991)
14. Atanassov, K.: On Generalized Nets Theory. "Prof. M. Drinov" Academic Publishing House, Sofia (2007)
15. Zoteva, D., Krawczak, M.: Generalized nets as a tool for the modelling of data mining processes. A survey. Issues in Intuitionistic Fuzzy Sets and Generalized Nets, vol. 13, pp. 1–60 (2017)

Generalized Net Model of Biometric Authentication System Based on Palm Geometry and Palm Vein Matching

Zhelyana Ivanova and Veselina Bureva(✉)

"Prof. Dr. Assen Zlatarov" University, 1 "Prof. Yakimov" Blvd., Burgas 8010, Bulgaria
zhelyana.ivanova@abv.bg, vbureva@btu.bg

Abstract. The introduction of technological safeguards is more important than ever, but with the increase in networks, devices, connections and applications, this it is extremely complicated. For many people, biometric security is the answer - the technology that can serve and support all industries. The biometric systems are large-scale technology implementations that have to provide capabilities in time. The aim of the current investigation is to construct a Generalized net (GN) model of biometric authentication system based on palm geometry and palm vein matching. It presents the idea of reading biometric data through a censor and camera. In the former version, via a censor placed in the housing of the device, the palm of the user is scanned. In the latter, the data is collected via a scan of the user's veins with the camera of the device. Both variants are run simultaneously. The aims are to ease the use of smart devices, to increase the security and guarantee the protection of personal data.

Keywords: Biometric system · Generalized net · Palm Vein authentication · Palm geometry · Security

1 Introduction

In today's digitally active world, where cybercrime, theft and fraud are on the rise, organizations are demanding the most secure, accurate and reliable methods, for authentication, data access, physical access and general security. The security of the systems has become a challenge with attacks against all fronts - not only externally by outsiders, but also by internal manipulation. The introduction of technological safeguards is more important than ever, but with the increase in networks, devices, connections and applications, this it became extremely complicated. For many people, biometric security is the answer - the technology of the platform that can serve and support all industries. The biometric characteristics have to provide the following properties: robustness, distinctiveness, availability, accessibility. Robustness is a measure of the characteristic that should not change over time. Distinctiveness is a measure of the differences in the biometric pattern among the general population. Availability describes the case when the whole population should contain the characteristic. Accessibility is the case when the characteristic

S. S. Sotirov et al. (Eds.): BioInfoMed 2020, LNNS 374, pp. 121–130, 2022.
https://doi.org/10.1007/978-3-030-96638-6_13

should be easy to acquire. Biometrics combines methods for uniquely recognizing people based on their inherent physical or behavior traits. Physiological biometrics is face, finger and palm prints, hand/finger geometry, iris, DNA, odor, earlobe, sweat pore, lips. Behavioral biometrics includes keystroke, signature, gait and speech recognition. The biometric technologies are used in the fields of law enforcement and public security, military, travel/border control, civil identification, healthcare, physical and logical access, and commercial applications. In information technology in particular, biometrics is used to identify people as an access control. In the recent years the topic of pattern recognition and related scientific fields are investigated intensively [8, 13, 15, 16, 18–20, 22].

2 Generalized Net Model of Biometric Authentication System Based on Palm Geometry and Palm Vein Matching

The aim of the current investigation is to construct a Generalized net (GN) model of biometric authentication system based on palm geometry and palm vein matching. GNs are suitable instrument for describing parallel processes in their dynamic development. The theory of GNs is presented in [1, 6, 7]. GN models for pattern recognition processes are published [3–5]. GN model of biometric access-control system and GN model of biometric identification system presenting the classification of multimodal biometric systems are constructed [2, 9]. The processes of iris recognition [12, 21], face recognition [17], fingerprints recognition [10] and signature verification [11] are already modeled using the apparatus of GNs. In the current research work the idea of receiving biometric data through a mobile phone sensor and camera is discussed. The palm of the user is scanned via a sensor placed in the housing of the device. The user's veins data is collected via a scan with the camera of the device. Both processes run simultaneously. The Generalized net model of biometric authentication system based on palm geometry and palm vein matching is constructed using GNDraw software [14]. It contains 6 transitions and 31 places (Fig. 1).The set of transitions A has the following form:

$$A = \{Z_0, Z_1, Z_2, Z_3, Z_4, Z_5, Z_6\},$$

where the transitions describe the following processes:

- Z_0 - "user for recognition";
- Z_1 - "preliminary processing (preprocessing)";
- Z_2 - "palm geometry extraction";
- Z_3 - "palm veins extraction";
- Z_4 - "storing data in database".
- Z_5 - "comparison of biometric characteristics";
- Z_6 - "password authentication";

The users veins and palm geometry is scanned (transition Z_0) and the received images are preprocessed (transition Z_1). Thereafter the palm veins and the palm geometry are extracted from the input images in parallel (transition Z_2 and transition Z_3). The result is written in the database in the case if it does not contain the data for the selected user (transition Z_4). If the database contains the palm vein template and palm geometry template for the given user the new templates are used for comparison with the current user templates (transition Z_5). Additionally the user has to write its password which is compared with the previously stored password in the database (transition Z_6). Depending of the received results from the comparison the user can be recognized or not. The presented generalized net model of biometric authentication system based palm geometry and palm vein matching can be used to describe and simulate the work process with smart devices. The processes allow for a significant improvement in the quality of work by increasing the security of the device and protection of the user's personal data.

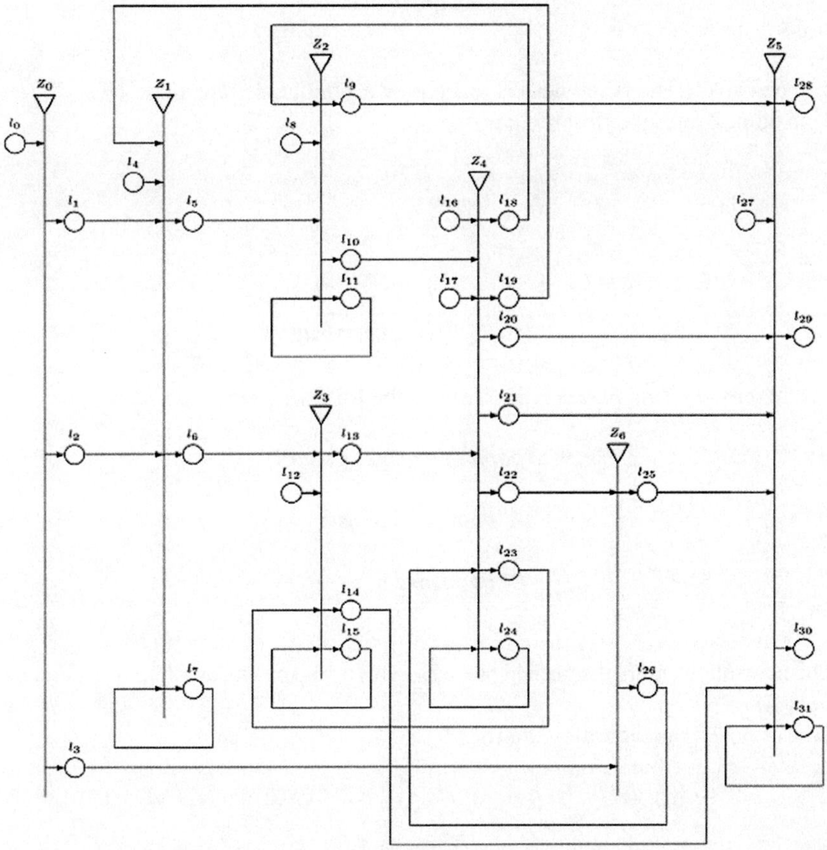

Fig.1. Generalized net model of biometric authentication system based on palm geometry and palm vein matching

Initially the generalized net model has the following tokens:

In place l_{24} there is one α-token. It will be in its own place during the whole functioning time of the GN. It has the following characteristic: "database". The token in place l_{24} generates tokens at certain time moments which will move to places $l_{18}, l_{19}, l_{20}, l_{21}, l_{22}, l_{23}$ with the following characteristics:

„*information about the palm*"

in place l_{18},

„*preliminary information*"

in place l_{19},

„*information for comparison*"

in place l_{20},

„*templates for comparison*"

in place l_{21},

„*password information*"

in place l_{22},

„*information about the veins*"

in place l_{23}.

The new token enters the net via place L_0 with initial characteristic: "*user*".

Transition Z_0 has the following form:

$$Z_0 = \langle \{l_0\},\ \{l_1, l_2, l_3\}, R_0 \rangle$$

where:

$$R_0 = \begin{array}{c|ccc} & l_1 & l_2 & l_3 \\ \hline l_0 & true & true & true \end{array}$$

The tokens entering places l_1, l_2, l_3 have the following characteristics:

"*palm geometry of user*"

in place l_1,

"*palm vein of user*"

in place l_2,

"*password of user*"

in place l_3.

The new token enters the net via place L_4 with initial characteristic: "*parameters for preprocessing*".

Transition Z_1 has the following form:

$$Z_1 = \langle \{l_1, l_2, l_4, l_7, l_{19}\},\ \{l_5, l_6, l_7\}, R_1 \vee (\wedge(l_1, l_2, l_4, l_{19}), l_7) \rangle$$

where:

$$R_1 = \begin{array}{c|ccc} & l_5 & l_6 & l_7 \\ \hline l_1 & false & false & true \\ l_2 & false & false & true \\ l_4 & false & false & true \\ & W_{7,5} & W_{7,6} & W_{7,7} \\ l_{19} & false & false & true \end{array}$$

and
$W_{7,5}$ = "there is preprocessed palm geometry information";
$W_{7,6}$ = "there is preprocessed palm vein data";
$W_{7,7} = \neg(W_{7,5} \wedge W_{7,6})$.
The tokens entering places l_4 and l_5 obtain the following characteristics:

„*preprocessed palm geometry information*"

in place l_4,

„*preprocessed palm vein data*"

in place l_5.

The new token enters the net via place L_8 with initial characteristic: "*parameters for palm geometry features extraction*".
Transition Z_2 has the form:

$$Z_2 = \langle \{l_5, l_8, l_{11}, l_{18}\}, \{l_9, l_{10}, l_{11}\}, R_2 \vee (\wedge(l_5, l_8, l_{18}), l_{11} \rangle,$$

where

$$R_2 = \begin{array}{c|ccc} & l_9 & l_{10} & l_{11} \\ \hline l_5 & false & false & true \\ l_8 & false & false & true \\ l_{11} & W_{11,9} & W_{11,10} & W_{11,11} \\ l_{18} & false & false & true \end{array}$$

and
$W_{11,9}$ = "there is a palm geometry templates (patterns) for comparison";
$W_{11,10}$ = "there is a palm geometry templates (patterns) for storing in database";
$W_{11,11} = \neg(W_{11,9} \wedge W_{11,10})$.
The tokens entering places l_9 and l_{10} obtain the following characteristics:

„*palm geometry templates for comparison*"

in place l_9 and

„*palm geometry templates for storing*"

in place l_{10}.

The new token enters the net via place L_{12} with initial characteristic: *"parameters for palm vein features extraction"*.

Transition Z_3 has the following form:

$$Z_3 = \langle\{l_6, l_{12}, l_{15}, l_{23}\}, \{l_{13}, l_{14}, l_{15}\}, R_3 \vee (\wedge(l_6, l_{12}, l_{23}), l_{15})\rangle$$

where:

$$R_3 = \begin{array}{c|ccc} & l_{13} & l_{14} & l_{15} \\ \hline l_6 & \mathit{false} & \mathit{false} & \mathit{true} \\ l_{12} & \mathit{false} & \mathit{false} & \mathit{true} \\ l_{15} & W_{15,13} & W_{15,14} & W_{15,15} \\ l_{23} & \mathit{false} & \mathit{false} & \mathit{true} \end{array}$$

and

$W_{15,13}$ = "there is a vein templates (patterns) for storing";

$W_{15,14}$ = "there is a vein templates (patterns) for comparison";

$W_{15,15} = \neg(W_{15,13} \wedge W_{15,14})$.

The tokens entering places l_{13} and l_{14} obtain the following characteristics:

„vein templates (patterns) for storing"

in place l_{13} and

„vein templates (patterns) for comparison"

in place l_{14}.

The two new tokens enter the net via places l_{16} and l_{17} with initial characteristics:

"new password information" in place l_{16} and
"new templates" in place l_{17}.

Transition Z_4 has the following form:

$$Z_4 = \left\langle \begin{array}{l} \{l_{10}, l_{13}, l_{16}, l_{17}, l_{24}, l_{26}\}, \{l_{18}, l_{19}, l_{20}, l_{21}, l_{22}, l_{23}, l_{24}\}, \\ R_4, \vee(\wedge(l_{10}, l_{13}, l_{16}, l_{17}, l_{26}), l_{24}) \end{array} \right\rangle$$

where:

$$R_4 = \begin{array}{c|ccccccc} & l_{18} & l_{19} & l_{20} & l_{21} & l_{22} & l_{23} & l_{24} \\ \hline l_{10} & false & false & false & false & false & false & true \\ l_{13} & false & false & false & false & false & false & true \\ l_{16} & false & false & false & false & false & false & true \\ l_{17} & false & false & false & false & false & false & true \\ l_{24} & W_{24,18} & W_{24,19} & W_{24,20} & W_{24,21} & W_{24,22} & W_{24,23} & W_{24,24} \\ l_{26} & false & false & false & false & false & false & true \end{array}$$

and
$W_{24,18}$ = "there is information about the palm";
$W_{24,19}$ = "there is preliminary information";
$W_{24,20}$ = "there is information for comparison";
$W_{24,21}$ = "there are templates for comparison";
$W_{24,22}$ = "there is password information";
$W_{24,23}$ = "there is information about the veins";
$W_{24,24} = \neg(W_{24,18} \wedge W_{24,19} \wedge W_{24,20} \wedge W_{24,21} \wedge W_{24,22} \wedge W_{24,23})$.

The tokens entering places $l_{18}, l_{19}, l_{20}, l_{21}, l_{22}$ and l_{23} obtain the following characteristics:

„*information about the palm*"

in place l_{18},

„*preliminary information*"

in place l_{19},

„*information for comparison*"

in place l_{20},

„*templates for comparison*"

in place l_{21},

„*password information*"

in place l_{22},

„*information about the veins*"

in place l_{23}.

The new token enters the net via place L_{27} with initial characteristic: *"parameters for comparison (pattern matching)"*.

Transition Z_5 has the following form:

$$Z_5 = \left\langle \begin{array}{l} \{l_9, l_{14}, l_{20}, l_{21}, l_{25}, l_{27}, l_{31}\}, \{l_{28}, l_{29}, l_{30}, l_{31}\}, R_5, \\ \vee(\wedge(l_9, l_{14}, l_{20}, l_{21}, l_{25}, l_{27}), l_{31}) \end{array} \right\rangle$$

where:

$R_5 =$

	l_{28}	l_{29}	l_{30}	l_{31}
l_9	*false*	*false*	*false*	*true*
l_{14}	*false*	*false*	*false*	*true*
l_{20}	*false*	*false*	*false*	*true*
l_{21}	*false*	*false*	*false*	*true*
l_{25}	*false*	*false*	*false*	*true*
l_{27}	*false*	*false*	*false*	*true*
l_{31}	$W_{31,28}$	$W_{31,29}$	$W_{31,30}$	$W_{31,31}$

and
$W_{31,28} =$ "there is a result from palm match";
$W_{31,29} =$ "there is a result from password match";
$W_{31,30} =$ "there is a result from palm veins match";
$W_{31,31} = \neg(W_{31,28} \wedge W_{31,29} \wedge W_{31,30})$.
The tokens entering places l_{28}, l_{29} *and* l_{30} obtain the following characteristics:

„*result from palm match*"

in place l_{28},

„*result from password match*"

in place l_{29}

„*result from palm veins match*"

in place l_{30}.

Transition Z_6 has the following form:

$$Z_6 = \langle \{l_3, l_{22}\}, \{l_{25}, l_{26}\}, R_6, \vee(l_3, l_{22}) \rangle$$

where:

$R_6 =$

	l_{25}	l_{26}
l_3	$W_{3,25}$	$W_{3,26}$
l_{22}	*true*	*false*

and
$W_{3,19} =$ "there is a confirmed password",
$W_{3,20} =$ "there is a password for storing";
The tokens entering places l_{25}*and* l_{26} obtain the following characteristics:

„*confirmed password*"

in place l_{25} and

„*password for storing*"

in place l_{20}.

3 Conclusion

The presented generalized net model of a system for biometric authentication can be used to describe and simulate the work with smart devices. The processes allow for a significant improvement in the quality of work by increasing the security of the device and protection of the user's personal data. The biometric systems are large-scale technology implementations that have to provide capabilities in time.

Acknowledgment. The authors are grateful for the support provided by the Bulgarian Ministry of Education and Science under the National Research Programme "Information and Communication Technologies for a Digital Single Market in Science, Education and Security" approved by DCM # 577/ 17.08.2018.

References

1. Atanassov, K.: Generalized Nets. World Scientific, Singapore, New Jersey, London (1991)
2. Atanassov, K., Boumbarov, O., Gluhchev, G., Hadjitodorov, S., Shannon, A., Vassilev, V.: A generalized net model of biometric access-control system. In: Proceedings of the 9th WSEAS International Conference on Automatic Control, Modeling & Simulation, Istanbul, Turkey, 27–29 May 2007, pp 78–81 (2007)
3. Atanassov, K., et al.: Generalized nets decision making and pattern recognition. Warsaw School of Information Technology, Warszawa (2006)
4. Atanassov, K., Gluhchev, G., Hadjitodorov, S., Shannon, A., Vasilev, V.: Generalized nets in image processing and pattern recognition. In: Sixth International Workshop on GNs, Sofia, pp. 47–60 (2005)
5. Atanassov, K., Gluhchev, G., Hadjitodorov, S., Shannon, A., Vassilev, V.: Generalized Nets and Pattern Recognition. KvB Visual Concepts Pty Ltd., Monograph No. 6, Sydney (2003)
6. Atanassov, K., Sotirova, E.: Generalized Nets. Prof. M. Drinov Academic Publishing House, Sofia (2017). (in Bulgarian)
7. Atanassov, K.: On Generalized Nets Theory. "Prof. Marin Drinov" Academic Publishing House, Sofia (2007)
8. Bai, X., et al.: Explainable deep learning for efficient and robust pattern recognition: A survey of recent developments. Pattern Recogn. **120** (2021). IF: 7.196
9. Bureva, V., Sotirova, E., Bozov, H.: Generalized net model of biometric identification process. In: 2018 20th International Symposium on Electrical Apparatus and Technologies (SIELA), Bourgas, pp. 1–4 (2018)
10. Bureva, V., Yovcheva, P., Sotirov, S.: Generalized net model of fingerprint recognition with intuitionistic fuzzy evaluations. In: Kacprzyk, J., Szmidt, E., Zadrożny, S., Atanassov, K., Krawczak, M. (eds.) Advances in Fuzzy Logic and Technology 2017. IWIFSGN 2017, EUSFLAT 2017. Advances in Intelligent Systems and Computing, vol. 641, pp. 286–294. Springer, Cham (2018)
11. Gluhchev, G., Atanassov, K., Hadjitodorov, S., Szmidt, E.: A generalized net model for signature verification. In: Conference proceedings of Tenth International Workshop on Generalized Nets, Sofia, 5 December 2009, pp. 27–30 (2009)
12. Gocheva, E., Sotirov, S.: Modelling of the verification by iris scanning by generalized nets. In: Conference proceedings of Ninth International Workshop on Generalized Nets, Sofia, pp 9–13 (2008)

13. Huo, G., Zhang, Q., Zhang, Y., et al.: Multi-source heterogeneous iris recognition using stacked convolutional deep belief networks-deep belief network model. Pattern Recogn. Image Anal. **31**, 81–90 (2021)
14. Ikonomov, N.: GNDraw – software application for creating generalized nets. Issues Intuition. Fuzzy Sets Gener. Nets **13**, 61–71 (2017)
15. Li, G., Wu, Z., Liu, Y., et al.: 3D hand reconstruction from a single image based on biomechanical constraints. Vis. Comput. **37**, 2699–2711 (2021)
16. Misra S., Sridevi G., Laskar R., Modeling a virtual bare-hand interface system using a robust hand detection approach for HCI. In: Int. J. Pattern Recogn. Artif. Intell. **35**(05) (2021)
17. Parvathi, R., Gluhchev, G., Atanassov, K.: Generalized net model of face recognition. In: Conference proceedings of Ninth International Workshop on Generalized Nets, Sofia, 4 July 2008, pp 102–105 (2008)
18. Ross, A., Banerjee, S., Chowdhury, A.: Security in smart cities: a brief review of digital forensic schemes for biometric data. Pattern Recogn. Lett. **138**, 346–354 (2020). IF: 3.756
19. Sánchez-Lozano, E., Tzimiropoulos, G., Martinez, B., De la Torre, F., Valstar, M.: A functional regression approach to facial landmark tracking. IEEE Trans. Pattern Anal. Mach. Intell. **40**(9), 2037–2050 (2018). IF: 16.389
20. Shi, L., Wang, C., Jia, H., Hu, X.: EPS: robust pupil edge points selection with Haar feature and morphological pixel patterns. Int. J. Pattern Recogn. Artif. Intell. **35**(06), 2156002 (2021)
21. Sotirov, S.: Generalized net model of iris recognition using neural networks. Ann. Inform. Sect. Union Sci. Bulgaria **4**, 127–132 (2011)
22. Zhang, Z.: A gait recognition method for a moving target image in sports based on a decision tree. Int. J. Biometr. **13**(2/3), 165–179 (2021)

A Generalized Net Model of the Process of Obtaining and Diagnosing Convalescent Plasma from Patients with COVID-19

Nikolay Andreev[1], Simeon Ribagin[2,3(✉)], and Krassimir Atanassov[2,3]

[1] Transfusion Hematology Department, University Hospital "St. Anna", 1 D. Mollov Str., 1750 Sofia, Bulgaria
imuno_chem@abv.bg

[2] Institute of Biophysics and Biomedical Engineering, Bulgarian Academy of Sciences, Acad. G. Bonchev Str., Bl. 105, 1113 Sofia, Bulgaria
simribagin@gmail.com, krat@bas.bg

[3] Prof. D-r Asen Zlatarov University, 8010 Bourgas, Bulgaria

Abstract. In the present study we propose a GN-model of the process of obtaining and diagnosing convalescent plasma from patients with COVID-19. The presented GN-model describes in details the different steps during the evaluation of the convalescent plasma, which will find expression in process optimization.

Keywords: COVID-19 · Generalized net · Plasmapheresis · Convalescent blood plasma

1 Introduction

Humans infected with SARS-CoV-2 may develop COVID-19, which manifests across a wide spectrum of clinical severity ranging from a mild upper respiratory tract illness to a diffuse viral pneumonia causing acute respiratory failure, acute lung injury, multi organ dysfunction syndrome and death [1]. Although protection from COVID-19 infection or disease has yet to be directly correlated with levels of circulating antibodies against SARS-CoV-2 providing virus-neutralizing antibodies in the form of convalescent plasma may expedite disease resolution before the maturation of a patient's own humoral response [2]. The vast majority of patients who recover from COVID-19 illness develop circulating antibodies to various SARS-CoV-2 proteins 2-3 weeks following infection, which are detectable by enzyme-linked immunosorbent assay or other quantitative assays and often correlate with the presence of neutralizing antibodies. These antibodies appear to be protective, based on several primate studies showing animals could not be re-infected with SARS-CoV-2 weeks to months later. Multiple studies have now reported the use of COVID-19 convalescent plasma (CCP) to treat severely or critically ill COVID-19 patients, without unexpected or serious adverse events [3,6]. At the moment, much of the plasma collected by public blood services is

S. S. Sotirov et al. (Eds.): BioInfoMed 2020, LNNS 374, pp. 131–138, 2022.
https://doi.org/10.1007/978-3-030-96638-6_14

done via donations of whole blood, from which the plasma is then separated. This is a much less efficient collection method compared with plasmapheresis - a bedside process where plasma is taken from the donor whilst the other blood components are returned to the donor. In case of plasmapheresis, donors can donate higher volumes of plasma at one time and can donate once every 2 weeks compared with once every 3–4 months for whole blood donors [5]. In the Republic of Bulgaria there is a system of centers and wards at three hierarchical levels for collection, processing, diagnosis, storage and distribution of blood and blood products. The leading institution is the National Center for Transfusion Hematology (NCTH). At the next level are five Regional Centers for Transfusion Hematology (RCTH) in Plovdiv, Varna, Pleven and Stara Zagora and the Center at the Military Medical Academy, Sofia. At the third level, there are 23 Departments of Transfusion Hematology (DTH) at each Multiprofile Hospital for Active Treatment (MHAT). The present study aims to present the process of taking plasma or blood from survivors of Covid-19 in one of the Departments of Transfusion Hematology (DTH). It should be transported to the National Center for Transfusion Hematology (NCTH) for the diagnosis of transmissible infections. Depending on the need, the plasma will be distributed in the medical establishments of the country. Generalized nets (GNs) [4,7–9], are one of the tools for modeling parallel processes running in real time. By this reason, the whole process will be described through a GN model and the benefit of such a model for process optimization will be discussed.

2 The Generalized Net Model

The present GN (see Fig. 1) contains 12 transitions, 26 places and 5 types $(\beta, \gamma, \varepsilon, \zeta, \rho)$ of tokens that represent, respectively:

- β - blood for processing,
- γ - blood for returning to the donor,
- ε - blood for transmissible diseases (HIV, HBV, HCV, Wass),
- ζ - blood for evaluation of blood group, Rh and antibodies screenings,
- ρ - request for blood product.

Tokens $\beta_i (i = 1, 2, ...)$ enter place l_1 with the initial characteristic

"identifier, name address and telephone of i-th person who will donate blood;

time-moment (date, hour, minute), quantity of the blood, diagnostic for

antibody for COVID-19",

For brevity, the index $i = 1, 2, ...$ will be omitted, but when we discuss token β- ε-, etc. tokens, we will have in mind β_i- ε_i-, etc. tokens.

$$Z_1 = \langle \{l_1\}, \{l_2, l_3\}, \frac{\quad \quad \begin{array}{c|cc} & l_2 & l_3 \end{array}}{\begin{array}{c|cc} l_1 & W_{1,2} & W_{1,3} \end{array}} \rangle.$$

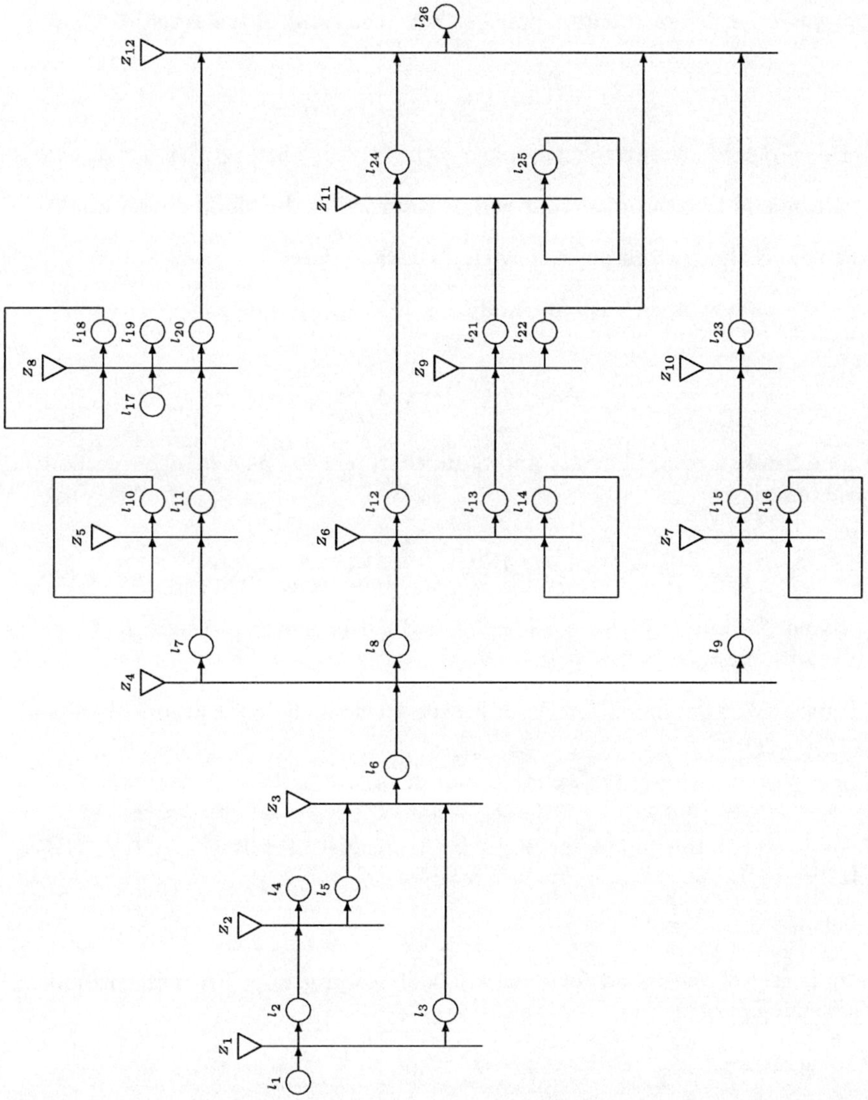

Fig. 1. GN model of the process of obtaining and diagnosing convalescent plasma from patients with COVID-19

where
$W_{1,2}$ = "in the DTH there is a plasmapheresis machine",
$W_{1,3} = \neg W_{1,2}$
where $\neg P$ is the negation of the predicate P.

If $W_{1,2} = true$, token β enters place l_2 with a characteristic

"blood, quality".

If $W_{1,3} = true$, token β enters place l_3 with the same characteristic.

$$Z_2 = \langle \{l_2\}, \{l_4, l_5\}, \begin{array}{c|cc} & l_4 & l_5 \\ \hline l_2 & true & true \end{array} \rangle.$$

Token β splits to two tokens: token γ that enters place l_4 with a characteristic

"plasma and erythrocites that will be returned to the blood donor, quality"

and token β that enters place l_5 with a characteristic

"plasma with antibody for COVID-19, quality".

$$Z_3 = \langle \{l_3, l_5\}, \{l_6\}, \begin{array}{c|cc} & l_4 & l_5 \\ \hline l_2 & true & true \end{array} \rangle.$$

The β-token from places l_3 and l_5 unite in place l_6 as a β-token without any characteristic.

$$Z_4 = \langle \{l_6\}, \{l_7, l_8, l_9\}, \begin{array}{c|ccc} & l_7 & l_8 & l_9 \\ \hline l_6 & true & true & true \end{array} \rangle.$$

Token β splits to three tokens β, ε and ζ that enter places l_7, l_8, l_9, respectively, with characteristics

"quantity of the blood for Fresh Frozen Plasma (FFP), thrombocytes and erythrocytes"

for token β,

"quantity of the blood for tests for transmissible diseases (HIV, HBV, HCV, VDRL)"

for token ε,

"quantity of the blood for evaluation of blood group, Rh and antibodies screenings"

for token ζ.

$$Z_5 = \langle \{l_7, l_{10}\}, \{l_{10}, l_{11}\}, \begin{array}{c|cc} & l_{10} & l_{11} \\ \hline l_7 & true & false \\ l_{10} & W_{10,10} & W_{10,11} \end{array} \rangle,$$

where
$W_{10,10}$ = "there are no tests results for the i-th person",
$W_{10,11} = \neg W_{10,10}$.

Token β from place l_7 enters place l_{10} and from this place, it enters place l_{11} with a characteristic

"duration of the mandatory quarantine period".

$$Z_6 = \langle \{l_8, l_{14}\}, \{l_{12}, l_{13}, l_{14}\}, \begin{array}{c|ccc} & l_{12} & l_{13} & l_{14} \\ \hline l_8 & false & false & true \\ l_{14} & W_{14,12} & W_{14,13} & false \end{array} \rangle,$$

where
$W_{14,12}$ = "The results of the transmissible diseases tests are negative",
$W_{14,13} = \neg W_{14,12}$,
i.e., there is at least one positive test for transmissible disease. Token ε from place l_8 enters place l_{14} with a characteristic

"results of the transmissible diseases tests".

On the next time step, token ε enters one of the other output places with a characteristic

"the blood of the i-th person is suitable for use"

in place l_{12} and

"the blood of the i-th person must be tested again"

in place l_{13}.

$$Z_7 = \langle \{l_9, l_{16}\}, \{l_{15}, l_{16}\}, \begin{array}{c|cc} & l_{15} & l_{16} \\ \hline l_9 & false & true \\ l_{16} & true & false \end{array} \rangle.$$

Token ζ from place l_9 enters place l_{16} with a characteristic

"results of the evaluation of blood group and Rh".

On the next time step, token ζ enters place l_{15} with a characteristic

"results of the evaluation of antibodies screenings".

In some time-moments of the GN-functioning, tokens ρ enter place l_{17} with the initial characteristic

"request for blood product for patient with COVID-19".

$$Z_8 = \langle \{l_{11}, l_{17}, l_{18}\}, \{l_{18}, l_{19}, l_{20}\}, \begin{array}{c|ccc} & l_{18} & l_{19} & l_{20} \\ \hline l_{11} & true & false & false \\ l_{17} & false & true & false \\ l_{18} & W_{18,18} & W_{18,19} & W_{18,20} \end{array} \rangle,$$

where
$W_{18,18}$ = "the FFP is in the safekeeping deadline (2 years) and there has not been a request for it" or "transmissible diseases tests of the i-th person are positive" or "the blood of the i-th person contains antibodies",
$W_{18,19}$ = "the FFP is in the term of limitation and there is a request for it",
$W_{18,20}$ = "the FFP is already outside the term of limitation".

Token β from place l_{11} enters place l_{18} with characteristics

"information for i-th person's blood group and Rh; the FFP must be preserved at $-25\,^{\circ}$C".

When $W_{18,19} = true$, token β enters place l_{19} together with token ρ and the two tokens unite with a characteristic

"name of the person who will obtain the FFP; the identifier(s) of the bank(s) with FFP and antibody for COVID-19".

When $W_{18,20} = true$, token β enters place l_{20} with a characteristic:

"the amount of blood to be disposed of".

$$Z_9 = \langle\{l_{13}\}, \{l_{21}, l_{22}\}, \begin{array}{c|cc} & l_{21} & l_{22} \\ \hline l_{13} & W_{13,21} & W_{13,22} \end{array}\rangle,$$

where:
$W_{13,21} =$ "The results of the second tests for transmissible diseases are negative",
$W_{13,22} = \neg W_{13,21}$.

Token ε from place l_{13} enters place l_{21} with a characteristic

"the blood of the i-th person is suitable for use"

and in place l_{22} with a characteristics

"the blood of the i-th person must be disposed of".

$$Z_{10} = \langle\{l_{15}\}, \{l_{23}\}, \begin{array}{c|c} & l_{23} \\ \hline l_{15} & W_{15,23} \end{array}\rangle.$$

where
$W_{15,23} =$ "the evaluation of antibodies screenings, does not met the requrements for the i-th person",

Token ζ from place l_{15} enters places l_{23} with a characteristic

"the blood of the i-th person's must be disposed of".

$$Z_{11} = \langle\{l_{12}, l_{21}, l_{25}\}, \{l_{24}, l_{25}\}, \begin{array}{c|cc} & l_{24} & l_{25} \\ \hline l_{12} & false & true \\ l_{21} & false & true \\ l_{25} & W_{25,24} & W_{25,25} \end{array}\rangle,$$

where
$W_{25,24} =$ "the preservation period of the tests results has finished",
$W_{25,25} = \neg W_{25,24}$.

Token ε from place l_{12} or from place l_{21} enters place l_{25} with a characteristic

"the i-th person's tests results, time-moment (date, hour, minute)"

and it enters places l_{24} with a characteristic

"the safekeeping deadline of the i-th person's tests results has expired".

$$Z_{12} = \langle \{l_{20}, l_{22}, l_{23}, l_{24}\}, \{l_{26}\}, \begin{array}{c|c} & l_{26} \\ \hline l_{20} & true \\ l_{22} & true \\ l_{23} & true \\ l_{24} & true \end{array} \rangle.$$

Each of the tokens from the input places of the transition enters place l_{26} with a characteristic

"information about the disposed biological material".

3 Conclusion

The presented GN-model describes in details the different steps during the evaluation of the convalescent plasma, which will find expression in process of optimization and knowledge extraction. The model is in a way an opportunity to more fully meet the needs of patients positive for Covid-19 in their treatment. It allows in NCTH and in each RCTH and DTH with a plasmapheresis equipment to obtain convalescent plasma from those who have already recovered with a high levels of antibodies. In this way, all patients' needs for convalescent plasma treatment will be met and fewer of them will reach intensive care units.

Acknowledgements. The present research has been supported by the Bulgarian National Science Fund under Grant Ref. No. DN-02/10 "New Instruments for Knowledge Discovery from Data and Their Modelling".

References

1. Greenland, J.R., et al.: COVID-19 infection: implications for perioperative and critical care physicians. Anesthesiology **132**(6), 1346–1361 (2020). https://doi.org/10.1097/ALN.0000000000003303
2. Liu, S.T.H., Lin, H., Baine, I., et al.: Convalescent plasma treatment of severe COVID-19: a propensity score-matched control study. Nat. Med. **26**, 1708–1713 (2020). https://doi.org/10.1038/s41591-020-1088-9
3. Shen, C., Wang, Z., Zhao, F., et al.: Treatment of 5 critically ill patients with COVID-19 with convalescent plasma. JAMA **323**(16), 1582–1589 (2020). https://doi.org/10.1001/jama.2020.4783
4. Zoteva, D., Krawczak, M.: Generalized nets as a tool for the modelling of data mining processes. A survey. Issues Intuit. Fuzzy Sets Gener. Nets **13**, 1–60 (2017)
5. https://ec.europa.eu/commission/presscorner/detail/en/ip_20_1435

6. https://www.hematology.org/covid-19/covid-19-and-convalescent-plasma
7. Alexieva, J., Choy, E., Koycheva, E.: Review and bibliography on generalized nets theory and applications. In: Choy, E., Krawczak, M., Shannon, A., Szmidt, E. (eds.) A Survey of Generalized Nets. Raffles KvB Monograph No. 10, pp. 207–301 (2007)
8. Atanassov, K.: Generalized Nets. World Scientific, Singapore, New Jersey, London (1991)
9. Atanassov, K.: On Generalized Nets Theory. "Prof. M. Drinov" Academic Publishing House, Sofia (2007)

Possible Application of Generalized Nets in Telemedicine Screening of Corona Virus Disease 2019 (COVID-19)

Simeon Ribagin[1,2](✉)

[1] Department of Bioinformatics and Mathematical Modelling Institute of Biophysics and Biomedical Engineering, Bulgarian Academy of Sciences, Sofia, Bulgaria
sim_ribagin@mail.bg
[2] Department of Health and Social Care, Medical College, University "Prof. D-r Asen Zlatarov", Burgas, Bulgaria

Abstract. The purpose of the present paper is to present possible application of generalized nets with the development of a reduced GN-model as a telemedicine screening tool for COVID-19. The model describes a possible algorithm for remote symptom screening and the different transitions of the model are representing respectively the different criteria of the evaluation process.

Keywords: Telemedicine · COVID-19 · Generalized nets · GN-model

1 Introduction

The Novel Coronavirus 2019 was first reported in Wuhan, China in late December 2019. Since the beginning of the outbreak, there has been an exponential expansion of the infection and the WHO declared it as a public health emergency of international concern on January 30th [1]. The spread of this virus is now global with lots of casualties of all ages and social groups. The virus has been named SARS-CoV-2 and the disease it causes has become known as coronavirus disease 2019 (COVID-19). The rapid progression of the disease presents a real challenge for the whole world. Telemedicine has been thrust into the forefront during the COVID-19 outbreak [2–4], and has been recommended to screen suspected patients to limit risk of exposure and maximize medical staff protection [5]. Telehealth can be used to screen for COVID-19 symptoms and assess patients for potential exposure. The use of telehealth can slow transmission of the disease by keeping at-risk people out of waiting rooms and reducing their contact with healthcare facilities. It also enables patients who are not suffering from COVID-19 to continue to receive care. This is particularly important for elderly people, who are at the greatest risk during this pandemic. By improving staff productivity, it stretches the capacity of hospital staff so that more people can be treated. By reducing physical contact with infected patients, telehealth can protect the clinicians. Preventing the risk of contagion, especially via professionals, who are key assets that need to be taken care of in this context: avoiding direct physical contact, reducing the risk of exposure to respiratory secretions [6].

S. S. Sotirov et al. (Eds.): BioInfoMed 2020, LNNS 374, pp. 139–144, 2022.
https://doi.org/10.1007/978-3-030-96638-6_15

In a view of this, the development of specific screening algorithms and protocols based on the interactive telecommunication technologies and realized via mathematical tools for modeling will permit more efficient collection, transformation and evaluation of significantly large number of patient health data and to report the data into the required form. In general, the algorithm should include: triaging for COVID-19 symptoms, monitoring, contact tracing and data analysis. The purpose of the present paper is to present possible application of generalized nets [7–9] with the development of a reduced GN-model as a telemedicine screening tool for COVID-19. The model describes a possible algorithm for remote symptom screening and the different transitions of the model are representing respectively the different criteria of the evaluation process. Moreover, some of the structures of the proposed model represents database storage and processing of COVID-19 data.

2 Generalized Net Model for Screening of Corona Virus Disease 2019 (COVID-19)

The GN model (Fig. 1) has 4 transitions and 24 places with the following meaning:

- Transition Z_1 represents the personal data of the patients (age, gender, symptoms, etc.),
- Transition Z_2 represents the current symptoms of the patient,
- Transition Z_3 represents the severity of the current symptoms of the patient,
- Transition Z_4 represents the results from the Covid-19 testing procedures.

The net contains five types of tokens: α, β, μ, η and γ. Some of the model transitions contain a so called "special place" where a token stays and collect information about the specific parts of the screening process and serves as a database storage, which it represents as follows:

- In place l_3, token $\boldsymbol{\beta}$ stays permanently and collects the overall information obtained from the screening in the personal record (personal data),
- In place l_7, token $\boldsymbol{\mu}$ stays permanently and collects information about the current status of the patient obtained from the symptom evaluation,
- In place l_{10}, token $\boldsymbol{\eta}$ stays permanently and collects information about the severity of the symptoms,
- In place l_6, token $\boldsymbol{\gamma}$ stays permanently and collects information about the results from the Covid-19 tests

At the time of duration of the GN-functioning, some of these tokens can split, generating new tokens, that will transfer in the net obtaining respective characteristics, and also in some moments they will unite with some of tokens α, β, μ, η and γ.

Token α enters the net with additional characteristics "*person suspected of having a clinical features and epidemiological risk of Covid-19 infection*" in place $l1$.

The transition Z_1 of the GN-model has the following form:

$$Z_1 = \langle \{l_1, l_3\},\ \{l_2, l_3\}, r_1 \rangle,$$

where:

$$r_1 = \begin{array}{c|cc} & l_2 & l_3 \\ \hline l_1 & false & true \\ l_3 & W_{3,2} & true \end{array}$$

and $W_{3,2}$ = *"there is an established telecommunication connection (telephone, videoconference, online chat, etc.) between the person (user) and the medical specialist".*

On the next time-moment, token β splits to two tokens – the same token β and token α_1. When the predicate $W_{3,2}$ is true, token α_1 enters place l_2 and there it obtains a characteristic:

"start triage and evaluate for Covid-19 symptoms,".

The transition Z_2 has the following form:

$$Z_2 = \langle\{l_2, l_7\}, \{l_4\, l_5\, l_6\, l_7\}, r_2\rangle,$$

where:

$$r_2 = \begin{array}{c|cccc} & l_4 & l_5 & l_6 & l_7 \\ \hline l_2 & false & false & false & true \\ l_7 & W_{7,4} & W_{7,5} & W_{7,6} & true \end{array}$$

and,

$W_{7,4}$ = *"the patient reports fever or signs/symptoms of lower respiratory illness, smell or taste disturbances and close contact with laboratory confirmed COVID-19 patient within 14 days of symptom onset or history of travel from affected geographic areas (including risk from local epidemiological spread) within 14 days of symptoms";*
$W_{7,5}$ = *"the patient reports a history of travel from affected geographic areas (including risk from local epidemiological spread) with no signs/symptoms of lower respiratory illness";*
$W_{7,6}$ = "$\neg W_{7,4}$" *and* "$\neg W_{7,5}$"

The tokens from the two input places of transition Z_2 enter place l_7 and unite with token μ with the above mentioned characteristic. On the next time-moment, token μ splits to four tokens – the same token μ and tokens α_1, α_2 and α_3. When the predicate $W_{7,4}$ is true, token α_1 enters place l_4 and there it obtains a characteristic:

"high risk and potential COVID-19 infection, detailed evaluation of symptoms is needed".

When the predicate $W_{7,5}$ is true, token α_2 enters place $l_5$and there it obtains a characteristic:

"moderate risk of COVID-19 infection, recommend quarantine and self-monitoring for 14 days".

When the predicate $W_{7,6}$ is true, token α_2 enters place l_5 and there it obtains a characteristic:

"low risk of COVID-19 infection, no further evaluation is needed recommend social distancing, face mask and eye protection"

The transition Z_3 has the following form:

$$Z_3 = \langle\{l_4, l_5, l_{10}\}, \{l_8, l_9, l_{10}\}, r_3\rangle,$$

where:

$$r_3 = \begin{array}{c|ccc} & l_8 & l_9 & l_{10} \\ \hline l_4 & false & false & true \\ l_5 & false & false & true \\ l_{10} & W_{10,8} & W_{10,9} & true \end{array}$$

and,

$W_{10,8}$ = *"the patient reports: high fever and dry cough and shortness of breath or loss of taste and smell, sore throat and fatigue"*;
$W_{10,9}$ = *"the patient reports: extremely difficult breathing, severe constant dizziness or lightheadedness and acute pain"*;

The tokens from the three input places of transition Z_3 enter place l_{10} and unite with token η with the above mentioned characteristic. On the next time-moment, token η splits to three tokens – the same token η and tokens α_1 and α_2. When the predicate $W_{10,8}$ is true, token α_1 enters place l_8 and there it obtains a characteristic:

"Covid-19 testing indicated".

When the predicate $W_{10,9}$ is true, token α_2 enters place l_9 and there it obtains a characteristic:

"Urgent medical attention is needed".

The transition Z_4 has the following form:

$$Z_4 = \langle\{l_8\}, \{l_{11}, l_{12}, l_{13}\}, r_4,\rangle$$

where:

$$r_4 = \begin{array}{c|ccc} & l_{11} & l_{12} & l_{13} \\ \hline l_8 & false & false & W_{8,13} \\ l_{13} & W_{13,11} & W_{13,12} & true \end{array}$$

and,

$W_{8,13}$ = *"the patient has been tested for Covid-19 with RT-PCR or rapid antigen test"*
$W_{13,11}$ = *"the RT-PCR test is positive"*;
$W_{13,12}$ = "$\neg W_{13,11}$"

;

When the predicate $W_{8,13}$ is true, token α_1 enters place l_{13} and unites with token γ with the above mentioned characteristic. On the next time-moment, token γ splits to three tokens – the same token γ and tokens α_1 and α_2.

When the predicate $W_{13,11}$ is true, token α_1 enters place l_{11} and there it obtains a characteristic:

"the patient is positive for Covid-19 infection: inform regional health inspectorate, request details for contact tracing, quarantine the patient and prescribe home treatment and daily online visits for monitoring".

When the predicate $W_{10,9}$ is true, token α_2 enters place l_9 and there it obtains a characteristic:

"the patient is negative for Covid-19 infection: send patient for antibody test".

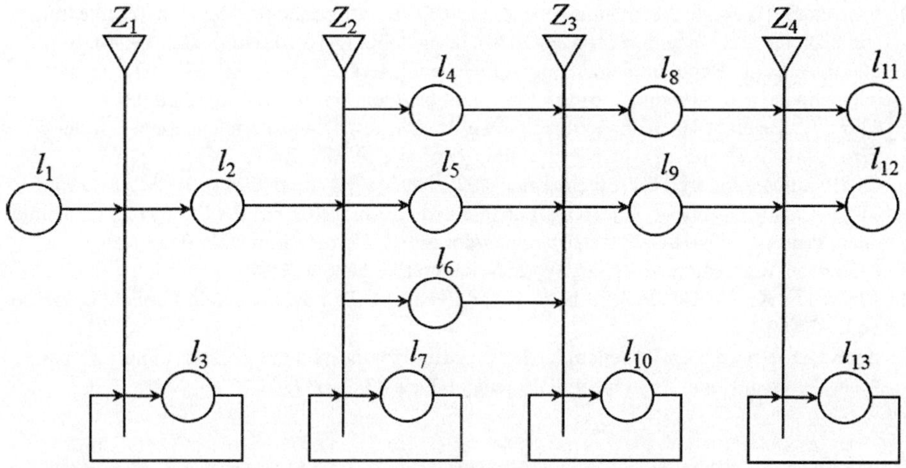

Fig. 1. Generalized net model for screening of corona virus disease 2019 (COVID-19).

3 Conclusions

The model describes a possible algorithm for remote symptom screening and the different transitions of the model are representing respectively the different criteria of the evaluation process. The GN-model may provide a framework that can be used by primary care practitioners to guide screening processes for patient suspected to have COVID-19, enabling more accurate and efficient identification of that condition and would assist in optimizing patient outcomes and more effective treatment. Moreover, some of the structures of the proposed model represents database storage and processing of COVID-19 data. This model can be complicated and detailed, which will significantly improve the accuracy of the evaluation of the patient and the reliability of the proposed algorithm.

Acknowledgements. The present research has been supported by the Bulgarian National Science Fund under Grant Ref. No. DN-02/10 "New Instruments for Knowledge Discovery from Data and Their Modelling and № НИХ - 442 /2020 "Exploring the possibilities of telemedicine, telerehabilitation and telepharmacy to improve patient care and medical education".

References

1. World Health Organization. World Health Organization. Statement on the second meeting of the International Health Regulations (2005) Emergency Committee regarding the outbreak of novel coronavirus (2019-nCoV). https://www.who.int/news/item/30-01-2020-statement-on-the-second-meeting-of-the-international-health-regulations-(2005)-emergency-committee-regarding-the-outbreak-of-novel-coronavirus-(2019-ncov). Accessed Oct 2020
2. Ahmed, S., Sanghvi, K., Yeo, D.: Telemedicine takes centre stage during COVID-19 pandemic. BMJ Innov. bmjinnov-2020-000440 (2020). https://doi.org/10.1136/bmjinnov-2020-000440
3. Bashshur, R., et al.: Telemedicine and the COVID-19 pandemic, lessons for the future, telemedicine and e-health, 571–573 (2020). https://doi.org/10.1089/tmj.2020.29040.rb
4. Dosaj, A., et al.: Rapid implementation of telehealth services during the COVID-19 pandemic telemedicine and e-Health, 116–120 (2021). https://doi.org/10.1089/tmj.2020.0219
5. Chou, E., Hsieh, Y.L., Wolfshohl, J., Green, F., Bhakta, T.: Onsite telemedicine strategy for coronavirus (COVID-19) screening to limit exposure in ED. Emerg. Med. J. **37**(6), 335–337 (2020). https://doi.org/10.1136/emermed-2020-209645. Epub 2020 May 3. PMID: 32366616
6. Vidal-Alaball, J., Acosta-Roja, R., et al.: Telemedicine in the face of the COVID-19 pandemic. Aten. Primaria **52**, 418–422 (2020). https://doi.org/10.1016/j.aprim.2020.04.003
7. Atanassov, K.: Generalized Nets. World Scientific, Singapore (1999)
8. Atanassov, K.: On Generalized Nets Theory. Prof. M. Drinov Academic Publishing House, Sofia (2007)
9. Zoteva, D., Krawczak, M.: Generalized nets as a tool for the modelling of data mining processes. A survey. Issues Intuition. Fuzzy Sets General. Nets **13**, 1–60 (2017)

Generalized Net Model of the Upper Limb Arterial Supply System

Valentin Vasilev[1], Krassimir Atanassov[2], and Evdokia Sotirova[1](✉)

[1] Faculty of Public Health and Health Care, Prof. Assen Zlatarov University, 1 Prof. Yakimov str., 8010 Burgas, Bulgaria
vailvasilev@mail.bg, esotirova@btu.bg

[2] Department of Bioinformatics and Mathematical Modelling, Institute of Biophysics and Biomedical Engineering, Bulgarian Academy of Sciences, Acad. G. Bonchev Str., Bl. 105, 1113 Sofia, Bulgaria
krat@bas.bg

Abstract. In a lot of papers, a different medical processes, different organs and systems in the human body are described by Generalized Nets. This models can be used for simulation of real processes and objects in the human body with the aim of using the information obtained in the process of medical diagnosis and research.

In the present investigation a Generalized Net model of the arterial supply of the upper limb as a part of the vascular system is presented. The vascular system ensures the circulation of blood and lymph in the body. It is composed of a CardioVascular System (CVS) which distributes blood, and a lymphatic system, which distributes lymph.

The transitions of the Generalized Net model of the arterial supply of the upper limb proposed in this paper have maximal simple form. The purpose of the construction of the model is to complement and enrich the knowledge about the CVS, or at least about a part of it.

Keywords: Generalized nets modelling · Vascular system · Arterial supply of the upper limb

1 Introduction

The modelling the vascular system if very important for understanding of vascular anomalies, and properly diagnose and treat [20]. The arterial supply with oxygenated blood to the upper limb is delivered via five main vessels: subclavian, axillary, brachial, radial, and ulnar arteries [13, 15, 16, 18]. The subclavian, axillary and brachial arteries are continuous with one another. The brachial artery bifurcating into the radial and ulnar arteries which later converge in the hand [11–13].

More than 750 Generalized Net models (GN, [1, 2]) of the different systems and organs in the human body are described in a series of paper and several monographs ([3–10, 14, 17]. The global GN model of the human body is presented in [3]. It can be used for investigation the behavior of the different systems of the human body and the relations between them. Each GN model of a particular system or organ can be detailed and made more complex.

S. S. Sotirov et al. (Eds.): BioInfoMed 2020, LNNS 374, pp. 145–153, 2022.
https://doi.org/10.1007/978-3-030-96638-6_16

In the present investigation a GN model of the arterial supply of the upper limb as a part of the vascular system is presented. The transitions of the GN model present the main arteries that carry oxygenated blood away from the heart to the upper limb are: subclavian artery, axillary artery, brachial artery, radial artery, ulnar artery. They divide into numerous branches.

The basic GN model of CVS is presented in [3]. In [14] this GN model is extended with intuitionistic fuzzy estimations. In [14] a GN model which describes the process of circulation of the blood and lymph in the upper limbs is constructed. In [19] a method for evaluation the blood supply of the lower limbs with peripheral artery disease is suggested.

2 Generalized Net Model of the Arterial Supply of the Upper Limb of the Vascular System

The GN-model contains 19 transitions and 62 places (Fig. 1). The following types of tokens are used:

- α-tokens - represent the blood in the arteries and capillaries, i.e., before the metabolism;
- β-tokens - represent the blood in the veins, i.e., after the metabolism;
- γ-token - represents the heart;
- χ-token - represents the brachiocephalic trunk;
- μ_1-, μ_2-, μ_3-, μ_4-, μ_5-, μ_6-, μ_7-, μ_8-tokens - represent the human body matter in which metabolism processes flow as follows:

  - μ_1- and μ_2-tokens represent the left subclavian and axillary arteries and the right subclavian and axillary arteries respectively
  - μ_3- and μ_4-tokens represent the left anterior artery and the right anterior artery respectively
  - μ_5- and μ_6-tokens represent the left brachial artery and the brachial artery respectively
  - μ_7- and μ_8-tokens represent the left radial artery and the right radial artery respectively.

Each one of these tokens obtain as a characteristic:
"blood, quantity, time-moment".
The transitions are as follow:

- Z_1 – the functions of the heart
- Z_2 – the functions of the aortic arch
- Z_3 – the functions of the brachiocephalic trunk
- Z_4/Z_5 – the functions of the left/right subclavian and axillary arteries
- Z_6/Z_7 – the functions of the branches of left/right anterior artery
- Z_8/Z_9 – the functions of the left/right brachial artery
- Z_{11}/Z_{12} – the functions of the left/right radial artery

- Z_{10}/Z_{13} – the functions of the left/right ulnar artery
- Z_{14}/Z_{17} – the functions of the left/right superficial palmar arch
- Z_{15}/Z_{16} – the functions of the left/right deep palmar arch
- Z_{18}/Z_{19} – the functions of the veins after metabolism

The transitions have the following forms:

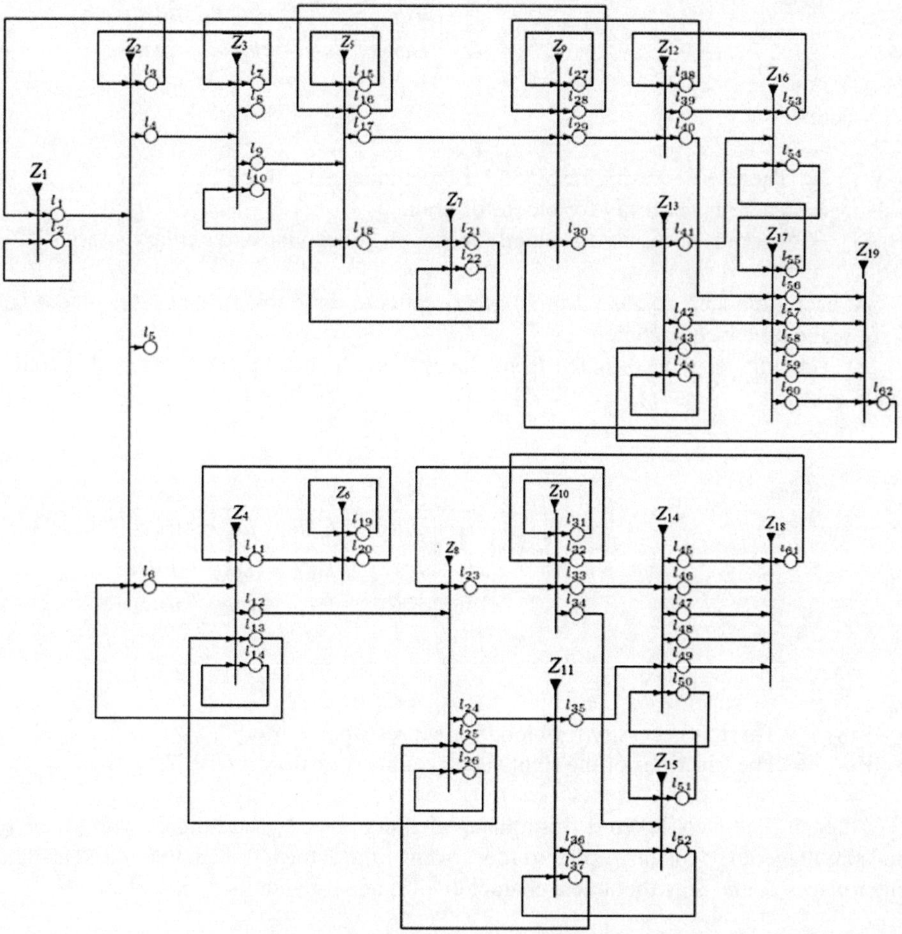

Fig. 1. Arterial supply of the upper limb of the vascular system.

$$Z_1 = \langle \{l_2, l_3\}, \{l_1, l_2\}, \begin{array}{c|cc} & l_1 & l_2 \\ \hline l_2 & true & true \\ l_3 & false & true \end{array} \rangle,$$

At each time-step, token from place l_3 enters place l_2 and unites with token γ and simultaneously, token γ splits to two tokens - the same token γ and a token α that enters place l_1.

$$Z_2 = \langle \{l_1, l_7, l_{13}\}, \{l_3, l_4, l_5, l_6\}, \begin{array}{c|cccc} & l_3 & l_4 & l_5 & l_6 \\ \hline l_1 & false & W_{1,4} & W_{1,5} & W_{1,6} \\ l_7 & true & false & false & false \\ l_{13} & true & false & false & false \end{array} \rangle,$$

where:

- $W_{1,4}$ = "There is necessity for blood for brachiocephalic trunk",
- $W_{1,5}$ = "There is necessity for blood for brain",
- $W_{1,6}$ = "There is necessity for blood for the left subclavian and axillary arteries".

At each time-step, token α from place l_1 splits to three tokens that enter places l_4, l_5, l_6 without new characteristics.

At each time-step, the β-tokens from places l_7 and l_{13} enter place l_3 where they unite in a β-token.

$$Z_3 = \langle \{l_4, l_{10}, l_{16}\}, \{l_7, l_8, l_9, l_{10}\}, \begin{array}{c|cccc} & l_7 & l_8 & l_9 & l_{10} \\ \hline l_4 & false & false & false & true \\ l_{10} & true & W_{10,8} & W_{10,9} & false \\ l_{16} & false & false & false & true \end{array} \rangle,$$

where:

- $W_{10,8}$ = "There is necessity for blood for suprascapular artery",
- $W_{10,9}$ = "The functions of the right subclavian and axillary arteries".

At each time-step, token α from place l_4 enters place l_{10} and unites with token χ and simultaneously, token χ splits to four tokens - the same χ-token, two α tokens that enter places l_8 and l_9 without new characteristics, and β-token that enter place l_7.

$$Z_4 = \langle \{l_6, l_{14}, l_{20}, l_{25}\}, \{l_{11}, l_{12}, l_{13}, l_{14}\}, \begin{array}{c|cccc} & l_{11} & l_{12} & l_{13} & l_{14} \\ \hline l_6 & false & false & false & true \\ l_{14} & true & true & true & true \\ l_{20} & false & false & false & true \\ l_{25} & false & false & false & true \end{array} \rangle,$$

At each time-step, token α from place l_6 enters place l_{14} and unites with token μ_1 and simultaneously, token μ_1 splits to four tokens - the same μ_1-token, two α tokens that enter places l_{11} and l_{12}, and β-token that enter place l_{13}.

$$Z_5 = \langle \{l_9, l_{15}, l_{21}, l_{28}\}, \{l_{15}, l_{16}, l_{17}, l_{18}\}, \begin{array}{c|cccc} & l_{15} & l_{16} & l_{17} & l_{18} \\ \hline l_9 & false & false & false & true \\ l_{15} & true & true & true & true \\ l_{21} & false & false & false & true \\ l_{28} & false & false & false & true \end{array} \rangle,$$

At each time-step, token α from place l_9 enters place l_{15} and unites with token μ_2 and simultaneously, token μ_2 splits to four tokens - the same μ_2-token, a α tokens that enters places l_{17} and l_{18}, and β-token that enter place l_{16}.

$$Z_6 = \langle \{l_{11}, l_{19}\}, \{l_{19}, l_{20}\}, \begin{array}{c|cc} & l_{19} & l_{20} \\ \hline l_{11} & true & false \\ l_{19} & true & true \end{array} \rangle,$$

At each time-step, token from place l_{11} enters place l_{19} and unites with token μ_3 and simultaneously, token μ_3 splits to two tokens - the same token μ_3 and a token β that enters place l_{20}.

$$Z_7 = \langle \{l_{18}, l_{22}\}, \{l_{21}, l_{22}\}, \begin{array}{c|cc} & l_{21} & l_{22} \\ \hline l_{18} & false & true \\ l_{22} & true & true \end{array} \rangle,$$

At each time-step, token from place l_{18} enters place l_{22} and unites with token μ_4 and simultaneously, token μ_4 splits to two tokens - the same token μ_4 and a token β that enters place l_{21}.

$$Z_8 = \langle \{l_{12}, l_{26}, l_{32}, l_{37}\}, \{l_{23}, l_{24}, l_{25}, l_{26}\}, \begin{array}{c|cccc} & l_{23} & l_{24} & l_{25} & l_{26} \\ \hline l_{12} & false & false & false & true \\ l_{26} & true & true & true & true \\ l_{32} & false & false & false & true \\ l_{37} & false & false & false & true \end{array} \rangle,$$

At each time-step, token α from place l_{12} enters place l_{26} and unites with token μ_5 and simultaneously, token μ_5 splits to four tokens - the same μ_5-token, a α tokens that enters places l_{23} and l_{24}, and β-token that enter place l_{25}.

$$Z_9 = \langle \{l_{17}, l_{27}, l_{38}, l_{43}\}, \{l_{27}, l_{28}, l_{29}, l_{30}\}, \begin{array}{c|cccc} & l_{27} & l_{28} & l_{29} & l_{30} \\ \hline l_{17} & true & false & false & false \\ l_{27} & true & true & true & true \\ l_{38} & true & false & false & false \\ l_{43} & true & false & false & false \end{array} \rangle,$$

At each time-step, token α from place l_{17} enters place l_{27} and unites with token μ_6 and simultaneously, token μ_6 splits to four tokens - the same μ_6-token, a α tokens that enters places l_{29} and l_{30}, and β-token that enter place l_{28}.

$$Z_{10} = \langle \{l_{23}, l_{31}, l_{61}\}, \{l_{31}, l_{32}, l_{33}, l_{34}\}, \begin{array}{c|cccc} & l_{31} & l_{32} & l_{33} & l_{34} \\ \hline l_{23} & true & false & false & false \\ l_{31} & true & true & true & true \\ l_{61} & true & false & false & false \end{array} \rangle,$$

At each time-step, token α from place l_{23} enters place l_{31} and unites with token μ_7 and simultaneously, token μ_7 splits to three tokens - the same μ_7-token, a α tokens that enters places l_{33} and l_{34}, and β-token that enter place l_{32}.

$$Z_{11} = \langle \{l_{24}, l_{52}\}, \{l_{35}, l_{36}, l_{37}\}, \begin{array}{c|ccc} & l_{35} & l_{36} & l_{37} \\ \hline l_{24} & true & true & false \\ l_{52} & false & false & true \end{array} \rangle,$$

At each time-step, α token from place l_{24} enter places l_{35} and l_{36} and β token from place l_{52} enters place l_{37}.

$$Z_{12} = \langle \{l_{29}, l_{53}\}, \{l_{38}, l_{39}, l_{40}\}, \begin{array}{c|ccc} & l_{38} & l_{39} & l_{40} \\ \hline l_{29} & false & true & true \\ l_{53} & true & false & false \end{array} \rangle,$$

At each time-step, α token from place l_{29} enter places l_{39} and l_{40} and β token from place l_{53} enters place l_{38}.

$$Z_{13} = \langle \{l_{30}, l_{44}, l_{62}\}, \{l_{41}, l_{42}, l_{43}, l_{44}\}, \begin{array}{c|cccc} & l_{41} & l_{42} & l_{42} & l_{44} \\ \hline l_{30} & false & false & false & true \\ l_{44} & true & true & true & true \\ l_{62} & false & false & false & true \end{array} \rangle,$$

At each time-step, token α from place l_{30} enters place l_{44} and unites with token μ_8 and simultaneously, token μ_8 splits to three tokens - the same μ_8-token, a α tokens that enters places l_{41} and l_{42}, and β-token that enter place l_{43}.

$$Z_{14} = \langle \{l_{33}, l_{35}, l_{51}\}, \{l_{45}, l_{46}, l_{47}, l_{48}, l_{49}, l_{50}\}, \begin{array}{c|cccccc} & l_{45} & l_{46} & l_{47} & l_{48} & l_{49} & l_{50} \\ \hline l_{33} & true & true & true & true & true & true \\ l_{35} & true & true & true & true & true & true \\ l_{51} & true & true & true & true & true & true \end{array} \rangle,$$

At each time-step, tokens α from places l_{33}, l_{35} and l_{51} enter places l_{45}, l_{46}, l_{47}, l_{48}, l_{49} and l_{50} without new characteristics.

$$Z_{15} = \langle \{l_{34}, l_{36}, l_{50}\}, \{l_{51}, l_{52}\}, \begin{array}{c|cc} & l_{51} & l_{52} \\ \hline l_{34} & true & true \\ l_{36} & true & true \\ l_{50} & true & true \end{array} \rangle,$$

At each time-step, tokens α from places l_{34}, l_{36}, and l_{50} unites into one α token that generates two new tokens - α token that enter place l_{51} and β-token that enter place l_{52}.

$$Z_{16} = \langle \{l_{39}, l_{41}, l_{54}\}, \{l_{53}, l_{54}\}, \begin{array}{c|cc} & l_{53} & l_{54} \\ \hline l_{39} & true & true \\ l_{41} & true & true \\ l_{54} & true & true \end{array} \rangle,$$

At each time-step, tokens α from places l_{39}, l_{41}, and l_{54} unites into one α token that generates two new tokens - α token that enter place l_{54} and β-token that enter place l_{53}.

$$Z_{17} = \langle \{l_{40}, l_{42}, l_{54}\}, \{l_{55}, l_{56}, l_{57}, l_{58}, l_{59}, l_{60}\}, \begin{array}{c|cccccc} & l_{55} & l_{56} & l_{57} & l_{58} & l_{59} & l_{60} \\ \hline l_{40} & true & true & true & true & true & true \\ l_{42} & true & true & true & true & true & true \\ l_{54} & true & true & true & true & true & true \end{array} \rangle,$$

At each time-step, tokens α from places l_{40}, l_{42} and l_{54} enter places l_{55}, l_{56}, l_{57}, l_{58}, l_{59} and l_{60} without new characteristics.

$$Z_{18} = \langle \{l_{45}, l_{46}, l_{47}, l_{48}, l_{49}\}, \{l_{61}\}, \begin{array}{c|c} & l_{61} \\ \hline l_{45} & true \\ l_{46} & true \\ l_{47} & true \\ l_{48} & true \\ l_{49} & true \end{array}$$

$$Z_{19} = \langle \{ l_{56}, l_{57}, l_{58}, l_{59}, l_{60}\}, \{l_{62}\}, \begin{array}{c|c} & l_{62} \\ \hline l_{56} & true \\ l_{57} & true \\ l_{58} & true \\ l_{59} & true \\ l_{60} & true \end{array} .$$

3 Conclusions

In the present research, a GN model of the of the arterial supply of the upper limb of the vascular system is presented. The purpose of the construction of the GN model of the CVS is to complement and enrich the knowledge about the CVS, or at least about a part of it. Objective criteria and parameters can be chosen which through the GN model would give us a clearer understanding about the state of the CVS, presence and weight of the pathology, the risk degree and a relatively correct prognosis for the development of one disease or another.

Acknowledgements. The authors are grateful for the support provided by the Bulgarian Ministry of Education and Science under the National Research Programme "Information and Communication Technologies for a Digital Single Market in Science, Education and Security" approved by DCM # 577 / 17.08.2018".

References

1. Atanassov, K.: Generalized Nets. World Scientific, Singapore (1991)
2. Atanassov, K.: On Generalized Nets Theory. Sofia, Prof. M. Drinov Publishing House, Sofia, Bulgaria (2007)
3. Atanassov, K., Chakarov, V., Shannon, A., Sorsich, J.: Generalized Net Models of the Human Body. Prof. M. Drinov Publishing House, Sofia, Bulgaria (2008)
4. Atanassov, K., Shannon, A.: Principal generalized net model of an excretory system. In: Proceedings of the Fifth International Workshop on Generalized Nets, Sofia, 10 November 2004, pp. 13–18 (2004)

5. Atanassov, K., Chakarov, V., Shannon, A.: Principal generalized net model of the human muscle-sceletal system. In: Proceedings of Scientific Session of Burgas Free University, Burgas, 9–11 June 2006, vol. III, pp. 392–398 (2006)
6. Chakarov, V., Shannon, A., Atanassov, K.: Principal generalized net model of endocrine system. Comptes rendus de l'Academie Bulgare des Sciences **61**(6), 705–712 (2008)
7. Chakarov, V., Shannon, A., Atanassov, K.: Stress principal I generalized net model with intuitionistic fuzzy estimations of human nervous system. Notes Intuitionistic Fuzzy Sets **13**(3), 17–23 (2007)
8. Chakarov, V., Shannon, A., Atanassov, K.: Generalized net model of human hematopoetic system. Int. Electron. J. Bioautomation **2**, 107–114 (2005)
9. Chakarov, V., Shannon, A., Atanassov, K.: Generalized net model with intuitionistic fuzzy estimations of human nervous system. Notes Intuitionistic Fuzzy Sets **13**(3), 17–23 (2007)
10. Chakarov, V., Atanassov, K., Shannon, A.: Generalized net model of the human muscular-sceletal system. A survey of generalized nets. In: Choy, E.Y.H., Krawczak, M., Shannon, A., Szmidt, E. (eds.) Raes KvB Monograph, Sydney, no. 10, pp. 127–140 (2007)
11. Efstratios, G., Aliki, F.: Issues in teaching vascular anatomy: thoughts and suggestions from the clinician's point of view. Anat. Sci. Educ. (2020). https://doi.org/10.1002/ase.2040
12. Carroll, M.A., Blandino, J., Flynn, A., Laughran, R., Pennella, S.: Neurovascular axillary variations: superficial brachial artery and single-corded brachial plexus. Anat. Sci. Int. **96**(1), 161–167 (2020). https://doi.org/10.1007/s12565-020-00563-x
13. Olinger, A.: Upper limb arteries. Bergman's Comprehensive Encyclopedia of Human Anatomic Variation, pp. 583–618 (2016)
14. Ribagin, S., Chakarov, V., Atanassov, K.: Generalized net model of the upper limb vascular system. In: Proceedings of IEEE 6th Conference on Intelligent Systems, Sofia, 6–8 September 2012, vol. 2, pp. 229–233 (2012)
15. Rodríguez-Niedenführ, M., Vázquez, T., Nearn, L., Ferreira, B., Parkin, I., Sañudo, J.: Variations of the arterial pattern in the upper limb revisited: a morphological and statistical study, with a review of the literature. J. Anatomy **199**(5), 547–566 (2001)
16. Rodriguez-Niedenfuhr, M., Vazquez, T., Parkin, I., Sanudo, J.: Arterial patterns of the human upper limb: update of anatomical variations and embryological development. Eur. J. Anat. **7**, 21–28 (2003)
17. Sorsich, J., Shannon, A., Atanassov, K.: Generalized net model of the cardiovascular system (an intuitionistic fuzzy approach). In: Kacprzyk, J., Atanassov, K. (eds.) Proceedings of the Fourth International Conference on Intuitionistic Fuzzy Sets, vol. 2, Notes on Intuitionistic Fuzzy Sets, vol. 6, no. 4, pp. 59–63 (2000)
18. Thiele, C., Thornburg, D., Van Nuland, S., Langley, N.: Axillary artery variation: the rule not the exception. Natl. J. Clin. Anatomy **9**(3), 82–89 (2020)
19. Vasilev, V., Sotirova, E., Atanassov, K., Sotirov, S.: Intuitionistic fuzzy assessments of the abdominal aorta and its branches. In: Kahraman, C., Cevik Onar, S., Oztaysi, B., Sari, I.U., Cebi, S., Tolga, A.C. (eds.) INFUS 2020. AISC, vol. 1197, pp. 26–31. Springer, Cham (2021). https://doi.org/10.1007/978-3-030-51156-2_4
20. Upton, J., Coombs, C.J., Mulliken, J.B., Burrows, P.E., Pap, S.: Vascular malformations of the upper limb: a review of 270 patients. J. Hand Surg. **24**(5), 1019–1035 (1999). https://doi.org/10.1053/jhsu.1999.1019

Generalized Net Model of Ant Lion Optimizer

Elena Chorukova[1,2], Olympia Roeva[1], and Krassimir Atanassov[1](✉)

[1] Institute of Biophysics and Biomedical Engineering, Bulgarian Academy of Sciences, Acad. G. Bonchev Str. Bl. 105, Sofia 1113, Bulgaria
olympia@biomed.bas.bg, krat@bas.bg

[2] The Stephan Angeloff Institute of Microbiology, Bulgarian Academy of Sciences, Acad. G. Bonchev Str. Bl. 26, Sofia 1113, Bulgaria
elena@microbio.bas.bg

Abstract. In the presented paper, the functioning and the results of the work of a metaheuristic algorithm, known as Ant Lion Optimizer (ALO) is described using the apparatus of generalized nets (GNs). The developed GN-model mimic the optimization processes based on the behaviour of antlions and their prey in nature. To model such interactions, ants are required to move over the search space, and antlions are allowed to hunt them and become fitter using traps. The proposed GN-model executes the ALO algorithm conducting abovementioned basic steps and thus performing an optimal search. ALO GN-model considered here is another contribution to the open, namely "to present each of the artificial intelligence areas by GNs".

Keywords: Generalized nets · Ant lion optimizer · Metaheuristic algorithm

1 Introduction

One of the contemporary fields of artificial intelligence is the field of metaheuristic algorithms – a scientific method to problem-solving that extends the idea of heuristic algorithms, where "meta" denotes "beyond" or "on a higher level" [1]. Metaheuristic algorithms can be divided into three main groups [2, 3]: (i) *evolutionary algorithms* (genetic algorithms, genetic programming, differential evolution, evolutionary strategies, evolutionary programming and harmony search); (ii) *swarm-based algorithms* (artificial bee colony, particle swarm optimization, firefly algorithm, cuckoo search algorithm, ant colony optimization, antlion optimizer, etc.); (iii) *trajectory-based algorithms* (tabu search, simulated annealing, hill climbing).

The antlion optimizer (ALO) is a new *swarm-based metaheuristic algorithm* for optimization, which mimics the hunting mechanism of antlions in nature [4]. To model such interactions, ants are required to move over the search space, and antlions are allowed to hunt them and become fitter using traps. Since ants move stochastically in nature when searching for food, a random walk models ants' movement. Antlions are able to build traps proportional to their fitness and ants are required to move randomly. The final stage of hunt is when an ant reaches the bottom of the pit and is caught in the antlion's jaw. After this stage, the antlion pulls the ant inside the sand and consumes its body.

S. S. Sotirov et al. (Eds.): BioInfoMed 2020, LNNS 374, pp. 154–162, 2022.
https://doi.org/10.1007/978-3-030-96638-6_17

The ALO is used to solve various problems as control of a multi-area system [5], parameter identification of photovoltaic cell model [6], solving optimal reactive power dispatch problem [7], optimal sizing of renewable energy resources [8], optimal cluster analysis [9], influenza prediction [10], state based object oriented testing [11], etc. The presented results show that the ALO is an efficient and robust algorithm [12].

The aim of this paper is to describe the functioning and results of ALO algorithm using the apparatus of generalized nets (GNs) [13]. A detailed description of any algorithm in terms of GNs can lead to a better understanding of the inner workings of the algorithm and consequently to an improvement of its performance [14, 15].

The GNs are proven as an appropriate and efficient tool for describing the essence of various optimization methods, such as metaheuristic algorithms. Up to now, GNs have been successfully used as an instrument for modelling of ant colony optimization [16], genetic algorithms [17], cuckoo search algorithm [14], etc.

The paper is organized as follows: background notes on ALO algorithm are presented in Sect. 2. The developed generalized net model of ALO algorithm is discussed in Sect. 3. Concluding remarks and further work directions are given in Sect. 4.

2 The ALO Algorithm

The ALO algorithm [4] is defined as a three-tuple function that approximates the global optimum for optimization problems as follows: *ALO*(*ABC*) where *A* is a function that generates the random initial solutions, *B* manipulates the initial population provided by the function *A*, and *C* returns true when the end criterion is satisfied. The functions *A*, *B*, and *C* are defined as follows:

$$0 \xrightarrow{A} \{M_{Ant}, M_{OA}, M_{Antlion}, M_{OAL}\} \tag{1}$$

$$\{M_{Ant}, M_{Antlion}\} \xrightarrow{B} \{M_{Ant}, M_{Antlion}\} \tag{2}$$

$$\{M_{Ant}, M_{Antlion}\} \xrightarrow{C} \{true, false\} \tag{3}$$

In (1), random initialization of the antlion and ant matrices $\{M_{Ant}, M_{OA}, M_{Antlion}, M_{OAL}\}$ are made using the function *A*, where M_{Ant} is the matrix of the position of ants, $M_{Antlion}$ includes the position of antlions, M_{OA} contains the corresponding fitness of ants, and M_{OAL} has the fitness of antlions.

The ALO algorithm starts with an initial population of *n* ants x_i, $i = 1, 2, \ldots, n$. An initial solution is generated based on:

$$\mathbf{x} = Lb + (Ub - Lb)rand\ (size(Lb)), \tag{4}$$

where *rand* is a random number generator, uniformly distributed in the space [0;1] and *Ub* and *Lb* are *d*-dimensional vectors, the *upper* and *lower ranges* of the attributes of the optimization problem, respectively.

Of each iteration of (2), the function *B* updates the position of each ant with respect to an antlion selected by the roulette wheel operator and the elite. It is assumed that every ant randomly walks around a selected antlion by the roulette wheel and the elite simultaneously as follows:

$$Ant_i^t = \frac{R_A^t + R_E^t}{2}, \tag{5}$$

where R_A^t is the random walk around the antlion selected by the roulette wheel at t-th iteration, R_E^t is the random walk around the elite at t-th iteration, and Ant_i^t indicates the position of i-th ant at t-th iteration.

Since ants move stochastically in nature when searching for food, a random walk is chosen for modelling ants' movement as follows:

$$X(t) = [0, cumsum(2r(t_1) - 1), cumsum(2r(t_2) - 1), cumsum(2r(t_m) - 1)], \tag{6}$$

where *cumsum* calculates the cumulative sum, m is the maximum number of iteration, t shows the step of random walk (iteration in this study), and $r(t)$ is a stochastic function defined as follows:

$$r(t) = \begin{cases} 1 & if\ rand > 0.5 \\ 0 & if\ rand \leq 0.5 \end{cases}, \tag{7}$$

where *rand* is a random number generated with uniform distribution in the interval of [0; 1].

The updating position is then accomplished by two random walks around the selected antlion and elite.

When all the ants randomly walk, they are evaluated by the fitness function. The quality of the new solution is evaluated as f_i^{new}. If any of the ants become fitter than any other antlions, we assume it was caught by the antlion and the antlion update goes to its position to build the trap (i.e., $f_i^{new} < f_i$), their positions are considered as the new positions for the antlions in the next iteration. The best solutions are kept when fraction p_a of best positions is updated as elite. At the last step of each iteration, the solutions are ranked and the current best solution is stored. The best antlion is compared to the best antlion found during optimization (elite) and substituted if it is necessary. These steps iterative until the function C returns false, i.e., a predefined number of iterations (generations) is reached, or a stop criterion is satisfied.

3 Generalized Net Model of ALO Algorithm

The proposed here GN-model of the workflow of the ALO algorithm applied for an arbitrary problem, is shown on Fig. 1. GN-transitions are associated with the main steps of the ALO algorithm.

The first transition of the GN-model is related to the initialization phase of the ALO algorithm, where the initial values of the ALO parameters are set, along with the specifics of the problem being solved. The initial set of solutions is generated as a result of the transition's functioning. The GN-transition Z_1 has the form.

$$Z_1 = \langle \{l_1, l_2\}, \{l_3, l_4, l_5\}, r_1, \wedge(l_1, l_2) \rangle,$$

$$r_1 = \begin{array}{c|ccc} & l_3 & l_4 & l_5 \\ \hline l_1 & T & T & F \\ l_2 & T & T & F \end{array}$$

where T denotes "true" and F denotes "false". The transition can be activated when there are tokens in both of the input places l_1 and l_2, i.e., the parameters of the ALO algorithm and the problem are loaded.

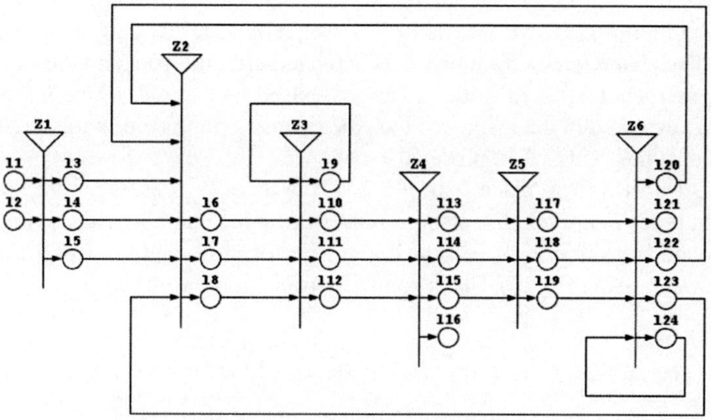

Fig. 1. ALO algorithm: a generalized net model.

The token χ enters the net through place l_1 with an initial characteristic "ALO parameters: n (number of ants), *Max_iteration* (maximum number of generations), M_{Ant} (positions of ants), $M_{Antlion}$ (positions of antlions), M_{OA} (fitness of ants), and M_{OAL} (fitness of antlions); position of elite, sorted antlions, convergence curve.

From the implementation point of view, each token in the GN has a list of characteristics. Each characteristic is an object with a name, type of the stored data and its value. For example, the characteristic *number of ants* should be described as (n, Integer, 40). The values, set as initial characteristics of the token χ, are the optimal parameter values of the ALO algorithm with respect to the specific problem considered during the current functioning of the GN.

The token κ enters the net via place l_2 with an initial characteristic "Problem parameters: objective function $f(\boldsymbol{x})$, $x = (x_1, x_2,..., x_d)$ in a d-dimensional space (number of parameters whose optimal values are sought); upper range (Ub) and lower range (Lb) of the ants".

The form of the objective function (cost function, loss function, fitness function, etc.), as well as the constraints on the values of the searched parameters depend entirely on the specific problem that is being considered. For example, the most commonly used objective function in fitting data problem is the least mean square. The name of the specific objective function (e.g., "least_mean_square"), which should be used during the current GN functioning, and two vectors with the numeric values of the upper and the lower bounds of the searched parameters are set as initial characteristics to the token κ by the characteristic function X. The GN functioning can be started multiple times for different problems at consideration and therefore with different initial settings (parameters of the ALO algorithm and the specific problem, as well as a particular objective function).

The token χ can be transferred to the output places l_3 and l_4. The token κ can be transferred to the output places l_3 and l_5. In case a token can be moved to more than one output place, the algorithm of transition's functioning allows the token to be split in as many identical copies as the number of output places to which it can be transferred.

When the transition Z_1 is activated, the tokens χ and κ unite into a new token τ in place l_3. The merging of two tokens in its essence is a unification of the information they carry. The characteristic function Φ is activated after the consolidation of the initial tokens. It generates the list of initial solutions, based on Eq. (4), which list is added to the information stored in the token τ. Thus, the token τ obtains the characteristic "Initial population of n host ants x_i" in place l_3.

Upon its transfer from place l_1 to place l_4, the token χ preserves its characteristic. Similarly, in place l_5 the token κ also preserves the same characteristic as in place l_2.

The second transition Z_2 is responsible for performing random walk for modelling ants' movement; using Eq. (6).The second GN-transition has the form.

$$Z_2 = \langle \{l_3, l_4, l_5, l_{20}, l_{22}, l_{23}\}, \{l_6, l_7, l_8\}, r_2, \vee(\wedge(l_3, l_4, l_5), \wedge(l_{20}, l_{22}, l_{23}))\rangle,$$

$r_2 =$

	l_6	l_7	l_8
l_3	T	F	F
l_4	T	T	F
l_5	T	F	T
l_{20}	T	F	F
l_{22}	T	T	F
l_{23}	T	F	T

The transition can be activated when there are tokens in the input places l_3, l_4 and l_5, or in the input places l_{20}, l_{22} and l_{23}, simultaneously. For the first iteration of the ALO algorithm, the current position of elite and sorted antlions, the algorithm's parameters and the considered problem parameters are stored in the tokens τ, χ and κ in the input places l_3, l_4 and l_5, respectively. As for the rest of the iterations of the ALO algorithm, the data comes from the tokens in the input places l_{20}, l_{22} and l_{23}.

In place l_6, the tokens τ, χ and κ merge into a new token, υ. Performing random walk (Eq. (6)), the new token υ obtains the characteristic "Updated position $X(t)$ of each ant with respect to an antlion".

Transferring from either place l_4 or place l_{22} to place l_7, the token χ obtains the characteristic "ALO parameters and current computational time". The token κ (from place l_5 or place l_{23}) obtains the characteristic "Problem parameters and current evaluations of $f_i(x)$" in place l_8.

The form of the third GN-transition is

$$Z_3 = \langle \{l_6, l_7, l_8, l_9\}, \{l_9, l_{10}, l_{11}, l_{12}\}, r_3, \vee(\wedge(l_6, l_7, l_8), l_9) \rangle,$$

$$r_3 = \begin{array}{c|cccc} & l_9 & l_{10} & l_{11} & l_{12} \\ \hline l_6 & T & F & F & F \\ l_7 & T & F & T & F \\ l_8 & T & F & F & T \\ l_9 & F & T & F & F \end{array}$$

The third transition is responsible for the evaluation of the newly generated solutions. The decision whether to update antlion positions and fitness based of the ants is made during the transition's functioning.

The transition can be activated when there are tokens in the input places l_6, l_7 and l_8, simultaneously, or in the input place l_9. The functioning of the transition Z_3 is performed in two steps. On the first step, in place l_9, the tokens υ, χ and κunite into a new token φ, which obtains the characteristic "Evaluated f_i". The newly generated solutions are evaluated after merging the information of the three tokens. When there is a token in the input place l_9, the transition Z_3 can be activated again. The token φ is moved then from place l_9 to place l_{10}. The characteristic function Φ, applied to the token φ in place l_{10}, performs the comparison between the evaluations of the newly generated solutions and solutions in the ants. As a result, the token φ obtains the characteristic "Chosen ants based on f_i" in place l_{10}.

When the token χ moves from place l_7 to place l_{11}, only the computational time in its characteristic is updated. When the token κ moves from place l_8 to place l_{12}, only the sum of f_i evaluations is updated.

The transition Z_4 is where is update the position of elite if any antlions becomes fitter than it. The fourth transition of the GN-model has the form.

$$Z_4 = \langle \{l_{10}, l_{11}, l_{12}\}, \{l_{13}, l_{14}, l_{15}, l_{16}\}, r_4, \wedge(l_{10}, l_{11}, l_{12}) \rangle,$$

$$r_4 = \begin{array}{c|cccc} & l_{13} & l_{14} & l_{15} & l_{16} \\ \hline l_{10} & T & F & F & T \\ l_{11} & T & T & F & T \\ l_{12} & T & F & T & T \end{array}$$

The transition Z_4 can be activated when there are tokens in the input places l_{10}, l_{11} and l_{12}, simultaneously.

The token φ in place l_{10} can be transferred to the output places l_{13} and l_{16}. The token $\chi(l_{11})$ can be transferred to the output places l_{13}, l_{14} and l_{16}. The token κ (from l_{12}) can be transferred to the output places l_{13}, l_{15} and l_{16}.

In place l_{13}, the token φ is merged with the tokens χ and κ into a new token ψ. The characteristic function Φ uses the value of the parameter p_a, in the original token χ in order to determine which of the elite have to be updated. Applied to the token ψ in place l_{10}, Φ performs the local random walk to generate the new solutions which should replace the abandoned ones. As a result, the token ψ obtains the characteristic "New elite population". In place l_{16}, the token φ merges with the tokens χ and κ into a new token ν, which obtains the characteristic "Abandoned worst elite".

When the token χ moves from l_{11} to l_{14} and the token κ moves from l_{12} to l_{15}, only the computational time and the number of evaluations of the f_i are respectively updated in their characteristics.

The form of the fifth transition of the GN-model is

$$Z_5 = \langle\{l_{13}, l_{14}, l_{15}\}, \{l_{17}, l_{18}, l_{19}\}, r_5, \wedge(l_{13}, l_{14}, l_{15})\rangle,$$

$$r_5 = \begin{array}{c|ccc} & l_{17} & l_{18} & l_{19} \\ \hline l_{13} & T & F & F \\ l_{14} & T & T & F \\ l_{15} & T & F & T \end{array}$$

During the functioning of this transition, all solutions (ants) are ranked on the basis of f_i. In place l_{17} new token ω (resulting from the merging of the tokens ψ, χ and κ) obtains the characteristic "Rated solutions with corresponding f_i".

When the token χ moves from l_{14} to l_{18} and the token κ moves from l_{15} to l_{19}, only the computational time and the sum of f_i evaluations are updated in their characteristics, respectively.

The last transition determines whether the end of the ALO algorithm is reached, i.e., if the *MaxGeneration* count is reached or the stopping criterion is met. The sixth transition of the GN-model has the form.

$$Z_6 = \langle\{l_{17}, l_{18}, l_{19}, l_{24}\}, \{l_{20}, l_{21}, l_{22}, l_{23}, l_{24}\}, r_6, \wedge(l_{17}, l_{18}, l_{19}, l_{24})\rangle,$$

$$r_6 = \begin{array}{c|ccccc} & l_{20} & l_{21} & l_{22} & l_{23} & l_{24} \\ \hline l_{17} & W_1 & W_2 & F & F & F \\ l_{18} & W_1 & W_2 & T & F & F \\ l_{19} & W_1 & W_2 & F & T & F \\ l_{24} & F & F & F & F & T \end{array}$$

where W_1 = "The ALO algorithm has not finished" and $W_2 = \neg W_1$.

Data information, in the form "Best solution, corresponding objective function value, computational time and number of evaluations of the objective function for each iteration of the algorithm", is collected throughout the whole GN-model functioning and is stored in the token ξ that stays in place l_{24}.

The transition Z_6 can be activated when there are tokens in the input places l_{17}, l_{18}, l_{19} and l_{24}, simultaneously. If the end of the algorithm has not been reached, the tokens ω, χ and κ merge into a new token, ς, which acquires the characteristic "Rated solutions, corresponding f_i value for the current iteration" in place l_{20}. At the end of the ALO algorithm, in place l_{21}, the token ς obtains the characteristic "Best solution, corresponding f_i value, total computational time and complete number of f_i evaluations".

When the token χ moves from l_{18} to l_{22} and the token κ moves from l_{19} to l_{23}, again only the computational time and the sum of f_i evaluations are respectively updated in their characteristics.

The best solution with the corresponding value of f_i, the computational time and the sum of f_i evaluations are added to the characteristic of the token ξ in place l_{24} at the end of each loop of the ALO algorithm.

4 Conclusion

In this paper, the generalized net theory is used to model the metaheuristic algorithm Ant lion optimizer. The developed GN-model mimic the optimization processes based on the behaviour of antlions and their prey in nature. The GN-model execute the ALO algorithm conducting its basic steps and thus performing an optimal search.

The GN-model of ALO allows the possibility to store information during its operation as a database. The collected data could be used in a case of comparable optimization problems as *a priori* information when choosing the ALO parameters and/or for an approximate prediction of the outcomes. The use of the collected information in the database could serve to enhance the algorithm's performance.

ALO GN-model considered here is another contribution to the open problem specified in [15], namely "to present each of the artificial intelligence areas by GNs". Future work will include verification of the feasibility of converting the universal GN-model proposed in [14] into the ALO GN-model herewith presented using the appropriate hierarchical operators. The results will serve to confirm the hypothesis that all concepts from the artificial intelligence can be described by one uniform mathematical formalism – generalized net theory.

Acknowledgements. The work presented here is partially supported by the National Scientific Fund of Bulgaria under Grant DN02/10 "New Instruments for Knowledge Discovery from Data, and their Modelling".

References

1. Jamil, M., Yang, X.-S.: A literature survey of benchmark functions for global optimization problems. Int. J. Math. Model. Numer. Optim. **4**(2), 150–194 (2013)
2. Fister, I., Fister, I., Jr., Yang, X.-S., Brest, J.: A comprehensive review of firefly algorithms. Swarm Evol. Comput. **13**, 34–46 (2013)
3. Shehab, M., Khader, A.T., Al-Betar, M.A.: A survey on applications and variants of the cuckoo search algorithm. Appl. Soft Comput. **61**, 1041–1059 (2017)
4. Mirjalili, S.: The ant lion optimizer. Adv. Eng. Softw. **83**, 80–98 (2015)
5. Raju, M., Saikia, L.C., Sinha, N.: Automatic generation control of a multi-area system using ant lion optimizer algorithm based PID plus second order derivative controller. Int. J. Electr. Power Energy Syst. **80**, 52–63 (2016)

6. Wu, Z., Yu, D., Kang, X.: Parameter identification of photovoltaic cell model based on improved ant lion optimizer. Energy Convers. Manag. **151**, 107–115 (2017)
7. Mouassa, S., Bouktir, T., Salhi, A.: Ant lion optimizer for solving optimal reactive power dispatch problem in power systems. Eng. Sci. Technol. Int. J. **20**(3), 885–895 (2017)
8. Dinakara, P.R.P., Veera, R.V.C., Gowri, M.T.: Ant lion optimization algorithm for optimal sizing of renewable energy resources for loss reduction in distribution systems. J. Electr. Syst. Inf. Technol. **5**(3), 663–680 (2018)
9. Majhi, S.K., Biswal, S.: Optimal cluster analysis using hybrid K-means and ant lion optimizer. Karbala Int. J. Mod. Sci. **4**(4), 347–360 (2018)
10. Hu, H., Li, Y., Bai, Y., Zhang, J., Liu, M.: The improved antlion optimizer and artificial neural network for Chinese influenza prediction. Complexity 2019, Article ID 1480392 (2019). https://doi.org/10.1155/2019/1480392
11. Sharma, R., Saha, A.: Ant lion optimizer for state based object oriented testing. J. Inf. Optim. Sci. **40**(2), 219–232 (2019). https://doi.org/10.1080/02522667.2019.1578085
12. Abualigah, L., Shehab, M., Alshinwan, M., Mirjalili, S., Elaziz, M.A.: Ant lion optimizer: a comprehensive survey of its variants and applications. Arch. Comput. Methods Eng. **28**(3), 1397–1416 (2020). https://doi.org/10.1007/s11831-020-09420-6
13. Atanassov, K.T.: Generalized Nets. World Scientific, Singapore (1991)
14. Roeva, O., Zoteva, D., Atanassova, V., Atanassov, K., Castillo, O.: Cuckoo search and firefly algorithms in terms of generalized net theory. Soft. Comput. **24**(7), 4877–4898 (2019). https://doi.org/10.1007/s00500-019-04241-7
15. Atanassov, K.T.: Generalized Nets in Artificial Intelligence. Volume 1: Generalized Nets and Expert Systems. Academic Publishing House "Prof. M. Drinov", Sofia (1998)
16. Fidanova, S., Atanassov, K., Marinov, P.: Generalized nets in artificial intelligence: generalized nets and ant colony optimization, Vol 5. "Prof. M. Drinov" Academic Publishing House, Sofia (2011)
17. Roeva, O., Shannon, A., Pencheva, T.: Description of simple genetic algorithm modifications using generalized nets. In: Proceedings of 6th IEEE International Conference on Intelligent Systems, Bulgaria, pp. 178–183 (2012)

Implementation of Expanding Hierarchical Operators in GN IDE

Dafina Zoteva[1,2](✉)

[1] Institute of Biophysics and Biomedical Engineering, Bulgarian Academy of Sciences, 105 Acad. Georgi Bonchev Str., 1113 Sofia, Bulgaria
[2] Faculty of Mathematics and Informatics, Sofia University "St. Kliment Ohridski", 5 Blvd James Bourchier, 1164 Sofia, Bulgaria
dafinaz@fmi.uni-sofia.bg

Abstract. Generalized Nets (GNs) are an instrument for modelling and optimization of parallel and concurrent processes. Numerous operators have been introduced over the years to extend the GNs' modelling power. Generalized Net Integrated Development Environment (GN IDE) – a software tool for creating GN-models and running simulations on such, has been developed to support the research in the GNs' theory and applications. Recent enhancements in GN IDE involve the integration of GN operators.

The implementation in GN IDE of two of the GNs' hierarchical operators, defined to replace a specified transition or place in a GN-model with an entire GN, is discussed in the present paper. These new features of GN IDE make it possible to easily refine existing GN-models in details.

Keywords: Generalized nets · Hierarchical operators · GN IDE

1 Introduction

Generalized Nets (GNs) [3,4,7] are an extension of Petri nets (PNs) [11] and all other PNs' extensions and modifications. Standard GNs, however, are not simply an automatic collection of the components of the other types of nets. New components were added to their definition so that all other PNs' extensions and modifications can be described by standard GNs. GNs are used as a means to describe in a systematic and structured way reusable models of complex systems which involve parallel or concurrent processes [12].

Generalized Net Integrated Development Environment (GN IDE) has been developed in the past few years in order to facilitate the constant research in the GNs' theory and applications [1,8]. Since then, GN IDE has been successfully applied for comparing different mathematical models of the *E. coli* fed-batch cultivation process [9], as well as for simulating a typical process of wastewater treatment plant using real experimental data [10].

Operators in the GNs' framework have been defined since the very beginning of the development of the GNs' theory [3,4,7]. When applied to a particular GN,

S. S. Sotirov et al. (Eds.): BioInfoMed 2020, LNNS 374, pp. 163–173, 2022.
https://doi.org/10.1007/978-3-030-96638-6_18

each of the operators results in a new GN with specific properties. The diversity of the GNs' operators has motivated their different classifications.

According to the GNs' components to which they might be applied, the operators can be divided into six different types: *global*, *local*, *hierarchical*, *reducing*, *extending* and *dynamical* [3,4,7].

Initially, the main functionality implemented in GN IDE involved creating GN-models and running their simulations, with a simplified algorithm of GN's functioning integrated into the system. Further, this algorithm in GN IDE has been enhanced by implementing merging and splitting of tokens. This can be regarded as an integration of the relevant dynamical operators into the software system.

Two features related to the reducing operators over GNs have been recently implemented in GN IDE [2]. The first one allows a reduction of GN-components without any loss of information. This way, different test scenarios can be explored by creating a single GN-model. The action of applied reducing operators can be reverted. The second feature allows identifying the class of reduced GNs to which a particular GN-model can be referred to.

In order to enhance GN IDE and to open up new possibilities for applications of the software tool, the consistent implementation of GNs' operators in GN IDE should continue.

Decomposition is a common approach when modelling complex systems. Different subprocesses or states of such a system can be described first by simpler models and integrated later into one composite model. This is often easier than creating directly a detailed version of the model.

In one rough simplified GN-model the states or the subprocesses of a system are usually depicted by GN-places or GN-transitions. When they are presented in more details, entire GNs are commonly used. To include the detailed models into the overall model of the system, two expanding hierarchical operators $\mathcal{H}_1(E, l^*, E')$ and $\mathcal{H}_3(E, Z, E')$ [3,4,7] can be exploited. The implementation of these hierarchical operators in GN IDE and their possible applications are discussed in the present paper.

Some brief theoretical notes regarding the definitions and the restrictions of the two hierarchical operators ($\mathcal{H}_1$ and $\mathcal{H}_3$) are presented in Sect. 2. Their integration in GN IDE is discussed in Sect. 3. Concluding remarks are given is Sect. 4.

2 Notes on Hierarchical Operators over Generalized Nets

There are seven *hierarchical operators* introduced in the GNs' theory [3,4], divided into two groups according to the results of their action:

- expanding a given GN ($\mathcal{H}_1$, $\mathcal{H}_3, \mathcal{H}_6, \mathcal{H}_7$ and in some cases – $\mathcal{H}_5$);
- shrinking a given GN ($\mathcal{H}_2$, $\mathcal{H}_4$ and in some cases – $\mathcal{H}_5$).

Other three groups are formed when the object of their action is considered:

- acting upon or giving as a result of their work a place ($\mathcal{H}_1$ and $\mathcal{H}_2$);

– acting upon or giving as a result of their work a transition ($\mathcal{H}_3$ and $\mathcal{H}_4$);
– acting upon or giving as a result of their work a subnet ($\mathcal{H}_5$, $\mathcal{H}_6$ and $\mathcal{H}_7$).

They are named hierarchical operators since they are related to the degree of details of the GN-models achieved after their application.

The expanding operators can be seen as a means of considering the structure of a modelled process in more details, while the shrinking operators – as a means of integrating and ignoring irrelevant details of the process.

The focus of the present paper is on the two expanding hierarchical operators $\mathcal{H}_1(E, l^*, E')$ and $\mathcal{H}_3(E, Z, E')$. Therefore, only their definitions will be given and discussed in details.

2.1 Hierarchical Operator $\mathcal{H}_1(E, l^*, E')$

The hierarchical operator $\mathcal{H}_1$ has the form

$$\mathcal{H}_1(E, l^*, E') = E'',$$

where l^* is a fixed place of the GN E, i.e., $l^* \in L_E$ (the set of all places of E), E' is a GN that functions similarly to the place l^* and E'' is the resulting GN.

The place l^* of the given GN E is replaced by an entire GN E'. Thus, the resulting GN E'' is a more detailed version of E. Moreover, $\mathcal{H}_1$ operator does not change the functioning of the GN E, the expected results of the work of the GN E'' are the same as those of the original GN E.

Let $\Box\,[l_1, l_2]$ denote a transition type $\Box$, where all occurrences of the variable l_1 are substituted by the variable l_2.

Let $A \div B = (A - B) \cup (B - A)$.

Let $l, l', l'' \notin L_E \cup L_{E'}$, where L_E and $L_{E'}$ are the sets of all places of E and E', respectively.

Let Q_E^I and Q_E^O are the sets of all input and output places of E, respectively.

Let $Q_E = Q_E^I \cup Q_E^O$, i.e., Q_E is therefore the set of all border places of E.

1. If $l^* \notin L_E$: $E'' = E$.
If the chosen GN-place l^* does not belong to the GN E, then the operator does not change the input GN E.

2. If $l^* \in L_E - Q_E$,
i.e., l^* is an input place for one transition and an output place for another transition in E (not an input or an output place for the entire GN E), the resulting GN E'' is then defined as follows:

$$E'' = \langle\langle(A_E \div (\{\langle L_1', L_1'', t_1^1, t_2^1, r_1, M_1, \Box_1\rangle, \langle L_1', L_1'' \div \{l^*, l'\}, t_1^1, t_2^1, r_1, M_1,$$
$$\Box_1[l^*, l']\rangle | l^* \in L_1''\} \cup \{\langle L_2', L_2'', t_1^2, t_2^2, r_2, M_2, \Box_2\rangle,$$

$$\langle L_2' \div \{l^*, l''\}, L_2'', t_1^2, t_2^2, r_2, M_2, \Box_2[l^*, l'']\rangle | l^* \in L_2''\})) \cup A_{E'} \cup$$
$$\{\langle\{l'\}, Q_{E'}^I, t_1^1, 0, r', M', \vee(l')\rangle, \langle Q_{E'}^O, \{l''\}, t_1^2, 0, r'', M'', \vee(Q_{E'}^O)\rangle\},$$
$$\pi_A \cup \pi_{A'}, \pi_L \cup \pi_{L'}, c_L \cup c_{L'}, f \cup f', \theta_1^A \cup \theta_1^{A'}, \theta_2^A \cup \theta_2^{A'}\rangle,$$

$$\langle K_E \cup K_{E'}, \pi_K \cup \pi_{K'}, \theta_K \cup \theta_{K'} \rangle, \langle T, t_E^o, t_E^* \frac{t_E^o}{t_{E'}^o} \rangle,$$

$$\langle X_E \cup X_{E'}, \Phi_E \cup \Phi_{E'}, b_E \cup b_{E'} \rangle \rangle,$$

where

$$r' = \begin{array}{c|cccc} & & & Q_{E'}^I & \\ \hline l' & W_1 & W_2 & \ldots & W_{card(Q_{E'}^I)} \end{array}$$

and for $1 \leq i \leq card(Q_{E'}^I)$
$W_i =$ "the token is an element of the set K_i",
where

$$K_E = \bigcup_{i=1}^{card(Q_{E'}^I)} K_i \ .$$

The index matrices (IMs) [5] for the capacities of the transitions' arcs and the transitions' conditions are:

$$M' = \begin{array}{c|cccc} & & & Q_{E'}^I & \\ \hline l' & \infty & \infty & \ldots & \infty \end{array}, \qquad r'' = \begin{array}{c|c} & l'' \\ \hline & true \\ Q_{E'}^O & \vdots \\ & true \end{array}, \qquad M'' = \begin{array}{c|c} & l'' \\ \hline & \infty \\ Q_{E'}^O & \vdots \\ & \infty \end{array}.$$

The GN E' is to replace l^*. Two auxiliary transitions are therefore included in the resulting GN E'', surrounding E':

$$Z_L = \langle \{l'\}, Q_{E'}^I, t_1^1, 0, r', M', \vee(l') \rangle,$$

$$Z_R = \langle Q_{E'}^O, \{l''\}, t_1^2, 0, r'', M'', \vee(Q_{E'}^O) \rangle.$$

They serve as a link between E' and the elements of the original GN E. The left transition Z_L has exactly one input place l' and all the input places of E' as its output places. The right auxiliary transition Z_R has as input places the output places of the GN E' and exactly one output place l''.

3. If $l^* \in Q_E^I$,
i.e. l^* is an input place of the GN E, the resulting E'' is then defined as follows:

$$E'' = \langle \langle (A_E - \{\langle L', L'', t_1, t_2, r, M, \square \rangle, \langle L' \div \{l^*, l'\}, L'', t_1, t_2, r, M, \square \rangle | l^* \in L'\})$$

$$\cup \{\langle Q_{E'}^O, \{l\}, t_1^2, 0, r''', M''', \vee(Q_{E'}^O) \rangle\} \cup A_{E'}, \pi_A \cup \pi_{A'}, \pi_L \cup \pi_{L'}, c_L \cup c_{L'},$$

$$f \cup f', \theta_1^A \cup \theta_1^{A'}, \theta_2^A \cup \theta_2^{A'} \rangle, \langle K_E \cup K_{E'}, \pi_K \cup \pi_{K'}, \theta_K \cup \theta_{K'} \rangle,$$

$$\langle T, t_E^o, t_E^* \frac{t_E^o}{t_E'^o} \rangle, \langle X_E \cup X_{E'}, \Phi_E \cup \Phi_{E'}, b_E \cup b_{E'} \rangle \rangle,$$

where

$$r''' = \begin{array}{c|c} & l \\ \hline & true \\ Q^O_{E'} & \vdots \\ & true \end{array}, \qquad M''' = \begin{array}{c|c} & l \\ \hline & \infty \\ Q^O_{E'} & \vdots \\ & \infty \end{array}.$$

One auxiliary transition $\langle Q^O_{E'}, \{l\}, t^2_1, 0, r''', M''', \vee(Q^O_{E'}) \rangle$ is included in the resulting GN E'' to link E' and the remaining elements of E. It has as input places the output places of E' and exactly one output place l. l will replace l^* of E in the resulting GN E''.

4. If $l^* \in Q^O_E$,
i.e., l^* is an output place of the GN E, the resulting GN E'' is defined as follows:

$$E'' = \langle\langle(A_E \div (\{\langle L'_1, L''_1, t_1, t_2, r, M, \square\rangle, \langle L', L'' \div \{l^*, l'\}, t_1, t_2, r, M, \square[l^*, l']\rangle | l^* \in L'_2\} \cup \{\langle\{l\}, Q^I_{E'}, t^2_1, 0, r'''', M'''', \vee(l)\rangle\})) \cup A_{E'},$$
$$\pi_A \cup \pi_{A'}, \pi_L \cup \pi_{L'}, c_L \cup c_{L'}, f \cup f', \theta^A_1 \cup \theta^{A'}_1, \theta^A_2 \cup \theta^{A'}_2\rangle,$$
$$\langle K_E \cup K_{E'}, \pi_K \cup \pi_{K'}, \theta_K \cup \theta_{K'}\rangle, \langle T, t^o_E, t^*_E . \frac{t^o_E}{t^o_{E'}}\rangle,$$
$$\langle X_E \cup X_{E'}, \Phi_E \cup \Phi_{E'}, b_E \cup b_{E'}\rangle\rangle,$$

where

$$r'''' = \begin{array}{c|cccc} & & & Q^I_{E'} & \\ \hline l & W_1 & W_2 & \dots & W_{card(Q^I_{E'})} \end{array}$$

and for $1 \leq i \leq card(Q^I_{E'})$
$W_i =$ "the token is an element of the set K_i",
where

$$K_E = \bigcup_{i=1}^{card(Q^I_{E'})} K_i,$$

$$M'''' = \begin{array}{c|cccc} & & & Q^I_{E'} & \\ \hline l & \infty & \infty & \dots & \infty \end{array}.$$

An auxiliary transition $\langle\{l\}, Q^I_{E'}, t^2_1, 0, r'''', M'''', \vee(l)\rangle$, with one input place l and the input places of E' as its output places, is included in the resulting GN E'' to link E' with the elements of E. l will replace l^* of E in the resulting GN.

2.2 Hierarchical Operator $\mathcal{H}_3(E, Z, E')$

The operator $\mathcal{H}_3(E, Z, E')$ is similar to $\mathcal{H}_1(E, l^*, E')$, except that it interchanges a fixed transition Z of a given GN E with an entire GN E', which functions in the same way and has the same input and output places as the transition Z, $pr_1 Z = Q^I_{E'}$ and $pr_2 Z = Q^O_{E'}$.

The operator is defined as follows:

$$\mathcal{H}_3(E, Z, E') = E'',$$

where

1. if $Z \notin A$: $E'' = E$.

If the chosen transition Z does not belong to E, then the input GN E is not affected by the operator $\mathcal{H}_3$.

2. if $Z \in A$ and $Z = G_2(E')$, where G_2 is the second global operator [3,4,7], then:

$$E'' = \langle\langle (A_E - \{Z\}) \cup A_{E'}, \pi_A \cup \pi_{A'}, \pi_L \cup \pi_{L'}, c_L \cup c_{L'}, f \cup f', \\ \theta_1^A \cup \theta_1^{A'}, \theta_2^A \cup \theta_2^{A'} \rangle, \langle K_E \cup K_{E'}, \pi_K \cup \pi_{K'}, \theta_K \cup \theta_{K'} \rangle, \\ \langle T, t^o_E, t^*_E \frac{t^0_E}{t^o_{E'}} \rangle, \langle X_E \cup X_{E'}, \Phi_E \cup \Phi'_E, b_E \cup b'_E \rangle\rangle.$$

The global operator G_2 changes the form and the structure of a given GN by narrowing it down to a single transition. Only the input and output places of the given GN remain in the model, as if the resulting GN is regarded as a "black box". The statement $Z = G_2(E')$ should therefore be interpreted as: the GN E' has the same input and output places as the transition Z, moreover, E' and Z function in the same way.

3 Implementation of Hierarchical Operators in GN IDE

GN IDE as a platform independent software tool developed for creating and running simulations on GN-models. GN IDE provides essential features related to the GN functioning. The multilayered architecture of the system has been designed in a way that components with weak non-cyclic dependencies to be formed and new ones to be easily integrated [8].

The integration of the reducing operators in GN IDE has been carried out in a new independent layer of the system [2]. The implementation of the hierarchical operators in GN IDE extends this same layer with three new classes: the abstract class *HierarchicalOperator* and its derived *H1HierarchicalOperator* and *H3HierarchicalOperator*.

The abstract class *HierarchicalOperator* has a data member:

- *private GeneralizedNet mainGN*, which represents the original GN E whose place or transition will be replaced.

The following methods are part of the interface of the abstract super class *HierarchicalOperator*.

- *public HierachicalOperator(GeneralizedNet inputGN)*{} is the constructor of the class. It has one input argument of type *GeneralizedNet*, used to initialize the data member of the class. The class cannot be instantiated, no objects can be created.

- *public abstract List*⟨Command⟩ *changeElemWithGN*(*GeneralizedNet gn*){}. This method has one input argument, the GN E' which should change one particular element of the original GN E. The method is *abstract.* It is therefore overridden in the derived classes.

The derived class *H1HierarchicalOperator* implements the behaviour of the $\mathcal{H}_1$ hierarchical operator. It extends the abstract class *HierarchicalOperator.* In addition to the inherited data member, the derived class has one of its own:

- *private Place placeToBeReplaced*, representing the place l^* of the original GN E which should be replaced by the GN E'.

The constructor of this class has two input arguments, one to initialize the inherited data member *mainGN* and the other, of type *Place*, to initialize its own data member. The validity of the input data is checked in the constructor. The place should be an element of *mainGN*, the comparison is based on the identifiers of the places.

The abstract method *changeElemWithGN* is overridden in the derived class. Certain restrictions on the form of its input argument (the GN E') have been considered:

- if the selected element is an inner place for the original GN E, the GN E', chosen to replace it, should have exactly one input and one output place;
- if the selected element is an input place for the original GN E, the GN E', chosen to replace it, should have exactly one output place;
- if the selected element is an output place for the original GN E, the GN E', chosen to replace it, should have exactly one input place.

When the resulting GN E'' is built upon the original GN E, new elements like places, transitions, tokens and functions are brought into the model. The sets of places, transitions, tokens and functions of the original model E are extended with the relevant elements of E'.

A number of changes are entailed further by the fact that l^* is to be replaced by:

- l' and l'' in the left and right transitions of l^*, respectively, in case l^* is an inner place of E;
- l, in case l^* is either an input or an output place of E.

When l^* is an inner place of E and it is replaced by l' and l'' in its left and right transitions, respectively:

- all occurrences of l^* in the label sets of the index matrices of the left and right transitions of l^* are replaces by l' and l'', respectively;
- the characteristic function of the place l^* itself is set as a characteristic function of the place l'. All occurrences of l^* in the characteristic functions of the model are replaced by l';
- all occurrences of l^* in the predicates, used to describe the left and right transitions' conditions, are replaces by l' and l'', respectively.

When l^* is either an input or an output place of E, it is replaced by l. Then:

- all occurrences of l^* in the index matrices of the right or left transitions of l^*, respectively, are replaces by l;
- the characteristic function of the place l^* itself is set as a characteristic function of the place l. All occurrences of l^* in the characteristic functions of the model are replaced by l;
- all occurrences of l^* in the predicates, used to describe the right or left transitions' conditions, respectively, are replaces by l.

Since the elements of E' and the original GN E will be combined in a new model E'', in order to avoid possible conflicts, the identifiers of the elements of the GN E' (transition, places, tokens, etc.) should be different from the identifiers of the elements of E. This also applies to the characteristic functions and the predicates for the transitions' conditions in the GN-models.

The class *H3HierarchicalOperator* is another derived class of the abstract *HierarchicalOperator*. It implements the behaviour of the $\mathcal{H}_3$ hierarchical operator. The class has a data member of the following type:

- *private Transition transitionToBeReplaced*, representing the transition Z of the original GN E which should be replaced by the GN E'.

The constructor of this class also requires two input arguments, one to initialize the inherited data member *mainGN* and the other, of type *Transition*, to initialize its own data member. The validity of the input data is always checked in the constructor of the class. The transition should be an element of *mainGN*, the comparison is based on the identifiers of the transitions.

In the overridden method *changeElemWithGN* of this derived class the number of input and output places of the input argument (the GN E') is required to be equal to the number of input and output places of the transition Z of E, which should be replaced. Moreover, a mapping between the input and output places of Z and the GN E' is requested from the user. This mapping is essential in order to determine which characteristic functions will be modified, also which are the left and/or right transitions whose index matrices, predicates for transitions' conditions and positions will be affected.

The restrictions and requirements discussed above, as well as the actions that follow them, are performed in auxiliary methods of the *HierarchicalOperator* class.

- *private void moveTokens(Place oldHost, Place newHost)*{}. The method has two input parameters, the old host and the new host where the tokens should be moved to.
- *private void changeLeftTransitionIndexMatrices(Place, Transition)*{}. The method has two input parameters – the new output place and the left transition. It is responsible for changing the related labels of the index matrices, the capacities and the function references of the left transition.

- *private void changeRightTransitionIndexMatrices(Place, Transition)*{}. This method has two input parameters, similarly to the one for the left transition – the new input place and the right transition. All changes are related to the right transition.

In all cases, the time-step of the entire GN E'' should be consistent with the time components of the two GNs E and E'. The greatest common devisor of the time-steps of E and E' is an appropriate value.

The changes of the time components of the resulting GN E'' are carried out in an auxiliary method

- *private void changeTimeComponents(GeneralizedNet, int)*{}. It has two input parameters: the GN, whose time components should be changed, and an integer parameter, representing the time-step of the GN E'.

Depending on the position of the GN-element which should be replaced in the original GN E, a certain number of the remaining GN-elements are affected when it comes to visualization of the resulting GN E'' in GN IDE. The original positions of these GN-elements has to be changed. For example, new places and transitions are added, therefore, some of the elements (arcs, places, transitions) of the original model are translated.

A utility class *DrawableObjects* has been created to perform the changes in the positions of the GN-elements. The main methods of this class are:

- *public static List⟨Command⟩*
  translateObjectWithPosition(GnObjectWithPosition, int dx, int dy){}.

  The method translates an object with a position (an object of class *Point*, *Place* or *Transition*), specified as an input parameter, by a definite offset, a translation vector.
- *public static void translateArc(List⟨Point⟩ arc, int dx, int dy)*{}.
  The method translates an arc, specified in the input parameters, with a certain translation vector.
- *public static void resizeArc(List⟨Point⟩ arc, int length, int height)*{}.
  The method resizes a given arc by a specified length and height, set as input parameters.
- *public static List⟨Command⟩ connectPlacesToTransition*
  (List⟨PlaceReference⟩ places, Transition t, int gnLen,
  Point lowestPoint, int gridStep){}.
  The method adds new connecting arcs between a given transition and a list of places.

The two hierarchical operators $\mathcal{H}_1$ and $\mathcal{H}_3$ are accessible in GN IDE through the *GN* menu, *Apply operator* menu item (Fig. 1). The related menu items are active only before or after a simulation.

The main or input GN-model should be loaded first in GN IDE. The first step of the operators' execution is to check if the relevant GN-element is selected. The operator $\mathcal{H}_1$ requires a GN-place to be chosen, while the operator $\mathcal{H}_3$ expects

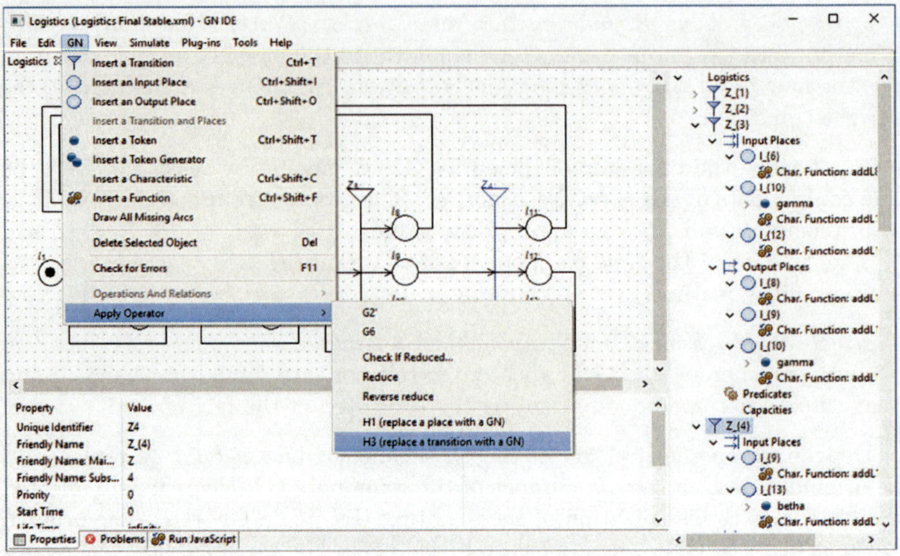

Fig. 1. H_3 hierarchical operator menu item in GN IDE

a GN-transition. A dialogue box with an error message is shown in case the selected GN-element is of the wrong type.

The restrictions on the form and the results of the work of the GN E' which should replace the selected GN-element are checked next. The GN E' should not change the functioning of the GN E.

When all prerequisites are met, the particular hierarchical operator can be applied.

Reverting operators to restore the GN-model to its original state before applying the hierarchical operators have not been implemented in GN IDE. However, the model of the input GN will not be changed unless the modifications of the operators $\mathcal{H}_1$ or $\mathcal{H}_3$ are stored in the corresponding XML file with its description. If the resulting GN is stored in a new XML file, the two GNs (original and resulting) remain independent.

4 Conclusion

GN IDE, as a software tool for creating GN-models and running simulations on such, has been a subject of development in the recent years. GN IDE has been created to assist the constant research in the GNs' theory and applications. In order to expand the possibilities of its applications, GN IDE has been extended by integrating two of the hierarchical operators of GNs' theory.

The specifics of their implementation in GN IDE have been discussed in the present paper. Using these new features a GN-model can be refined in details after replacing a GN-place or a GN-transition with an entire GN.

The hierarchical operators implemented in GN IDE can be conveniently employed for modelling the hierarchically embedded systems and organs of the human body. These new application possibilities will be explored in near future. Each of the transitions in the GN-model of the human body, published in [6], will be replaced by the GN-model of the respective human system (nervous system, cardiovascular system, respiratory system, etc.), also found in [6]. Further, the transitions in these models related to human organs will be replaced by entire GN-models using the relevant expanding hierarchical operators.

Acknowledgement. The work presented here is partially supported by the National Science Fund of Bulgaria under the Grant DN02/10 "New Instruments for Knowledge Discovery from Data, and their Modelling".

References

1. Angelova, N., Todorova, M., Atanassov, K.: GN IDE: implementation, improvements and algorithms. Comptes rendus de l'Academie Bulgare des Sciences **69**(4), 411–420 (2016)
2. Angelova, N., Zoteva, D.: Implementation of the reducing operators over generalized nets in GN IDE. Issues Intuition. Fuzzy Sets General. Nets **12**, 80–92 (2016)
3. Atanassov, K.: Generalized Nets. World Scientific, Singapore (1991)
4. Atanassov, K.: On generalized nets theory. "Prof. Marin Drinov" Academic Publishing House, Sofia (2007)
5. Atanassov, K.: Index Matrices: Towards an Augmented Matrix Calculus. Springer, Cham (2014). https://doi.org/10.1007/978-3-319-10945-9
6. Atanassov, K., Chakarov, V., Shannon, A., Sorsich, J.: Generalized Net Models of the Human Body. "Prof. M. Drinov" Academic Publishing House, Sofia (2008)
7. Atanassov, K., Sotirova, E.: Generalized Nets. "Prof. Marin Drinov" Academic Publishing House, Sofia (2017)
8. Dimitrov, D.G.: GN IDE: a software tool for simulation with generalized nets. In: Proceedings of the Tenth International Workshop on Generalized Nets, pp. 70–75 (2009)
9. Dimitrov, D., Roeva, O.: Development of generalized net for testing of different mathematical models of *E. coli* cultivation process. In: Angelov, P., et al. (eds.) Intelligent Systems'2014. AISC, vol. 322, pp. 657–668. Springer, Cham (2015). https://doi.org/10.1007/978-3-319-11313-5_58
10. Georgieva, V., Angelova, N., Roeva, O., Pencheva, T.: Simulation of parallel processes in wastewater treatment plant using generalized net integrated development environment. Comptes rendus de l'Academie bulgare des Sciences **69**(11), 1493–1502 (2016)
11. Petri, C.A.: Kommunication mit Automaten. Ph.D. thesis, University of Bonn, 1962; Schriften des Inst. fur Instrument. Math., No. 2, Bonn (1962)
12. Zoteva, D., Krawczak, M.: Generalized nets as a tool for the modelling of data mining processes. A survey. Issues Intuition. Fuzzy Sets General. Nets **13**, 1–60 (2017)

Artificial Intelligence in Biomedicine and Healthcare

Fuzzy U-Net Neural Network Design for Image Segmentation

Mark Kirichev, Todor Slavov, and Galina Momcheva(✉)

Department of Computer Science, Varna Free University "Chernorizets Hrabar", Yanko Slavchev str 84, Chayka resort, 9007 Varna, Bulgaria
galina.momcheva@vfu.bg

Abstract. One of the problems in biomedical image analysis is the problem of nuclei segmentation. The annotation of nuclei by hand has proven itself very time consuming with varying results depending on many factors. More recently convolutional neural networks have made the problem of automatic image segmentation easier, faster and more reliable. In this article, an extension to the standard U-Net is proposed which aims at improving the quality of biomedical image segmentation. By integrating Fuzzy computations in the standard U-Net architecture we have achieved even better accuracies than the ones reached by the base architecture. The Fuzzy Layers resemble a sense of uncertainty, which is already seen in the real world. This allows for a more precise detection and instance segmentation of cellular nuclei. When the original U-Net was first developed one of its focuses was to be trainable with a small dataset. This quality has been proved useful when undertaking the task of Biomedical Image Segmentation. The model was trained with the Kaggle 2018 Dataset. The segmentation process comes right after, including the following two steps: 1) creating a prediction matrix, 2) thresholding the matrix to attain a visual result. The second step exhausts the following techniques: Manual Thresholding; Adaptive Thresholding; Gaussian Thresholding, and Otsu Thresholding. The results point out which thresholding technique, combined with a certain set of Fuzzy Layers, yields the best possible results in terms of accuracy of the models.

Keywords: Deep learning · Fuzzy layer · U-net · Semantic segmentation · Otsu thresholding · Hyperparameter optimization · Keras tuner · Grid search · Automated ML

1 Image Segmentation

1.1 Description

In digital image analysis, image segmentation is the process of dividing an image into multiple segments (pixel clusters). The goal of segmentation is to give a better perspective over an image making it easier to analyze it. It can also be used for quantitative measurements. The technique can be used for object recognition and tracking. On a more basic level the process of assigning a value to every pixel in an image, such that pixels that share the same value have some qualities or characteristics in common [1].

S. S. Sotirov et al. (Eds.): BioInfoMed 2020, LNNS 374, pp. 177–184, 2022.
https://doi.org/10.1007/978-3-030-96638-6_19

1.2 Problems

Manual image segmentation is a very time consuming task – a single image can take minutes of an experienced professional, which is not desirable in the field of medicine [2]. Automated techniques significantly reduce this time – a standard CNN takes less than a second [3]. The segmentations of microscopic images are highly sensitive to experimental conditions (such as human error, level of expertise, fatigue and subjectivity), which often leads to increased variability of the final images even in slides processed in one batch. To reduce this variance we introduce Fuzzy Logic in image segmentation.

2 Fuzzy U-Net Model Architecture

2.1 The Standard U-Net

The U-Net is a convolutional neural network designed for fast and precise semantic segmentation of images. It was released in 2015 [4] and has proven itself repeatedly to be one of the best currently available methods for image segmentation. The aim of the design is to perform well even with a minimal amount of training data and still yield satisfactory results (Fig. 1).

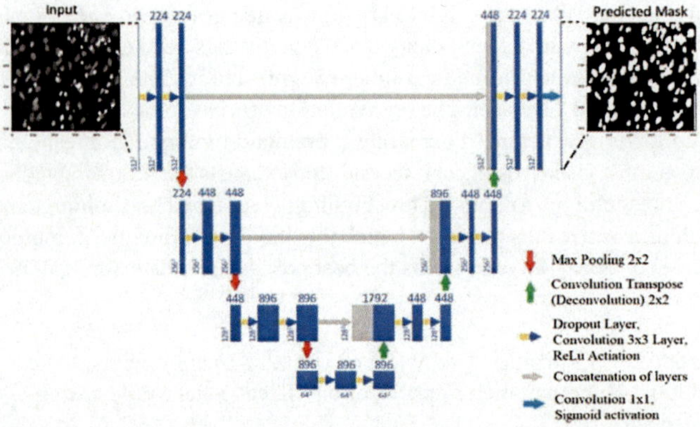

Fig. 1. Graphical representation of the design of our U-Net model, used for obtaining the hereinafter results

2.2 Initial Cross Validation

For the optimization of the parameters of the model used for predictions and for its testing, the Kaggle Dataset was divided in a ratio 9:1 [5]. The larger part has been introduced to the model in the training phase, while the smaller was used in the process of validation and comparison of true results versus predicted ones.

In order for this overfitting problem to be avoided, we have implemented the following two approaches: 1) using dropout for every layer, ensuring a relative randomness in the model. The dropout values vary with every layer, rising with the increase of depth in terms of the convolution layers. Moreover, the dropout values represented an arithmetic progression model with a first term of 0.1 and a common difference of 0.1; 2) the 90% fraction of the original dataset is kept constant, however, the particular images varied with each iteration of the training process. This random split into training and testing sets is easily achieved using the Sci-kit library and the train_test_split function. The 90% selection of the original 700 segmented nuclei images, on which the training is performed, coupled with the dropout ensure that a set of different enough inputs is introduced to the model in order for it to be optimal. Through multiple epochs, this method proves to be a key element in evaluating the best possible weights in the neural network architecture.

2.3 Integration of Fuzzy Layers into the Standard U-Net

Concatenating Fuzzy Layers with the standard U-Net architecture has proven to yield better results both accuracy-wise and visually [6]. Adding Fuzzy Layers in some key points of the model allows for a level of uncertainty, which better represents a nature-like behavior [7]. Therefore, the placement of the Fuzzy Layers in the model serves for a way for it to display different results when the Layers are differently arranged. Depending on the positioning of the Fuzzy Layers, they are taken with a different weight in the final calculations, which means there is room for optimization in the placement of said layers [8].

Table 1. Summary of trained models with 1–5 Fuzzy Layers concatenated to the U-Net.

Number of layers	1	2	3	4	5
Number of such Fuzzy architectures	15	30	30	15	3
Average accuracy	0.854288	0.852531	0.849569	0.849902	0.837194
Maximum accuracy	0.871554	0.8838943	0.877656	0.869461	0.851802
Minimum accuracy	0.823202	0.8133097	0.82635	0.829388	0.812643

In Table 1 shown above is shown the summary of all the trained models consisting of 1, 2 3, 4 or 5 integrated Fuzzy Layers. We can see that in the average case best accuracy is achieved when 1 FL is used but out of all trained models, the one that performed the best had two Fuzzy Layers.

Table 2. Summary of all the models consisting of the first through fifth Fuzzy Layer

FL activated	A	B	C	D	E
Number of such architectures trained	48	48	48	48	48
Average accuracy	0.852984	0.848939	0.85071	0.850761	0.846178
Maximum accuracy	0.877656	0.883894	0.883894	0.873607	0.877313
Minimum accuracy	0.81264	0.81264	0.81264	0.81264	0.81264

In Table 2 layers A, B, C, D, and E correspond to the Fuzzy Layers, shown on Fig. 2 starting from the layer which is visually represented as the highest one and continuing in the down direction. In Table 2. all models that have with the aforementioned (A-E) layers are summarized. The conclusion that can be drawn is that the first Fuzzy Layer has the best impact on the overall accuracy of the model. In contrast, models consisting of the fifth layer usually perform worse.

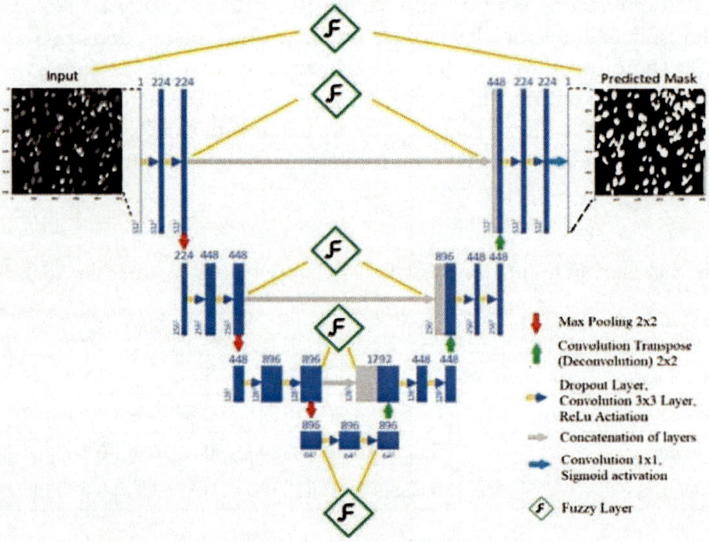

Fig. 2. Graphical representation of all the possible placements of the Fuzzy Layers around the original U-Net

The position of the Fuzzy Layers in the U-Net model in Fig. 2 has been determined to be optimal in between the different sets of convolution-dropout layers and deconvolution-dropout layers. This is due to the fact that these sets of layers form a frame of images, carrying the same number of pixels and, therefore, the same quantity of information for the neural network architecture. The Fuzzy Layers, in this case, represent an alternative path in the model, which exploits the use of the Fuzzy membership function instead of the usual Conv2D or Conv2DTranspose transformations. By concatenating the results from both paths the neural network has information of both the normal U-Net, as well as the Fuzzy Layers, which surround it.

3 Thresholding

3.1 Thresholding Methods Tested

To produce as good as possible bitmasks from the acquired predictions (Fig. 3.1) we tried different thresholding techniques such as manual thresholding, Global isodata thresholding, Adaptive Gaussian, Adaptive mean and Otsu thresholding [10]. The first method would return satisfying results only in some cases, while in others there would be no distinguishable segmentation between some nuclei. In the average scenario, the second method would yield good results but in some extreme cases would also not perform well. The third and fourth approach would also perform well but we found out that the final approach proved to be effective for all occasions. In conclusion, global methods of thresholding appear to give good bitmasks overall but not as good as Otsu's method [11] (Fig. 3.2).

Finally, a thresholding was implemented, using the following mathematical expression:

$$\begin{cases} 1, P(x, y) \geq \sigma_{\omega}(x, y) + G(x, y) * W \\ 0, P(x, y) < \sigma_{\omega}(x, y) + G(x, y) * W \end{cases},$$

where $P(x, y)$ is the value of the current pixel, $\sigma_{\omega}(x, y)$ is the Otsu value, $G(x, y)$ is the Gaussian blur factor and W is a weight, used to negate biases (Figs. 4, 5 and 6).

After experimenting with different values of W, we found out that this method gave us the most visually desirable output, which we accomplished to derive from the input. The following results prove it by highlighting better segmentation, better nuclei separation and sharper object boundaries.

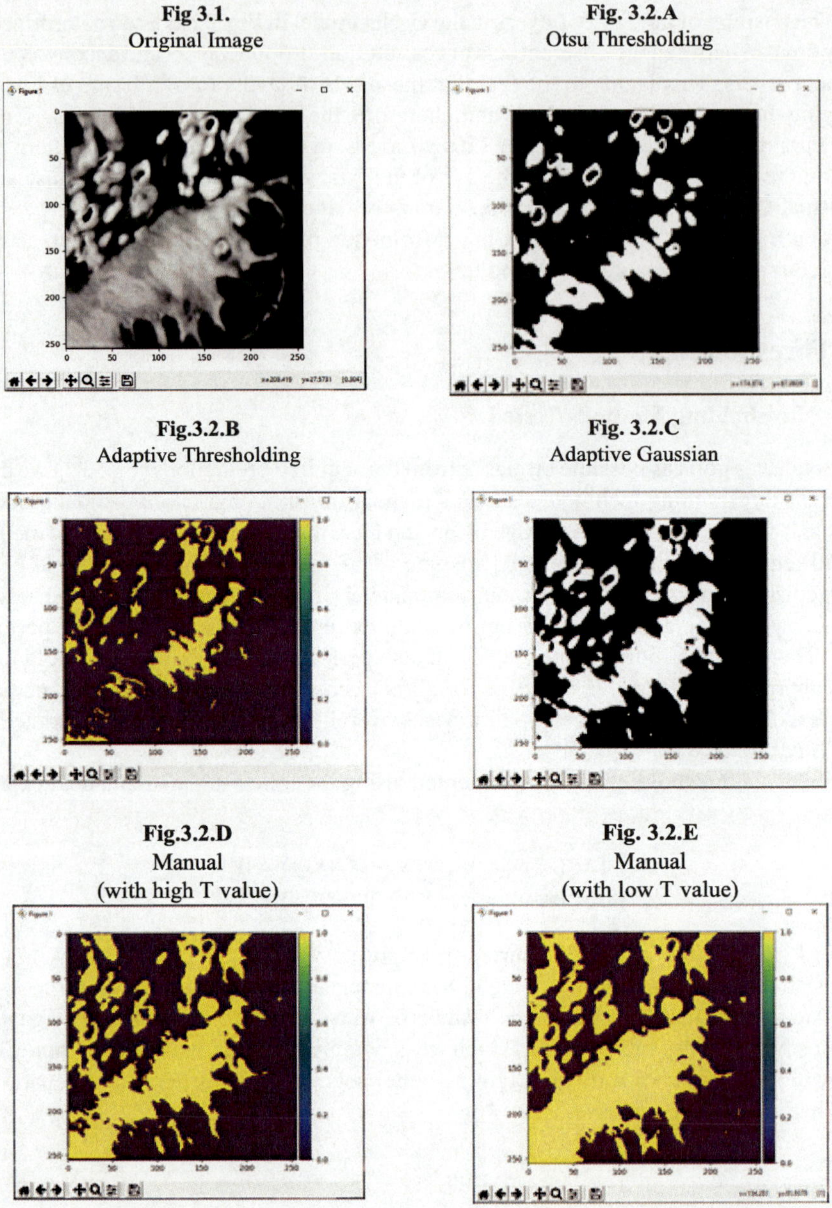

Fig. 3. Visual representation of the Original image and the types of thresholding performed

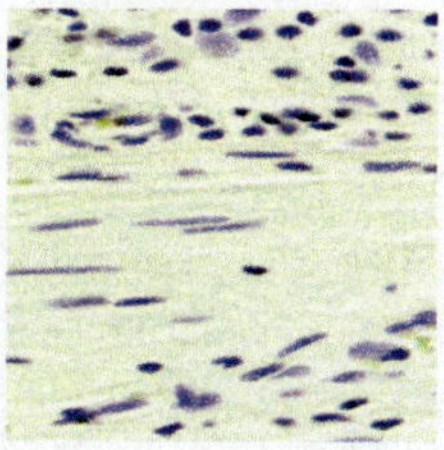

Fig. 4. Original image

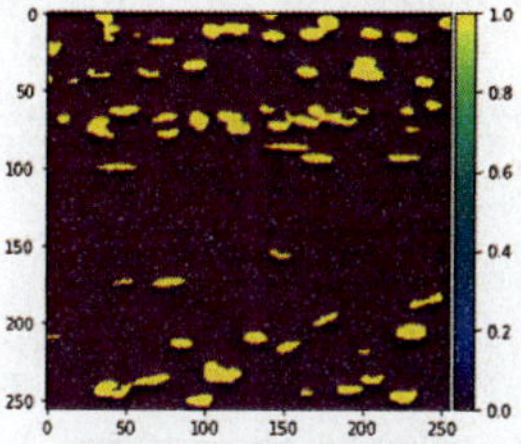

Fig. 5. Otsu method

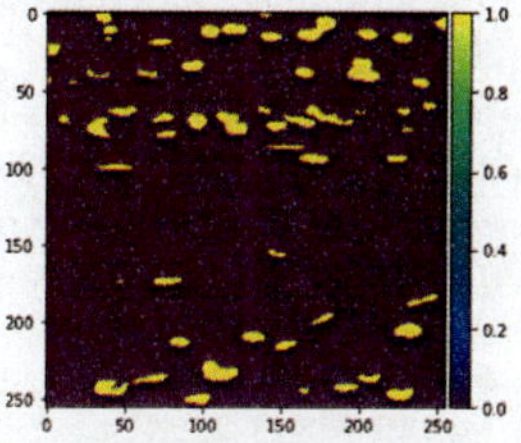

Fig. 6. Our method

3.2 Conclusion

The different placement of the Fuzzy Layers results in different model accuracies and yields varying visual representations as a result. The Fuzzy Logic helps to differentiate the groups of nuclei sharing a common part and makes those differences more visible. The accuracy values for the training of Fuzzy Layers implemented inside the U-Net architecture range from 0.81 (approx.) to 0.88 (approx.). This proves that the models depend on the location of the Fuzzy Layers and the segmentation can be further developed through the addition of a proper set of them. Moreover, the number of included Fuzzy Layers also has a major impact on the results with just above 7% accuracy difference amongst the number of Fuzzy Layers incorporated in the model.

References

1. Haralick, R., Shapiro, L.: Image segmentation techniques. Comput. Vis. Graph. Image Process. **29**(1), 100–132 (1985)

2. Rosas González, S., Birgui Sekou, T., Hidane, M., Tauber, C.: 3D automatic brain tumor segmentation using a multiscale input U-net network. In: Crimi, A., Bakas, S. (eds.) BrainLes 2019. LNCS, vol. 11993, pp. 113–123. Springer, Cham (2020). https://doi.org/10.1007/978-3-030-46643-5_11
3. Falcao, A., Udupa, A., Miyazawa, J.: An ultra-fast user-steered image segmentation paradigm: live wire on the fly. IEEE Trans. Med. Imaging **19**(1), 55–62 (2000)
4. Ronneberger, O., Fischer, P., Brox, T.: U-Net: convolutional networks for biomedical image segmentation. In: Navab, N., Hornegger, J., Wells, W.M., Frangi, A.F. (eds.) MICCAI 2015. LNCS, vol. 9351, pp. 234–241. Springer, Cham (2015). https://doi.org/10.1007/978-3-319-24574-4_28
5. Kaggle Data Science Bowl 2018 Dataset. https://www.kaggle.com/c/data-science-bowl-2018/data. Accessed 10 Oct 2020
6. Korshunova, K.: A convolutional fuzzy neural network for image classification. In: 3rd Russian-Pacific Conference on Computer Technology and Applications (RPC), Vladivostok, pp. 1–4 (2018)
7. Xi, Z., Panoutsos, G: Interpretable machine learning: convolutional neural networks with RBF fuzzy logic classification rules. In: International Conference on Intelligent Systems (2018)
8. Li, L., Jamieson, K., DeSalvo, G., Rostamizadeh, A., Talwalkar, A.: Hyperband: a novel bandit-based approach to hyperparameter optimization. J. Mach. Learn. Res. **18**, 6765–6816 (2018)
9. Springenberg, J., Klein, A., Falkner, S., Hutter, F.: Bayesian optimization with robust Bayesian neural networks. In: Advances in Neural Information Processing Systems, vol. 29, pp. 4134–4142. Curran Associates, Inc. (2016)
10. Goh, T., Basah, S., Yazid, H., Safar, M., Saad, F.: Performance analysis of image thresholding: Otsu technique. Meas. J. Int. Meas. Confed. **114**, 298–307 (2018)
11. Gong, J., Li, L., Chen, W.: Fast recursive algorithms for two-dimensional thresholding. Pattern Recogn. **31**(3), 295–300 (1998)

Effects of Variable Impulsive Perturbations on the Stability of Fractional-Order Cohen–Grossberg Neural Networks with Respect to Functions

Ivanka Stamova[1], Sotir Sotirov[2](✉), Stanislav Simeonov[2], and Gani Stamov[1]

[1] College of Science, University of Texas at San Antonio, One UTSA Circle, San Antonio, TX 78249, USA
{ivanka.stamova,gani.stamov}@utsa.edu

[2] Department of Computer Systems and Technologies, Burgas "Prof. Dr. Assen Zlatarov" University, Prof. Yakimov 1 str., 8010 Burgas, Bulgaria
ssotirov@btu.bg

Abstract. In is paper, we study the problems of stability of the equlibrium with respect to the manifold defined by a function for impulsive fractional-order Cohen-Grossberg neural networks. The effects of variable impulsive perturbations are investigated. The impulses are realized as continuous functions and can be considered as a control. The main results are obtained by employing the Lyapunov method and comparison principle.

Keywords: σ-manifold · Mittag–Leffler stability · Variable impulses · Cohen–Grossberg neural networks · Lyapunov method

1 Introduction

The class of Cohen–Grossberg neural networks (CGNNs) introduced in 1983 [1] has been intensively studied in the existing literature because of the great opportunities for applications in emerging areas as associative memory, parallel computing, pattern recognition, signal and image processing, etc. that are important to science, engineering, biology and medicine.

On the other side, because of the boundless opportunities for modelling applications and generalizations, fractional-order derivatives have been effectively introduced in numerous real-world models [2–5]. In fact, it has been theoretically and experimentally proven that fractional-order derivatives can capture non-local relations as well as memory and hereditary properties the processes. However, the results on fractional-order CGNNs are still very few [6–8] and the theory of this important class of models is still under the development.

S. S. Sotirov et al. (Eds.): BioInfoMed 2020, LNNS 374, pp. 185–194, 2022.
https://doi.org/10.1007/978-3-030-96638-6_20

In addition, in the existing results on fractional CGNNs impulsive effects have been ignored so far. Note that, the behavior of integer-order impulsive CGNNs have been studied by some authors [9–12]. However, corresponding results on impulsive CGNNs with fractional derivatives do not exist, and this is the main goal of our research.

In is paper, we study a stability problem of the equilibrium with respect to manifolds defined by a function for impulsive fractional-order Cohen–Grossberg neural networks. The effects of variable impulsive perturbations are investigated. The impulses are realized as continuous functions and can be considered as a control. Indeed, since impulsive phenomena exist in numerous fields of science and engineering [13–15] and the impulsive control methods are find to be more efficient than other control strategies, there has been an increasing research interest on the impulsive generalizations of different neural network systems.

In addition, the uncertain model, corresponding to the introduced impulsive fractional CGNN system is presented and the robust stability result is established. The main tools in our study are the Lyapunov function method and the comparison principle.

2 Preliminaries

We will use the following notations: $\mathbb{R}^n$ denotes the n-dimensional Euclidean space with the norm $||x|| = \sum_{i=1}^{n} |x_i|$ and $\mathbb{R}_+ = [0, \infty)$. In our research we will apply the Caputo-type fractional-order derivatives defined by the following definition.

Definition 1. [4] *For any $t \geq t_0$, $t_0 \in \mathbb{R}_+$, the Caputo fractional derivative of order q, $0 < q < 1$ with the lower limit t_0 for a function $l \in C^1[[t_0, b], \mathbb{R}]$, $b > t_0$, is defined as*

$$ {}_{t_0}^{C}D_t^q l(t) = \frac{1}{\Gamma(1-q)} \int_{t_0}^{t} \frac{l'(s)}{(t-s)^q} ds, $$

where Γ is the Gamma function given by

$$ \Gamma(q) = \int_0^{\infty} e^{-t} t^{q-1} dt. $$

A specific type of the fractional-order CGNN models introduced in [6–8] is the following system

$$ {}_{t_0}^{C}D_t^q x_i(t) = -a_i(x_i(t)) \left[b_i(x_i(t)) - \sum_{j=1}^{n} c_{ij} f_j(x_j(t)) + I_i \right], \tag{1} $$

where $t \geq 0$, $x = (x_1, x_2, \ldots, x_n)^T$, $x_i(t)$ denotes the state variable of the ith neuron at time t, the number of neurons $n \geq 2$, the model parameters $c_{ij}, I_i \in \mathbb{R}$, the functions $a_i, b_i, f_i \in C[\mathbb{R}, \mathbb{R}]$, $i, j = 1, \ldots, n$. Particularly, a_i represents the amplification function for the ith neuron, b_i is a well behaved function, c_{ij} are the synaptic strengths, I_i represents a constant external input, $i, j = 1, \ldots, n$.

In our research we will extend the model (1) to an impulsively controlled model.

Let $x_0 = (x_{10}, x_{20}, \ldots, x_{n0}) \in \mathbb{R}^n$ and let $x(t)$, where

$$x(t) = x(t; t_0, x_0) = (x_1(t; t_0, x_0), x_2(t; t_0, x_0), \ldots, x_n(t; t_0, x_0))^T$$

be the solution of system (1) that satisfies the initial condition $x(t_0) = x_0$.

We will consider the case when at some moments of time the integral curve of the solution $x(t)$ of (1) meets the hypersurfaces

$$\sigma_k = \big\{(t, x) \in [t_0, \infty) \times \mathbb{R}^n : \ t = \tau_k(x)\big\}, \ k = 1, 2, \ldots,$$

where $\tau_0(x) \equiv t_0$, the functions $\tau_k : \mathbb{R}^n \to \mathbb{R}_+$, $k = 1, 2, \ldots$.

The solutions of the equations $t = \tau_k(x(t))$, $k = 1, 2, \ldots$, will be denoted by t_{l_k}. Let us note that, in general, $k \neq l_k$. The points $t_{l_1}, \ t_{l_2}, \ldots \ (t_0 < t_{l_1} < t_{l_2} < \ldots)$ will be called impulsive moments.

The corresponding impulsively controlled system is designed as

$$\begin{cases} {}_{t_0}^{C}D_t^q y_i(t) = -a_i(y_i(t)) \left[b_i(y_i(t)) - \sum\limits_{j=1}^{n} c_{ij} f_j(y_j(t)) + I_i \right], \quad t \neq \tau_k(y(t)), \\ \\ \Delta y_i(t) = y_i(t^+) - y_i(t) = P_{ik}(y_i(t)), \ t = \tau_k(y(t)), \end{cases} \tag{2}$$

where $t \geq t_0$, $\Delta y_i(t)$ is the state jumping of the ith neuron at an impulsive moment which is a solution of one of the equations $t = \tau_k(y(t))$, $y_i(t^-) = y_i(t)$ and $y_i(t^+)$ are, respectively, the states of the ith neuron before and after the sudden changes at the impulsive times, the functions $P_{ik} \in C[\mathbb{R}, \mathbb{R}]$, $i = 1, 2, \ldots, n$, $k = 1, 2, \ldots$.

The initial condition, associated with the impulsive model (2) is

$$y(t_0^+) = y_0, \ y_0 \in \mathbb{R}^n. \tag{3}$$

By means of the impulsively controlled response system (2) we can study and manipulate the behavior of the fractional-order CGNN (1). The solutions $y(t)$ of problems of the type (2), (3) are [5,13–15] piecewise continuous functions with points of discontinuity of the first kind at which they are left continuous and satisfy the following relations:

$$y_i(t_{l_k}^-) = y_i(t_{l_k}), \ y_i(t_{l_k}^+) = y_i(t_{l_k}) + P_{il_k}(y_i(t_{l_k})).$$

Let $\sigma = \sigma(t, y)$, $\sigma : [t_0, \infty) \times \mathbb{R}^n \to \mathbb{R}$ be a function in $[t_0, \infty) \times \mathbb{R}^n$. Define the sets:

$$M_t = \{y \in \mathbb{R}^n : \ \sigma(t, y) = 0, \ t \in [t_0, \infty)\},$$
$$M_t(\varepsilon) = \{y \in \mathbb{R}^n : \ ||\sigma(t, y)|| < \varepsilon, \ t \in [t_0, \infty)\}, \ \varepsilon > 0.$$

In order to investigate the stability behavior of the solutions of the fractional CGNN system (2) with respect to the manifolds defined by the function σ, we will need the following assumptions:

H2.1. The function a_i, $i = 1, 2, \ldots, n$ are continuous and bounded and there exist positive constants $\underline{a}_i$ and $\overline{a}_i$ such that $1 < \underline{a}_i \leq a_i(t) \leq \overline{a}_i$ for $t \in \mathbb{R}$.

H2.2. For the continuous functions b_i there exist positive constants B_i such that

$$\frac{b_i(\chi_1) - b_i(\chi_2)}{\chi_1 - \chi_2} \geq B_i,$$

for any $\chi_1, \chi_2 \in \mathbb{R}$, $\chi_1 \neq \chi_2$ and $i = 1, 2, \ldots, n$.

H2.3. There exist constants $L_i > 0$ such that the activation functions f_i satisfy

$$|f_i(\chi_1) - f_i(\chi_2)| \leq L_i|\chi_1 - \chi_2|$$

for all $\chi_1, \chi_2 \in \mathbb{R}$, $\chi_1 \neq \chi_2$, $i = 1, 2, \ldots, n$.

H2.4. The function σ is continuous on $[t_0, \infty) \times \mathbb{R}^n$ and the set M_t, is an $(n-1)$-dimensional manifold in $\mathbb{R}^n$.

H2.5. Each solution $y(t)$ of the initial value problem (2), (3) satisfying

$$||\sigma(t, y(t; t_0, y_0))|| \leq H < \infty$$

is defined on $[t_0, \infty) \times \mathbb{R}^n$.

H2.6. The functions $\tau_k(y)$, $k = 1, 2, \ldots$, are continuous, and the following relations hold

$$t_0 < \tau_1(y) < \tau_2(y) < \ldots, \ \tau_k(y) \to \infty \text{ as } k \to \infty$$

uniformly on $y \in \mathbb{R}^n$.

Note that condition H2.6 is very important for impulsive control systems with variable impulsive perturbations. For such systems, different solutions may have different impulsive moments, and phenomena as "merging" the solutions after an impulsive event, or "beating" of solutions are, in general, possible. Condition H2.6 guarantees the absence of such phenomena, as well as, the existence of an equilibrium state of the model (2) $y^* = (y_1^*, y_2^*, \cdots, y_n^*)^T$, $y_i^* = const$, $i = 1, 2, \ldots, n$ given as a solution of the system

$$\begin{cases} b_i(y_i^*) - \sum_{j=1}^{n} c_{ij} f_j(y_j^*) + I_i = 0, \\ P_{ik}(y_i^*) = 0, \ k = 1, 2, \ldots. \end{cases} \tag{4}$$

By definition, the state y^* is an equilibrium for the model (1), too. For some efficient sufficient conditions for the existence of such states for fractional-order CGNNs we refer to [6–8].

We will introduce a definition of global Mittag–Leffler stability of the equilibrium y^* of (2) with respect to the function σ (or with respect to the manifold defined by this function), which generalizes definitions given in [16].

Definition 2. *The equilibrium y^* of the fractional CGNN system (2) is said to be globally Mittag–Leffler stable with respect to the function σ if for $t_0 \in \mathbb{R}_+$ and $y_0 \in \mathbb{R}^n$ there exists a constant $\mu > 0$ such that*

$$y(t) - y^* \in M_t\big(m(y_0 - y^*)E_q(-\mu(t-t_0)^q)\big), \quad t \geq t_0,$$

where E_q is the corresponding Mittag–Leffler function given by

$$E_q(z) = \sum_{\kappa=0}^{\infty} \frac{z^\kappa}{\Gamma(q\kappa+1)},$$

$m(0) = 0$, $m(y)$ is Lipschitz continuous with respect to $y \in \mathbb{R}^n$, and $m(y) \geq 0$.

To translate the stability analysis of the equilibrium y^* to this of the origin, we set $u_i(t) = y_i(t) - y_i^*$, $i = 1, 2, \ldots, n$, $t \geq t_0$, and consider the following system

$$\begin{cases} {}_{t_0}^{C}D_t^q u_i(t) = -\alpha_i(u_i(t))\left[\beta_i(u_i(t)) - \sum_{j=1}^{n} c_{ij}\hat{f}_j(u_j(t))\right], \; t \neq \tau_k(u(t)+y^*), \\ \Delta u_i(t) = u_i(t^+) - u_i(t) = \hat{P}_{ik}(u_i(t)), \; t = \tau_k(u(t)+y^*), \\ u(t_0^+) = y_0 - y^* = u_0, \end{cases} \tag{5}$$

where $t \geq t_0$, $\alpha_i(u_i(t)) = a_i(u_i(t) + y_i^*)$, $\beta_i(u_i(t)) = b_i(u_i(t) + y_i^*) - b_i(y_i^*)$, $\hat{f}_j(u_j(t)) = f_j(u_j(t)+y_j^*) - f_j(y_j^*)$, $\hat{P}_{ik}(u_i(t)) = P_{ik}(u_i(t)+y_i^*)$, $i, j = 1, 2, \ldots, n$, $k = 1, 2, \ldots$.

Denote by $u(t; t_0, u_0)$, $u \in \mathbb{R}^n$, the solution of system (5), and define the following sets

$$\mathcal{G}_k = \big\{(t,u) \in [t_0, \infty) \times \mathbb{R}^n : \tau_{k-1}(u+y^*) < t < \tau_k(u+y^*)\big\}, \; k = 1, 2, \ldots,$$

$$\mathcal{G} = \bigcup_{k=1}^{\infty} \mathcal{G}_k$$

and in the future considerations, we will adopt the Lyapunov functions approach. That is why in the proofs of our main results we will use the class of piecewise continuous auxiliary functions given as:

$V_0 = \big\{V : [t_0, \infty) \times \mathbb{R}^n \to \mathbb{R}_+ : V \in C[G, \mathbb{R}_+]$, V is locally Lipschitz with respect to its second argument on each of the sets $\mathcal{G}_k$; $V(t, 0) = 0$ for $t \geq t_0$, $V(\tau_0^{*-}, u) = \lim_{\substack{t \to \tau_0^* \\ t < \tau_0^*}} V(t,u)$ and $V(\tau_0^{*+}, u) = \lim_{\substack{t \to \tau_0^* \\ t > \tau_0^*}} V(t,u)$, exist, and $V(\tau_0^{*-}, u) = V(\tau_0^*, u)$ for any $\tau_0^* \geq t_0$ and $u \in \mathbb{R}^n\big\}$.

Let for simplicity denote

$$F(t,u) = -\alpha_i(u_i(t))\left[\beta_i(u_i(t)) - \sum_{j=1}^{n} c_{ij}\hat{f}_j(u_j(t))\right].$$

For $V \in V_0$. For $(t,u) \in \mathcal{G}$ we define the following derivative of V with respect to system (5) defined by

$$ {}^{C}D^{q}_{+}V(t,u) = \lim_{\chi\to 0^+} \sup \frac{1}{\chi^q}\big[V(t,u) - V(t-\chi, u - \chi^q F(t,u))\big], $$

where $0 < q < 1$.

The following lemma from [5] will be also important.

Lemma 1. *Assume that the function $V \in V_0$ is such that for $t \in [t_0, \infty)$,*

$$ V(t^+, u + \Delta(u)) \leq V(t,u), \quad t = \tau_k(u + y^*), \quad k = 1,2,\ldots $$
$$ {}^{C}D^{q}_{+}V(t,u) \leq -\mu V(t,u), \ t \neq \tau_k(u+y^*), \ k = 1,2,\ldots. $$

Then

$$ V(t, u(t; t_0, u_0)) \leq V(t_0^+, u_0) E_q(-\mu(t-t_0)^q) $$

for $t \in [t_0, \infty)$.

3 Main Results

Theorem 1. *Assume that:*

1. *Assumptions H2.1–H2.6 are fulfilled.*
2. *There exists a positive number*

$$ \mu^* = \underline{a} \min_{1\leq i\leq n} \frac{\overline{a}_i}{\underline{a}_i}\Big(B_i - \sum_{j=1}^{n} |c_{ji}| L_j\Big) $$

where $\underline{a} = \min_{1\leq i\leq n} a_i$.

3. *The functions $\hat{P}_{ik}$ are such that*

$$ \hat{P}_{ik}(u_i(t)) = -\gamma_{ik} u_i(t), \ t \neq \tau_k(u+y^*), \ 0 < \gamma_{ik} < 2, $$

and

$$ \underline{a}|1-\gamma_{ik}| \leq |1 - \hat{\gamma}_k| \leq \frac{\underline{a}}{\overline{a}}, $$

where $\hat{\gamma}_k = \min_{1\leq i\leq n} \gamma_{ik}$, $i = 1,2,\ldots,n$, $k = 1,2,\ldots$, $\overline{a} = \max_{1\leq i\leq n} a_i$.

4. *For the function $\sigma(t,u)$ the next inequalities hold*

$$ ||\sigma(t,u)|| \leq \sum_{i=1}^{n} \int_0^{u_i(t)} \frac{sgn(s)}{\alpha_i(s)} ds \leq \Lambda(H)||\sigma(t,u)||, \quad t \in [t_0, \infty), $$

where $u = (u_1, u_2, \ldots, u_n)^T$, $\Lambda(H) \geq 1$ exists for any $0 < H \leq \infty$.

Then for any $\mu \in (0, \mu^*)$ the equilibrium y^* of the impulsive fractional CGNN (2) is globally Mittag–Leffler stable with respect to the function σ.

Proof. Let $y(t) = (y_1(t), y_2(t), \dots, y_n(t))^T$ be any solution of (2) with an initial value $y_0 \in \mathbb{R}^n$ and $u(t) = y(t) - y^*$, $t \geq t_0$.

Consider the Lyapunov-type function

$$V(t, u(t)) = \sum_{i=1}^{n} \int_0^{u_i(t)} \frac{sgn(s)}{\alpha_i(s)} ds,$$

where $\alpha_i(u_i(t)) = a_i(u_i(t) + y_i^*)$.

It is easy to prove that

$$\frac{1}{\overline{a}}||u(t)|| \leq V(t, u(t)) \leq \frac{1}{\underline{a}}||u(t)||, \ t \in \mathbb{R}_+. \tag{6}$$

Then, for $t = \tau_k(u + y^*)$, $k = 1, 2, \dots$, condition 3 of Theorem 1 and (6) yield

$$\begin{aligned} V(t^+, u(t^+)) &\leq \frac{1}{\underline{a}}||u(t^+)|| = \frac{1}{\underline{a}}\sum_{i=1}^{n}|u_i(t^+)| = \frac{1}{\underline{a}}\sum_{i=1}^{n}(1 - \gamma_{ik})|u_i(t)| \\ &< \sum_{i=1}^{n}|1 - \gamma_{ik}||u_i(t)| \leq \frac{|1 - \hat{\gamma}_k|}{\underline{a}}\sum_{i=1}^{n}|u_i(t)| \\ &\leq \frac{1}{\overline{a}}\sum_{i=1}^{n}|u_i(t)| = \frac{1}{\overline{a}}||u(t)|| \leq V(t, u(t)). \end{aligned} \tag{7}$$

Let $t \geq t_0$, $\tau_{k-1}(u + y^*) < t < \tau_k(u + y^*)$, $k = 1, 2, \dots$. Then, we have

$$\begin{aligned} {}^{C}D_{+}^{q}V(t, u(t)) &= -\sum_{i=1}^{n} {}_{t_0}^{C}D_t^q\Big(\int_0^{u_i(t)} \frac{sgn(s)}{\alpha_i(s)} ds\Big) \\ &= \sum_{j=1}^{n} \frac{1}{\Gamma(1-q)} \int_{t_0}^{t} \Big[\int_0^{u_i(\eta)} \frac{sgn(s)}{\alpha_i(s)} ds\Big]' \frac{1}{(t-\eta)^q} d\eta \\ &= \sum_{j=1}^{n} \frac{1}{\Gamma(1-q)} \int_{t_0}^{t} \frac{sgn(u_i(\eta))}{\alpha_i(y_i(\eta))} \frac{[u_i(\eta)]'}{(t-\eta)^q} d\eta \\ &\leq \sum_{i=1}^{n} sgn(y_i(t)) \frac{\overline{a}_i}{\underline{a}_i}\Big(- \beta_i(u_i(t)) + \sum_{j=1}^{n} c_{ij}\hat{f}_j(u_j(t))\Big). \end{aligned}$$

Using (6) and conditions 1 and 2 of Theorem 1, we get

$$\begin{aligned} {}^{C}D_{+}^{q}V(t, u(t)) &\leq -\sum_{i=1}^{n} \frac{\overline{a}_i}{\underline{a}_i}\Big(B_i|u_i(t)| - \sum_{j=1}^{n}|c_{ij}|L_j|u_j(t)|\Big) \\ &= -\sum_{i=1}^{n} \frac{\overline{a}_i}{\underline{a}_i}\Big(B_i - \sum_{j=1}^{n}|c_{ji}|L_j|\Big)|u_i(t)| \leq -\mu V(t, u(t)), \ t \neq \tau_k(u + y^*). \end{aligned} \tag{8}$$

Then by virtue of (7), -(8), (3.3) and Lemma 1, we have

$$V(t, u(t)) \leq V(t_0^+, u_0)\big(E_q(-\mu(t-t_0)^q\big), \ t > t_0.$$

Hence, from condition 4 of Theorem 1 we obtain

$$\begin{aligned}||\sigma(t, u(t, t_0, u_0)|| &\leq V(t, u(t)) \leq V(t_0^+, u_0)\big(E_q(-\mu(t-t_0)^q\big) \\ &\leq \Lambda(H)||\sigma(t_0^+, y_0 - y^*)||\big(E_q(-\mu(t-t_0)^q\big), \quad t \geq t_0.\end{aligned}$$

Then

$$||\sigma(t, y(t, t_0, y_0) - y^*|| \leq \mathcal{M}\big(E_q(-\mu(t-t_0)^q\big), \quad t \geq t_0,$$

where $\mathcal{M} = \mathcal{M}(y_0 - y^*) > \Lambda(H)||\sigma(t_0^+, y_0 - y^*)|| \geq 0$ and $\mathcal{M} = 0$ holds only if $\sigma(t_0^+, y_0 - y^*) = 0$.

From the last estimate it follows that the equilibrium y^* (2) is globally Mittag–Leffler stable with respect to the function σ and the proof is complete.

Remark 1. Theorem 1 offers an impulsive control rule for the stability behavior of the fractional-order CGNN system (1). The result is a generalization of the stability and stabilization results established in [6–8] to the impulsive case.

Remark 2. The stability with respect to function σ notion extends the Lyapunov-type stability notions. When the function $\sigma(t, u) = ||u||$, $t \geq t_0$, the obtained results will be reduced to Lyapunov stability of the equilibrium y^* of the impulsive control fractional CGNN system (2).

Finally, to evaluate the effects of uncertain parameters in the impulsive control system (2), we will consider the uncertain case corresponding to (2) given by

$$\begin{cases} {}_{t_0}^{C}D_t^q y_i(t) = -a_i(y_i(t))\Big[b_i(y_i(t)) - \sum\limits_{j=1}^{n}\big(c_{ij} + \tilde{c}_{ij}\big) f_j(y_i(t)) \\ \qquad\qquad + I_i + \tilde{I}_i\Big], \quad t \neq \tau_k(y(t)), \\ y_i(t^+) = y_i(t) + \Big(P_{ik}(y_i(t)) + \tilde{P}_{ik}(y_i(t))\Big), \ t = \tau_k(y(t)), \end{cases} \tag{9}$$

where the constants $\tilde{c}_{ji}$, $\tilde{I}_i$ and functions $\tilde{P}_{ik} \in C[\mathbb{R}, \mathbb{R}]$, $i, j = 1, \ldots, n$, $k = 1, 2, \ldots$, represent the uncertain parameters in the fractional CGNN system [17, 18].

Remark 3. The study of the effects of uncertain parameters that represent uncertain factors on the behavior of neural networks is of a great interest to researchers. Robust stability analysis for integer-order CGNNs with uncertain parameters has been conducted by several authors. See, for example, [19–21] and the references therein. However, the investigations of uncertain CGNNs of fractional order still do not exist.

Definition 3. *The equilibrium y^* of (2) is said to be globally robustly Mittag-Leffler stable with respect to the function σ if for $t_0 \in \mathbb{R}_+$, $y_0 \in \mathbb{R}^n$ and for any $\tilde{c}_{ji}$, $\tilde{I}_i$, $\tilde{P}_{ik}$ $i, j = 1, \ldots, n$ the equilibrium of (9) is globally Mittag–Leffler stable with respect to the function σ.*

The proof of the next theorem follows directly from Theorem 1.

Theorem 2. *Assume that:*

1. *The conditions of Theorem 1 hold.*
2. *There exists a positive number*

$$\mu^{**} = \underline{a} \min_{1 \le i \le n} \frac{\overline{a}_i}{\underline{a}_i} \Big(B_i - \sum_{j=1}^{n} |c_{ji} + \tilde{c}_{ji}| L_j \Big)$$

   where $\underline{a} = \min_{1 \le i \le n} a_i$.
3. *The unknown functions $\tilde{P}_{ik}$ are such that $\tilde{P}_{ik}(u_i) = \tilde{\gamma}_{ik} u_i$ and $0 < \tilde{\gamma}_{ik} < 2 - \gamma_{ik}$, $i = 1, 2, \ldots, n$, $k = 1, 2, \ldots$.*

Then, for any $\mu \in (0, \mu^{**})$ the equilibrium y^* of system (2) is globally robustly Mittag–Leffler stable with respect to the function σ.

4 Conclusion

In this paper, an impulsive control CGNN system of fractional order is developed. The notion of stability of the equlibrium with respect to the manifold defined by a function is adopted to the introduced model. The effects of variable impulsive perturbations on the stability behavior are investigated. Our results are new and extend the results on fractional CGNN models [6–8] to the impulsive case. In addition, the uncertain parameters are also considered and a robust stability result is established. The Lyapunov function method and comparison results are the main tools in the investigation.

Acknowledgements. This research was funded in part by the European Regional Development Fund through the Operational Program "Science and Education for Smart Growth" under contract UNITe No BG05M2OP001–1.001–0004 (2018–2023).

References

1. Cohen, M.A., Grossberg, S.: Absolute stability of global pattern formation and parallel memory storage by competitive neural networks. IEEE Trans. Syst. Man Cybern. **13**, 815–826 (1983)
2. Baleanu, D., Diethelm, K., Scalas, E., Trujillo, J.: Fractional Calculus: Models and Numerical Methods. World Scientific, Hackensack (2017)
3. Hilfer, R.: Applications of Fractional Calculus in Physics. World Scientific, Hackensack (2001)
4. Podlubny, I.: Fractional Differential Equations. Academic Press, San Diego (1999)

5. Stamova, I.M., Stamov, G.T.: Functional and Impulsive Differential Equations of Fractional Order: Qualitative Analysis and Applications. CRC Press/Taylor and Francis Group, Boca Raton (2017)
6. Ke, Y., Miao, C.: Stability analysis of fractional-order Cohen-Grossberg neural networks with time delay. Int. J. Comput. Math. **92**(6), 1102–1113 (2015)
7. Pratap, A., Raja, R., Cao, J., Lim, C.P., Bagdasar, O.: Stability and pinning synchronization analysis of fractional order delayed Cohen-Grossberg neural networks with discontinuous activations. Appl. Math. Comput. **359**, 241–260 (2019)
8. Rajivganthi, C., Rihan, F.A., Lakshmanan, S., Muthukumar, P.: Finite-time stability analysis for fractional-order Cohen-Grossberg BAM neural networks with time delays. Neural Comput. Appl. **29**, 1309–1320 (2018). https://doi.org/10.1007/s00521-016-2641-9
9. Bai, C.: Stability analysis of Cohen-Grossberg BAM neural networks with delays and impulses. Chaos, Solitons Fractals **35**, 263–267 (2008)
10. Li, K., Zeng, H.: Stability in impulsive Cohen-Grossberg-type BAM neural networks with time-varying delays: a general analysis. Math. Comput. Simul. **80**, 2329–2349 (2010)
11. Li, X.: Exponential stability of Cohen-Grossberg-type BAM neural networks with time-varying delays via impulsive control. Neurocomputing **73**, 525–530 (2009)
12. Stamov, G., Stamova, I., Simeonov, S., Torlakov, I.: On the stability with respect to h-manifolds for Cohen-Grossberg-type bidirectional associative memory neural networks with variable impulsive perturbations and time-varying delays. Math. (MDPI) **8**, 335 (2020)
13. Haddad, W.M., Chellaboina, V.S., Nersesov, S.G.: Impulsive and Hybrid Dynamical Systems, Stability, Dissipativity, and Control. Princeton University Press, Princeton (2006)
14. Liu, X., Zhang, K.: Impulsive Systems on Hybrid Time Domains. Springer, Cham (2019). https://doi.org/10.1007/978-3-030-06212-5
15. Stamova, I.M., Stamov, G.T.: Applied Impulsive Mathematical Models. Springer, Cham (2016). https://doi.org/10.1007/978-3-319-28061-5
16. Li, Y., Chen, Y.Q., Podlubny, I.: Stability of fractional-order nonlinear dynamic systems: Lyapunov direct method and generalized Mittag-Leffler stability. Comput. Math. Appl. **59**, 1810–1821 (2010)
17. Liu, B., Liu, X., Liao, X.: Robust stability of uncertain impulsive dynamical systems. J. Math. Anal. Appl. **290**, 519–533 (2004)
18. Stamov, G.T., Stamova, I.M., Cao, J.: Uncertain impulsive functional differential systems of fractional order and almost periodicity. J. Franklin Inst. **355**(12), 5310–5323 (2018)
19. Chen, T., Rong, L.: Robust global exponential stability of Cohen-Grossberg neural networks with time delays. IEEE Trans. Neural Netw. **15**(1), 203–206 (2004)
20. Wan, Y., Cao, J., Wen, G., Yu, W.: Robust fixed-time synchronization of delayed Cohen-Grossberg neural networks. Neural Netw. **73**, 86–94 (2016)
21. Yuan, K., Cao, J., Li, H.-X.: Robust stability of switched Cohen-Grossberg neural networks with mixed time-varying delays. IEEE Trans Syst. Man Cybern. **36**(6), 1356–1363 (2006)

Visualization on Stability of Impulsive Cohen-Grossberg Neural Networks with Time-Varying Delays

Gani Stamov[1], Stanislav Simeonov[2], and Ivan Torlakov[2](✉)

[1] Department of Mathematics, Technical University of Sofia, Burgasko Shose Street, 8800 Sliven, Bulgaria
stamov@tu-sofia.bg

[2] Department of Computer Systems and Technologies, "Prof. Dr. Asen Zlatarov" University, Prof. Yakinov 1 Street, 8010 Burgas, Bulgaria

Abstract. Neural networks of type Cohen-Grossberg (CGNNs) were introduced by Cohen and Grossberg in [1]. Since it releases in 1983 many scientists were working for application of such models in different research fields. One of the most important roles of such systems is the stability it provides within the potential application therefore it's necessary to be investigated. In real-world, both biological and artificial neural networks have time delays due mainly to the limited speed of signal transmissions and amplifiers switching. Such time delays may affect the dynamic behavior of the network making it unstable or diverge.

Due to the countless different possibilities for modelling, behaviors of impulsive CGNNs have been investigated by many researchers [3–9, 11–13, 23]. Some of the examples, as theoretical results, given in those results [3, 5–8] are used in this paper as real-world neural networks in order to be viewed their graphical representation of a computer simulation. The effect of the time-varying delays over the stability of the system is investigated.

It is believed that these graphical results are useful for the design and exploration of impulsive CGNNs by researchers and students.

Keywords: Cohen-Grossberg neural networks · Delay · Stability · Time-varying delays · Graphical results

1 Introduction

Neural networks of type Cohen-Grossberg were introduced by Cohen and Grossberg in [1]. Since it releases in 1983 many scientists were working for application of such models [2–9, 12–18, 23] in different research fields [10, 14, 19–22, 24, 25] and had impacts on such as neurobiology, bioinformatics, etc. One of the most important roles of such systems is the stability it provides within the potential application therefore it's necessary to be investigated. In real-world, both biological and artificial neural networks have time delays due mainly to the limited speed of signal transmissions and amplifiers

S. S. Sotirov et al. (Eds.): BioInfoMed 2020, LNNS 374, pp. 195–201, 2022.
https://doi.org/10.1007/978-3-030-96638-6_21

switching. Such time delays may affect the dynamic behavior of the network making it unstable or diverge.

In this paper, we consider the following CGNN model with time-varying delays which is described by the following equations:

$$\begin{cases} \frac{dx_i}{dt} = -a_i(x_i(t))\left[b_i\left(x_i(t) - \sum_{j=1}^{2} c_{ij}f_j\left(x_j(t) - \tau_{ij}(t)\right)\right) + J_i\right], t \neq t_k, t \geq 0 \\ \Delta x_i = x_i\left(t_k^+\right) - x_i\left(t_k^-\right) = I_{ik}\left(x_i\left(t_k^-\right)\right),\ i = 1,\ 2,\ \ldots,\ n,\ k \in N \triangleq \{1,\ 2,\ \ldots\} \end{cases} \tag{1}$$

where $n \geq 2$ is the number of neurons in the network; $x_i(t)$ represents the state of the ith neuron at the time t; a_i is amplification function; b_i is an appropriately behaved function; J_i corresponds to the external input of the ith neuron; c_{ij} denotes to the connection strengths of the jth neuron on ith neuron; f_j represents the activation function; $\tau_{ij}(t)$ corresponds to the transmission delay along the axon of the j from the I neuron and satisfies $0 \leq \tau_{ij}(t) \leq \tau_{ij} \leq \tau(\tau = \max_{1 \leq i,j \leq n}\{\tau_{ij}\})$; t_k is impulsive moment and satisfies $0 < t_1 < t_2 < \ldots,\ \lim_{k\to\infty} t_k = +\infty$; $x_i(t_k^-)$ and $x_i(t_k^+)$ represents the left-hand and right-hand limits at t_k.

2 Results

2.1 Example 1 – Impulsive CGNN

The model defined by [3] has the following equation:

$$\begin{cases} x_i'(t) = a_i(x_i(t))\left[-b_i(t, x_i(t)) + \sum_{j=1}^{n} c_{ij}(t)f_j\left(x_j(t)\right) + I_i(t)\right] \\ x_i\left(t_k^+\right) - x_i\left(t_k^-\right) = y_{ik}x_i(t_k) \end{cases} \tag{2}$$

where $t \geq 0, t \neq t_k,\ i = 1, \ldots, n,\ k \in N = \{1, 2, \ldots\}$, $x_i(t)$ corresponds to the state of the ith unit in time t, $n \geq 2$ is the number of units in the network. It is assumed that $c_{ij}(t), I_i(t)$ are continuous and T-periodic functions. For the functions $a_i(u), b_i(t, u)$ and $f_j(u)$ are supposed that they are locally Lipschitz continuous with respect to u.

The following neural networks are used for the example with $n = 2$:

$$\begin{cases} x_1'(t) = a_1(x_1(t))\left[-b_1(t, x_1(t)) + \sum_{j=1}^{2} c_{1j}(t)f_j\left(x_j(t)\right) + I_1(t)\right], t \neq t_k \\ x_2'(t) = a_2(x_2(t))\left[-b_2(t, x_2(t)) + \sum_{j=1}^{2} c_{2j}(t)f_j\left(x_j(t)\right) + I_2(t)\right] \end{cases} \tag{3}$$

with impulses at t_k, such that $(0, 2\pi) \cap \{t_k\}_{k\in N} = \{t_1, t_2, t_3\}$. Taken

$$c_{11}(t) = \sin(t) \quad c_{12}(t) = \cos(t) \quad c_{21}(t) = -\cos(t) \quad c_{22}(t) = \frac{\sin(t)}{4} \tag{4}$$

and

$$b_1(t, u) = 4(\sin(t) + 2)u = \beta_1(t)u \quad b_2(t, u) = (3 + \cos(t))u = \beta_2(t)u \tag{5}$$

The following results were obtained (Fig. 1):

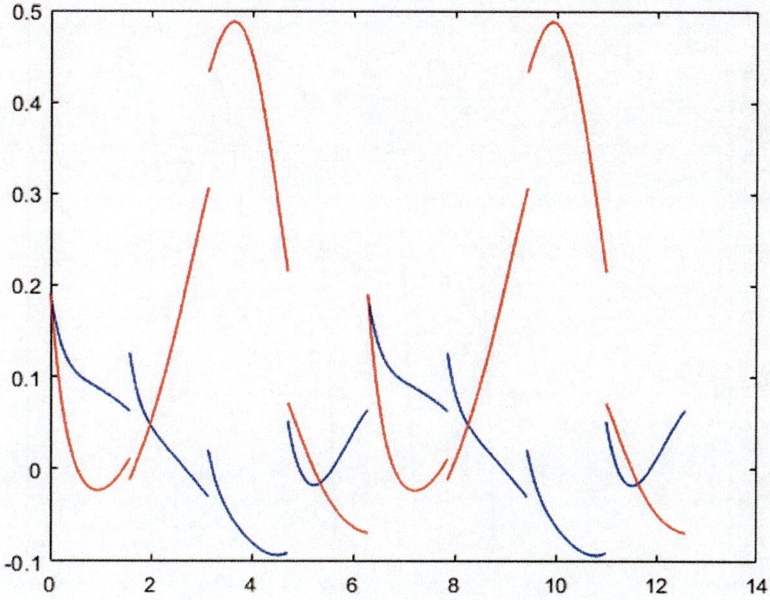

Fig. 1. x_1 and x_2 from the period of π and 4π.

2.2 Example 2 – Impulsive CGNN with Time-Varying Delays

Neural network according to [6] has the following equation:

$$\begin{cases} \frac{du_i(t)}{dt} = -a_i(u_i(t))\Big[b_i(u_i(t)) - \sum_{j=1}^{n} c_{ij}g_i\big(u_j(t)\big) \\ \qquad - \sum_{j=1}^{n} d_{ij}f_j\big(u_j(t-\tau_{ij}(t))\big) + I_j\Big], \quad t \neq t_k \\ u_i(t) = p_{ik}\big(u_1(t^-), \ldots, u_n(t^-)\big) + q_{ik}\big(u_1\big((t-\tau_{i1}(t))^-\big), \\ \qquad \ldots, u_n\big((t-\tau_{in}(t))^-\big)\big) + J_k, \quad t = t_k \end{cases} \tag{6}$$

for $i = 1, 2, \ldots, n$; $k = 1, 2, \ldots,$ where n corresponds to the number of units in the neural network.

The following model is used as example:

$$\begin{cases} \frac{dx_1(t)}{dt} = -(1 + 0.2\cos(x_1(t))) \\ \qquad [6x_1(t) - g_1(x_1(t)) + g_2(x_2(t)) - g_1(x_1(t-\tau(t))) + 2] \\ \frac{dx_2(t)}{dt} = -(1 + 0.2\sin(x_2(t))) \\ \qquad [4.5x_2(t) - g_2(x_2(t)) + g_1(x_1(t-\tau(t))) - g_2(x_2(t-\tau(t))) - 3] \end{cases} \tag{7}$$

Where $g_1(u) = g_2(u) = \frac{1}{2}(|u+1| - |u-1|), \quad \tau(t) = 3|\cos t| + 1.$

The following results were obtained (Fig. 2):

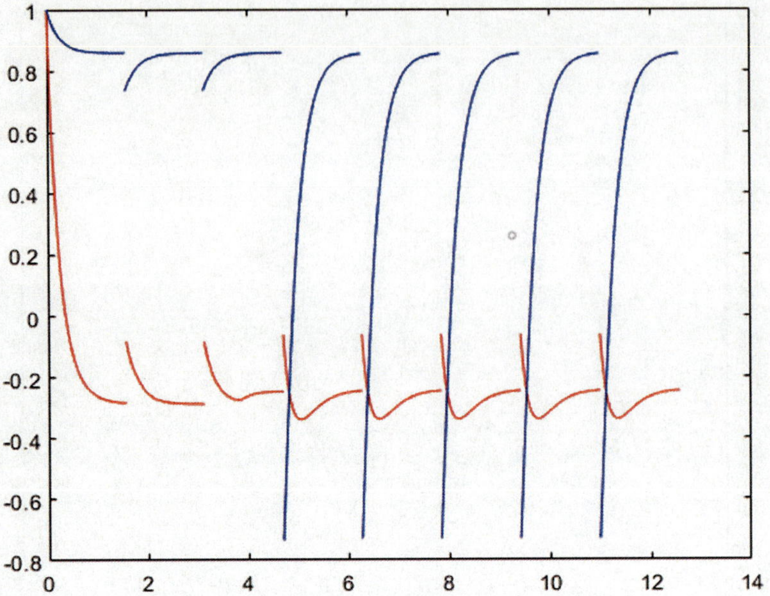

Fig. 2. x_1 and x_2 from the period of π and 4π.

2.3 Example 3 – Impulsive CGNN with Time-Varying Delays

The model used in [8] has the following equation:

$$\begin{aligned} Dx_i(t) = -a_i(x_i(t))\Big[b_i(x_i(t)) - \textstyle\sum_{j=1}^{n} c_{ij}f_j\big(x_j(t)\big)u_j \\ - \textstyle\sum_{j=1}^{n} d_{ij}f_j\big(x_j(t-\tau)\big)w_j + I_i\Big] \end{aligned} \tag{8}$$

where

$$\begin{aligned} x = (x_1, x_2, \ldots, x_n)^T \in R,\ a_i(x_i) > 0,\ u_j = 1 + \textstyle\sum_{k=1}^{+\infty} \beta_{jk}\delta(t - t_k), \\ w_j = 1 + \textstyle\sum_{k=1}^{+\infty} \gamma_{jk}\delta(t - t_k),\ t_0 < t_1 < t_2 < \ldots,\ \lim_{k\to+\infty} t_k = +\infty, \end{aligned} \tag{9}$$

$\delta(t)$ is the Dirac impulsive function.
The following equation is used for the example:

$$\begin{cases} \dot{x}_1(t) = -\left(6 - \frac{\cos t}{1+x_1^2(t)}\right)\{x_1(t) - [\frac{1}{10}\sin t\sin(x_1(t)) + \frac{1}{12}\cos t\cos(x_2(t))] \\ \quad -[\frac{1}{21}\sin t\sin(x_1(t-\tau_1(t))) + \frac{1}{30}\cos t\cos(x_2(t-\tau_2(t)))] + \sin t\} \\ \dot{x}_2(t) = -\left(5 + \frac{\sin t}{2+x_2^2(t)}\right)\{x_2(t) - [\frac{1}{12}\cos t\sin(x_1(t)) + \frac{1}{18}\sin t\cos(x_2(t))] \\ \quad -\left[\frac{1}{18}\cos t\sin(x_1(t-\tau_1(t))) + \frac{1}{30}\sin t\cos(x_2(t-\tau_2(t)))\right] + \cos t\Big\} \\ \qquad x_1(t_k) = (1 + b_{ik})x_1\left(t_k^-\right) \\ \qquad x_2(t_k) = (1 + b_{ik})x_2\left(t_k^-\right) \end{cases} \tag{10}$$

where

$$b_{ik} = -1 + (-1)^k \textit{ for } i = 1, 2, \ k \in N \quad \tau_i(t) = \tfrac{1}{2}\sin t + e^t + 1, \quad i = 1, 2$$
$$t_{2k+1} = t_1 + 2k\pi \quad t_{2k} = t_2 + 2(k-1)\pi \quad 0 < t_1 < t_2 < 2\pi, \ k \in N \tag{11}$$

The following results were obtained (Fig. 3):

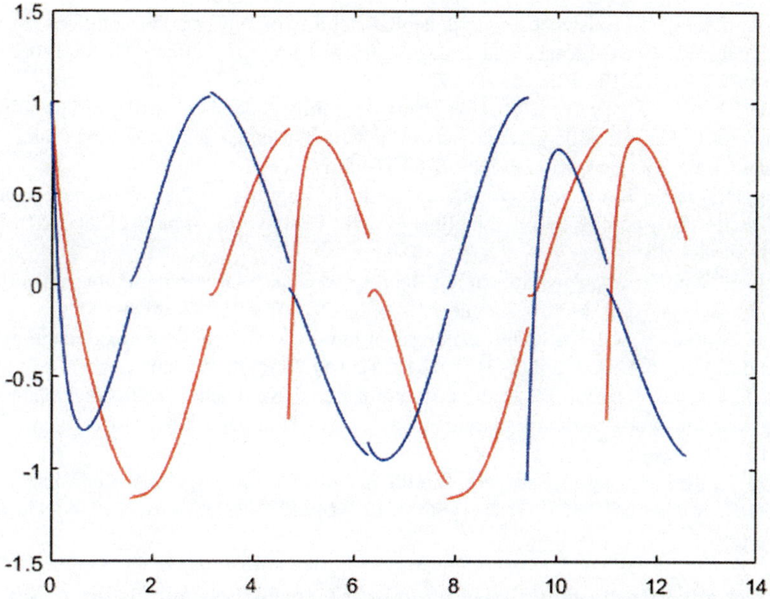

Fig. 3. x_1 and x_2 from the period of π and 4π.

It's visible that the given model has a unique periodic solution which is globally exponentially stable.

3 Conclusion

The given models and examples are believed to be useful for the design and further exploration of impulsive Cohen-Grossberg neural networks with time-varying delays by researchers and students. Deeper understanding of the time-varying delays over the stability of systems would benefit building and developing of such systems in the future and their use.

Funding. This research was funded in part by the European Regional Development Fund through the Operational Program "Science and Education for Smart Growth" under contract UNITe No. BG05M2OP001–1.001–0004 (2018–2023).

References

1. Cohen, M.A., Grossberg, S.: Absolute stability of global pattern formation and parallel memory storage by competitive neural networks **42**, 288–308 (1987). ISSN 0166-4115
2. Guo, S., Huang, L.: Stability analysis of Cohen-Grossberg neural networks. IEEE Trans. Neural Netw. **17**(1), 106–117 (2006). ISSN 1045-9227 (print), 1941-0093 (web)
3. Lisena, B.: Dynamical behavior of impulsive and periodic Cohen–Grossberg neural networks. Nonlinear Anal.: Theory Methods Appl. **74**(13), 4511–4519 (2011). ISSN 0362-546X
4. Xu, C., Zhang, Q.: Existence and exponential stability of anti-periodic solutions for a high-order delayed Cohen–Grossberg neural networks with impulsive effects. Neural Process. Lett. **40**(3), 227–243 (2013). ISSN 1370-4621
5. Li, Y., Zhao, L., Zhang, T.: Global exponential stability and existence of periodic solution of impulsive Cohen–Grossberg neural networks with distributed delays on time scales. Neural Process. Lett. **33**(1), 61–81 (2011). ISSN 1370-4621
6. Song, Q., Zhang, J.: Global exponential stability of impulsive Cohen–Grossberg neural network with time-varying delays. Nonlinear Anal.: Real World Appl. **9**(2), 500–510 (2006). ISSN 1468-1218
7. Yang, Z., Xu, D.: Impulsive effects on stability of Cohen–Grossberg neural networks with variable delays. Appl. Math. Comput. **177**(1), 63–78 (2006). ISSN 0096-3003
8. Chen, Z., Ruan, J.: Global stability analysis of impulsive Cohen–Grossberg neural networks with delay. Phys. Lett. A **345**(1–3), 101–111 (2005). ISSN 0375–9601
9. Pan, J., Liu, X., Zhong, S.: Stability criteria for impulsive reaction–diffusion Cohen–Grossberg neural networks with time-varying delays. Math. Comput. Model. **51**(9–10), 1037–1050 (2010). ISSN 0895-7177
10. Kang, B., Koo, N.: Stability properties in impulsive differential systems of non-integer order. J. Korean Math. Soc. (대한수학회지) **56**(1), 127–147 (2019). ISSN 0304-9914 (print), 2234-3008 (web)
11. Zhang, L., Yang, Y., Xu, X.: Synchronization analysis for fractional order memristive Cohen–Grossberg neural networks with state feedback and impulsive control. Physica A: Stat. Mech. Appl. **506**, 644–660 (2018). ISSN 0378-4371
12. Song, Q., Zhang, J.: Global exponential stability of impulsive Cohen–Grossberg neural network with time-varying delays. Nonlinear Anal.: Real World Appl. **9**(2), 500–510 (2008). ISSN 1468-1218
13. Meng, Y., Huang, L., Guo, Z., Hu, Q.: Stability analysis of Cohen–Grossberg neural networks with discontinuous neuron activations. Appl. Math. Model. **34**(2), 358–365 (2010). ISSN: 0307-904X
14. Wan, A., Wang, M., Peng, J., Mao, W.: Global exponential stability analysis of Cohen-Grossberg neural networks. Math. Appl. (Wuhan) **19**, 381–387 (2006)
15. Gan, Q.: Adaptive synchronization of Cohen–Grossberg neural networks with unknown parameters and mixed time-varying delays. Commun. Nonlinear Sci. Numer. Simul. **17**, 3040–3049 (2012). ISSN 1007-5704
16. Yuan, K., Cao, J., Li, H.-X.: Robust stability of switched Cohen–Grossberg neural networks with mixed time-varying delays. IEEE Trans. Syst. Man Cybern. Part B (Cybern.) **36**(6), 1356–1363 (2006). ISSN 1941-0492
17. Ozcan, N.: Stability analysis of Cohen–Grossberg neural networks of neutral-type: multiple delays case. Neural Netw. **113**, 20–27 (2019). ISSN 0893-6080
18. Pratap, K.A., Raja, R., Cao, J., Lim, C.P., Bagdasar, O.: Stability and pinning synchronization analysis of fractional order delayed Cohen–Grossberg neural networks with discontinuous activations. Appl. Math. Comput. **359**, 241–260 (2019). ISSN 0096-3003

19. Benchohra, M., Henderson, J., Ntouyas, S.: Impulsive Differential Equations and Inclusions. Hindawi Publishing Corporation, New York (2006). ISBN 977-5945-50-X
20. Haddad, W.M., Chellaboina, V.S., Nersesov, S.G.: Impulsive and Hybrid Dynamical Systems, Stability, Dissipativity, and Control, 1st ed. Princeton University Press, Princeton (2006). ISBN 9780691127156
21. He, W., Qian, F., Cao, J.: Pinning-controlled synchronization of delayed neural networks with distributed-delay coupling via impulsive control. Neural Netw. **85**, 1–9 (2017). ISSN 0893-6080
22. Stamova, I.M., Stamov, T., Simeonova, N.: Impulsive control on global exponential stability for cellular neural networks with supremums. J. Vibr. Control **19**(4), 483–490 (2013). ISSN: 1077-5463 (print) 1741-2986 (web)
23. Li, X.: Existence and global exponential stability of periodic solution for impulsive Cohen–Grossberg-type BAM neural networks with continuously distributed delays. Appl. Math. Comput. **215**(1), 292–307 (2009). ISSN 0096-3003
24. Benchohra, M., Henderson, J., Ntouyas, S.K., Ouahab, A.: Impulsive functional differential equations with variable times. Comput. Math. Appl. **47**, 1659–1665 (2004). ISSN 0898-1221
25. Stamov, G.T., Stamova, I.M.: Integral manifolds for uncertain impulsive differential–difference equations with variable impulsive perturbations. Chaos Solitons Fractals **65**, 90–96 (2014). ISSN

Performance Prediction of a Microbial Fuel Cell Based on Artificial Neural Networks

Blagovesta Midyurova(✉), Aleksandar Dimitrov, Sotir Sotirov, and Todor Petkov

University "Prof. Dr. Assen Zlatarov", Burgas, Bulgaria
blagi77@abv.bg

Abstract. This study integrates artificial neural network processing with microbial fuel cell - based on the electricity generation. Innovative treatment of wastewater flows is being applied to the Microbial fuel cell. It is a typical bioelectrochemical system where chemical energy containing in organic matter is converted in electricity by bacteria activity.

The study is focused on monitoring the operational stability of a cell with a VITO air-cathode and an anode carbon brush. The process is controlled by the potential created through an open circuit potential (OCP) and closed circuit potential (CCP). The obtained voltage is in the range of 45.7 mV - 43.9 mV at CCP, while it is 202.5 - 201.5 mV at OCP. The extent of the chemical oxygen demand (COD) removal during the run is 78.6%.

The data obtained were used to train the Neural Network and to predict the generated voltage by the Microbial fuel cell.

Keywords: Neural network · Microbial fuel cell · Waste water

1 Introduction

1.1 Microbial Fuel Cells

The Microbial fuel cells (MFCs) are renewable bio electrochemical converters which transform chemical energy into electricity. MFCs provide possibility for simultaneous treatment of waste water and generation of energy from organic matter using microbial consortia as biocatalysts. Beside the treatment of anthropogenic wastes and waste waters, while generating rather than consuming electric energy, the MFCs have the ability to degrade toxic contaminants [1]. Many research teams focused their efforts on profound analyses of their characteristics and hence improvement of cell performance. Up to now, the studies of MFCs are based mainly on laboratory experiments and mathematical modelling of the design of the electrodes, reactor design, selection of materials, etc.

The artificial neural networks (ANNs) are flexible mathematical models which can identify complex non-linear relationships between the input and output data [2]. These models are quite useful when it is difficult to apply conventional mathematical equations. ANN work by transforming the input data into numerical values which are propagated through a network of neurons. ANN are combined with sensor arrays for discovering

S. S. Sotirov et al. (Eds.): BioInfoMed 2020, LNNS 374, pp. 202–209, 2022.
https://doi.org/10.1007/978-3-030-96638-6_22

analytes [3–7]. They can help predict data about biosensors, bioelectrochemical systems and MFCs by identifying specific water contaminants. The use of artificial neural network is a possibility which estimates these systems on the basis of different variable parameters. Tsompanas et al., 2019 [8], used back-propagation network with topology of 4-10-1 neurons which determined the voltage of each MFC. The neural network was trained with two different membranes and two different electrode configurations assembled and used in laborastory experiments for creation of a set of data on which the neural network was trained.

1.2 Artificial Neural Networks

The artificial neural networks are inspired by the biological neural network of the mammal brain [9–11]. The basic building block of the ANN is the artificial neuron which is shown in Fig. 1.

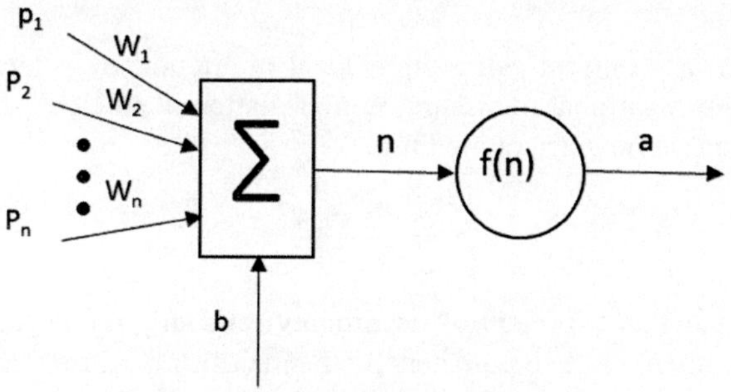

Fig. 1. Model of artificial neuron.

On the input of the neuron the input signals denoted by p that represents the data features are multiplied by the weight vector of the neuron denoted by w the result is summed with the bias value b. The result n enters into a functional block which applies the activation function of the neuron the actual output of the neuron is denoted by a.

In many cases a single neuron is not enough to perform complicated tasks. In order to achieve the goal the layer of the neurons are used, see Fig. 2, which can be extended into multilayer network.

The main benefit of using ANN is that it can solve different problems that conventional algorithms cannot. In order to do the training of the ANN the data examples must be provided.

On the other hand, artificial neural networks are a convenient tool for predicting the parameters of various processes. Here we will use a recurrent neural network to predict the loading of the cell. The network is trained with historical experimental data from the process and intuitionistic fuzzy estimates of the coefficient of affiliation, reflecting the degree of depletion of the nutrient medium-

$$\mu = \frac{a_i - A_k}{B - A} \tag{1}$$

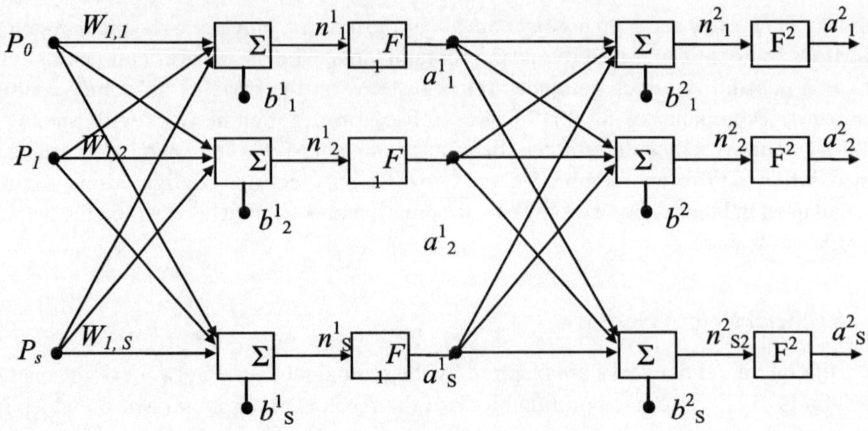

Fig. 2. The architecture of multilayer network.

where $\mu \in [0; 1]$; a_i – current value; A_k – local minimum; B_k – local maximum; B – global maximum; A - global minimum; non-affiliation coefficient – ν, reflecting the degree of saturation of the nutrient medium,

$$\nu = \frac{B_k - a_i}{B - A} \tag{2}$$

where $\nu \in [0; 1]$, and a coefficient of uncertainty reflecting the degree of inertia and inaccurate measurement of the parameters by the measuring instruments

$$\pi = 1 - \mu - \nu \tag{3}$$

These estimates are the basis on which cell loading is managed.

In the paper the membership function is perceived as something negative, and the non-membership function as something positive. Here we will connect the membership function with depletion of nutrient medium and the non-membership function with a degree of saturation of the nutrient medium. A part from our perception there is no other reason for these two expressions to be in this way, but if necessary, they can be exchanged.

2 Materials and Methods

Laboratory single-chamber, air-cathode MFC was constructed. The anode was graphite brush made from carbon fibers which are mechanically twisted around 14 cm long stainless steel wire with resistance of 1.4 Ω. Constructed MFC with air-cathode presented a single chambered plastic cylindrical system - volume V = 20 ml and carbon fiber brush anode (surface area of 200 cm^2) and air-cathode (VITO - Belgium) (surface area of 3.77 cm^2). A titanium wire was used for to connect the external circuit to the carbon electrodes. A schematic diagram of the laboratory MFC are presentedy in Fig. 3.

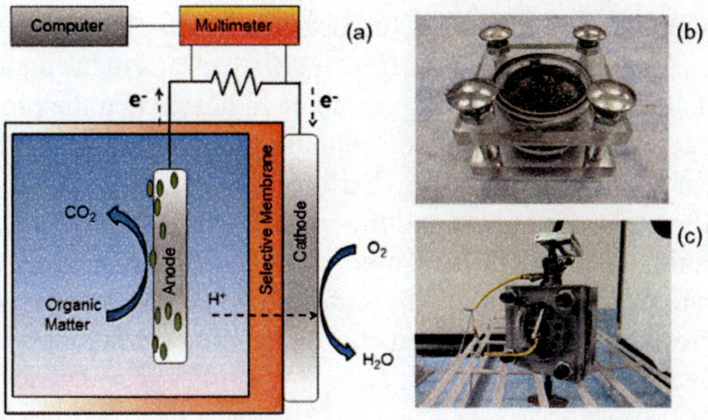

Fig. 3. (a) Basic components of a single-chamber MFC; (b) view of anode chamber and carbon fiber anode; (c) configuration of electrode connections and external resistor (adapted from Scott) [7]).

The air cathode is fixed by rubber gasket and separator of non-woven fabric was placed between the two electrodes to prevent the direct contact. The electrodes are connected to the external electric circuit through 1000 Ω resistor. To create conditions suitable for formation of biofilm, the anode chamber was charged with microorganisms isolated from bed sediments taken from the Yasna Polyana dam. A sample of the sediments was placed in a liquid nutritious medium with a composition of caseine tryptone – 10 g/dm^3; yeast extract – 5 g/dm^3; NaCl – 5 g/dm^3; glucose – 20 g/dm^3. After 72 h cultivation at temperature from 16 to 20 ºC, the obtained electrolyte was put in the anode chamber of the constructed MFC.

The operation stability of the cell with carbon brush anode and VITO air cathode was studied on the basis of measuring the voltage and analysis of the indicator chemical oxygen demand (COD). To monitor the content of organics in the medium, the tests Hach Lange LCK 514 and spectrophotometer Hach Lange DR 3900 were used. The voltage was measured during the experiments with Auto ranging digital Multimeter Model MY-66 and the values with open or closed circuit were measured by means of a switch. The following parameters were measured during the experiments: T (temperature), U (closed circuit voltage), OCV(open circuit voltage).

3 Results and Discussion

The values obtained for the voltages were in the range from 45.7 mV to 43.9 mV (closed electric circuit) and, respectively, from 202.5 mV to 201.5 mV (open electric circuit). The results obtained showed that after certain period of operation the voltages decreased which was most probably due to the exhaustion of the nutritious substances necessary for the growth of the microorganisms. This requires periodic replenishing of the anodic substrate with fresh nutrient. Thus, 6 cycles of system replenishing were done. The charging time was about 5 min since it is necessary to release the gaseous phase and homogenize the anodic electrolyte.

It is necessary to predict the voltage generated by the MFC during its period of operation. This is an essential indicator for the optimal conditions for the formation of biofilm of microorganisms on the anode (carbon brush) which paly the main role in the anaerobic oxidation of substrates and the transfer of electrons to the anode and the external electric circuit. The decrease of voltage values indicates that the bioelectrochemical system requires a new charge of the anode chamber with nutritious medium to maintain the vital activity of the microorganisms. Failure to supply fresh medium will lead to gradual death of the microorganisms and the system will stop functioning.

The input training data set is measured during a certain period of time and the measured results are used to obtain the input training vectors of the ANN.

The input vector consists of intuitionistic fuzzy values which are estimated by the equations as follows.

- formula (1) calculates the degree of membership, which is the first value of the input vector
- formula (2) calculates non-membership, which is the second value of the input vector
- formula (3) calculates the third value of the input vector.

The target vector consists of one value which can be 0 or 1. If the value is 1 then the MFC has to be loaded with a material in order to continue the voltage generation. If the value is 0 it is not necessarily to load with material the MFC.

The ANN consists of two layers of neurons the first layer applies 'sigmoid' activation function where

$$f(x) = \frac{1}{1 + e^{-x}} \tag{4}$$

The second layer consists of one neuron which applies linear activation function

$$f(x) = x \tag{5}$$

In Fig. 6 is shown the structure of the ANN. The algorithm goes through 339 iterations out of 1000 maximum, at each iteration the stopping parameters of the ANN are calculated. The input data set consists of 82 training samples each sample has the intuitionistic fuzzy estimation and the target value. The 76 of the input vectors were used for training the network, the rest six of 82 are excluded of the training set for the validation set of the ANN (Fig. 4).

In Fig. 5 is shown the calculation of the mean square error (MSE) at each iteration of the ANN. The learning trial stops at iteration 339 in accordance to the MSE.

When the learning trial finishes the measured data that was not in the training set was applied in order to validate the accuracy of the ANN. The model of the ANN predicts the time that the MFC has to be loaded in order to produce the output voltage. The 88 measurements of the output voltage were made each measurement is made 12 min from the previous one. In Fig. 6 are shown two graphics, the blue one represents the measured voltage value from the MFC, the brown one represents the time which has to be predicted from the ANN in order to load the MFC.

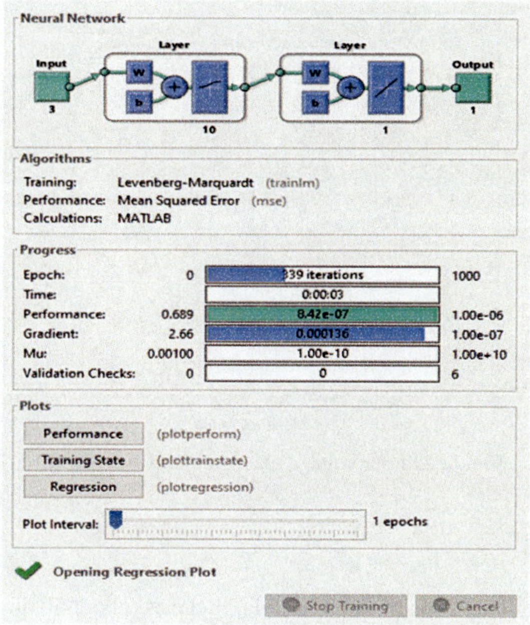

Fig. 4. The structure of ANN.

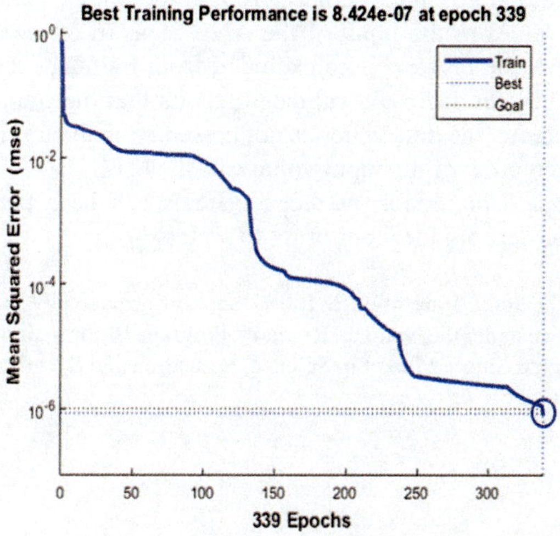

Fig. 5. The MSE at each iteration.

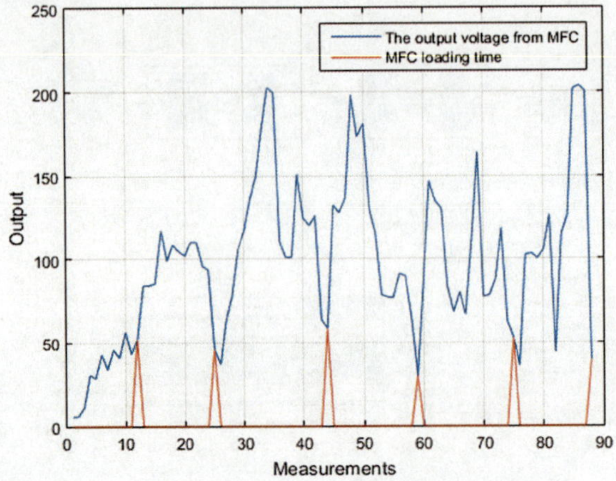

Fig. 6. Output voltage measurements from MFC

4 Conclusions

During the period of operation of the microbial fuel cell (about 50 days) cyclicity of the results obtained for the voltage generated was observed, a basic indicator for the partial assimilation of the organic matter.

The artificial neural network of supervised learning type was used in order to predict the loading time of the MFC with material. The artificial neural networks consisted with two layers of neurons and goes through 339 iterations of the input training data set. The training set that was used to build the model contained 82 input vectors, where the values of the vectors are time, temperature and the output voltage converted into intuitionistic fuzzy values as the input of the ANN. The 76 of the data set were used as the training vectors the rest six were excluded from training set for the purpose of validation. It could be seen from the validated results that the trained model of ANN could successfully predict the data which is not presented in the training set. The model could be used for prediction of the input voltage of the MFC.

The automation and prediction of these systems can be a prerequisite for their implementation as biosensors for monitoring of waters.

Acknowledgment. The authors are grateful for the support provided by the Bulgarian Ministry of Education and Science under the National Research Program "Information and Communication Technologies for a Digital Single Market in Science, Education and Security" approved by DCM # 577/ 17.08.2018.

References

1. Izadi, P., Rahimnejad, M., Ghoreyshi, A.: Power production and wastewater treatment simultaneously by dual-chamber microbial fuel cell technique. Biotechnol. Appl. Biochem. **62**(4), 483–488 (2015)

2. Li, B., Zhou, J., Zhou, X., Wang, X., Li, B., Santoro, C.: Surface modification of microbial fuel cells anodes: approaches to practical design. Electrochim. Acta **134**, 116–126 (2014)
3. Liao, Q., Zhang, J., Li, J., Ye, D., Zhu, X., Zhang, B.: Increased performance of a tubular microbial fuel cell with a rotating carbon-brush anode. Biosens. Bioelectron. **63**, 558–561 (2015)
4. Lovely, D.R.: Microbial fuel cells: novel microbial physiologies and engineering approaches. Curr. Opin. Biotech. **17**, 327–332 (2006)
5. Mohan, S.V., Velvizhi, G., Modestra, J.A., Srikanth, S.: Microbial fuel cell: critical factors regulating bio-catalyzed electrochemical process and recent advancements. Renew. Sustain. Energy Rev. **40**, 779–797 (2014)
6. Tursun, H., et al.: Carbon material optimized biocathode for improving microbial fuel cell performance. Front. Microbiol. **7**, 6 (2016)
7. King, S.T., Sylvander, M., Kheperu, M., Racz, L., Harper, W.F.: Detecting recalcitrant organic chemicals in water with microbial fuel cells and artificial neural networks. Sci. Total Environ. **497–498**, 527–533 (2014)
8. Tsompanas, M., You, J., Wallis, L., Greenman, J., Ieropoulos, I.: Artificial neural network simulating microbial fuel cells with different membrane materials and electrode configurations. J. Power Sources **436**, 226832 (2019)
9. Atanassov, K.: On Intuitionistic Fuzzy Sets Theory. Springer, Berlin (2012). https://doi.org/10.1007/978-3-642-29127-2
10. Bishop, C.M.: Neural networks for pattern recognition. Oxford University Press, Oxford (2000). ISBN 0 19 853864 2
11. Haykin, S.: Neural Networks: A Comprehensive Foundation. Prentice Hall, Hoboken (1999)

Functional Outcome Prediction of Operated Proximal Humerus Fractures by Means of Artificial Neural Networks

Stoyan Hristov[1], Asen Baltov[2], and Sotir Sotirov[3](✉)

[1] Department of Orthopedics and Traumatology, University Hospital for Active Treatment, 73 Stefan Stambolov, 8000 Burgas, Bulgaria
hristov@dr.com

[2] University Multiprofile Hospital for Active Treatment and Emergency Medicine "N.I. Pirogov", 21 Totleben str., 1606 Sofia, Bulgaria
asen_b@bitex.bg

[3] Burgas "Asen Zlatarov" University, 1 Y.Yakimov str., 8010 Burgas, Bulgaria
ssotirov@btu.bg

Abstract. The goal of our presentation is to predict functional outcome of operated Proximal Humeral Fractures by means of Artificial Neural Networks, so that we can compare anticipated results to real results recorded during scheduled follow-ups. We get the chance to assess above-mentioned method's reliability as a proper tool in our daily practice and to analyze it in details in order to improve it.

Keywords: Neural network · Proximal humerus fractures

1 Introduction

1.1 Proximal Humerus Fractures

Proximal humerus fractures (PHFx) are very common in patients above 65. PHFx are second most frequent upper extremity fracture accounting about 6 to 10% of fractures in this patient group [2, 5, 9, 10].

According Charles Court-Brown PHFx often occur in the fit elderly independent patient - a contributor to society, who might well be converted to a degree of social dependency. PHFx are mainly treated nonoperatively. Surgical treatment is becoming more frequent ranging between 30% to 60%.

Surgical treatment of such fractures with locking plate has a good functional outcome provided that the correct surgical technique is used. Medical records of patients' s demographic and baseline characteristics during hospitalization were created. Fracture classification acc. Neer Classification system based on plain radiographs and intraoperatively.

Follow-up evaluations on the third, sixth and twelfth month postoperatively- patient assessed with regard to pain, ROM (Forward Flexion, Medial and Lateral Rotation, Abduction), Constant and DASH score.

S. S. Sotirov et al. (Eds.): BioInfoMed 2020, LNNS 374, pp. 210–215, 2022.
https://doi.org/10.1007/978-3-030-96638-6_23

Above information is based on studies performed at University Hospital of Burgas by our research team in the period 2017–2019. The neural networks are one useful tools for object recognition and prognosis [1, 3, 7].

1.2 Artificial Neural Networks

Neural networks are a very well used tool for classification and clustering [1, 3, 4, 6–8, 11]. One of the most used neural networks are the ones with feed forward structure, using an algorithm for backpropagation of the error. An abbreviated description of the network is given in Fig. 1.

Figure 1 shows, in abbreviated notation, a classical three-layered neural network.

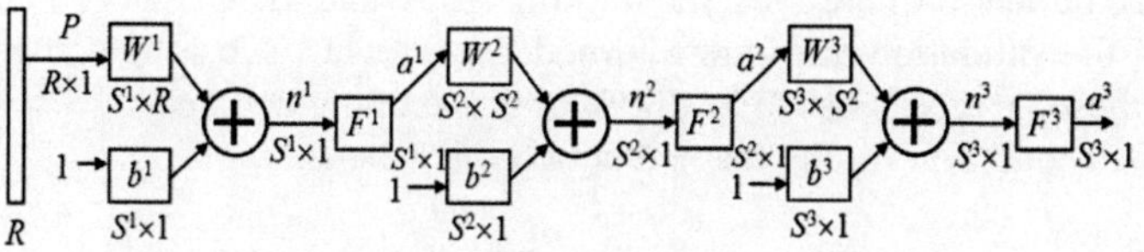

Block diagram of the feedforward neural network.

In multilayered networks, the exits of one layer become entries for the next one.

The equations describing this operation are:

$$a^{m+1} = f^{m+1}(w^{m+1}.a^m + b^{m+1})$$

for m $= 0, 1, 2, \ldots, M - 1$,

where:

P is a network input vector;

F^m is the transfer function of the exit of m-th layer.

m is the number of the layers in the network;

s^m is a current number of neutrons of a m-th layer of the neural network;

M is the number of the layers;

a^m is the output of the m-th layer of the neural network;

W is a matrix of the weight coefficients;

b is neuron's input bias.

The neuron in the first layer receives outside entries p. The neurons' exits from the last layer determine the neural network's exits a.

Since it belongs to the learning with teacher methods, to the algorithm are submitted training sets (an entry value and an achieving aim – on the network's exit)

$$\{p_1, t_1\}, \{p_1, t_2\}, \ldots, \{p_Q, t_Q\},$$

$Q \in (1, \ldots, n)$, n – numbers of learning couple, where p_Q is the entry value (on the network entry), and t_Q is the exit's value corresponding to the aim. Every network's entry is preliminary established and constant, and the exit has to correspond to the aim. The difference between the entry values and the aim is the error $e = t–a$.

The “back propagation” algorithm [3] uses mean-quarter error:

$$\hat{F} = (t - a)^2 = e^2.$$

In learning the neural network, the algorithm recalculates network’s parameters (W and b) in order to achieve mean-square error.

When the multilayer neural network is trained, usually the available data has to be divided into three subsets. The first subset is named “Training set” and is used for computing the gradient and updating the network weighs and biases. The second subset is named “Validation set”. The error of the validation set is monitored during the training process. The validation error normally decreases during the initial phase of training, so does the training set error. Sometimes, when the network begins to overfit the data, the error of the validation set typically begins to rise. When the validation error increases for a specified number of iterations, the training stops, and the weights and biases at the minimum of the validation error are returned [3]. The last subset is named “test set”. The sum of these three sets has to be 100% of the learning couples.

The classic condition for the learned network is when

$$e^2 < E_{\max},$$

where $E_{\max}$ is the maximum square error.

2 Realization and Discussion

The neural network has a structure of 9:66:12. At the input of the neural network is provided information about Age, According Neer’s Classification system, the measurements from the third month for Medial Rotation $\in [0 \div 360]$, Lateral Rotation $\in [0 \div 360]$, Forward Flexion $\in [0 \div 360]$, Abduction $\in [0 \div 360]$, Constant $\in [0 \div 100]$, and DASH Score assesment. The measured parameters for the sixth and twelfth month for Medial Rotation, Lateral Rotation, Forward Flexion, Abduction, Constant, DASH Score are fed to the output of the neural network. The aim is to predict the values of Medial Rotation, Lateral Rotation, Forward Flexion, Abduction, Constant and DASH Score without performing the corresponding measurements for the sixth and twelfth month. In the training and testing of the neural network, the data from real measurements of 63 women for 2019 who underwent this procedure at the University Hospital Burgas were used (Figs. 2, 3).

After the training process use 20 vectors for the testing and 10 vectors for verifying the neural network. The first 20 vectors are vectors that we take from the training set. The Mean Square Error (MSE) for this vectors is 0.024. For the vectors for verifying we use the vectors that we not used for the training process. The MSE for this vectors is 0.028.

This method gives us the opportunity to predict anticipated results on 6th and 12th month postoperatively based on recorded outcome 3 months postop. The main advantage of this method is the option to compare expected to achieved results and to analyze the reasons for disparity if any recorded. Another benefit of this method is its cost-effectiveness in terms of optimized waiting lists, tightness of doctors schedule and saved patients time.

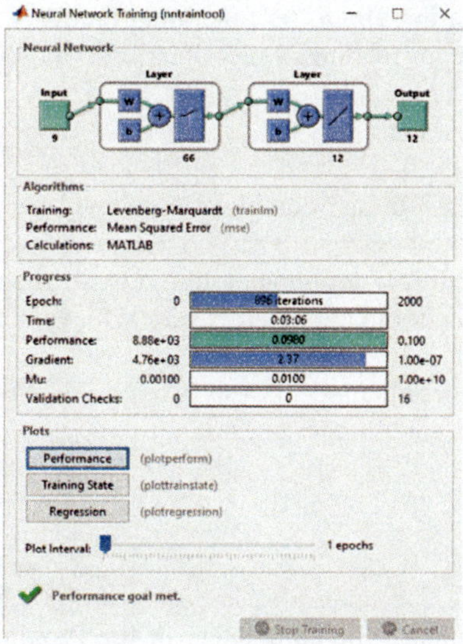

Fig. 1. Structure of the neural network

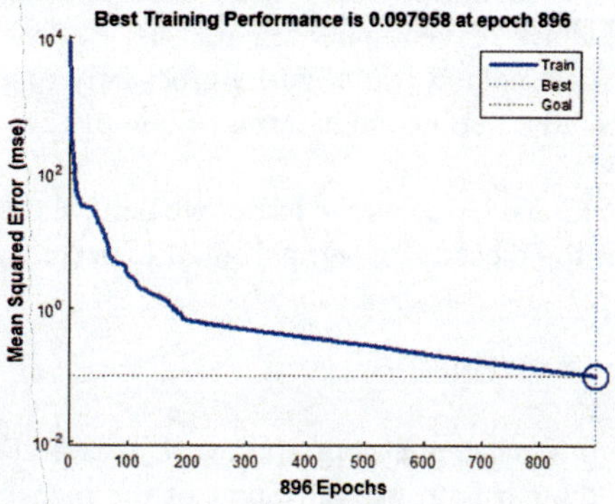

Fig. 2. Training process of the neural network

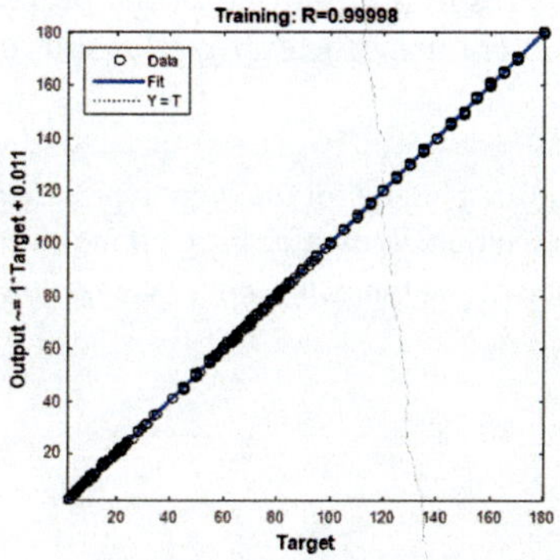

Fig. 3. The regression coefficient R

In ROM- measurements cases it's essential to take into consideration the interaction conditions within individual anatomic structures.

Single muscle groups with their respective insertion sites to adjacent bones and surrounding soft- tissue structures operate within functional units, mutually affecting each other, which complicates, even hinders precise measurements of indexes.

Individual parameter/Forward Flexion, Abduction, Medial Rotation& Lateral Rotation/ are measured as a possible range of active movements in standing/upright position, while rotation is measured in abducted upper limb position.

That's exactly what leads to a result, impacted by a certain percentage of subjectivity, caused by researches as such and by researche performer as well.

The measured DASH score gives us information about the functional recovery of operated limp at the moment, while Constant score readings reflect possible range of active movements and pain status as well.

Two main method disadvantages to be considered:

- The method doesn't allow timely screening of clinical complications, resulting in delayed revision treatment.
- The method could hardly efficiently replace clinical assessment and individual specialist's approach in terms of complete assessment of patient's recovery achievement.

Considering above issues we believe that a more detailed analysis is required, which has to be based on comparing of expected results for a longer period.

3 Conclusion

In this paper the neural network was presented. The main idea is during the process of diagnosis to use only part of the measurements of the parameters of the Proximal humerus fractures.

The method makes it possible to predict the expected results in the 6th and 12th month based on the recorded result 3 months after the procedure. The main advantage of this method is the ability to compare the expected results and analyze the reasons for non-compliance, if any.

Acknowledgment. The authors are grateful for the support provided by the Bulgarian Ministry of Education and Science under the National Research Programme "Information and Communication Technologies for a Digital Single Market in Science, Education and Security" approved by DCM # 577/ 17.08.2018.

References

1. Bishop, C.M.: Neural Networks for Pattern Recognition. Oxford university press, Oxford (2000). ISBN 0 19 853864 2
2. Karantonis, D.M., Narayanan, M.R., Mathie, M., Lovell, N.H., Celler, B.G.: Implementation of a real-time human movement classifier using a triaxial accelerometer for ambulatory monitoring. IEEE Trans. Inf. Technol. Biomed. **10**(1), 156–167 (2006)
3. Rumelhart, D., Hinton, G., Williams, R.: Training representation by back-propagation errors. Nature **323**, 533–536 (1986)
4. Zwe-Lee, G.: Wavelet-based neural network for power disturbance recognition and classification. IEEE Trans. Power Deliv. **19**(4), 1560–1568 (2004)
5. Gardner, M.J., et al.: The importance of medial support in locked plating of proximal humerus fractures. J. Orthop. Trauma **21**(3), 185–191 (2007)
6. Sawai, H., et al.: Parallelism, hierarchy, scaling in time-delay neural networks for spotting Japanese phonemes/CV-syllables. In: IJCNN, vol. 11, pp. 81−88 (1989)
7. Haykin, S.: Neural Networks: A Comprehensive Foundation. Prentice Hall, N.J. (1999)
8. Meissner, M., Schmuker, M., Schneider, G.: Optimized Particle Swarm Optimization (OPSO) and its application to artificial neural network training. BMC Bioinform. **7**(1), 125 (2006)

9. Plecko, M., Kraus, A.: Internal fixation of proximal humerus fractures using the locking proximal humerus plate. Oper. Orthop. Traumatologie **17**(1), 25–50 (2005)
10. Rusimov, L., et al.: Does supplemental intramedullary grafting increase stability of plated proximal humerus fractures? J. Orthop. Trauma **33**(4), 196–202 (2019)
11. Himavathi, S., Anitha, D., Muthuramalingam, A.: Feedforward neural network implementation in FPGA using layer multiplexing for effective resource utilization. IEEE Trans. Neural Netw. **18**(3), 880–888 (2007)

Application of Virtual Reality as a Tool for Structural Analysis of Molecules – Steroids, Pharmaceuticals and Pesticides

Todor Petkov[1(✉)], Stanislav Popov[1], Milen Todorov[1], Stanimir Surchev[1], Guy De Tré[2], and Maciej Krawczak[3]

[1] University "Prof. Dr. Assen Zlatarov", Burgas, Bulgaria
t-petkov@mail.bg
[2] Ghent University, Ghent, Belgium
[3] Polish Academy of Sciences, Warsaw, Poland

Abstract. The three dimensional nature of chemical interactions requires application of suitable technology for visualization of molecules in order to assess their reactivity. The molecular flexibility of organic chemicals allows those structures to exist in a set of different structural orientation (conformations) as a result of changes of structural variables such as bond rotation, valence angles and many other. Each spatial position may differ significantly in respect to chemical reactivity. Thus, in order to make an overall reactivity profile of a specific chemical all possible 3D positions should be considered and analyzed. In the current paper the focus is set on representative chemicals from groups of steroids, pharmaceuticals and pesticides. In order to analyze the reactivity of those chemicals the following steps are accomplished: 1) generation of initial 3D structures; 2) generation of possible 3D structural orientation; and 3) visualization of the 3D structures in a suitable virtual reality tool for education purposes.

Keywords: Virtual reality · Chemical molecules · Education

1 Introduction

Reality technologies include Virtual reality (VR), Augmented reality (AR) and Mixed reality (MR). From the three aforementioned, VR is proven to be the most suitable technology for demonstrating 3D objects for educational purposes [1]. However, VR systems have its own classification which includes the following types: full immersive, semi-immersive and non-immersive. Therefore, these types should be further analyzed by a making a comparison between them which will make the result more comprehensive. In this way, the choice of what type of VR technology should be used in different areas of education, and especially medicine, is even more specified.

From practical point of view, a representative set of most structurally distinctive conformers could be used for analysis of the chemical reactivity. The goal here is to create all conformers in 3D scene which will allow further visualization with an appropriate VR technology. In order to achieve the goal, the coordinates of conformers for each

S. S. Sotirov et al. (Eds.): BioInfoMed 2020, LNNS 374, pp. 216–223, 2022.
https://doi.org/10.1007/978-3-030-96638-6_24

molecule are generated and provided into a user friendly application, where the user could see each conformer and observe its 3D molecular structure.

2 Virtual Reality – Definition

The concept of Virtual reality (VR) refers to a whole simulated reality, which is built with computer systems by using digital formats. Building and visualizing this alternative reality requires hardware and software powerful enough to create a realistic immersive experience [2]. VR simulations provide a deep understanding of the material by a learner with its further application in real life.

The most important objective Virtual reality aims to achieve is total immersion. Total immersion means that the sensory experience feels so real, that the user forgets it is a virtual-artificial environment and begins to interact with it as he/she would naturally in the real world [3]. In a virtual reality environment, the generated artificial world may or may not have the properties of a real-world environment [4]. This means that the VR environment can mimic an everyday setting, or for example create a world in which the physical laws governing gravity, time and material properties no longer hold.

VR adds precision, and permits visualizing objects and processes otherwise impossible to show in a real environment. What is more, virtual technologies make it possible to be exposed to abstract ideas by using models that can be interacted, so it also eases exposing students to knowledge by following a constructivist approach. Therefore, virtual technologies encourage students to be active learners, because VR promote decision-taking when interacting with virtual environments, permitting autonomous exploration, understanding complex concepts, creating new experiences, and learning by doing [5].

3 Comparison Between Different Types of Virtual Reality Systems

The algorithm for implementing VR technology in education, proposed in [6] shows that VR is the best choice when it comes to demonstrating 3D objects in the field of chemistry or biology. It is based on different criteria: educational, technical, complexity of objects, number of students in a group. However, VR systems, in turn, have their own classification.

Although it is difficult to categorize all VR systems, most configurations fall into three main categories (non-immersive, semi-immersive and full-immersive) and each category can be ranked by the sense of immersion, or degree of presence it provides (see Table 1) [7]. Immersion or presence can be regarded as how powerfully the attention of the user is focused on the task in hand [8]. To further understand the difference between them, in this section a comparison of their main features is presented.

By combining the data from the table above and the criteria from the algorithm proposed in [6] we can determine the most appropriate type of VR system to be used in different educational scenarios (see Table 2).

Table 1. Qualitative performance of VR systems

Main features	Non-Immersive VR	Semi-Immersive VR	Full Immersive VR
Resolution	High	High	Low-Medium
Scale	Low	Medium-High	High
Sense of situational awareness	Low	Medium	High
Lag	Low	Low	Medium-High
Sense of immersion	None-Low	Medium-High	Medium-High

Table 2. Combination of qualitative performance and educational criteria

Object complexity	Number of students	Most appropriate VR System
High	≤10	Semi-Immersive
High	>10	Non-immersive
Medium	≤10	Full Immersive
Medium	>10	Semi-Immersive
Low	≤10	Full Immersive
Low	>10	Semi-Immersive

4 VR System for Structure of Chemical Molecules

The 3D models of steroids, pharmaceuticals and pesticides chemical structures consist mainly of atoms, the bonds between them and text showing their titles. Therefore, for educational purpose, from technical point of view, the resolution and lag needed for optimal experience should be of medium value. However, the scale and sense of situational awareness should be high. The complexity of the VR objects is medium and the number of students attending these types of classes is very often no more than ten in order individual comprehension to be achieved. Based on the above stated, Fig. 1 (see next page) shows that Full-immersive VR is the best choice when it comes to demonstrating objects from the field of chemistry.

The rest of the paper focuses on the programming construction of the molecules and their visualization in VR environment.

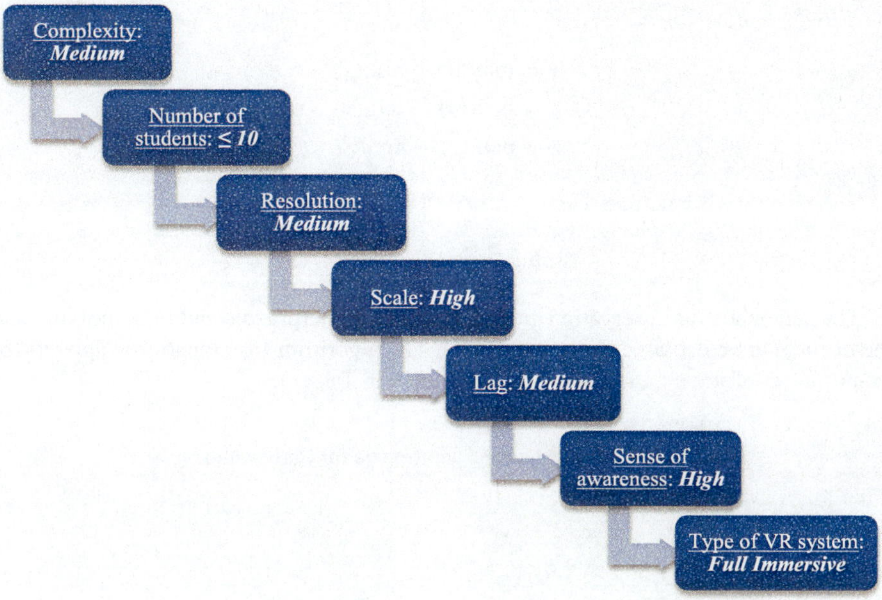

Fig. 1. Choosing the best VR system for visualizing 3D chemical molecules

5 Programming Construction

The ability to generate conformers of a molecule is useful if one needs to observe its 3D structure or when the molecule can be shown in multi-component screen based on its complementarity [9].

The Conformer Generator that is used provides the ability to minimize molecular conformations and to generate diverse conformer subsets [10, 11]. The methodology starts from an input 3D molecular structure with all hydrogen atoms. A final diverse set of conformers is clustered according to conformer similarity, each of the conformers is locally optimized in a torsion space [12].

The conformers have plenty of characteristics one of the important properties of the conformer is its planarity, the computation of the planarity can be performed in a several ways. The computation methodology of a planarity here is based on the coordinates of the atoms in 3D space for each conformer. The minimum and the maximum values of x, y and z coordinates of atoms of each conformer are calculated in order to compute the planarity. The most planar conformer is going to be the one that the computed value is less than the other conformers computed values.

Where

$$x = \max(x) - \min(x)$$
$$y = \max(y) - \min(y)$$
$$z = \max(z) - \min(z)$$

The planarity is calculated as

$$planarity = x * y * z$$

The chemicals for observation are Carbetamide, Omerprazole and Estradiol for each one of them are calculated four conformers. The algorithm for planarity is applied for each of the conformers and the results are shown in Table 3.

Table 3. The calculated conformers for Carbetamide

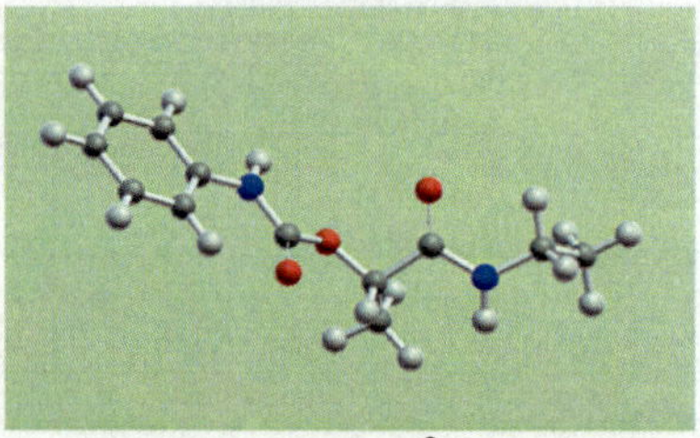

a) 504 m^3

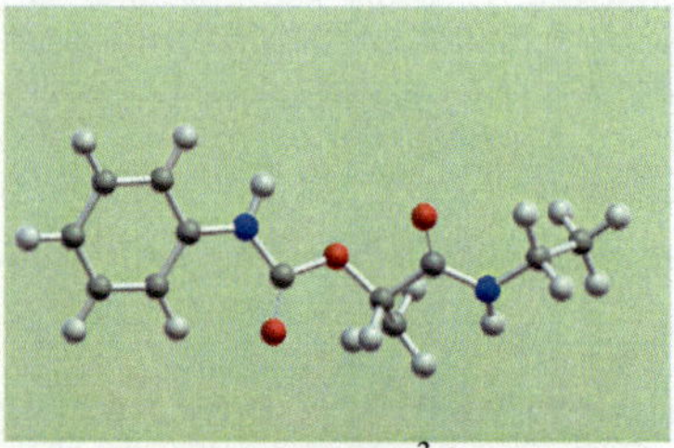

b) 294 m^3

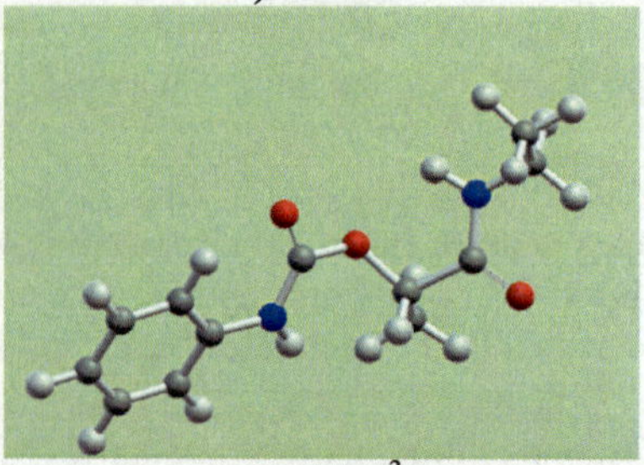

c) 437 m^3

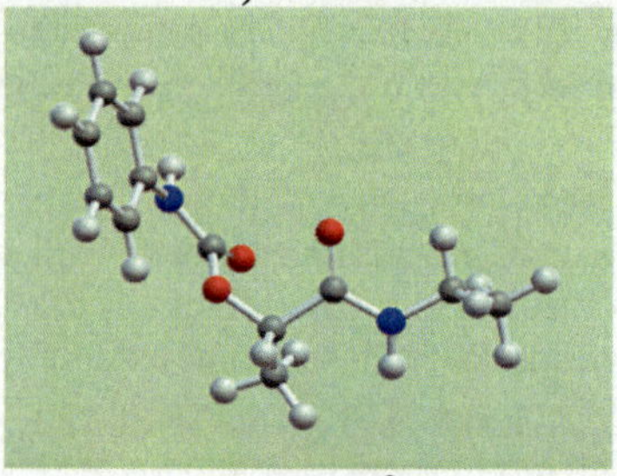

d) 459 m^3

From Table 3 it can be seen that the most planar conformer is b) image.
The generated conformers for the Omerprazole chemical are shown in Table 4.
From Table 3 it can be seen that the most planar conformer is c) image.
The next chemical is the estradiol its conformers can be seen in Table 5.
From Table 5 it can be seen that the most planar conformer is a) image.

Table 4. The calculated conformers for Omerpazole

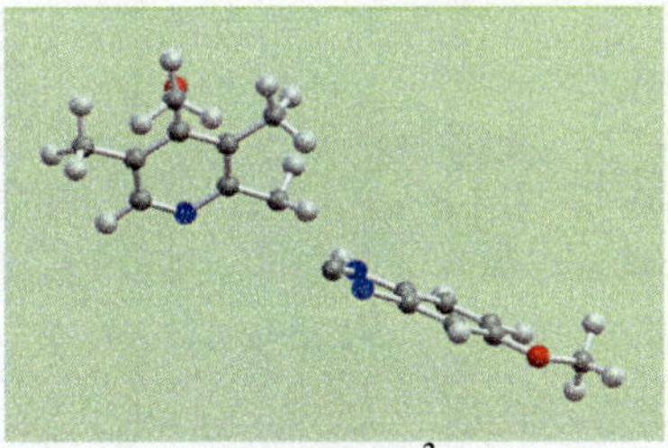

a) 1148 m^3

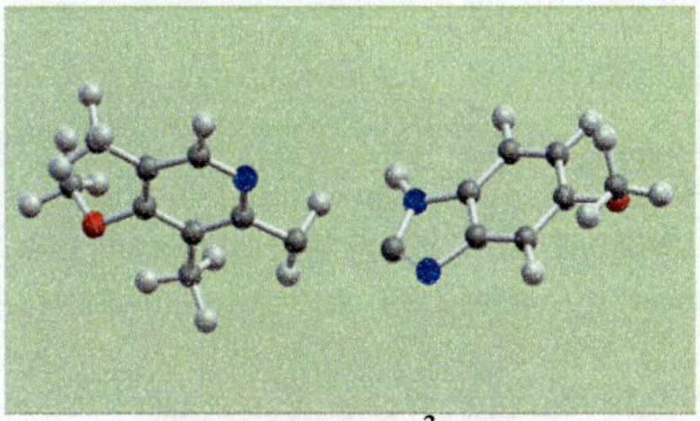

b) 611 m^3

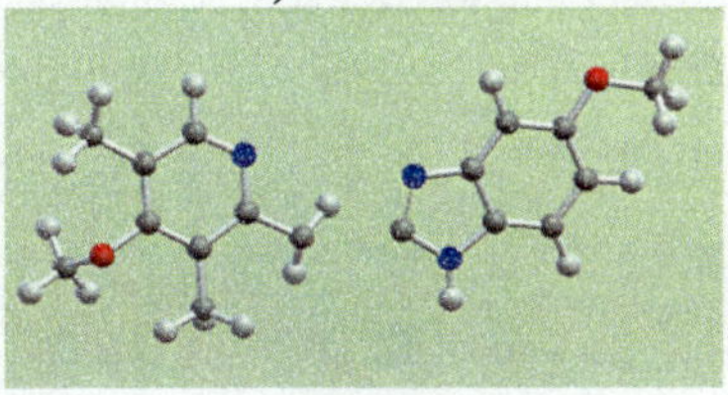

c) 293 m^3

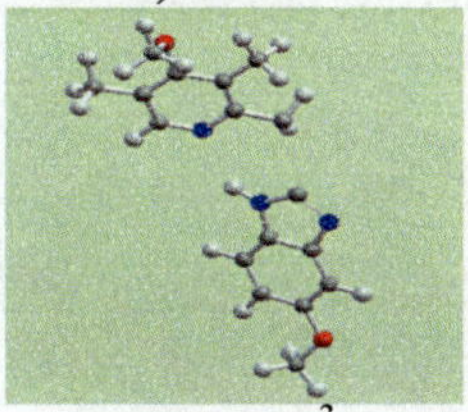

d) 891 m^3

Table 5. The calculated conformers for Estradiol

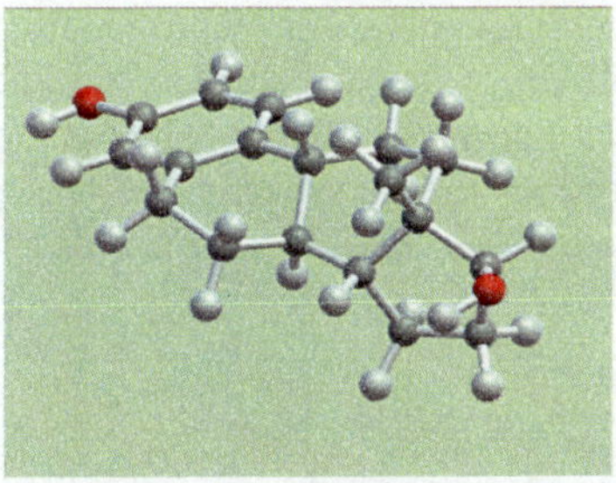

a) 412 m^3

b) 466 m^3

c) 442 m^3

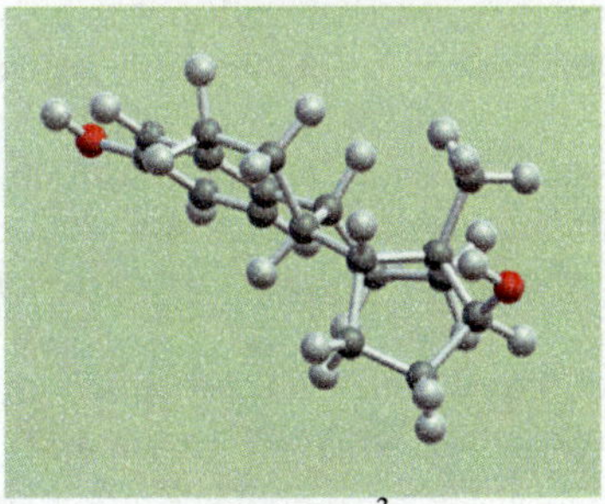

d) 471 m^3

Traditionally the analysis of molecular reactivity is based on a single representative 3D structure. Such simplifying assumptions could doom to failure the studies given the significant variation of the geometric and electronic characteristics across the multitude of energetically reasonable conformers representing the molecules [13]. Thus it is important to analyze the set of possible conformations. Recent technologies such as VR allows visual inspection of molecules by making use devices for virtual reality. For educational purposes this type of visualization is considered to be very helpful in understanding the reactivity profile of various molecules [14].

In the current study structural variations in molecules of three chemicals are represented. Among all chemical classes the focus is set on steroids (estradiol), pharmaceutical (omeprazole) and pesticide (Carbetamide). The current version of the developed VR module supports visualization of a set of conformers (Fig. 2).

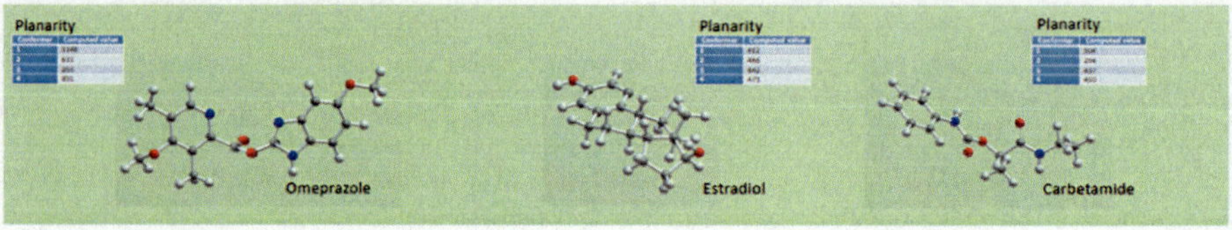

Fig. 2. Representative structures - pharmaceutical (omeprazole), steroids (estradiol) and pesticide (Carbetamide) and respective values for planarity.

The user can choose a conformers and could be able to get details for calculated planarity. The aim of the program is to give the user more realistic representation of the real models that are the subject for studying. The module can be extended in order to cover more chemical characteristics which will be very helpful also for educational purposes. It has to be noticed that using the ability of VR the students are going to learn more effectively and more important to apply their new knowledge in the practice.

6 Conclusion

The current study focuses on the usage of VR technology for creating and visualizing 3D structures of chemical molecules. Based on the comparison between different VR systems and combining the result with educational criteria, Full-Immersive VR proves to be the best solution for demonstrating chemical compounds in 3D scene for educational purposes. An algorithm for construction of 3D structures is proposed for visualization of different chemicals in Unity platform. Ultimately the user is able to analyze generated structures by making use different devices for virtual reality such as Oculus Rift, Samsung Gear or other.

The three dimensional representation of molecules form different classes - steroids, pharmaceutical and pesticides is useful for analyzing of their reactivity and other characteristics. For each molecule were extracted several conformers and the planarity for each of them were calculated.

References

1. Fromm, J., Radianti, J., et al.: More than experience? - On the unique opportunities of virtual reality to afford a holistic experiential learning cycle. Internet High. Educ. **50**, 100804 (2021). ISSN 1096-7516. https://doi.org/10.1016/j.iheduc.2021.100804
2. Li Z., Yue J., Jauregui D.: A new virtual reality environment used for e-learning. In: IEEE International Symposium on IT in Medicine & Education, pp. 445–449. IEEE (2009)
3. Kotranza, A., Lind, D.S., Pugh, C.M., Lok, B.: Real-time in-situ visual feedback of task performance in mixed environments for learning joint psycho-motor-cognitive tasks. In: 8th IEEE International Symposium on Mixed and Augmented Reality (ISMAR), Orlando, FL (2009)
4. Monahan, T., McArdle, G., Bertolotto, M.: Virtual reality for collaborative e-learning. Comput. Educ. **50**(4), 1339–1353 (2008)
5. Parong, P., Mayer, R.: Learning science in immersive virtual reality. J. Educ. Psychol. **110**(6), 785–797 (2018)
6. Popov, S., et al.: Virtual reality as educational technology. In: 2019, 29th Annual Conference of the European Association for Education in Electrical and Information Engineering (EAEEIE), pp. 1–6. Ruse, Bulgaria (2019). https://doi.org/10.1109/EAEEIE46886.2019.9000438
7. Kalawsky, R.: Science of Virtual Reality and Virtual Environments, 2nd edn. Addison-Wesley, Boston (2004)
8. Wismer, P., Lopez Cordoba, A., Baceviciute, S.: Immersive virtual reality as a competitive training strategy for the biopharma industry. Nat. Biotechnol. **39**, 116–119 (2021). https://doi.org/10.1038/s41587-020-00784-5
9. Mekenyan, O.: Dynamic QSAR techniques: applications in drug design and toxicology. Curr. Pharm. Des. **8**(17), 1605–1621 (2002)
10. Mekenyan, O., Dimitrov, D., Nikolova, N., Karabunarliev, S.: Conformational coverage by a genetic algorithm. J. Chem. Inf. Comput. Sci. **39**(6), 997–1016 (1999)
11. Pavlov, T., et al.: Conformational coverage by a genetic algorithm: saturation of conformational space. J. Chem. Inf. Model. **47**(3), 851–863 (2007)
12. Mekenyan, O., Pavlov, T., Grancharov, V., Todorov, M., Schmieder, P., Veith, G.: 2D–3D migration of large chemical inventories with conformational multiplication. Application of the genetic algorithm. J. Chem. Inf. Model. **45**, 283–292 (2005)
13. Li, J., Gramatica, P.: The importance of molecular structures, endpoints' values, and predictivity parameters in QSAR research: QSAR analysis of a series of estrogen receptor binders. Mol. Divers. **14**, 687–696 (2010)
14. Stadlinger, B., Jepsen, S., Chapple, I., Sanz, M., Terheyden, H.: Technology-enhanced learning: a role for video animation. Br. Dent. J. **230**, 93–96 (2021)

In-silico Investigation of Human Visual System

Petia Koprinkova-Hristova[1](✉), Simona Nedelcheva[1], and Nadejda Bocheva[2]

[1] Institute of Information and Communication Technologies, Bulgarian Academy of Sciences, Acad. G. Bonchev Str., Bl. 25A, 1113 Sofia, Bulgaria
petia.koprinkova@iict.bas.bg

[2] Institute of Neurobilogy, Bulgarian Academy of Sciences, Acad. G. Bonchev Str., Bl. 25, 1113 Sofia, Bulgaria
nadya@percept.bas.bg

Abstract. The paper presents in-silico investigations of the effects of brain lesions in different parts of human visual system. For this aim, a hierarchical spike timing neural network model reproducing performance of visual tasks with reinforcement learning by humans was implemented in NEST simulator. Its structure and connectivity is designed according to available information about corresponding brain areas organization. The model has two basic sub-structures: a perceptual part involved in visual information processing and perceptual decision making and basal ganglia that biases taken decisions according to received external reinforcement signal. The developed software includes also an option to introduce "damage" in each one of the model sub-structures thus allowing performance of in-silico investigations of effect of brain lesions. The simulations were performed feeding the model with a dynamic visual stimulus consisting of moving dots. Different scenarios including damage in single or several brain areas were prepared and model reactions were collected. Observed deterioration of visual task performance were summarized and commented.

Keywords: Spike-timing neural network · Visual system · Decision making · Reinforcement learning

1 Introduction

The studies of human nervous system and in particular of human brain has long history dated up to the ancient Egypt. Since then the brain research has gone through various phases such as philosophical, experimental, and theoretical ones. The modern neuroscience has already accumulated a lot of knowledge about the functional specialization of different brain areas as well as of individual neural cells (neurons) functionality. Nowadays, accumulated multi-scale brain data require the need for work on brain simulation [5].

S. S. Sotirov et al. (Eds.): BioInfoMed 2020, LNNS 374, pp. 224–235, 2022.
https://doi.org/10.1007/978-3-030-96638-6_25

The present work is focused on probably the most investigated part of the human brain - the visual system. The systematic investigations accumulated by far rich knowledge about its hierarchical structure performing the visual information processing starting from the eyes (light sensors) through the optic nerve to the visual cortex as well as about deeper brain areas related to perception and decision making.

Accounting for available information about structure and functionality of human visual system we developed an integrated spike timing neural network model that includes all major structures related to dynamic visual information processing starting from the retinal input and up to perceptual and decision-making areas including abilities for reinforcement learning [12,14].

In the present work, we conducted an in-silico investigation of the human visual system using our model. The aim of the study is to mimic lesions in different brain areas related to visual information processing and to observe their effect on the perceptual as well as biased by external reinforcement decision making.

The rest of the paper is organized as follows: Sect. 2 briefly describes the structure of our model; next, simulation experimental set-up was presented; results were discussed in the next Sect. 4; the paper finishes with concluding remarks and directions for future work.

2 Preliminary Research - Model Structure

Based on the available data about human brain structures playing role in visual motion information processing and decision making, as well as their connectivity, we developed a hierarchical spike timing neural network model consisting of two basic substructures: related to visual information perception and sensory-based decision making [12,13] shown in Fig. 1 and the basal ganglia that modulates the perceptual decision via external reinforcement [14,15], shown in Fig. 2. The overall model structure was implemented in NEST simulator [18].

Each coloured rectangle in both figures denotes a layer of neurons positioned in a regular two-dimensional grid. The names of the layers correspond to the brain structures they are representing.

Following the commonly accepted models from [26], the reaction of retinal ganglion cells (RGC) to luminosity changes was simulated by a spatio-temporal filter whose spatial component has a circular shape modeled by a difference of two Gaussians and the temporal component with bi-phasic profile determined by the difference of two Gamma functions. It generates a continuous signal by convolution of the spatio-temporal kernel with the visual stimuli (images falling on the retina). For details of our parallel implementation of that part of the model see [21].

The neurons in lateral geniculate nucleus (LGN) area were simulated by equations for integrate and fire (IAF) model whose parameters were determined from in-vivo experiments as in [2]. The rest of neurons in our model were simulated according to the IAF equations proposed in [27].

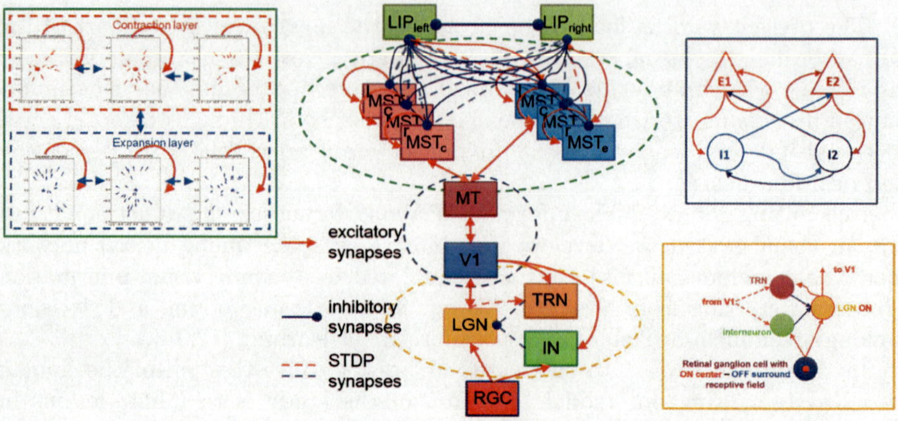

Fig. 1. Structure of perceptual part of the model as in [13].

The synapses (connections between neurons) are drawn by colour lines denoting their functionality as follows:

- Red arrows for excitatory connections; thick lines denote static while dashed lines - dynamic synapses;
- Blue dot ending lines for inhibitory connections; thick lines denote static while dashed lines - dynamic synapses;
- Violet arrows for dopamine dependent synapses;
- Dark blue arrows for anti-dopamine dependent synapses;
- Diamond ending dashed pink lines for dopamine releasing synapses.

We have two types of dynamic synapses: with spike timing-dependent plasticity (STDP) as in [24] and with dopamine-dependent plasticity as in [23]. Since our model also includes anti-dopamine synapses whose dynamics have to be opposite to that of dopaminergic ones, we've modified the model from [23] by converting the amplitudes A_+ and A_- of the dopamine eligibility trace dynamics from positive to negative.

The structure of the perceptual part of the model (Fig. 1) was developed and described in detail in our previous works [11,12,20].

The structure of the lateral geniculate nucleus (LGN) that transfers the signal generated by the sensory layer of our model (retina ganglion cells (RGC)) to the visual cortex (V1) was adopted from [7]. It includes also thalamic reticular nucleus (TRN) and interneurons (IN) that mediate the feedback/forward connectivity between LGN and V1 as shown in right bottom corner yellow rectangle of Fig. 1.

The structure and connectivity of the visual cortex (V1) and middle temporal (MT) areas was adopted from [16]. The neurons in these areas are orientation (V1) and direction (MT) sensitive and have elongated receptive fields defined by a Gabor probability function with orientation and phase parameters like in [8]. The 2D maps containing the neurons' orientations and phases at the corresponding 2D grid of the V1 and MT layers have typical for the mammalian brain"pinwheel-structure" generated by the approach from [25] (details are reported in [20]). The lateral connections in both V1 and MT areas follow the push-pull connectivity shown in right top blue rectangle of Fig. 1 as described in [16] while the connections from V1 to MT cells depend on the angle between orientation preferences of each pair of cells according to [4].

The medial superior temporal area (MST) was modeled like in [19] by two layers sensitive to expansion and contraction movement patterns having a circular shape on-center Gaussian receptive fields. Both layers have intra- and interlayer excitatory/inhibitory recurrent connections between cells having similar/different sensitivity as shown in left top green rectangle of Fig. 2 (for more details see [12]).

Lateral intra-parietal (LIP) area is the last layer of the perceptual part of the model that is responsible for making decisions based on accumulated sensory evidence. Since our model aims to decide whether the expansion center of moving dot stimulus is left or right from the stimulus center, in [12] we proposed a task-dependent design of excitatory/inhibitory connections as shown on left part of the Fig. 1. As in [13] the model has a complete set of feedback/forward connections between every two consecutive layers.

There is accumulated evidence (e.g. [1,6,10]) suggesting a significant role of the basal ganglia in learning by trial and error from external reinforcement signal telling us whether our perceptually based decisions are good or bad. The basal ganglia (BG) structure in our model is a combination of structures proposed in [9,17]. It is shown on the right part of Fig. 2. Following well known from literature structures (e.g. [17]), our model incorporates layers corresponding to the brain areas called Striatum (composed of D1 and D2 parts), Globus Pallidus externa (GPe), Subthalamic Nucleus (STN), and Substantia Nigra pars reticulata (SNr). Like in [9] it consists of two parallel structures connected via mutually inhibiting connections through their GPe areas that receive inputs from the left and right decision selecting LIP areas, respectively. The lateral connectivity in all layers was adopted from [17].

The striatum is divided into two sub-areas depending on the type of dopamine receptors they express (D1 and D2 on Fig. 2). Their lateral connections have short-range excitation and long-range inhibition characteristics like in [17]. The cortical input (coming from the LIP layer - the last of the sensory information processing part of our model) has dopaminergic synapses whose weights were randomly initialized and they are dynamically changed in dependence on the spiking activity of SNc area (considered as dopamine secreting structure).

The Globus Pallidus externa (GPe) and Subtalamic Nucleus (STN) form a common structure consisting of pairs of neurons connected one-to-one via glutamatergic (excitatory) and GABAergic (inhibitory) connections as shown on Fig. 2. The GPe layer has also lateral connections having negative center and positive surround shape as in [17].

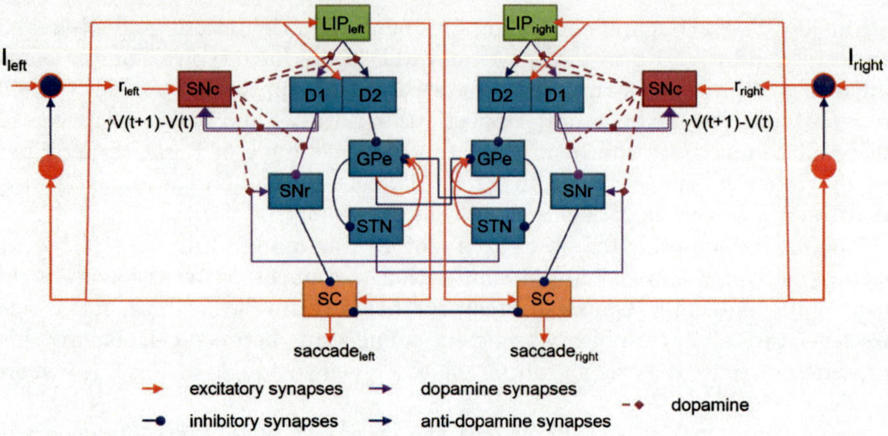

Fig. 2. Basal ganglia structure as in [14].

Substantia Nigra pars reticulata (SNr) has short-range excitatory and long-range inhibitory lateral connections like both Striatum layers.

The motor-reaction controlling structure superior colliculus (SC) receives inputs directly from the LIP area (involved in decision-making according to accumulated sensory information) as well as from the external reinforcement modulated output of BG (via SNr). The overall model connectivity is also enhanced by excitatory feedback connections from SC to their corresponding D1 and D2 areas of the Stratum as well as to LIP areas following recently reported findings [22,29,30]. Moreover, we introduced mutually inhibiting connections between the two SC groups.

Additionally, our model has a complete 2D layer of neurons corresponding to the Substantia Nigra pars compacta (SNc) area in the brain producing the neuromodulator (dopamine) proportional to the temporal difference error $\delta(t)$ (for more details see [1]) as follows:

$$\delta(t) = SNc = F(r(t) + \gamma D1(t+1) - D1(t)) \tag{1}$$

The value function is associated with D1 activity while the reinforcement signals on both left and right parts of the model r_{left} and r_{right} respectively are generated as the difference between SC left and right reactions and the external (teaching) input currents (I_{left} and I_{right} respectively). They affect both dopamine-producing areas (SNc left and right) (for details see [14]). The discount factor γ was set to 0.9.

3 Simulation Experiments

The visual stimulation (input of the model) consists of moving dot patterns with an imaginary expansion center to the left or right from the visual scene center as shown in Fig. 3. Detailed information about stimuli format is given in our previous works [12,21].

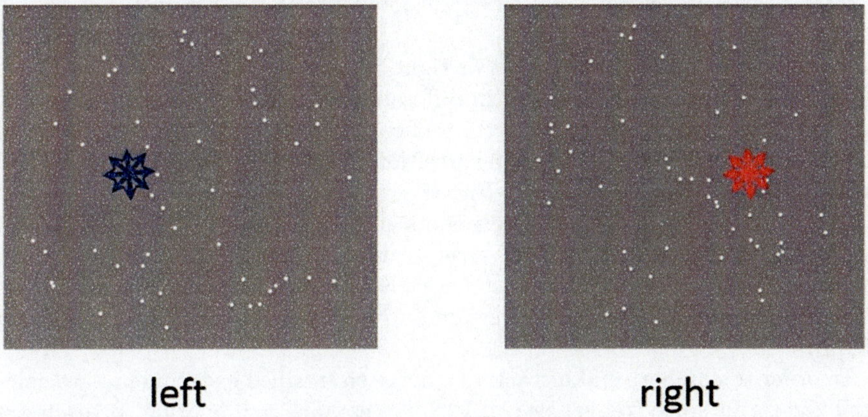

Fig. 3. Examples of stimuli.

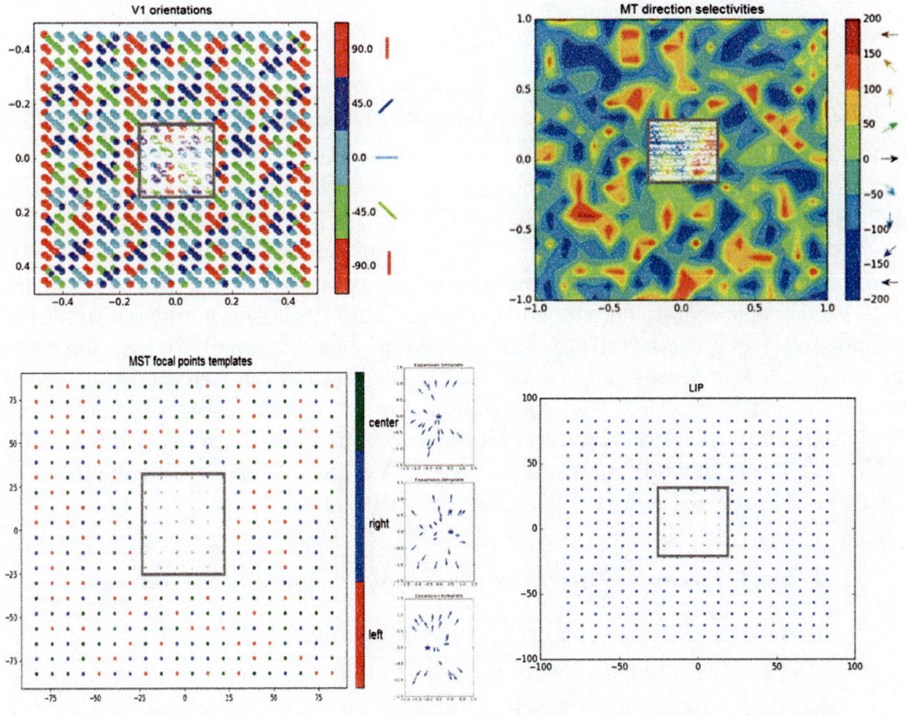

Fig. 4. Examples of lesions at centers of V1, MT, MST and LIP areas.

The model, like human test subjects in our experiments reported in [12], has to react to the perceived stimulus expansion center with a perceptual decision (increased spiking activity in LIP left or right area) as well as on the external reinforcement in corresponding left or right SC areas.

In order to introduce "lesions" in different areas, the software was upgraded with graphical tool projecting structures of consecutive model layers and allowing the user to chose manually "lesion" position. Examples of lesions in V1, MT, MST, and LIP areas are shown in Fig. 4.

We performed simulation experiments inducing lesions of different size in single and multiple areas (model layers) and recorded spiking activity of all neurons. For the sake of comparison we also performed simulations with and without reinforcement signal corresponding to the perceptual decision like in [13,14].

In order to compare spiking activity between in-silico experiments, we calculated the mean spike frequency (MSF) for a complete simulation for each area (layer) as in [28] as follows:

$$MSF_{layer} = n_{spikes}/t_{simulation} \tag{2}$$

Here n_{spikes} is the total number of spikes per neuron:

$$n_{spikes} = (\sum_{i=1}^{N_{layer}} n_{spikes_i})/N_{layer} \tag{3}$$

of a given *layer* consisting of N_{layer} neurons and $t_{simulation}$ is simulation time in seconds.

4 Results and Discussion

The moving dot stimulus in all simulations has expansion center to the left from the visual scene center, i.e. the correct perceptual decision should be increased spiking activity in the left LIP and SC areas. Its duration was 1670 ms. Since the duration of spatio-temporal kernel of RCG cells was 150 ms (see [21]) the overall simulation time was 1520 ms. The model dimensions as reported in [14] were 7400 spike timing neurons totally (2400 neurons in thalamic relay composed by LGN, TRN and IN; 1000 neurons in each of V1 and MT areas; 800 neurons in each of MST and LIP areas; 1400 neurons in BG area).

Table 1. Mean spike frequency per group of neurons without lesions.

RL	SC_{left}	SC_{right}	LIP_{left}	LIP_{right}	$MSTe$	MT	$V1$
Yes	303.5987	1.9934	207.3372	14.1053	8.1546	10.9191	14.3007
No	302.2303	0.9408	206.4589	13.9161	8.1414	11.0020	14.2993

Table 2. Mean spike frequency per group of neurons in case of lesions in V1.

Damage, %	*RL*	SC_{left}	SC_{right}	LIP_{left}	LIP_{right}	*MSTe*	*MT*	*V1*
6.8	Yes	303.5263	1.9934	207.3339	14.1020	8.1595	11.7309	13.4178
6.8	No	302.1513	0.9408	206.4391	13.9178	8.1398	11.0954	12.7829
30.2	Yes	307.0568	1.9605	210.1941	12.8766	8.3026	9.3303	9.6118
30.2	No	301.7434	0.9013	205.9342	14.2484	8.1431	9.4138	9.8039

Table 3. Mean spike frequency per group of neurons in case of lesions in MT.

Damage, %	*RL*	SC_{left}	SC_{right}	LIP_{left}	LIP_{right}	*MSTe*	*MT*	*V1*
2.4	Yes	303.6316	1.9934	207.3059	14.1036	8.1546	10.4717	14.3914
2.4	No	302.1711	0.9408	206.4556	13.9145	8.1414	10.2789	14.3592
6.2	Yes	303.5724	1.9934	207.3289	14.1053	8.1546	10.6053	14.3783
6.2	No	302.1250	0.9408	206.4819	13.9161	8.1398	10.5480	14.3625
9	Yes	303.5658	1.9934	207.3372	14.1036	8.1513	9.5158	14.4151
9	No	302.0395	0.9408	206.4622	13.9161	8.1382	9.5026	14.4230

Table 4. Mean spike frequency per group of neurons in case of lesions in MSTe.

Damage, %	*RL*	SC_{left}	SC_{right}	LIP_{left}	LIP_{right}	*MSTe*	*MT*	*V1*
5	Yes	3.4671	298.0197	1.1299	217.9062	6.4655	11.0026	14.3046
5	No	302.2368	1.3553	206.2928	14.1414	7.8701	10.9822	14.3145
6	Yes	3.3947	298.2763	1.1250	217.7105	6.0329	10.9967	14.3033
6	No	3.7566	296.6053	1.1168	217.1743	6.1184	10.9967	14.3046

Table 5. Mean spike frequency per group of neurons in case of lesions in LIP.

Damage, %	*RL*	SC_{left}	SC_{right}	LIP_{left}	LIP_{right}	*MSTe*	*MT*	*V1*
6.2	Yes	3.3421	298.9605	1.0921	218.2434	7.0592	11.0000	14.3079
6.2	No	3.8224	296.3947	1.1316	217.2434	7.0592	10.9941	14.3039
9	Yes	3.5461	297.4211	1.1316	217.5263	6.9342	11.0013	14.3026
9	No	3.9539	295.5329	1.1414	216.8026	6.9293	10.9500	14.3178

Table 6. Mean spike frequency per group of neurons in case of lesions in V1, MT and MSTe.

Damage, %	*RL*	SC_{left}	SC_{right}	LIP_{left}	LIP_{right}	*MSTe*	*MT*	*V1*
16	Yes	303.3421	0.7303	207.0280	13.1809	6.9112	8.7112	11.9197
16	No	305.0855	0.8092	208.5033	12.5970	7.0033	8.7368	11.9132
25	Yes	302.6776	1.4803	206.1875	13.8174	5.9539	7.2789	10.4447
25	No	302.5263	0.6776	205.8158	13.5905	5.9326	7.2250	10.5493

Tables 1, 2, 3, 4, 5 and 6 present MSFs for all observed during simulation layers having different percentage of damage (lesions' sizes) at the center of the corresponding area with and without reinforcement signal.

Table 1 shows MFS in case of model simulation without lesions. Next, Table 2 demonstrates that with the increase of lesion's size in V1 area the MSF in V1 decreases. The decrease in case of presence of reinforcement signal supporting perceptual decision is a little bit smaller for smaller lesion's sizes. The activity in the next MT area also decreases with the increase of lesions' size but here the effect of reinforcement signal is bigger for smaller lesion size. The reinforcement, however, provokes an increase of MST activity in an attempt to overcome the effect of damage in V1. We also observed that the reinforcement signal helps to overcome damage even more in the perceptual decision area (LIP) as well as the final decision in SC.

Results from simulation with lesions in MT area in Table 3 demonstrate that the model increases slightly MSF in the V1 area in an attempt to overcome the damage effect but over some size of the lesion (6.2%) the activity in the damaged MT area decreases. The model however manages to support approximately the same level of MSF in both LIP and SC areas. The presence of reinforcement signal here provokes higher MSF in both perceptual (LIP) and final decision (SC) areas in comparison with simulation without reinforcement.

Results in case of damages in MST area in Table 4 demonstrate a faster decrease in MST activity with the increase of lesion's size in case of simulation with reinforcement signal in comparison with simulation without reinforcement. The model tries to increase the activity in the MT and V1 area in order to compensate for damage in the upper MST area. In this case, however, the lesions above 5% prevent the model to take the proper decision (left) so the taken perceptual and final decision is the wrong one (right).

In case of damage in the perceptual decision area (LIP), simulation results in Table 5 demonstrate that even small lesions prevent the model to take the proper decision like in previous case (lesions in MST area). The reinforcement signal in this case increases activity in V1 and MT areas in an attempt to provoke proper decision while in the MST area the MSF decreases in both cases (with or without reinforcement).

In the case of equal damage size in all V1, MT, and MST areas (Table 6) we observed that MSF decreases in all three damaged areas. The MSF in decision areas (LIP and SC) however is different in the case of presence and absence of external reinforcement signal. For smaller damage size (16%) the model without reinforcement tries to overcome its effect with increased activity in the correct (left) areas of LIP and SC but for bigger lesions (25%) the reinforcement signal manages to support a little bit higher MSF in both areas. In contrast, reinforcement signal tries to decrease the activity in the wrong (right) part of the model output but for bigger lesions the model tends to make the wrong decision again.

5 Conclusion

In conclusion, damage in a given structure has a significant effect on its activity; damages in the earlier motion processing areas (V1 and MT) have a smaller effect on the model output structures (LIP and SC) while damages in the higher-level structures (LIP and MST) have more significant effect on the model ability to take proper decision while it has almost insignificant effect on the activity in earlier motion processing areas (e.g. V1).

An explanation could be found in the "pin-wheel" structure of orientation and direction selectivity in V1 and MT layers as their neurons are organized in space clusters having various selectivity properties that make them much more robust to the damages.

In contrast, in MST layer we have only three template types having focal points at the center, left and right positions respectively. We suppose that if we enrich that area with a bigger number of templates having more diverse focal points, it would become also more robust to damages like V1 and MT areas. However, in contrast to the receptive fields of neurons in the V1 and MT areas, the connection templates in the MST gather information from neurons having very diverse characteristics that probably make them more vulnerable to the damages.

Concerning the LIP area, its neurons do not have such diverse specialization and their activity depends on the exciting/inhibiting connectivity from the MST area that is artificially designed to be task-dependent in our model.

The effect of the reinforcement signal is more significant in the areas close to the BG (i.e. LIP and SC areas) while it is almost neglectful on the areas far from it (i.e. V1 and MT). This might be due to the fact that the damaging effect is concentrated mainly in the area it belongs to. The relatively stronger effect of reinforcement signal on the MST area might also be explained by the fact that all simulated damages were positioned at the center of the visual field, affecting the center of visual stimulus, which is of primary importance for the proper model response.

In the future we plan to enrich the templates type in the MST area in order to test its robustness. We also plan to design task-independent connectivity from MST to LIP areas thus to allow both left and right areas to compete for the final decision. This could be achieved by the introduction of other spike timing dependent plasticity rules like Clopath synapses [3] that were recently implemented in NEST simulator.

Acknowledgements. The present research has been supported by the Bulgarian National Science Fund under Grant Ref. No. DN02/3/2016 "Modelling of voluntary saccadic eye movements during decision making".

References

1. Barto, A.G.: Adaptive critics and the basal ganglia. In: Houk, J.C., Davis, J.L., Beiser, D.G. (eds.) Models of Information Processing in the Basal Ganglia, pp. 215–232. MIT Press, Cambridge (1995)
2. Casti, A., Hayot, F., Xiao, Y., Kaplan, E.: A simple model of retina-LGN transmission. J. Comput. Neurosci. **24**, 235–252 (2008)
3. Clopath, et al.: Connectivity reflects coding: a model of voltage-based STDP with homeostasis. Nat. Neurosci. **13**(3), 344–352 (2010)
4. Escobar, M.-J., Masson, G.S., Vieville, T., Kornprobst, P.: Action recognition using a bio-inspired feedforward spiking network. Int. J. Comput. Vis. **82**, 284–301 (2009)
5. Fan, X., Markram, H.: A brief history of simulation neuroscience. Front. Neuroinform. **13** (2019). https://doi.org/10.3389/fninf.2019.00032
6. Frank, M.J., Seeberger, L.C., O'Reilly, R.C.: By carrot or by stick: cognitive reinforcement learning in parkinsonism. Science **306, 5703**, 1940–1943 (2004). https://doi.org/10.1126/science.1102941
7. Ghodratia, M., Khaligh-Razavic, S.-M., Lehky, S.R.: Towards building a more complex view of the lateral geniculate nucleus: recent advances in understanding its role. Prog. Neurobiol. **156**, 214–255 (2017)
8. Gleeson, P., Martinez, R., Davison, A.: Network models of V1 (2016). http://www.opensourcebrain.org/projects/111
9. Igarashi, J., Shounob, O., Fukai, T., Tsujino, H.: Real-time simulation of a spiking neural network model of the basal ganglia circuitry using general purpose computing on graphics processing units. Neural Netw. **24**, 950–960 (2011)
10. Joel, D., Niv, Y., Ruppin, E.: Actor-critic models of the basal ganglia: new anatomical and computational perspectives. Neural Netw. **15**, 535–547 (2002)
11. Koprinkova-Hristova, P., Bocheva, N., Nedelcheva, S.: Investigation of feedback connections effect of a spike timing neural network model of early visual system. In: Innovations in Intelligent Systems and Applications (INISTA), Thessaloniki, Greece, 3–5 July 2018 (2018). https://doi.org/10.1109/INISTA.2018.8466292
12. Koprinkova-Hristova, P., Bocheva, N., Nedelcheva, S., Stefanova, M.: A model of self-motion perception developed in NEST. Front. Comput. Neurosci. (2019). https://doi.org/10.3389/fncom.2019.00020
13. Koprinkova-Hristova, P., et al.: STDP plasticity in TRN within hierarchical spike timing model of visual information processing. In: Maglogiannis, I., Iliadis, L., Pimenidis, E. (eds.) AIAI 2020. IAICT, vol. 583, pp. 279–290. Springer, Cham (2020). https://doi.org/10.1007/978-3-030-49161-1_24
14. Koprinkova-Hristova, P., Bocheva, N.: Brain-inspired spike timing model of dynamic visual information perception and decision making with STDP and reinforcement learning. In: Nicosia, G., et al. (eds.) LOD 2020. LNCS, vol. 12566, pp. 421–435. Springer, Cham (2020). https://doi.org/10.1007/978-3-030-64580-9_35
15. Koprinkova-Hristova, P., Bocheva, N.: Spike timing neural model of eye movement motor response with reinforcement learning. In: Georgiev, I., Kostadinov, H., Lilkova, E. (eds.) BGSIAM 2018. SCI, vol. 961, pp. 139–153. Springer, Cham (2021). https://doi.org/10.1007/978-3-030-71616-5_14
16. Kremkow, J., et al.: Push-pull receptive field organization and synaptic depression: mechanisms for reliably encoding naturalistic stimuli in V1. Front. Neural Circ. (2016). https://doi.org/10.3389/fncir.2016.00037
17. Krishnan, R., Ratnadurai, S., Subramanian, D., Chakravarthy, V.S., Rengaswamyd, M.: Modeling the role of basal ganglia in saccade generation: is the indirect pathway the explorer? Neural Netw. **24**, 801–813 (2011)

18. Kunkel, S., et al.: NEST 2.12.0. Zenodo (2017). https://doi.org/10.5281/zenodo.259534
19. Layton, O.W., Fajen, B.R.: Possible role for recurrent interactions between expansion and contraction cells in MSTd during self-motion perception in dynamic environments. J. Vis. **17**(5), 1–21 (2017)
20. Nedelcheva, S., Koprinkova-Hristova, P.: Orientation selectivity tuning of a spike timing neural network model of the first layer of the human visual cortex. In: Georgiev, K., Todorov, M., Georgiev, I. (eds.) BGSIAM 2017. SCI, vol. 793, pp. 291–303. Springer, Cham (2019). https://doi.org/10.1007/978-3-319-97277-0_24
21. Nedelcheva, S., Georgieva, K., Koprinkova-Hristova, P.: Parallel implementation of the model of retina ganglion cells layer. In: 2020 International Conference on Innovations in Intelligent Systems and Applications (INISTA), Novi Sad, Serbia, pp. 1–6 (2020). https://doi.org/10.1109/INISTA49547.2020.9194616
22. Plotkin, J.L., Goldberg, L.A.: Thinking Outside the Box (and Arrow): Current Themes in Striatal Dysfunction in Movement Disorders, The Neuroscientist (2018). https://doi.org/10.1177/1073858418807887
23. Potjans, W., Morrison, A., Diesmann, M.: Enabling functional neural circuit simulations with distributed computing of neuromodulated plasticity. Front. Comput. Neurosci. **4** (2010). https://doi.org/10.3389/fncom.2010.00141many
24. Rubin, J., Lee, D.D., Sompolinsky, H.: Equilibrium properties of temporally asymmetric Hebbian plasticity. Phys. Rev. Lett. **86**(2), 364–367 (2001)
25. Sadeh, S., Rotter, S.: Statistics and geometry of orientation selectivity in primary visual cortex. Biol. Cybern. **108**(5), 631–653 (2013). https://doi.org/10.1007/s00422-013-0576-0
26. Troyer, T.W., Krukowski, A.E., Priebe, N.J., Miller, K.D.: Contrast invariant orientation tuning in cat visual cortex: thalamocortical input tuning and correlation-based intracortical connectivity. J. Neurosci. **18**, 5908–5927 (1998)
27. Tsodyks, M., Uziel, A., Markram, H.: Synchrony generation in recurrent networks with frequency-dependent synapses. J. Neurosci. **20RC50**, 1–5 (2000)
28. Van Dijck, G., Van Hulle, M.M., Heiney, S.A., Blazquez, P.M., Meng, H., et al.: Probabilistic identification of cerebellar cortical neurones across species. Plos One **8**(3), e57669 (2013). https://doi.org/10.1371journal.pone.0057669
29. Wei, W., Rubin, J.E., Wang, X.-J.: Role of the indirect pathway of the basal ganglia in perceptual decision making. J. Neurosci. **35**(9), 4052–4064 (2015)
30. Yan, H., Wang, J.: Quantification of motor network dynamics in Parkinson's disease by means of landscape and flux theory. PLoS One **12**(3), e0174364 (2017)

Sequence-Based Prediction of Food-Originated ACE Inhibitory Peptides Using Deep Learning Algorithm

Margarita Terziyska[1(✉)], Ivelina Desseva[1], and Zhelyazko Terziyski[2]

[1] University of Food Technologies Plovdiv, Plovdiv, Bulgaria
m.terziyska@uft-plovdiv.bg

[2] University of Plovdiv Paisii Hilendarski, Plovdiv, Bulgaria

Abstract. In recent years, food originated bioactive peptides became a promising source of potential therapeutic agents. Predicting the biological activity of these peptides is crucial for the discovery and development of functional foods and effective peptides-based drugs. Antihypertensive peptides (AHTPs) are certainly the most reported food-derived peptides. These peptides inhibit a key enzyme in renin-angiotensin system, named angiotensin-converting enzyme (ACE), resulting in lowering of blood pressure. So far, AHTPs are obtained mainly by *in vitro* and *in vivo* protocols. This is a rather expensive and time-consuming procedure, and often require months of hard work, which is not always successful. To overcome this shortcoming, machine learning (ML) approaches are increasingly used.

In this study, a Long Short Term Memory (LSTM) is used for prediction of food-derived ACE inhibitory peptides. It was chosen this recurrent deep neural network as the most suitable for sequence-based prediction. The positive datasets are collected from the following food-derived peptide databases AHTPDB, FeptideDB, BIOPEP-UWM and BioPepDB, while the negative ones peptides without antihypertensive function were gathered. Then, the feature descriptors are generated via the Chou's pseudo amino acid composition method. They are inputted to the deep neural network classifier. Finally, the proposed LSTM approach is compared with Random Forest (RF) and Support Vector Macines (SVM) classifiers. It was demonstrated by 5-fold cross-validation that the deep learning algorithm has higher predictive accuracy than the other ML algorithms. This makes it suitable for identification of AHTPs.

Keywords: Deep learning · Machine learning · LSTM · Neural networks · Bioactive peptides · Antihypertensive peptides · Food-derived peptides

1 Introduction

Hypertension is a complex disease that affects billions of people around the world [1]. Its regulation depends mainly on a series of enzymes of the so-called renin - angiotensin - aldosterone system [2]. The commercial drugs used for lowering the blood pressure are in fact inhibitors of angiotensin-converting enzyme (ACE). Due to the many side effects

S. S. Sotirov et al. (Eds.): BioInfoMed 2020, LNNS 374, pp. 236–246, 2022.
https://doi.org/10.1007/978-3-030-96638-6_26

of these drugs, safer alternatives are constantly being sought. In this regard, research on peptides with such properties remains relevant.

Bioactive peptides are specific amino acids sequences that may exert beneficial properties on the body and have a potentially positive effect on human health that can be measurable [3]. They can be directly produced by the living cells or derived by the protein precursors. Food proteins possess a great potential to serve as sources of bioactive peptides, in addition to their nutritional value. The release of such peptides could be done by three general ways: microbial fermentation; gastrointestinal digestion or proteolytic hydrolysis. All of the three approaches consist of enzymatic proteolysis of proteins and their fragmentation to peptides and amino acids.

The antihypertensive peptides (AHTPs) are one of the most studied food-derived peptides. In general, the amino acid composition and sequence and the resulting molecular weight contribute significantly to the ACE-inhibitory properties of the peptides. It is found that the presence of hydrophobic (proline) and aliphatic amino acids (isoleucine and leucine) enhance the inhibitory activity [4]. So far, various protein sources of AHTPs are reported and summarized in many reviews [5, 6]. Certainly, milk proteins from different species are leading the ranking. The casein-derived three peptides IPP and VPP are the best characterized *in vitro* and *in vivo* [7, 8]. Recent studies even demonstrate that in addition to their antihypertensive function these peptides enhance insulin sensitivity and prevent insulin resistance [9].

Recently, there has been a growing interest in ML techniques used to predict the biological activity of peptides (BAPs). These techniques are an alternative to experimental *in vitro* and *in vivo* protocols, which are expensive and time consuming. In the scientific literature, numerous developments for identification peptides with different activity using ML techniques have been reported. Most of them used SVM [10, 11] and random forest techniques [12–14]. Concrete, for AHTPs, not much has been done in terms of their prediction with ML techniques. For the last 5 years, only a few predictive models have been proposed. In [15], the authors divided peptides into four groups depends on their length: tiny (2–3 residues), small (4–6 residues), medium (7–12 residues), and larger (>12 residues) peptides. For the tiny SVM-based regression models was used and for the others, an SVM classifier was employed. Three different feature representations, namely AAC, ATC and chemical descriptors, are used in this predictor, called AHTpin. The achieved average accuracy is 78.58%. A web server for predicting AHTPs, named PAAP and based on the RF classifier as a function of various combinations of amino acid, dipeptide and pseudoamino acid composition descriptors is presented in [16]. Classification models have been assessed via independent test dataset that demonstrated accuracy of 84.73%. Six different ML algorithms, namely, Adaboost, extremely randomized tree (ERT), gradient boosting (GB), k-nearest neighbor (k-NN), RF, and SVM using 51 feature sets derived from eight different feature encodings for the prediction of AHTPs underlie in the meta-predictor mAHTPred presented in [17]. Comparison of the three predictors discussed above shows that the performance of mAHTPred was the best and most robust, followed by those of PAAP and AHTpin [18]. A novel ML strategy for prediction of food-derived AHTPs is introduced in [19]. It is based on Extreme Gradient Boosting (XGBoost) algorithm composed of many single tree models. A PseAAC algorithm have been used as the method for feature encoding. This new model have been compared with k-NN, RF and SVM classifiers. The obtained results show that the XGBoost model achieve outstanding performance with the accuracy of 0.9841 and the area under the receiver operating characteristic curve of 0.9428, which is better than the other models.

In recent years, deep learning (DL) architectures such as DNNs, deep belief networks, recurrent neural networks and convolutional neural networks have become extremely popular. Many recent studies show the effectiveness of DL algorithms mainly in antimicrobial [20, 21] and anticancer [22] prediction. However, they are not currently used to predict antihypertensive peptides. Therefore, in this study a Long Short Term Memory (LSTM) is proposed for prediction of food-derived AHTPs. It was chosen LSTM since such structures achieved astounding results across a wide variety of application domains where data is sequential [23] (e.g. prediction of protein structure or protein function from the amino acid sequence which are classical bioinformatics problem etc.). The gathered from public databases positive and negative peptide sequences are transformed by the Chou's pseudo amino acid composition (PseAAC) method. The obtained features are fed as an input to train the LSTM model. To evaluate the performance of the proposed model, it was compared with two other ML algorithms, namely RF and SVM classifiers. It was demonstrated by 5-fold cross-validation that the deep learning algorithm has higher predictive accuracy than the other ML algorithms.

2 Methods

2.1 Dataset Constructing

As previous studies suggested, a new dataset have to be constructing that has to include both positive and negative samples. For the construction of the positive AHT peptide dataset, the FeptideDB [24], AHTPDB [25], BioPepDB [26] and BIOPEP-UWM [27] databases was used. The data was processing, the identical peptides was removed and finally, 3125 samples in the positive dataset was selected. According to [18] the negative samples should generally consist of the following peptides: (1) random peptides retrieved from the UniProt; (2) random shuffling of positive samples; or (3) peptides with different functions rather than the desired function. In this study, the option 3 was chosen and 2783 peptides without antihypertensive function were collected as negative samples. All the samples in the positive and negative datasets represent peptide sequence with length from 2 to 20 amino acids.

2.2 Peptides Representation

In order to be able to use the ML model, it is necessary to first transform the peptide sequences into fixed-length feature vectors. There are different approaches for protein and peptide representation, which can be categorized in two groups - sequence-based methods and non-sequential based methods. Generally, a peptide sequence can be represented as:

$$P = p_1 p_2 p_3 \dots p_L \tag{1}$$

where p_1 denotes the first residue in the peptide P, p_2 denotes the second residue in the peptide P and so forth. L denotes the length of P. Note that the residue p_i is an element of the standard amino acid alphabet {A, C, D, E, F, G, H, I, K, L, M, N, P, Q, R, S, T, V, W, Y}.

One of the most used and simple non-sequential method is Amino acid composition (ACC). In it, each peptide is represented with 20 features, each reflecting the occurrence of each amino acid in the sequence. Its main drawback is that it does not incorporate sequence-order information.

In this study, PseAAC, which is a modification on the classic AAC, was adopted as the feature encoding method of peptides.The concept of PseAAC was proposed by Kou-Chen Chou in 2001 [28] and nowadays it has been widely used in bioinformatics for prediction of bioactive peptides with different function. For example, PseAAC is applied in anticancer [29], antifungal [30], antihypertensive [23] and anti-HIV [31] peptides prediction.

The type II PseAAC was applied in present study. In this case, each peptide is represented by a vector with $20 + i\lambda$ dimensions, where i denotes the number of properties of the amino acid taken into consideration and λ is a coefficient that determines the distance of the interacted amino acids. To generate peptides representation it was used a web-based application PseAAC [32]. It was chosen the following properties of amino acids: hydrophobicity, hydrophilicity, mass, pK1 (alpha-COOH), pK2 (NH3) and pI (at 25 °C). The coefficient λ was set to 1 and the weight factor for the sequence order effect ω was set to 0.05. More details about the PseAAC algorithm and its parameters can be found in [33].

2.3 LSTM Model Architecture

It is well known that Neural Networks are a very powerful technique with many applications. One of their limitation is that, there is no memory associated with the model. To overcome this problem recurrent neural networks (RNNs) are widely used. These structures include a feedback loop, which serves as a kind of memory. In such a way the past inputs are taking to account in to the model. Long Short-Term Memory (LSTM) networks are a special kind of RNNs. LSTMs create both a short-term and a long-term memory component. Hence, LSTMs are great tool for application where has a sequence, such as Natural Language Processing (NLP), handwriting recognition, speech recognition, time series prediction and etc. LSTMs can also suitable to bioactive peptides prediction. That is the reason to choose LSTM in this study.

Many variants of the LSTM architecture have been proposed since their initial introduction in 1997 [34]. A large-scale comparative analysis of eight LSTM variants is presented in [35]. Here it was used the classical topology. Regardless of their variant, LSTMs represent a deep learning architecture based on the memory cell concept. The structure of the memory cell denoted C_j is shown on Fig. 1. The most important elements of LSTM cells are: *input gate* (gate that decides what should be forwarded to the next activation), *forget gate* (the gate that decides what information should be omitted), *output gate(*the gate that controls whether the information flow is passed on to the next LSTM unit*)* and *cell state* (the state of the cell passed in sequence after to the next steps). The three gates regulate the input-output information flow and thus control the state of the cell.

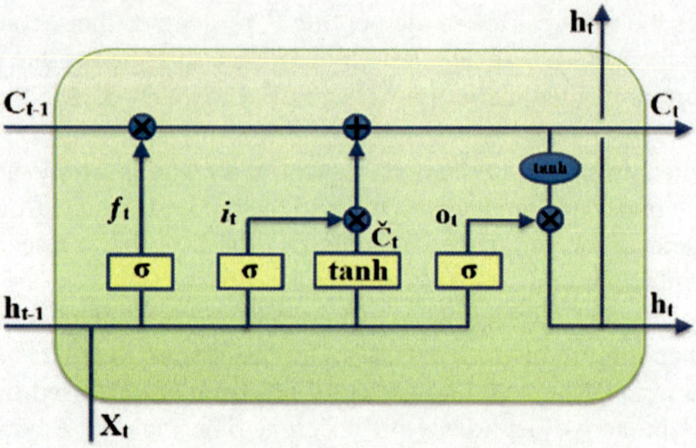

Fig. 1. Memory cell [36]

The **forget gate** take decision what information to throw away or kept from the previous cell. How do this? The forget gate takes the h_{t-1}, which is the output of the previous cell, and x_t, which represents the current cell input, then computes the output value between 0 and 1 by a sigmoid function. If the output is 1, all the information is kept; if the output is 0, all the information is discarded and for all other values it decides what part of the information from previous state is to be carried to the next state. Mathematically this can be described by the following expression:

$$f_t = \sigma\left(W_f \bullet \left[h_{t-1}, x_t\right] + b_f\right) \quad (2)$$

Then, a new sigmoid layer called, **input gate**, determine which information should enter the cell state. The output again is in the range of [0, 1]:

$$i_t = \sigma\left(W_i \bullet \left[h_{t-1}, x_t\right] + b_i\right) \quad (3)$$

Cell state is computed then by:

$$\tilde{C}_t = \tanh\left(W_C \bullet \left[h_{t-1}, x_t\right] + b_C\right) \quad (4)$$

where $\tilde{C}_t$ is a vector of new candidate values. Next step is to update the old cell state, C_{t-1}, into the new cell state C_t. It can be done by following expression:

$$C_t = f_t * C_{t-1} + i_t * \tilde{C}_t \quad (5)$$

Another sigmoid layer called, **output gate**, controls the extent to which the value in the cell is used to compute the output activation of the LSTM unit by:

$$o_t = \sigma\left(W_o \bullet \left[h_{t-1}, x_t\right] + b_o\right) \quad (6)$$

Finally, the output of the memory cell is obtain by:

$$h_t = o_t * \tanh(C_t) \quad (7)$$

The LSTM structure used in this study is shown in Fig. 2. It consists of LSTM layer with 60 memory cells, one dropout layer and one dense layer. A Dropout layer is used as a regularization method where input and recurrent connections to LSTM units are probabilistically excluded from activation and weight updates while training a network. This has the effect of reducing overfitting and improving model performance. A Dense layer feeds all outputs from the previous layer to all its neurons, each neuron providing one output to the next layer.

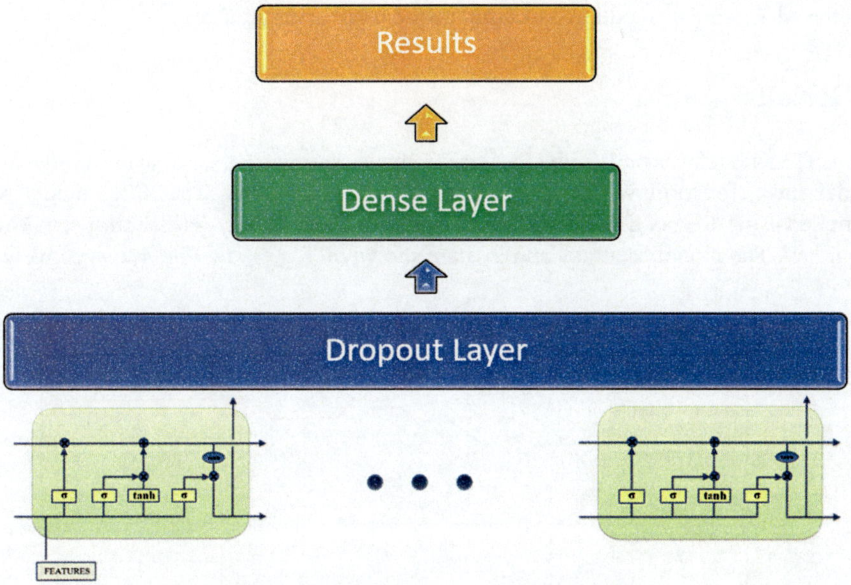

Fig. 2. Structure of the proposed LSTM

2.4 Prediction Performance Evaluation

In this study, it was used 5-fold cross-validation to evaluate the performance of the proposed deep learning LSTM model and comparison models. As performance metrics was chosen the most commonly used ones – accuracy (Acc), sensitivity (Sn), specificity (Sp), Precision (Prec), and Mattew correlation coefficient (MCC) which are computed by following expressions [18]:

$$Acc = \frac{TN + TP}{TN + TP + FN + FP} \tag{8}$$

$$Sn = \frac{TP}{TP + FN} \tag{9}$$

$$Sp = \frac{TN}{TN + FP} \tag{10}$$

$$\Pr ec = \frac{TP}{TP + FP} \tag{11}$$

$$MCC = \frac{TP * TN - FP * FN}{\sqrt{(TP + FN)(TP + FP)(TN + FP)(TN + FN)}} \tag{12}$$

where TP is the number of positive samples correctly classified in prediction and TN is the number of negative samples correctly classified by predictors. FP and FN represent the number of positive and negative samples misclassified, respectively. The ROC curve and the AUC were also adopted to evaluate the model performance.

3 Results

The LSTM model was built with the Keras framework (www.keras.io) using a sequential model and a TensorFlow [37] deep learning library back-end. The Keras model was compiled using 40 epochs with the 'adam' optimizer (loss: binary_crossentropy, metrics: accuracy). The model accuracy and loss are shown on Fig. 3 and Fig. 4, respectively.

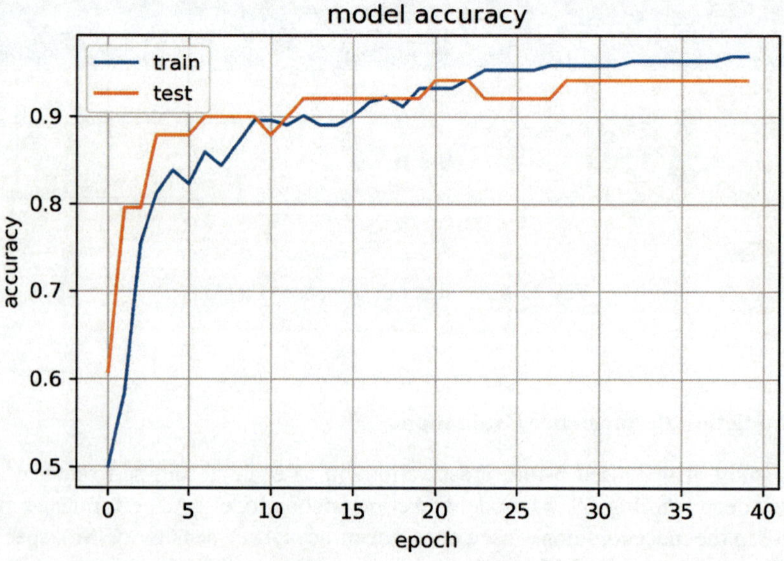

Fig. 3. Model accuracy

The LSTM classifier was compared with RF and SVM algorithms. For the implementation of these two ML models was used the library for the Python programming language Scikit-learn. They were tested with default parameters. The comparison results are summarized in Fig. 5.

Table 1 show the details of this comparison. The results show that average accuracy of 5-fold cross-validation is 85.42% with 5.79% standard deviation (SD), the average

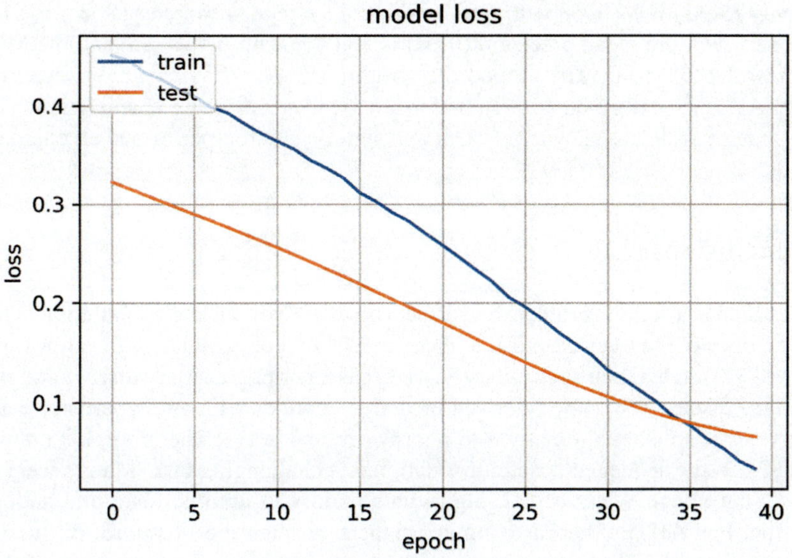

Fig. 4. Model loss

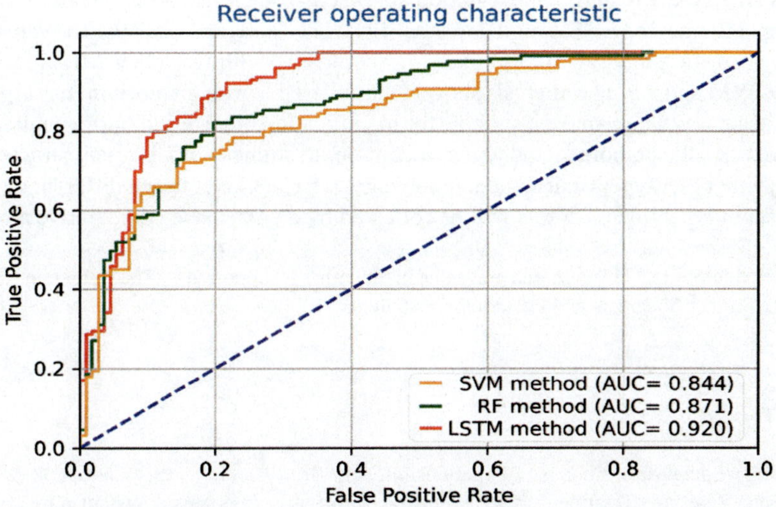

Fig. 5. The performance of the proposed LSTM and comparison models

Table 1. Performance of comparison models and LSTM

Model	Acc (%)	Prec (%)	Sen (%)	Spec (%)	Mcc (%)	AUC
SVM	72.5	68.57	76.92	63.64	54.15	0.84
RF	79.58	79.72	83.75	74.82	63.92	0.87
LSTM	85.42	85.04	89.17	81.18	71.12	0.92

sensitivity (Sens) is 89.17% with 6.07% SD, the average specificity (Spec) is 81.18% with 6.82% SD, the mean precision (Prec) is 85.04% with 6.12% SD and the MCC is 71.12% with 9.54% SD. The proposed algorithm showed an outstanding capability to identify AHTPs, performed an area under the receiver operating characteristic (ROC) curve (AUC) of 0.92, as shown in Fig. 5, and significantly outperformed all comparison methods.

4 Conclusions

Hypertension is a serious problem and affects up to 30% of the adult population in developed and developing countries. It is a major risk factor concomitant with cardiovascular disease (CVD) states such as coronary heart disease, peripheral artery disease and stroke, and kidney disease [38]. Food-derived bioactive peptides with antihypertensive activities serve as promising alternatives to pharmaceutical drugs. Therefore, the prediction of AHTPs on the sequence-based information is a challenging task. Many recent studies show the effectiveness of ML algorithms mainly in antimicrobial and anticancer prediction. For AHTPs there is a limited papers in scientific literature. In this study, it was developed LSTM model for sequence-based prediction of food-originated ACE inhibitory peptides. The Chou's pseudo amino acid composition (PseAAC) method was used as an efficient feature representation approach. Then, the obtained features are fed as an input to train the LSTM model to identify AHTPs and non- AHTPs. To evaluate the capability of the proposed algorithm, it was compared with two other ML algorithms, namely SVM and RF. The analysis shows that the deep learning algorithm has high predictive accuracy, which make it suitable for identification of ACE inhibitory peptides. As a future work, the authors intend to conduct in-depth studies with different structures of LSTM, with other types of deep neural networks (eg CNN) and to test different methods for presenting peptides. All this will be followed by an extensive comparative analysis.

Acknowledgements. This research presented in this paper is supported by The Bulgarian National Program "Young Scientists and Postdoctoral Students".

References

1. Mill, J.: Social determinants of hypertension. Arq. Bras. Cardiol. **113**(4), 696–698 (2019)
2. Daien, V., et al.: Treatment of hypertension with renin–angiotensin system inhibitors and renal dysfunction: a systematic review and meta-analysis. Am. J. Hypertens. **25**(1), 126–132 (2012)
3. Sánchez, A., Vázquez, A.: Bioactive peptides: a review. Food Qual. Saf. **1**(1), 29–46 (2017)
4. Aslam, M.Z., et al.: Role of bioactive peptides in reducing the severity of hypertension with the inhibition of ACE. Int. J. Pept. Res. Ther. **25**(4), 1639–1649 (2019). https://doi.org/10.1007/s10989-018-09806-y
5. Balgir, P., Kaur, T., Sharma, M.: Antihypertensive peptides derived from food sources. MOJ Food Process. Technol. **2**(1), 1–6 (2016)
6. Martínez-Maqueda, D., Miralles, B., Recio, I., Hernández-Ledesma, B.: Antihypertensive peptides from food proteins: a review. Food Funct. **3**(4), 350–361 (2012)

7. Yamaguchi, N., Kawaguchi, K., Yamamoto, N.: Study of the mechanism of antihypertensive peptides VPP and IPP in spontaneously hypertensive rats by DNA microarray analysis. Eur. J. Pharmacol. **620**(1–3), 71–77 (2009)
8. Turpeinen, A., Ehlers, P., Kivimäki, A., Järvenpää, S., Filler, I., et al.: Ile-Pro-Pro and Val-Pro-Pro tripeptide-containing milk product has acute blood pressure lowering effects in mildly hypertensive subjects. Clin. Exp. Hypertens. **33**(6), 388–396 (2011)
9. Chakrabarti, S., Wu, J.: Milk-derived tripeptides IPP (Ile-Pro-Pro) and VPP (Val-Pro-Pro) promote adipocyte differentiation and inhibit inflammation in 3T3-F442A cells. PloS One **10**(2), e0117492 (2015)
10. Xu, L., Liang, G., Wang, L., Liao, C.: A novel hybrid sequence-based model for identifying anticancer peptides. Genes **9**(3), 158 (2018)
11. Wei, L., Zhou, C., Chen, H., Song, J., Su, R.: ACPred-FL: a sequence-based predictor using effective feature representation to improve the prediction of anti-cancer peptides. Bioinformatics **34**(23), 4007–4016 (2018)
12. Manavalan, B., Shin, T., Kim, M., Lee, G.: AIPpred: sequence-based prediction of anti-inflammatory peptides using random forest. Front. Pharmacol. **9**, 276 (2018)
13. Bhadra, P., Yan, J., Li, J., Fong, S., Siu, S.W.: AmPEP: sequence-based prediction of antimicrobial peptides using distribution patterns of amino acid properties and random forest. Sci. Rep. **8**(1), 1–10 (2018)
14. Laengsri, V., Nantasenamat, C., Schaduangrat, N., Nuchnoi, P., Prachayasittikul, V., Shoombuatong, W.: TargetAntiAngio: a sequence-based tool for the prediction and analysis of anti-angiogenic peptides. Int. J. Mol. Sci. **20**(12), 2950 (2019)
15. Kumar, R., et al.: An in silico platform for predicting, screening and designing of antihypertensive peptides. Sci. Rep. **5**, 12512 (2015)
16. Win, T., Schaduangrat, N., Prachayasittikul, V., Nantasenamat, C., Shoombuatong, W.: PAAP: a web server for predicting antihypertensive activity of peptides. Future Med. Chem. **10**(15), 1749–1767 (2018)
17. Manavalan, B., Basith, S., Shin, T., Wei, L., Lee, G.: mAHTPred: a sequence-based meta-predictor for improving the prediction of anti-hypertensive peptides using effective feature representation. Bioinformatics **35**(16), 2757–2765 (2019)
18. Basith, S., Manavalan, B., Hwan Shin, T., Lee, G.: Machine intelligence in peptide therapeutics: a next-generation tool for rapid disease screening. Med. Res. Rev. (2020)
19. Wang, L., Niu, D., Wang, X., Shen, Q., Xue, Y.: A Novel Machine Learning Strategy for Prediction of Antihypertensive Peptides Derived from Food with High Efficiency. BioRxiv (2020)
20. Yan, J., et al.: Deep-AmPEP30: improve short antimicrobial peptides prediction with deep learning. Mol. Ther.-Nucleic Acids (2020)
21. Veltri, D., Kamath, U., Shehu, A.: Deep learning improves antimicrobial peptide recognition. Bioinformatics **34**(16), 2740–2747 (2018)
22. Youmans, M., Spainhour, C., Qiu, P.: Long short-term memory recurrent neural networks for antibacterial peptide identification. In: 2017 IEEE International Conference on Bioinformatics and Biomedicine (BIBM), pp. 498–502. IEEE (2017)
23. Sherstinsky, A.: Fundamentals of recurrent neural network (RNN) and long short-term memory (LSTM) network. Physica D: Nonlinear Phenomena **404**, 132306 (2020)
24. Panyayai, T., et al.: FeptideDB: a web application for new bioactive peptides from food protein. Heliyon **5**(7), e02076 (2019)
25. Kumar, R., et al.: AHTPDB: a comprehensive platform for analysis and presentation of antihypertensive peptides. Nucleic Acids Res. **43**(D1), D956–D962 (2015)
26. Li, Q., et al.: BioPepDB: an integrated data platform for food-derived bioactive peptides. Int. J. Food Sci. Nutr. **69**(8), 963–968 (2018)

27. Minkiewicz, P., Iwaniak, A., Darewicz, M.: BIOPEP-UWM database of bioactive peptides: current opportunities. Int. J. Mol. Sci. **20**(23), 5978 (2019)
28. Chou, K.: Prediction of protein cellular attributes using pseudo-amino acid composition. Proteins: Struct. Funct. Bioinform. **43**(3), 246–255 (2001)
29. Hajisharifi, Z., Piryaiee, M., Beigi, M., Behbahani, M., Mohabatkar, H.: Predicting anticancer peptides with Chou's pseudo amino acid composition and investigating their mutagenicity via Ames test. J. Theor. Biol. **341**, 34–40 (2014)
30. Mousavizadegan, M., Mohabatkar, H.: Computational prediction of antifungal peptides via Chou's PseAAC and SVM. J. Bioinform. Comput. Biol. **16**(04), 1850016 (2018)
31. Poorinmohammad, N., Mohabatkar, H., Behbahani, M., Biria, D.: Computational prediction of anti HIV-1 peptides and in vitro evaluation of anti HIV-1 activity of HIV-1 P24-derived peptides. J. Pept. Sci. **21**(1), 10–16 (2015)
32. Du, P., Wang, X., Xu, C., Gao, Y.: PseAAC-Builder: a cross-platform stand-alone program for generating various special Chou's pseudo-amino acid compositions. Anal. Biochem. **425**(2), 117–119 (2012)
33. Chou, K.C.: Prediction of protein cellular attributes using pseudo-amino acid composition. Proteins: Struct. Funct. Bioinform. **43**(3), 246–255 (2001)
34. Hochreiter, S., Schmidhuber, J.: Long short-term memory. Neural Comput. **9**(8), 1735–1780 (1997)
35. Greff, K., Srivastava, R., Koutník, J., Steunebrink, B., Schmidhuber, J.: LSTM: a search space odyssey. IEEE Trans. Neural Netw. Learn. Syst. **28**(10), 2222–2232 (2016)
36. Zhao, Y., Yang, R., Chevalier, G., Xu, X., Zhang, Z.: Deep residual bidir-LSTM for human activity recognition using wearable sensors. Math. Probl. Eng. (2018)
37. Abadi, M., Barham, P., Chen, J., Chen, Z., Davis, A., et al.: Tensorflow: a system for large-scale machine learning. In: 12th USENIX Symposium on Operating Systems Design and Implementation (OSDI 2016), pp. 265–283 (2016)
38. Kearney, P., Whelton, M., Reynolds, K., Whelton, P., He, J.: Worldwide prevalence of hypertension: a systematic review. J. Hypertens. **22**(1), 11–19 (2004)

Cascading Approach for Automatic ICD-10 Codes Association To Diseases in Bulgarian

Boris Velichkov[1,2], Simeon Gerginov[1], Panayot Panayotov[1], Sylvia Vassileva[1], Gerasim Velchev[1], Ivan Koychev[1,2], and Svetla Boytcheva[3(✉)]

[1] Faculty of Mathematics and Informatics, Sofia University "St. Kliment Ohridski", Blvd "James Bourchier" 5, 1164 g.k. Lozenets, Sofia, Bulgaria
koychev@fmi.uni-sofia.bg
[2] GATE Institute, Sofia University "St. Kliment Ohridski", 125 Tsarigradsko shose Blvd., Block 2, Floor 3, Sofia, Bulgaria
[3] Institute of Information and Communication Technologies, Bulgarian Academy of Sciences, Acad. G. Bonchev Str., Bl. 25, 1113 Sofia, Bulgaria

Abstract. ICD-10 is the 10th revision of the International Classification of Diseases, a medical ontology for the encoding of diseases and related health problems provided by the World Health Organization. This encoding is used by physicians to be able to describe diseases in a standardized way. Since this is currently performed manually by medical professionals, the ability to automate this task would save time and allow doctors to focus more on patient care. The task of automatic association of ICD-10 codes to a textual description is an extreme scale multi-class multi-label classification task, due to the huge number of classes – 11000, and the possibility to assign multiple valid ICD-10 codes to a diagnosis. Moreover, for the application of machine learning algorithms for this task, a large training data set is required. This task is even a bigger challenge for low resource languages such as the Bulgarian language. We created semi-automatically a dataset from linked open data and public documents. The corpora contain about 350,000 diagnoses in the Bulgarian language labeled with 3-sign and 4 sign ICD-10 codes. The paper presents a cascading approach for automatic classification of ICD-10 codes to diagnosis, which uses the hierarchical nature of the ICD-10 classification, to improve the accuracy of classification. This approach is tested and compared with the flat classification approach on the above-mentioned date set. Different machine learning algorithms are tested, including those based on deep learning transformers like BERT models. The results from the conducted experiments provide evidence that the proposed approach which takes into account the hierarchical structure of the ICD-10 codes outperforms the ones that ignore it.

Keywords: ICD–10 codes · Health informatics · Big data · Machine learning · Deep learning · Text-based classification

S. S. Sotirov et al. (Eds.): BioInfoMed 2020, LNNS 374, pp. 247–260, 2022.
https://doi.org/10.1007/978-3-030-96638-6_27

1 Introduction

For the application of Artificial Intelligence in healthcare, standardization and inseparability have significant importance. Several medical nomenclatures and statistical classifications are widely used for the transformation of diagnoses, procedures, drugs, etc. into standard codes. Such codes are of primary importance because they are used for the determination of different diagnoses, statistics about mortality, cause-of-death reports, and investigations of morbidity. The most widely used nomenclature in healthcare practice is SNOMED CT[1] and the statistical classification International Classification of Diseases 9th revision (ICD-9[2]) and 10th revision (ICD-10[3]). In this paper, we will address the problem of automatic association of diagnostic codes to a textual description of diagnosis in Bulgarian. In our study, we will explore ICD-10, because the ICD-10 official translation in Bulgarian[4] is used in all primary clinical documents sent to the Bulgarian National Health Insurance Fund.

The task for automatic association of ICD-10 codes to a textual description is computationally expensive because it requires big training corpora due to the huge number of classes - about 11,000, and there is high-variability of the textual descriptions of diagnoses. On the other hand, this is a multi-class, multi-label classification that is considered an extreme scale classification task.

We propose a cascading approach for the automatic association of ICD-10 codes to diagnosis in the Bulgarian language. In this approach, the core functionality is the use of the hierarchical nature of the ICD-10 classification that allows to narrow down the process and to reduce the search space. It is done at levels, starting with defining the first character in the encoding and so on until we define the entire code. As a result, individual classifications are combined into one common classification. The approach is compared to different machine learning approaches, including those based on deep learning [9] transformers like BERT models. The experimental results show that the proposed approach is more efficient than classical classification approaches, and some translation deep learning models like BERT [13].

2 Related Work

The task for automatic association of ICD-10 codes to diagnosis was addressed by several pieces of research for many languages. Unfortunately, due to the limited resources like annotated corpora and the high complexity of the task, it was not explored to its full scale. The task is language-dependent and different solutions were proposed addressing the specific language features. Thus the evaluation results vary between different languages, due to the different training corpora sizes and a different number of distinct classes (ICD-10 codes) used.

[1] http://www.snomed.org/.

[2] https://www.cdc.gov/nchs/icd/icd9.htm.

[3] https://www.who.int/classifications/icd/icdonlineversions/en/.

[4] https://ncpha.government.bg/bg/2019-02-19-23-22-18/icd-10.

Bagheri et al. [7] apply several classification methods for the Dutch language case for annotated corpora of 5,548 records and 1,195 (4-sign) ICD-10 codes and 608 (3-sign) ICD-10 codes. The best results for the classification task are achieved using average word embedding SVM - F1 54.9%. Almagro et al. [1] present results for a dataset that consists of 7,254 Spanish hospital discharged reports, with approx. 7,000 ICD-10 codes (4-sign). The authors investigate the performance of different classification algorithms over the number of classes (ICD-10) codes and show that the best results are achieved for many classes by Dependency-LDA, for a relatively smaller number of classes - by SVM, and for just a few classes - the most appropriate methods are Multi-Layer Perceptrons (MLPs) and Gradient Boost (P10 about (40%)). Chen et al. [12] explore the longest common subsequence problem for this task for diagnosis in Chinese and report F1 score of 81.1% for a dataset that contains about 1,000 textual descriptions of hepatitis-related diagnoses (about 100 codes). A hierarchical approach applied for the Chinese language was proposed by Ning et al. [15] with F1 score of 91.08%, the experiments were performed over a dataset with 16,330 manually coded instances and a subset of ICD-10 codes. They first apply a similarity search to determine the most probable branch of the ICD-10 hierarchy and to reduce the search space. Ceylan et al. [11] propose a method based on information retrieval techniques for Turkish, applied for about 51,955 records and 486 ICD-10 codes with the best score of 0.765. Another research for the Turkish language is presented by Arifoglu et al. [4] - CodeMagic - a recommendation system that proposes top K possible ICD-10-AM codes associated with a patient record. The authors in [6] explore machine learning techniques and report results for a subset of 1500 ICD-10 classes with achieved 92% precision for the primary disease. Another approach is to view the problem as a multi-label classification task and use neural networks like CNN, LSTM/BiLSTM, and HA-GRU [18], or applying BERT which has shown good results on this task in German [3]. Hybrid approaches, that combine different models show a slight improvement in the results [3].

The major problem with training datasets - the high imbalance of 4-character codes - is addressed in [16], where they propose methods for imbalance effects reduction, by techniques like splitting feature spaces or compressing label dimension.

In previous research for the Bulgarian language, [8], it was shown that the multi-class SVM method achieves promising results (F1 score 84.5%) for a dataset of 26,826 diagnoses and 448 different ICD-10 codes. Inspired by the promising results of hierarchical approaches presented in [15] and [10], we propose a cascading approach for ICD-10 codes association and compare it to other approaches for flat classification for diagnoses in Bulgarian. The proposed approach is relatively language-independent and can be easily adapted for other languages.

3 Dataset

3.1 Labeled Dataset of Diagnosis in Bulgarian with ICD-10 codes

The ICD-10 classification is a statistical hierarchical classification that organizes the diagnoses into 4 different levels of categories. Figure 1 illustrates the hierarchical nature of the ICD-10 classification. In addition, each diagnosis can be associated to one or more 4-sign categories as the diagnosis can be associated with the primary disease category or the category of diseases affecting a particular part of the body.

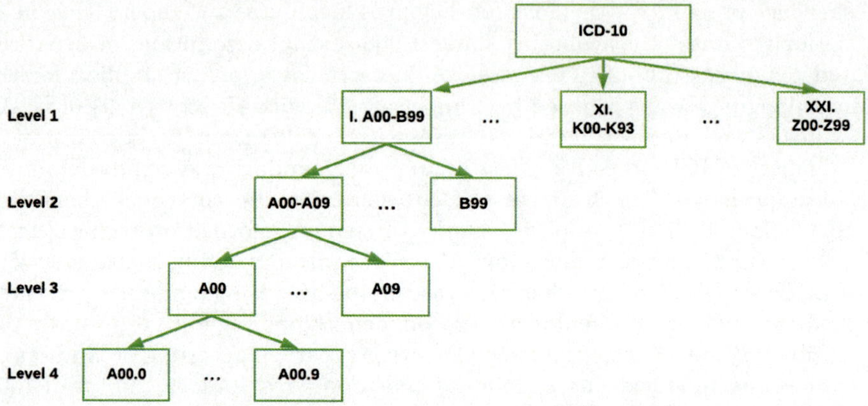

Fig. 1. The ICD-10 Hierarchical Classification.

We use the public dataset[5] with text descriptions of diagnoses in Bulgarian language and labeled with the corresponding ICD-10 codes, generated by Boytcheva et al. [9]. The dataset contains 189,254 diagnoses labeled with 3-character ICD-10 codes in 2,035 unique categories and 377,939 samples labeled with 4-character ICD-10 codes in 11,064 unique categories. The data was assembled from several different resources including official ICD-10 classification documents in Bulgarian as well as WikiData[6] disease information translated from English to Bulgarian. In this dataset, each diagnosis is labeled with the corresponding 4-character code as well as with the related 3-character parent code in order to retain information on the hierarchical structure. The dataset contains text descriptions of diagnoses in Bulgarian, Latin, and transliterated Latin into Cyrillic. We will call this dataset the source data (SD).

[5] https://github.com/BorisVelichkov/ICD10-Medical-Data.
[6] https://www.wikidata.org/.

3.2 Augmented Dataset

Due to the fact that the dataset is extremely imbalanced, we also use an augmented version of that dataset described in Velichkov et al. [17]. The augmented dataset contains 440,273 diagnoses samples labeled with 4-character codes in 5,854 unique categories. The augmentation procedure takes advantage of the hierarchical nature of the ICD-10 classification. In order to guarantee sufficient representation of each category, several operations are applied to underrepresented classes, which leads to some categories reduction: (i) For 4-character categories all diagnoses are generalized to the parent 3-character category; (ii) For 3-character categories augmentation is applied by swapping or deleting a character from the diagnosis text which adds 361 samples to the dataset; (iii) For certain groups of categories that involve causes or factors influencing the health, the authors have rolled up all sub-categories to the top-level category following ICD-10 classification groupings and pruning the corresponding 4-character categories. In particular, categories in the range V01-Y98 are classified as VY and Z00-Z99 as Z, respectively. Thus each category in the augmented dataset has at least 5 diagnosis samples.

We split the dataset using stratification and assign 80% for training, 10% for hyper-parameter tuning, and 10% for testing using scikit-learn library[7]. If a class has less than 3 samples, we augment the dataset using the nlpaug library by swapping random words or letters in the diagnosis[8]). The resulting dataset will be called the augmented dataset (AD). We also generate a more balanced subset of AD that contains up to 10,000 examples for each top-level class that contains 176,186 text descriptions of diagnoses, which will be denoted as AD-10.

3.3 Preprocessed Dataset

The following standard preprocessing pipeline was applied on the textual description of the diagnosis in the dataset:

- Tokenization - WordPiece[9] is used for tokenization;
- Stop words removal - removing any stop words from the BTB list[10];
- Stemming - applying the BulStem stemmer[11];

3.4 Features Selection

There are 56,013 unique words in the AD dataset. Feature selection is applied to the one-hot vector representations using the Information Gain metric. The feature selection produces a space with 18,479 features (words) and slightly

[7] Scikit-Learn https://scikit-learn.org/stable/.
[8] nlpaug library https://pypi.org/project/nlpaug/.
[9] https://huggingface.co/transformers/tokenizer_summary.html.
[10] BTB stopword list in Bulgarian: http://bultreebank.org/wp-content/uploads/2017/04/BTB-StopWordList.zip.
[11] BulStem stemmer: https://github.com/mhardalov/bulstem-py.

reduces the number of training examples to 422,672 distinct one-hot vectors. The size of the generated file with one-hot vectors is 14.5 GB.

For the AD-10 dataset, the number of selected features is 11,934 and the number of distinct one-hot vectors is 176,186, and the file size is 6.06 GB.

4 Cascade Approach for ICD-10 codes Association

The huge space dimensionality and the high imbalance make the task too complex. We can benefit from the hierarchical structure of ICD-10 classification, by splitting the task into several subtasks and reducing the number of classes for each of them.

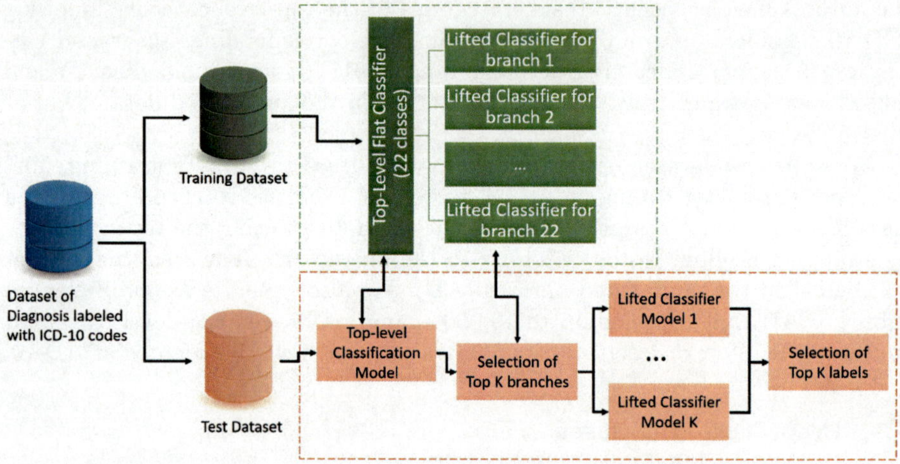

Fig. 2. Cascade classification.

This method explores information about a local hierarchical structure such as parent-child and sibling relationships during model learning. In the proposed cascade approach (Fig. 2) we first start with local classification at the top-level, i.e. 22 categories. Then, after identification of the top-level parent class, local classifiers per node are trained for each branch for the corresponding parent (Fig. 3).

In the Local Classifier per Node approach, a binary classifier is trained for each node in the hierarchy except the root. The purpose of this approach is to train the classifiers that can effectively distinguish between sibling nodes in the hierarchy. Usually, for the training of a node, we use examples belonging to this node and its descendants as positive examples, and those belonging to the siblings of the node and its descendants as negative examples.

In the Local Classifier per Level approach, a multi-class classifier is trained for each level in the hierarchy. Training is done level by level, where the classifier

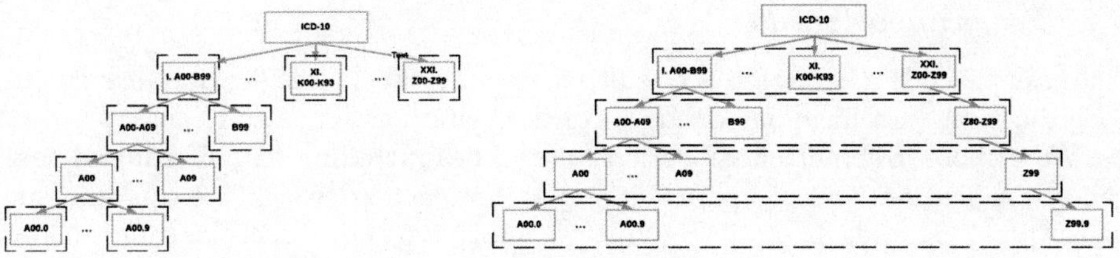

Fig. 3. Local classification per Node (left) and per Level (right).

uses the examples from nodes within the same level and its descendants. Different nodes in the same level correspond to different sibling classes. The prediction consists of the best node at each level in the hierarchy. Since the predictions at each level are independent, there is a possibility for vertical inconsistency in the prediction.

In the proposed method we use local classification for the top-level only, combined with local node classification for the lifted classes, which are a result of pruning underrepresented classes and substituting them with their parent.

First, we apply the trained model for the top level on the input (text description of diagnosis), then the top K classes are selected. For each of those top K classes, the corresponding local node classifier is applied. All classification results are merged and the top K candidate labels (ICD-10 codes) are selected and are presented as an output of the classification.

5 Experiments

5.1 Experiment Objectives

The aim of the experiments is to study and compare different approaches for ICD-10 code association to free-text diagnoses. The task is very challenging due to the following factors:

- high complexity of medical texts
- large number of classes
- multi-class multi-label classification - one text description of a diagnosis can be associated with more than one code.
- multi-level categories - a diagnosis can be associated not only with code on the lowest level (4-character code) but with a code on a higher level (3-character code).
- imbalanced nature of the training dataset
- class ambiguity - the same text can be associated with different codes

The main hypothesis that we evaluate in this paper is that a cascade approach, i.e. top-down classifications level by level is more efficient than flat classification.

5.2 Experiment Setup

The task for code association is a multi-class multi-label classification task where each diagnosis can have one or more correct classes assigned to it.

We explore different classification approaches including flat, lifted, and cascade classification, and compare the results. For each approach, we run different algorithms including Logistic Regression, FastAI[12] and a deep learning model using BERT (Bidirectional Encoder Representations from Transformers) [13].

5.3 Evaluation Metrics

We compare the accuracy, and precision of the best models from each algorithm.

For the correct class, we accept any class which is valid for a diagnosis.

Accuracy is defined as the number of diagnosis samples associated with the correct class.

The following evaluation metrics are used:

$$Precision = \frac{tp}{tp + fp}$$

and

$$Accuracy = \frac{tp + tn}{tp + fp + tn + fn}$$

Where tp (true positive) is the number of correctly associated samples per class; fp (false positive) - the number of samples incorrectly associated to belong to the class.

5.4 Flat Classification

Using the augmented dataset by Velichkov et al. [17], we train a direct multi-class classification model that uses all 4,274 4-character leaf categories in the ICD-10 classification. For the purposes of flat classification, we have removed all 3-character parent codes and use only the 4-character code classification information. This approach ignores the hierarchical structure of the classes and attempts to assign the most appropriate class using a single classification model. Figure 4 shows the usage of the flat classification approach for ICD-10 codes. The number of samples in our dataset is reduced to 171,035 in 4,274 categories.

We use BERT for text classification in order to build the multi-class classifier and we take the label with the highest probability for our evaluation. BERT (Bidirectional Encoder Representations from Transformers) [13] is a context-aware language model pre-trained on a large corpus of data. It has shown excellent results in a variety of NLP tasks including text classification. Multiple BERT models exist for different languages and/or domains.

[12] https://www.kaggle.com/nikkisharma536/fastai-toxic.

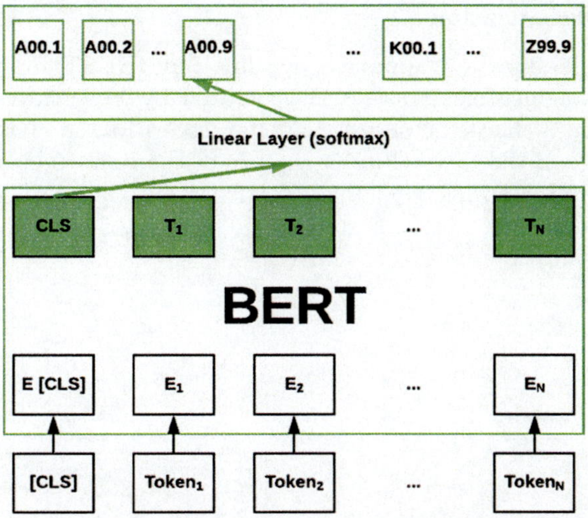

Fig. 4. Flat multi-class classification of ICD-10 codes using BERT.

Based on the results by Velichkov et al. [17], the fine-tuned Multilingual BERT model has shown the best performance for Bulgarian diagnosis texts in comparison with other models based on BioBERT [14], ClinicalBERT [2] and SlavicBERT [5]. Therefore, we focus on that model in our experiments.

The Multilingual BERT model is based on BERT-base and pre-trained using Wikipedia articles in 104 languages incl. Bulgarian[13].

In addition, the authors of [17] apply fine-tuning of the language model using 10,000 medical articles in Bulgarian scraped from several different websites. The fine-tuning is performed by running the masked language task [13] using 15% masked words that the model is trained to recognize.

Using the fine-tuned Multilingual BERT language model, we modify it further for the text classification task.

The BERT model takes the tokenized input and pre-pends and appends the [CLS] and [SEP] special tokens. The [CLS] token represents the diagnosis concept embedding and is used for text classification. The text classification model consists of a linear layer that is trained to classify the diagnosis concept embedding into the correct ICD-10 category using the softmax function. It is trained for a maximum of 10 epochs and produces the ICD-10 code with the highest probability to match the diagnosis. We report the accuracy, precision, and recall metrics for this classification.

The best results for flat classification applied over the reduced SD dataset, with the MultilingualBERT-Fine-tuned model - 79% accuracy and precision 67%.

[13] Multilingual BERT https://github.com/google-research/bert/blob/master/multilingual.md.

5.5 Lifted Classification

For the lifted classification approach, we fine-tune BERT for text classification using the augmented dataset AD generated by Velichkov et al. [17]. We include both the 3-character and 4-character codes in the classification, thus taking advantage of the hierarchical structure of the ICD-10 classification. The augmented dataset contains 345,591 diagnosis samples labeled with 4-character codes in 5854 unique categories. Figure 5 illustrates the differences between Flat and Lifted classification approaches for ICD-10.

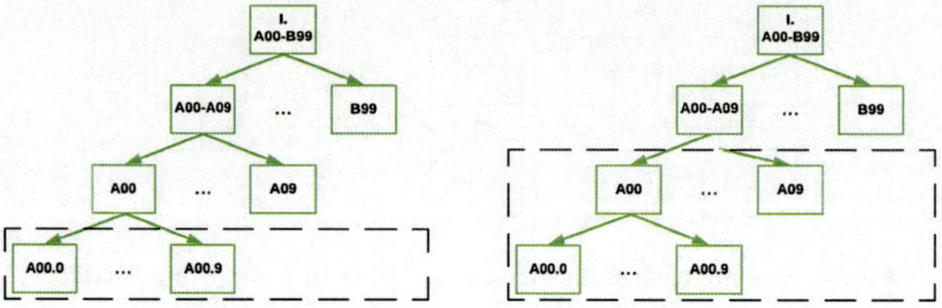

Fig. 5. Flat classification (left) vs Lifted classification (right).

We use the fine-tuned Multilingual BERT model as a starting point for the training. The resulting modified text classification model supports classification for 5,854 unique classes that have at least 5 samples in the dataset. We apply the same procedures as in flat classification for pre-processing the samples including stemming and removing stop words. Similar to the flat classification approach, we split the dataset and assign 80% for training, 10% for hyper-parameter tuning, and 10% for testing. The model architecture follows the same pattern as the flat classification case.

We display the results of all evaluated algorithms in Table 3. The best results are achieved by the Lifted Fine-tuned Multilingual BERT classifier trained with the following configuration: learning rate - 1e-4, batch size - 64, best epoch - 10.

Within the group of Lifted classification algorithms, BERT does the best job of classifying 3- and 4-character diagnoses with significantly higher accuracy, precision, and recall.

When comparing the Flat and Lifted approaches, the Flat BERT model performs worse than its Lifted counterpart. The additional hierarchical information in the Lifted approach makes a significant difference in the precision and recall metrics of the classifier given the same model architecture. In addition, the Lifted approach classifies across 5,854 classes, while the Flat one uses only 4,274 classes.

5.6 Cascade Classification

The idea of the cascade classification proposed in the current article is to build a basic classifier at the top level to determine which group the disease belongs to. Below the top level is the first character of the ICD-10 code (A, B, C, etc.). The total number of classes at this level is 22. Once it is determined to which group the disease belongs, it is submitted to a classifier, which is trained only on the respective group of codes and which must say what the exact code of the disease is. We used logistic regression for a top-level classification algorithm. For its training, we created a dataset based on AD using a one-hot vector. Logistic regression showed the best result with 27% accuracy. As a result, we balanced the number of examples to a maximum of 10,000 examples per class, but the results did not improve - the accuracy dropped to 16%. This experiment serves as a baseline. As a result, we decided to experiment in more detail using deep learning algorithms for text classification from FastAI. As a source of how to run FastAI's algorithms for our problem, we used an example from Kaggle[14]. After modifying the example so it fits our problem, we trained the algorithm and managed to reach more than 2 times better results than the logistic regression - 59% (see Table 1). The training of the algorithm consists of two main phases - first, training of an AWD-LSTM[15] type language model for recognizing the text context and second, training of the final model, which recognizes the ICD-10 codes at the top level. For local node classification for the different branches, we used some classical classification algorithms as:

- Logistic regression (LR) - Lasso (L1), C = 1;
- Random Forest (RF) - Number of trees: 15, Maximal number of considered features: 3, Replicable training, unlimited Maximal tree splitting nodes with maximum instances: 3;
- Stochastic Gradient Descent (SGD);
- Neural network (NN) - Hidden layers: 100, Activation: ReLu, Solver: Adam, Alpha: 0.0001, Max iterations: 300, Replicable training.

The results in Table 2 show that once the most probable branches are predicted and the number of candidates is reduced, the process for label prediction is more accurate.

Table 1. Classification results for top-level classes

Approach	Dataset	Algorithm	Accuracy	Avg. Precision
Cascade	AD	Logistic Regression	27%	81%
Cascade	AD	FastAI	59%	59%
Cascade	AD-10	Logistic Regression	16%	89%

[14] https://www.kaggle.com/nikkisharma536/fastai-toxic.
[15] https://docs.fast.ai/text.models.awdlstm.

Table 2. Classification results for branches

Approach	Dataset	Algorithm	Accuracy	Avg. Precision
Cascade	AD	Logistic Regression	87%	72%
Cascade	AD	Random Forest	91%	64%
Cascade	AD	NN	96%	76%
Cascade	AD	SGD	74%	67%

Table 3. Classification results using different algorithms.

Approach	Dataset	Algorithm	Accuracy	Avg. Precision
Lifted	AD	MultilingualBERT-Fine-tuned	81%	87%
Cascade	AD	FastAI+SGD	44%	40%
Cascade	AD	FastAI+NN	57%	45%
Cascade	AD	FastAI+LR	51%	42%
Cascade	AD	FastAI+RF	53%	38%
Cascade	AD	FastAI+FastAI	18%	18%

6 Conclusion and Further Work

The paper presents a cascading approach for the automatic association of ICD-10 codes to diagnosis. The approach uses the hierarchical nature of the ICD-10 classification to narrow down the classification task. The approach is top-down, starting with the prediction of the first character in the encoding (the top category) and then for the sub-categories, we predict the entire code. As a result, individual classifications of sub-categories are combined into one common classification. The approach is compared with different machine learning algorithms for flat multi-class, multi-label classification, including those based on deep learning transformers like BERT models. The results from experiments show that the proposed approach is more efficient than flat classification approaches, including the ones based on pre-trained deep neural network models such as BERT, when the process is narrowed down and the most probable branches are selected. Future works include the exploration of an approach that recursively walks down the hierarchy until the entire code is discovered.

Acknowledgments. This research is partially funded by the Bulgarian Ministry of Education and Science, grant DO1-200/2018 'Electronic health care in Bulgaria' (e-Zdrave) and the Bulgarian National Science Fund, grant DN-02/4-2016 'Specialized Data Mining Methods Based on Semantic Attributes' (IZIDA).

References

1. Almagro, M., Unanue, R.M., Fresno, V., Montalvo, S.: ICD-10 coding of Spanish electronic discharge summaries: an extreme classification problem. IEEE Access **8**, 100073–100083 (2020)
2. Alsentzer, E., et al.: Publicly available clinical bert embeddings. arXiv preprint arXiv:1904.03323 (2019)
3. Amin, S., Neumann, G., Dunfield, K., Vechkaeva, A., Chapman, K., Wixted, M.: Mlt-dfki at clef ehealth 2019: Multi-label classification of ICD-10 codes with bert (September 2019)
4. Arifoğlu, D., Deniz, O., Aleçakır, K., Yöndem, M.: CodeMagic: semi-automatic assignment of ICD-10-AM codes to patient records. In: Czachórski, T., Gelenbe, E., Lent, R. (eds.) Information Sciences and Systems 2014, pp. 259–268. Springer, Cham (2014). https://doi.org/10.1007/978-3-319-09465-6_27
5. Arkhipov, M., Trofimova, M., Kuratov, Y., Sorokin, A.: Tuning multilingual transformers for language-specific named entity recognition. In: Proceedings of the 7th Workshop on Balto-Slavic Natural Language Processing, pp. 89–93 (2019)
6. Atutxa, A., Pérez, A., Casillas, A.: Machine learning approaches on diagnostic term encoding with the icd for clinical documentation. IEEE J. Biomed. Health Inform. **22**(4), 1323–1329 (2017)
7. Bagheri, A., Sammani, A., Van der Heijden, P.G., Asselbergs, F.W., Oberski, D.L.: Automatic icd-10 classification of diseases from dutch discharge letters. In: BIOINFORMATICS 2020–11th International Conference on Bioinformatics Models, Methods and Algorithms, Proceedings; Part of 13th International Joint Conference on Biomedical Engineering Systems and Technologies, BIOSTEC 2020, vol. 13, pp. 281–289. SciTePress (2020). https://doi.org/10.5220/0009372602810289
8. Boytcheva, S.: Automatic matching of ICD-10 codes to diagnoses in discharge letters. In: Proceedings of the Second Workshop on Biomedical Natural Language Processing, pp. 11–18. Association for Computational Linguistics, Hissar, Bulgaria (September 2011). https://www.aclweb.org/anthology/W11-4203
9. Boytcheva, S., Velichkov, B., Velchev, G., Koychev, I.: Automatic generation of annotated corpora of diagnoses with icd-10 codes based on open data and linked open data. In: 2020 15th Conference on Computer Science and Information Systems (FedCSIS), pp. 163–167. IEEE (2020)
10. Catling, F., Spithourakis, G.P., Riedel, S.: Towards automated clinical coding. Int. J. Med. Inform. **120**, 50–61 (2018)
11. CEYLAN, N.M., ALPKOÇAK, A., ESATOGLU, A.E.: Tıbbi kayıtlara icd-10 hastalık kodlarının atanmasına yardımcı akıllı bir sistem (2012)
12. Chen, Y., Lu, H., Li, L.: Automatic icd-10 coding algorithm using an improved longest common subsequence based on semantic similarity. PLoS ONE **12**(3), e0173410 (2017)
13. Devlin, J., Chang, M.W., Lee, K., Toutanova, K.: Bert: Pre-training of deep bidirectional transformers for language understanding. arXiv preprint arXiv:1810.04805 (2018)
14. Lee, J., et al.: Biobert: a pre-trained biomedical language representation model for biomedical text mining. Bioinformatics **36**(4), 1234–1240 (2020)
15. Ning, W., Yu, M., Zhang, R.: A hierarchical method to automatically encode Chinese diagnoses through semantic similarity estimation. BMC Med. Inform. Decis. Mak. **16**(1), 1–12 (2016)

16. Parlak, B., Uysal, A.K.: On feature weighting and selection for medical document classification. In: Rocha, Á., Reis, L.P. (eds.) Developments and Advances in Intelligent Systems and Applications. SCI, vol. 718, pp. 269–282. Springer, Cham (2018). https://doi.org/10.1007/978-3-319-58965-7_19
17. Velichkov, B., et al.: Automatic icd-10 codes association to diagnosis: bulgarian case. In: CSBio 2020: Proceedings of the Eleventh International Conference on Computational Systems-Biology and Bioinformatics, pp. 46–53 (2020). https://doi.org/10.1145/3429210.3429224
18. Wang, Q., et al.: A study of entity-linking methods for normalizing Chinese diagnosis and procedure terms to icd codes. J. Biomed. Inform. **105**, 103418 (2020). https://doi.org/10.1016/j.jbi.2020.103418

Intuitionistic Fuzzy Sets with Applications in Bioinformatics and Biomedicine

Interval-Valued Intuitionistic Fuzzy Estimations of an Ultrasonic Image for Recognition Purposes

Todor Kostadinov and Veselina Bureva(✉)

Department of Computer Systems and Technology, Faculty of Technical Science, Prof. "Assen Zlatarov" University, Prof. Yakimov Street 1, 8010 Burgas, Bulgaria
{kostadinov_todor,vbureva}@btu.bg

Abstract. In the current research work the authors investigate the process of pattern recognition in terms of an object of interest, obtained by a radar image technique by applying the tools of Interval-Valued Intuitionistic Fuzzy Sets (IVIFS). The process of image acquisition by using the remote sensing approach is usually performed by a fast Fourier transform of the reflected signal. Due to the nature of waves, the reflection signal is comprised of direct and reflected signals, causing image blurring and smudging. The approach presented in this research work aims to propose an analytical evaluation of the processed image in comparison to preliminary defined images such as images of internal organs, tissues or bones. By applying the technique in question, it is possible to apply the method for recognition of any king of objects, and in particular – difference between organs with normal structure and organs with anomalies. The algorithm itself uses an interval-valued intuitionistic fuzzy estimations (IVIF-estimations) representing the dependencies between the target image and the selected image. The resulted IVIF-estimations represent the similarities between the selected images. According to the quantity of the processed key pixels in the images, there is a tradeoff between performance and quality of recognition. In order to verify the algorithm correctness, a numerical experiment on liver images with a different structure has been carried out.

Keywords: Interval-valued intuitionistic fuzzy estimation · Pattern recognition · Ultrasonic image · Internal organs

1 Image Acquisition

In the recent years different pattern recognition approaches are discussed in the literature [6–8, 11, 12, 14, 15]. In the current investigation the authors present an imaging technique called bistatic forward scattering inverse synthetic aperture radar (BFISAR) which is used to depict the object of interest. It uses the movement of the transmitter in order to observe objects of interest represented in the current case as a healthy human liver and one with cirrhosis. The geometry depicted in Fig. 1 is comprised of moving transmitter and stationary receiving station, situated in a Cartesian coordinate system, as depicted in the figure below.

S. S. Sotirov et al. (Eds.): BioInfoMed 2020, LNNS 374, pp. 263–268, 2022.
https://doi.org/10.1007/978-3-030-96638-6_28

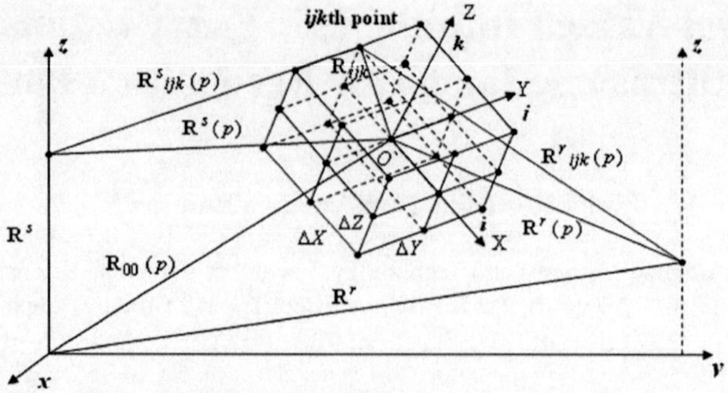

Fig. 1. BFISAR kinematics

where

$$\mathbf{R}^s(p) = \mathbf{R}^s - \mathbf{R}_{00}(p) = \mathbf{R}^s - \mathbf{R}_{00}(0) - \mathbf{V}\left(\frac{N}{2} - p\right)T_p \tag{1}$$

Is the range distance from the transmitter to the mass center of the object,

$$\mathbf{R}^r(p) = \mathbf{R}^r - \mathbf{R}_{00}(p) = \mathbf{R}^r - R_{00}(0) - \mathbf{V}\left(\frac{N}{2} - p\right)T_p \tag{2}$$

Is the Range distance vector from the transmitter to the ijk-th point of interest.

This imaging approach represents the object of observation as an assembly of point scatterers, observed by a bistatic system. Every point scatterer can be interpreted as a pixel of the object image, corresponding to the module of the range distance transmitter-object-receiver. The BFISAR technique can be regarded as a reception of broadband signal reflected by each point scatterer of the object which in this case is consisted of a series of linear frequency modulated pulses, and the signal is a superposition of the reflected pulses:

$$\begin{aligned} \dot{S}(p,t) &= \sum_{ijk} \dot{S}_{ijk}(p,t) \\ &= \sum_{ijk} a_{ijk}\, \mathbf{rect}\frac{t - t_{ijk}(p)}{T} \mathbf{exp}\left\{ -j\left[\begin{matrix} \omega\left(t - t_{ijk}(p)\right) + \\ + b\left(t - t_{ijk}(p)\right)^2 \end{matrix} \right] \right\} \end{aligned} \tag{3}$$

The range resolution of the image is resulted by the displacement of the observed object, as the azimuth resolution is achieved by the resolution properties of the signal, depending on the bandwidth.

The images acquisition is a result of a process called image reconstruction. In the complex signal, the time delays of the signals reflected by the object point scatterers, referred as range dimension is extracted by inverse Fourier transform known as range compression. The Doppler frequencies of the signals received, referred as azimuth dimension can also be extracted by inverse Fourier transform

$$\hat{I}(\hat{p},\hat{k}) = \sum_{p=1}^{N}\left[\sum_{k=1}^{K} \hat{S}(k,p).\exp\left(j2\pi\frac{k\hat{k}}{K}\right)\right]\exp\left(j2\pi\frac{p\hat{p}}{N}\right). \tag{4}$$

The objects in question – a human liver with normal structure a size and a human liver with cirrhosis are depicted in Fig. 2(a) and (b).

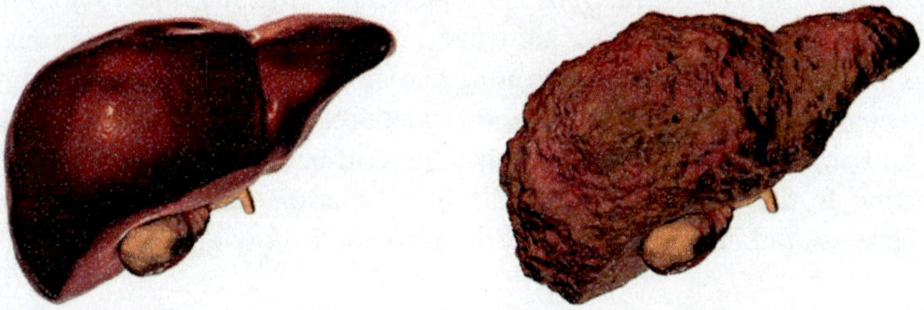

a) b)

Fig. 2. Objects, aim of imaging

After the remote imaging technique has been applied to the objects, the resulting images are represented in Fig. 3(a) and (b) respectively:

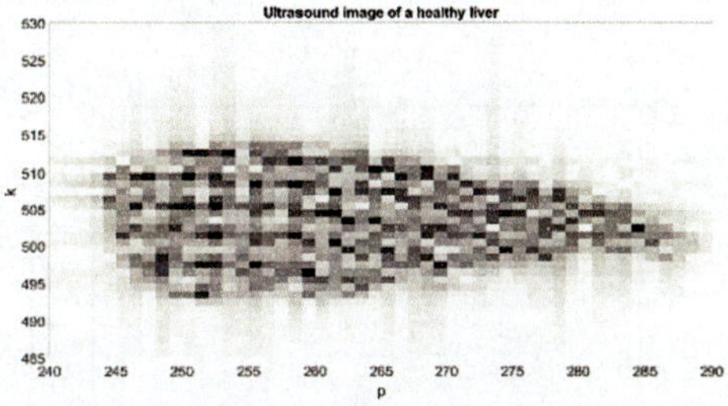

a)

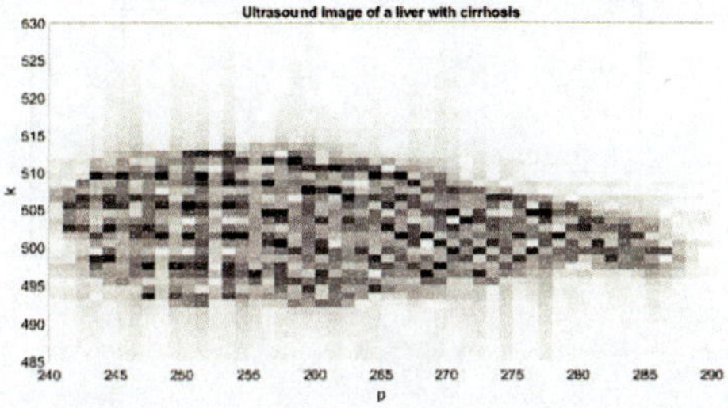

b)

Fig. 3. Objects images – healthy liver with normal structure (a) and liver with cirrhosis (b)

2 Interval-Valued Intuitionistic Fuzzy Estimations

The theory of intuitionistic fuzzy sets is presented in [1, 3, 4]. The intuitionistic fuzzy evaluations are used for assessment of objects based on several criteria. The intutionistic fuzzy evaluation has the form of intuitionistic fuzzy pair $\langle a, b\rangle$ where $a, b \in [0, 1]$ and $a + b \leq 1$. The investigations using intuitionstic fuzzy evaluations are presented [2, 5, 9, 10]. In the current paper a method for image classification using interval-valued intuitionistic fuzzy estimations is proposed [8, 13]. The images are obtained by the BFISAR technique using ultrasonic waves. The interval-valued intuitionistic fuzzy estimations (IVIF-estimations) representing the similarities between the target image and the selected images in the comparison procedure. According to the quantity of the processed key pixels in the images, there is a tradeoff between performance and quality of recognition. In the current paper the IVIF-estimations are defined to evaluate the similarity between the images. The procedure has the following form:

1. Target image is juxtaposed to the second image. The original pattern *ABCD* and the other pattern *AGHI* are given.
2. Sections *BC* and *HI* are fuzzified so that they are modified to the regions *BCEF* and *HIKJ*.

   The following regions are determined (Fig. 4):

   $s = \#AGHOCD$, $a = \#AELK$, $b = \#EBML$, $c = \#BGJM$, $d = \#MJHO$, $e = \#KLNI$, $f = \#INFD$, $g = \#NOCF$ and $h = \#LMON$.

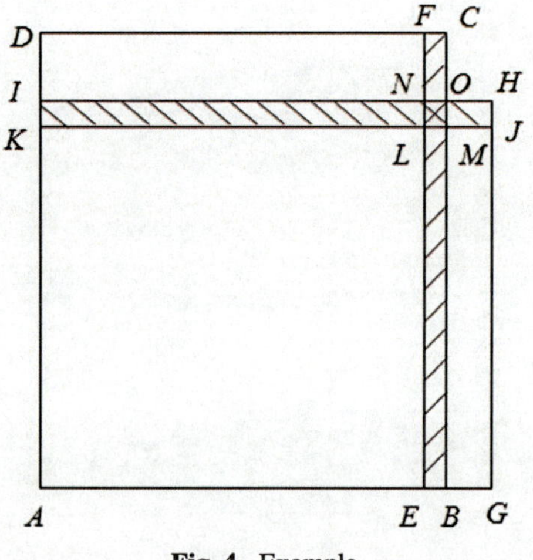

Fig. 4. Example

   Thereafter $s = a + b + c + d + e + f + g + h$ (Fig. 4).
3. The IVIF-degree of coincidence of the second pattern with the original are defined:

$$\langle M, N\rangle = \left\langle \left[\frac{a}{s}, \frac{a+b+e+h}{s}\right], \left[\frac{c}{s}, \frac{c+d}{s}\right]\right\rangle \tag{5}$$

   The degree of uncertainty is determined as the interval $\left[0, \frac{f+g}{s}\right]$.

For the purposes of images classification, an intuitionistic fuzzy estimation has been used. In order to assess the resulted images, an approach using the number of correlating and non-correlating image pixels is used. The number of pixels common for the both images (correlating pixels) is at a total of 628 pixels. The number of pixels (non-correlating) of the image from Fig. 3(a) is 87 at total and the number of pixels (non-correlating) of the image from Fig. 3(b) is 136. The total number of pixels for the image of the healthy liver – Fig. 3(a) is 715 and for the liver with cirrhosis is 764. The border pixels are 223.The Interval-Valued Intuitionistic Fuzzy Estimation for the images from Figs. 2 and 3 is:

$$M = \langle 0.74;\ 0.80\rangle,\quad N = \langle 0.16;\ 0.20\rangle.$$

The degree of uncertainty is in the interval $\langle 0;\ 0.13\rangle$.

The result presents the not enough similarity between the images. The pictures can not to be considered as the exactly the same. They can have close relationships and they can be considered to be in one group in the classification problem.

3 Conclusion

The presented image acquisition technique and interval-valued intuitionistic fuzzy estimations method can be successfully applied in the fields of image classification and pattern recognition. The proposed approach will be extended in the future research work: the algorithm will be applied to the set of images. The resulting interval-valued intuitionistic fuzzy estimations will be used to apply InterCriteria analysis to detect the correlations.

Acknowledgment. The authors are grateful for the support provided by the Bulgarian Ministry of Education and Science under the National Research Programme "Information and Communication Technologies for a Digital Single Market in Science, Education and Security" approved by DCM # 577/17.08.2018.

References

1. Atanassov, K., Szmidt, E., Kacprzyk, J.: On intuitionistic fuzzy pairs. Notes Intuit. Fuzzy Sets **19**(3), 1–13 (2013)
2. Atanassov, K.: Intuitionistic fuzzy logics as tools for evaluation of data mining processes. Knowl.-Based Syst. **80**, 122–130 (2015)
3. Atanassov, K: Interval-Valued Intuitionistic Fuzzy Sets. Springer, Switzerland (2020). https://doi.org/10.1007/978-3-030-32090-4
4. Atanassov, K.: On Intuitionistic Fuzzy Sets Theory. Springer, Berlin (2012)
5. Bureva, V., Yovcheva, P., Sotirov, S.: Generalized net model of fingerprint recognition with intuitionistic fuzzy evaluations. In: Kacprzyk, J., Szmidt, E., Zadrożny, S., Atanassov, K.T., Krawczak, M. (eds.) IWIFSGN/EUSFLAT -2017. AISC, vol. 641, pp. 286–294. Springer, Cham (2018). https://doi.org/10.1007/978-3-319-66830-7_26
6. Huo, G., Zhang, Q., Zhang, Y., et al.: Multi-source heterogeneous iris recognition using stacked convolutional deep belief networks-deep belief network model. Pattern Recogn. Image Anal. **31**, 81–90 (2021)

7. Hu, X., Li, X., Li, Y., Tang Y., Zhang, D.: A new method and implementation of blind restoration algorithm for moving fuzzy license plate image based on frequency-domain characteristics. Int. J. Pattern Recogn. Artif. Intell. **35**(08), 2154024 (2021)
8. Kim, T., Sotirova, E., Shannon, A., Atanassova, V., Atanassov, K., Jang, L.: Interval valued intuitionistic fuzzy evaluations for analysis of a student's knowledge in university e-learning courses. Int. J. Fuzzy Log. Intell. Syst. **18**(3), 190–195 (2018)
9. Kostadinov, T., Bureva, V.: Pattern recognition with intuitionistic fuzzy estimations. Notes Intuit. Fuzzy Sets **23**(2), 88–94 (2017)
10. Michalıkova, A.: Intuitionistic fuzzy sets and their use in image classification. Notes Intuit. Fuzzy Sets **25**(2), 60–66 (2019)
11. Mukherjee, S.: Denoising with discrete Morse theory. Vis. Comput. **37**, 2883–2894 (2021). IF: 2.601
12. Sogi, N., Zhu, R., Xue, J., Fukui, K.: Constrained mutual convex cone method for image set based recognition. Pattern Recogn. **121**, 108190 (2019). IF:7.196
13. Sotirova, E., Shannon, A., Atanassova, V., Atanassov, K., Bureva, V.: Interval valued intuitionistic fuzzy evaluations for analysis of students' knowledge. In: Intuitionistic and Type-2 Fuzzy Logic Enhancements in Neural and Optimization Algorithms (2020)
14. Wang, Y., Qi, Y., Yao, H., Gong, D., Wu, Q.: Image editing with varying intensities of processing. Comput. Vis. Image Underst. **211**, 103260 (2021). IF: 3.876
15. Xiao, G., Wang, H., Ma, J., et al.: Segmentation by continuous latent semantic analysis for multi-structure model fitting. Int. J. Comput. Vis. **129**, 2034–2056 (2021). IF: 7.410

Intuitionistic Fuzzy Representation of Uncertainty in Biomedical Operations

Stoyan Poryazov[1], Velin Andonov[1,2](✉), and Emiliya Saranova[1]

[1] Institute of Mathematics and Informatics, Bulgarian Academy of Sciences, Acad. G. Bonchev Str., Block 8, 1113 Sofia, Bulgaria
{stoyan,velin_andonov,emiliya}@math.bas.bg

[2] Institute of Biophysics and Biomedical Engineering, Bulgarian Academy of Sciences, Acad. G. Bonchev Str., Block 21, 1113 Sofia, Bulgaria

Abstract. The paper studies the problem of representation of the uncertainty in biomedical operations. Each operation is considered as a request (e.g., a patient) offered to a virtual service device, representing service of a manual, semiautomatic or automatic action. Some concepts of Service Systems Theory are justified and/or extended, especially traffic characterization in the International Telecommunication Union recommendations [4]. This allows three service qualifications to be made: (i) traffic characterization (e.g., requests' flow intensity (rate), probability of the next service direction, service duration, traffic intensity and volume); (ii) causal classification (offered, blocked, repeated, served, carried, lost, parasitic [14], effective, etc.); (iii) proposed intuitionistic fuzzy service characterization (successful (belonging to a fully completed successful service), uncertain (uncompleted, due to any reason, target service) and unsuccessful (a service not belonging to the desired successful service)). A conceptual model of the causal service structure of a virtual service system is presented, including graphical representation. A notation system of the devices and the devices' parameters is proposed based on the qualifiers, corresponding to each of the traffic types considered. For every virtual service device, using the causal conceptual representation, an intuitionistic fuzzy pair [7] is defined which characterizes the uncertainty in the process of servicing of the requests. The degrees of membership, non-membership and uncertainty [6] are expressed as ratios of the parameters of the virtual devices. The problem of representing the uncertainty in a more complex system consisting of two consecutively composed services is also studied. The qualitative difference between the causal classification and the proposed intuitionistic fuzzy characterization is shown.

Keywords: Conceptual modelling · Intuitionistic fuzzy pairs · Service systems · Uncertainty

S. S. Sotirov et al. (Eds.): BioInfoMed 2020, LNNS 374, pp. 269–278, 2022.
https://doi.org/10.1007/978-3-030-96638-6_29

1 Introduction

Biomedical operations are considered in many aspects in institutions such as Biomedical Operation and Research Office, Kennedy Space Center, USA and UNESCO [1]. Usually in biomedical operations complex devices are used, including mobile microrobots [3] and micro/nanorobots [5]. A device, that is used for biomedical operation or safety-critical applications like point-of-care health assessment, massive parallel DNA analysis, automated drug discovery, air-quality monitoring and food-safety testing, must have the attributes like reliability, dependability and correctness [8].

The paper considers biomedical operations in service systems such as hospital units, telecare/telehealth centers, etc. [9,10,16]. More precisely, the focus of the paper is on the problem for representation of uncertainty in the conceptual and analytical models of the operations. The apparatus of the service systems theory is used. Some concepts are justified and extended, for example traffic characterization in the definitions of the International Telecommunication Union (ITU) [4].

The biomedical operations are represented as service of requests (patients, sensor signal, etc.) in a virtual device. This approach allows easier conceptual and analytical modelling of various compositions of services. This is illustrated in [14], where the problem for traffic quality composition in service compositions is studied. The causal structure is presented through virtual devices corresponding to parasitic, carried and served traffic. The causal composition and decomposition of traffic quality is presented both graphically and analytically. A naming system for the virtual devices is proposed which takes into account the levels of inclusion of the base virtual devices in the comprise virtual devices.

Using the conceptual modeling approach from [12,15] and the intuitionistic fuzzy approach to the modeling of uncertainty (see [6]), here we propose an intuitionistic fuzzy characterization of uncertainty in biomedical operations. It is based on the causal service classification developed by the authors but is essentially qualitatively different from it.

The results presented in the paper can be applied to the measurement, prediction and partial management in real service systems.

2 Base Virtual Devices

In the conceptual models of service networks we use base virtual devices [12]. A general representation of a base virtual device such as server, buffer, switch, etc., is shown in Fig. 1. Every such device named x has the following parameters:

- Fx – Intensity or incoming rate (frequency) of the flow of requests (i.e., the number of requests per time unit) to device x;
- Px – Probability of directing the requests towards device x;
- Tx – Service time (duration of servicing of a request) in device x;
- Yx – Traffic intensity [Erlang];
- Vx – Traffic volume [Erlang - time unit];
- Nx – Number of lines (service resources, positions, capacity) of device x.

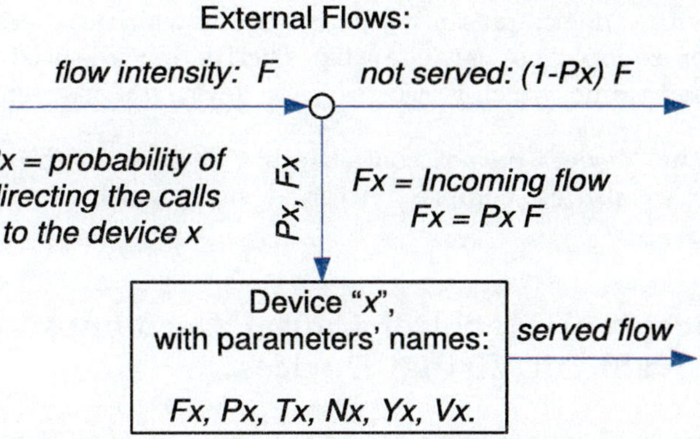

Fig. 1. Graphical representation of a base virtual device (see [17]).

Different types of base virtual devices are used in the models. The ones used in the paper together with their graphical representations are shown in Fig. 2.

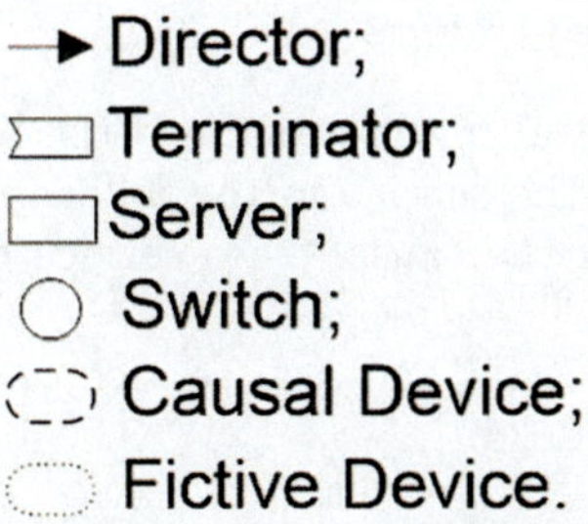

Fig. 2. Base virtual device types and their graphical representations.

The devices of each type have specific functions (see [12]):

- Director – this device points unconditionally to the next device, which the request shall enter but without transferring, changing or delaying it.
- Terminator – this device eliminates every request entered (so it leaves the model without any traces).
- Server – this device models the delay (service time, holding time) of requests in the corresponding device without their generation or elimination. It models also traffic and time characteristics of the requests processing.
- Switch (Transition) – this device selects one of its possible exits for each request entered, thus determining the next device where this request shall go to.
- Causal device – virtual device defined for presentation of causes of service ending, e.g., successful (carried) or not (interrupted, abandoned, etc.).

- Fictive device – device presenting fictive traffic which is necessary for engineering. For example, not carried traffic is fictive, but it is used for calculating of the offered traffic, which is necessary for device dimensioning.

The base virtual devices do not contain other devices. In the models of service networks, we also use comprise virtual devices, which contain base virtual devices.

3 A Conceptual Model of Causal Decomposition of the Traffic in Virtual Devices

We group causes of service ending, and the corresponding causal devices, in generalized comprising causal devices. If a device has unlimited capacity, three causal generalizations are enough: 'parasitic', 'carried' and 'served' [4].

The *parasitic traffic* in a pool of resources is the traffic which was unsuccessfully served in the pool. The parasitic traffic occupies real resources but not for a useful service execution.

The *carried traffic* in a pool of resources is the traffic which was successfully served in the pool (and carried to the next service device). We distinguish between two types of carried traffic:

- 'zero service', e.g., zero waiting in a buffer if the buffer is empty and there is free requested place for the service in the following device. The requests are receiving zero service in the causal 'zero service device' and may be served without delay.
- 'genuine service' – successful and real service of requests in the pool. The service time is noticeable.

The *served traffic* in a pool of resources is any traffic occupying (using) resources in the pool. The served traffic is a sum of the carried and the parasitic traffic.

The *not served traffic* is the traffic not occupying the recourses in the pool, due to any reason, e.g., 'blocking' [4], in the case of lack of service places in the pool. The intensity of the not served traffic is a multiplication of the not served requests' flow intensity and mean service time of the served requests.

The *offered traffic* is a sum of the served and not served traffic intensities in the pool.

Every Causal Device Parameter's Name is a concatenation:

Causal name = <qualifier><qualifier>.<Parameter's Symbol>.<Device Name>.

Some of the qualifiers used for characterization of the traffic are (see [15]):

- crr. = carried;
- nsr. = not served;
- ofr. = offered;
- prs. = parasitic;
- srv. = served.

'Parameter's symbol' is one of the letters P, F, T, Y, V, N, as they are described in Sect. 2. Qualifiers are used to characterize the parameters of the devices [12]. The used qualifiers may be two, one or none. If parameter's symbol is omitted, the causal name is a name of a device (see Fig. 3). The device name is in small or subscript letters. For instance, $crr.Fx$ is the intensity of the carried flow of requests of the device x, $prs.Fx$ is the intensity of the parasitic flow of requests of the device x (see Fig. 3). In the figures, only the names of causal devices may be present. The names of the device parameters are implicit.

A presentation of a service network portion 'x' is shown in Fig. 3. It contains a virtual device $srv.x$ with limited capacity. This may cause requests rejection due to lack of service places (call attempts blocking).

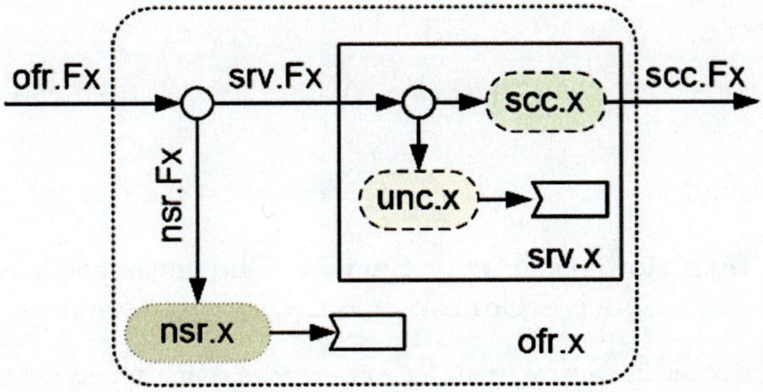

Fig. 3. Causal decomposition of the traffic inside a virtual service device x.

The device $nsr.x$ is not a part of the service device but is a part of the comprise fictive device $ofr.x$ which is necessary for dimensioning and traffic estimation of the operations. The qualifier *unc.* is used to characterize the traffic for which the reasons for the parasitic load cannot be determined explicitly. There might be several reasons as to why it cannot be determined explicitly. Some of them are:

- Abandoning/Failure by the users.
- Unsatisfying terms of service according to the users' opinion.
- Technical (objective) reasons – electricity failure, equipment failure, etc.
- Preempted – overtaken by higher priority request.

An important assumption must be stated about the device $unc.x$. It is assumed that in the case of uncertain service the service of the requests is terminated.

The qualifier *scc.* corresponds to the carried traffic.

4 An Intuitionistic Fuzzy Service Characterization

Using the conceptual model of the causal decomposition of the traffic in a virtual service device, described in the previous section, and the Intuitionistic Fuzzy Sets (IFSs, see [6]) theory and, more specifically, the notion of Intuitionistic Fuzzy Pair (IFP), we propose an intuitionistic fuzzy representation of the uncertainty of the services.

IFP is an object with the form $\langle a, b\rangle$ where $a, b \in [0, 1]$ and $a + b \leq 1$ which is used as an evaluation of some object or process. The components (a and b) are interpreted as degrees of membership and non-membership, or degrees of validity and non-validity, or degrees of correctness and non-correctness, etc. [7].

For a service device named x, we define an IFP $\langle \mu_x, \nu_x\rangle$, where:

$$\mu_x = \frac{scc.Yx}{ofr.Yx}\,; \nu_x = \frac{nsr.Yx}{ofr.Yx}\,. \tag{1}$$

From the conceptual model (see Fig. 3), it is clear that the degree of uncertainty is given by:

$$\pi_x = \frac{unc.Yx}{ofr.Yx}\,. \tag{2}$$

5 An Intuitionistic Fuzzy Service Characterization in a Consecutive Composition of two Operations

In [14], traffic quality aggregations for some service compositions are proposed. Here, we obtain intuitionistic fuzzy characterization of uncertainty in a consecutive composition of two operations. The graphical representation of the composition is shown in Fig. 4.

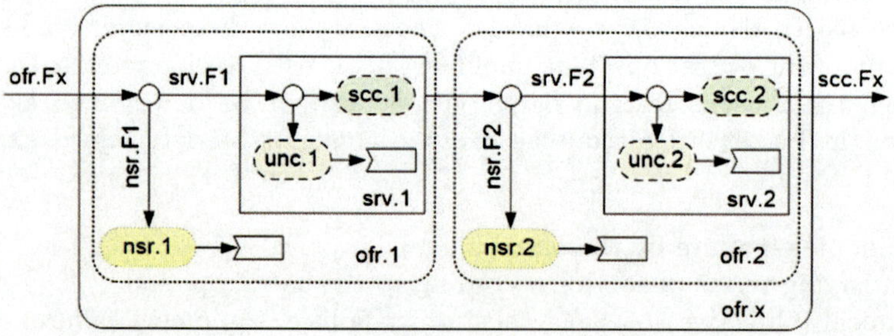

Fig. 4. A consecutive composition of two operations.

Using the graphical representation and elementary probability theory, for the flow intensity of the served requests in the comprise virtual device x, we obtain:

$$srv.Fx = srv.F_2 = ofr.Fx(1 - nsr.P_1)(1 - unc.P_1)(1 - nsr.P_2)\,. \tag{3}$$

The mean service time of the requests in the comprise device x is obviously a sum of the mean service times in the first and the second service devices, i.e.:

$$srv.Tx = srv.T_1 + srv.T_2\,. \tag{4}$$

The Little's theorem (see [2]) applied to the comprise service device x gives:

$$srv.Yx = srv.F_2(srv.T_1 + srv.T_2) = srv.Fx\; srv.Tx\,, \tag{5}$$

$$ofr.Yx = ofr.Fx\; srv.Tx\,, \tag{6}$$

and

$$scc.Yx = scc.Tx\; scc.Fx\,. \tag{7}$$

The mean service time for the successively served requests by the comprise device x ($scc.T_x$) is obviously a sum of the mean service times of the successively served requests in the causal devices $scc.1$ and $scc.2$:

$$scc.Tx = scc.T_1 + scc.T_2\,. \tag{8}$$

The flow intensity $scc.Fx$ can be expressed through the intensity of the offered flow to the comprise device and the probabilities of directing the requests towards the fictive devices $nsr.1$ and $nsr.2$, and the causal devices $unc.1$ and $unc.2$, in the following way:

$$scc.Fx = ofr.Fx(1 - nsr.P_1)(1 - unc.P_1)(1 - nsr.P_2)(1 - unc.P_2)\,. \tag{9}$$

Using the definition of μ_x (see (1)) and Eqs. (3)–(9), for μ_x we have:

$$\mu_x = \frac{scc.Yx}{ofr.Yx} =$$

$$\frac{(scc.T_1 + scc.T_2)(1 - nsr.P_1)(1 - unc.P_1)(1 - nsr.P_2)(1 - unc.P_2)}{srv.T_1 + srv.T_2}\,. \tag{10}$$

In order to obtain expression for ν_x, first we derive expressions for the flow intensities $nsr.F_1$, $nsr.F_2$ and $nsr.Fx$. Here, it is important to note that the classification of the service as uncertain in the first base virtual device ($unc.1$), in the composition is classified as not served ($nsr.$). This is because the device $unc.1$ deflects the flow of requests and they do not enter the second base virtual device, according to the assumption made in Sect. 3, and thus these requests are not serviced by the comprise service device.

$$nsr.F_1 = ofr.Fx\; nsr.P_1\,. \tag{11}$$

$$nsr.F_2 = ofr.Fx(1 - nsr.P_1)(1 - unc.P_1)nsr.P_2\,. \tag{12}$$

$$\begin{aligned} nsr.Fx = {} & ofr.Fx[nsr.P_1 + (1 - nsr.P_1)unc.P_1] \\ & + ofr.Fx(1 - nsr.P_1)(1 - unc.P_1)nsr.P_2\,. \end{aligned} \tag{13}$$

Using the Little's theorem, for the mean service time of the not served requests in the comprise device ($nsr.Tx$) we have:

$$nsr.Tx = \frac{nsr.Yx}{nsr.Fx}. \tag{14}$$

The traffic intensity of the not served requests of the comprise device ($nsr.Yx$) is a sum of the traffic intensities of the not served requests in the first ($nsr.Y_1$) and the second ($nsr.Y_2$) embedded devices:

$$nsr.Yx = nsr.Y_1 + unc.Y_1 + nsr.Y_2. \tag{15}$$

$$nsr.Y_1 = ofr.Fx\, nsr.P_1\, nsr.T_1. \tag{16}$$

$$nsr.Y_2 = ofr.Fx\,(1 - nsr.P_1)(1 - unc.P_1) nsr.P_2\, nsr.T_2. \tag{17}$$

$$unc.Y_1 = ofr.Fx\,(1 - nsr.P_1) unc.P_1\, nsr.T_1. \tag{18}$$

From Eqs. (11)–(18), we obtain:

$$\nu_x = \frac{nsr.Yx}{ofr.Yx} =$$

$$\frac{nsr.T_1\, nsr.P_1 + (1 - nsr.P_1)[unc.P_1 unc.T_1 + (1 - unc.P_1) nsr.P_2 nsr.T_2]}{srv.T_1 + srv.T_2}. \tag{19}$$

Similarly, in order to obtain expression for π_x, first we derive analytical expressions for the flow intensities $unc.F_1$, $unc.F_2$ and $unc.Fx$:

$$unc.F_1 = ofr.Fx(1 - nsr.P_1) unc.P_1. \tag{20}$$

$$unc.F_2 = ofr.Fx(1 - nsr.P_1)(1 - unc.P_1)(1 - nsr.P_2) unc.P_2. \tag{21}$$

$$unc.Fx = ofr.Fx(1 - nsr.P_1)(1 - unc.P_1)(1 - nsr.P_2) unc.P_2. \tag{22}$$

Using the Little's theorem and Fig. 4, for the traffic intensities $unc.Y_1$, $unc.Y_2$ and $unc.Yx$ we have:

$$unc.Y_1 = unc.F_1\, unc.T_1. \tag{23}$$

$$unc.Y_2 = unc.F_2\, unc.T_2. \tag{24}$$

$$unc.Yx = unc.Y_2. \tag{25}$$

From (2) and (20)–(25), we obtain analytical expression for π_x:

$$\pi_x = \frac{(1 - nsr.P_1)(1 - unc.P_1)(1 - nsr.P_2) unc.P_2\, unc.T_2}{srv.T_1 + srv.T_2}. \tag{26}$$

6 Conclusions and Future Work

In compositions consisting of a large number of embedded devices it may be difficult to obtain IFPs, characterizing the comprise devices. In such compositions, an evaluation of the uncertainty can be obtained through aggregation of the IFPs of the embedded service devices.

In the proposed intuitionistic fuzzy characterization of a composition of two consecutive services, a change in the classification of the flow of requests is made depending on the location of the devices. The flow of requests classified as uncertain (*unc.*) in the first base virtual device is classified as not served (*nsr.*) with respect to the comprise service device. This is a significant difference between the causal classification and the proposed intuitionistic fuzzy classification. It allows a more precise intuitionistic fuzzy estimation of uncertainty in service composition, unlike the approximate estimations from above and below.

The proposed intuitionistic fuzzy characterization can be used to define new Quality of Service (QoS) indicators.

Acknowledgements. The work of Stoyan Poryazov is partially supported by the joint research project "Symbolic-Numerical Decision Methods for Algebraic Systems of Equations in Perspective Telecommunication Tasks" of IMI-BAS, Bulgaria and JINR, Dubna, Russia.

The work of Velin Andonov is supported by the 2020 National Scientific Program for post doctoral and young scientists of the Ministry of Education and Science of Bulgaria.

The work of Emiliya Saranova was supported by the Task 1.2.5. "Prediction and Guaranteeing of the Quality of Servicein Human-Cyber-Physical Systems" of the National Scientific Program "Information and Communication Technologies for a Single Digital Market in Science, Education and Security (ICT in SES)" financed by the Bulgarian Ministry of Education and Science.

References

1. Birth of the Universal Declaration on the Human Genome and Human Rights. Division of the Ethics of Science and Technology of UNESCO, p. 166 (1999). http://www.unesco-chair-bioethics.org/wp-content/uploads/2015/05/119390eo.pdf
2. Haverkort, B.R.: Performance of Computer Communication Systems: A Model-Based Approach. John Wiley & Sons, New York, NY, USA (1998)
3. Ceylan, H., Giltinan, J., Kozielski, K., Sitti, M.: Mobile microrobots for bioengineering applications. Lab Chip **17**, 1705–1724 (2017)
4. ITU-T Recommendation E.600 (03/93): Terms and definitions of traffic engineering
5. Li, J., De Àvila, B.E.F., Gao, W., Zhang, L., Wang, J.: Micro/nanorobots for biomedicine: delivery, surgery, sensing, and detoxification. Sci. Robot. **2**(eaam6431), 9 (2017). http://robotics.sciencemag.org
6. Atanassov, K.: On Intuitionistic Fuzzy Sets Theory. Springer-Verlag, Berlin, Heidelberg (2012)
7. Atanassov, K., Szmidt, E., Kacprzyk, J.: On intuitionistic fuzzy pairs. Notes Intuit. Fuzzy Sets **19**(3), 1–13 (2013)

8. Majumder, M., Das, N., Saha, S.K.: A novel technique for multiple faults and their locations detection and start electrode selection in microfluidic digital biochip. J. Innov. Opt. Health Sci. **6**(4), 1350032 (2013)
9. Stefanova-Pavlova, M., Andonov, V., Tasseva, V., Gateva, A., Stefanova, E.: Generalized nets in medicine: an example of telemedicine for people with diabetes. In: Angelov, P., Sotirov, S. (eds.) Imprecision and Uncertainty in Information Representation and Processing. SFSC, vol. 332, pp. 327–357. Springer, Cham (2016). https://doi.org/10.1007/978-3-319-26302-1_20
10. Stefanova-Pavlova, M., et al.: Modeling telehealth services with generalized nets. In: Sgurev, V., Yager, R.R., Kacprzyk, J., Atanassov, K.T. (eds.) Recent Contributions in Intelligent Systems. SCI, vol. 657, pp. 279–290. Springer, Cham (2017). https://doi.org/10.1007/978-3-319-41438-6_16
11. D'Urso, P.: Informational paradigm, management of uncertainty and theoretical formalisms in the clustering framework: a review. Inf. Sci. **400–401**, 30–62 (2017)
12. Poryazov, S., Saranova, E.: Models of Telecommunication Networks with Virtual Channel Switching and Applications. Prof. Marin Drinov Academic Publishing House (2012). (In Bulgarian)
13. Ganchev, I., van der Mei, R.D., van den Berg, H. (eds.): Autonomous Control for a Reliable Internet of Services. LNCS, vol. 10768. Springer, Cham (2018). https://doi.org/10.1007/978-3-319-90415-3
14. Poryazov, S., Andonov, V., Saranova, E.: Different Traffic Quality Aggregations for a Service Composition. In: 2019 Big Data, Knowledge and Control Systems Engineering (BdKCSE), pp. 1–5. Sofia, Bulgaria (2019)
15. Poryazov, S., Andonov, V., Saranova, E.: Traffic quality aggregations of a queuing system. In: 2020 IEEE 10th International Conference on Intelligent Systems (IS), pp. 102–110. Varna, Bulgaria (2020)
16. Andonov, V., Stephanova, D., Esenturk, M., Angelova, M., Atanassov, K.: Generalized net model of telemedicine based on body temperature sensors. In: Proceedings of the 14th International workshop on Generalized Nets, Burgas, pp. 78–89 (2013)
17. Andonov, V., Poryazov, S., Saranova, E.: Analytical model of overall telecommunication system with queuing. Int. J. Inf. Models Anal. **8**(3), 212–230 (2019)

Intuitionistic Fuzzy Evaluations for Analysis of the Proximal Humerus Fractures

Stoyan Hristov[1], Asen Baltov[2], Evdokia Sotirova[3](✉), and Hristo Bozov[3]

[1] Department of Orthopedics and Traumatology, University Hospital for Active Treatment, 73 Stefan Stambolov, 8000 Burgas, Bulgaria
hristov@dr.com

[2] University Multiprofile Hospital for Active Treatment and Emergency Medicine "N.I. Pirogov", 21 Totleben str., 1606 Sofia, Bulgaria
asen_b@bitex.bg

[3] Faculty of Public Health and Health Care, Prof. Assen Zlatarov University, 1 Prof. Yakimov str., Burgas, Bulgaria
esotirova@btu.bg

Abstract. The paper is devoted of the assessment with intelligent instrument the very often happened Proximal humerus fractures. For our research 42 patients divided into two groups are analysed: *Group 1* with 19 patients, operated via angular-stable locking plate and polymethylmetacrylate augmentation, and *Group 2* with 23 patients, where iliac crest bone graft has been used. The analyzed data have been collected in the period 2017–2019 in Department of orthopedics and traumatology in University hospital for active treatment in Burgas.

A method for evaluation the degree of recovery of the patients with Proximal humerus fractures have been proposed. The assessment of the patient' recovery is formed on the basis of a set of intuitionistic fuzzy estimations $\langle \mu, \nu \rangle$ of real numbers from the set $[0, 1] \times [0, 1]$. These intuitionistic fuzzy estimations reflect the degree of successful recovery μ, or non-successful recovery ν, for each patient. The degree of uncertainty $\pi = 1 - \mu - \nu$ represents such cases wherein the treatment is not finished or we cannot provide information on the patient.

Keywords: Proximal humerus fractures evaluation · Intuitionistic fuzzy assessment · Augmentation

1 Introduction

Proximal humerus fractures are fractures occurring at or proximal to the surgical neck of humerus happen very often. In patients above the age of 65 years, Proximal humerus fractures (PHFx) are second most frequent upper extremity fracture, and the third most common nonvertebral osteoporotic fractureafter proximal femur and distal radius fractures [5]. Most of the patients with PHFx are treated nonoperatively. However, surgical treatment is becoming more frequent, with fracture reconstruction increasing at a higher rate than prosthetic replacement [6, 7]. Surgical treatment of displaced proximal humeral fractures with use of the locking proximal humeral plate that was evaluated in our and

S. S. Sotirov et al. (Eds.): BioInfoMed 2020, LNNS 374, pp. 279–284, 2022.
https://doi.org/10.1007/978-3-030-96638-6_30

other studies, can lead to a good functional outcome provided that the correct surgical technique is used [3, 4].

In the paper is proposed a method for evaluation of patients with Proximal humerus fractures (PHFx) obtained in the in the period 2017–2019 in Department of orthopedics and traumatology in University hospital for active treatment in Burgas. For the assessment of the PHFx the theory of Intuitionistic Fuzzy Sets is used (IFSs, [1, 2]). Intuitionistic fuzzy sets extend both concepts of classical set theory and fuzzy set theory by allowing the assessment of the elements by two functions: for membership and for non-membership, which belong to the real unit interval [0, 1] and whose sum belongs to the same interval.

2 Purpose

The purpose of the research is to perform operative treatment of Proximal Humeral Fractures in elderly patients via angular-stable locking plate (ASLP/PHILOS) and polymethylmetacrylate (PMMA) augmentation, find out type and frequency of complications arising during such treatment and to compare achieved functional result to results of a similar group of patients, where bone autograft augmentation has been used by us.

The analyzed processes and functional results are very complex and complete information for them is not always available. In this case, mathematical models will be developed to handle data for patients' Range of Motion (ROM) containing elements of uncertainty. An approach for evaluation the degree of recovery of the patients with Proximal humerus fractures will be suggested.

3 Material and Methods

19 patients (*Group 1*) have been operated via ASLP and PMMA augmentation and compared to 23 patients (*Group 2*), where iliac crest bone graft (ICBG) has been used. Average age respectively 70.1 and 71.6. All patients are older than 65, with low-energy trauma, in status of independency, with no records/findings for residual neurological symptomatic in injured extremity.

During hospitalization, the demographic and baseline characteristics of the patients were recorded. Fractures were classified according to Neer Classification system by the treating surgeon on the basis of plain radiographs and intraoperative fracture visualization. Scheduled follow-up evaluations were performed three, six, and twelve months postoperatively. At each follow-up visit, the patient was examined and interviewed with regard to pain, Forward Flexion, Abduction, Medial and Lateral Rotation, Constant and DASH score. True anteroposterior and transscapular Y-view radiographs were made postoperatively and at each follow-up visit. Radiographs were primarily evaluated by the treating surgeon for fracture-healing and possible complications.

For the evaluation the patients's condition and their ROM the theory of Intuitionistic Fuzzy Sets (IFSs) and interval valued sets is used. The IFSs are defined by Kr.Atanassov [1, 2]. They are an extension of the concept of fuzzy sets defined by L. Zadeh [8]. In the classical sense to each proposition one can assign its truth value: truth - denoted by 1, or falsity - 0. In the case of fuzzy logics this truth value is a real number in the interval [0, 1] and it is a "truth degree" of a particular proposition. In fuzzy logics the "falsity degree" is added, which will be in the interval [0, 1]. Thus, one assigns to the proposition p two real numbers $\mu(p)$ and $\nu(p)$ in interval [0, 1] with the following constraint to hold:

$$\mu(p) + \nu(p) \leq 1.$$

The suggested assessments will estimate the recovery process from PHFx to the respective patient. They reflect the degree of successful ROM μ, or non-successful ROM ν for each patient.

The degree of uncertainty $\pi = 1 - \mu - \nu$ represents the cases in which the treatment procedure has not been completed or the information for the current patient is not full.

4 Results

Here we discuss the proposed assessment model of analysing the patient's recovery process, which is realized on two phases.

First we determine the evaluation of the types of movement for three, six, and twelve months postoperatively for each patient. The next step is to determine the interval valued intuitionistic fuzzy evaluation of the types of movement by periods for in each group.

4.1 Determination the Patients' Assessments of the Different Types of Movement

Let us consider 2 groups of m_1 and m_2 patients respectively, the patients are labelled as follows $i_1 = 1, 2, \ldots, m_1$, $i_2 = 1, 2, \ldots, m_2$, and k types of movements, $l = 1, 2, \ldots, k$. The movements have to be evaluated for n periods after operation, $j = 1, \ldots, n$.

In current case we consider:

- 2 groups G_1 and G_2 of patients with $m = m_1 + m_2 = 19 + 23 = 42$ patients;
- 4 types of movements ($k = 4$): forward flexion (FF), abduction (A), medial rotation (MR), lateral rotation (LR);
- 3 time periods after operation ($n = 3$): three months, six months, twelve months.

Let the measured value for l-th type of movement for the j-th period for a patient i_g from *Group g*, has value $x_{l,j}^{G_g}(i_g)$, for $g = 1, 2$, $l = 1, 2, \ldots, k$, $j = 1, \ldots, n$.

Let $X_{l,j}^{min}$ is minimal requirement threshold value for the l-th type of movement and j-th period.

Let $X_{l,j}^{max}$ is maximal threshold value for the l-th type of movement and j-th period.

In this case the evaluations of the recovery process of the i_g-th patient $\left\langle \mu_{l,j}^{G_g}(i_g), \nu_{l,j}^{G_g}(i_g) \right\rangle$, where $\mu_{l,j}^{G_g}(i_g)$ and $\nu_{l,j}^{G_g}(i_g)$ corresponding to the degree of successful ROM and non-successful ROM, $\mu_{l,j}^{G_g}(i_g), \nu_{l,j}^{G_g}(i_g) \in [0, 1]$, $\mu_{l,j}^{G_g}(i_g) + \nu_{l,j}^{G_g}(i_g) \leq 1$ is:

$$\mu_{l,j}^{G_g}(i_g) = \frac{x_{l,j}^{G_g}(i_g)}{X_{l,j}^{max}},$$

$$\nu_{l,j}^{G_g}(i_g) = \frac{X_{l,j}^{min} \dot{-} x_{l,j}^{G_g}(i_g)}{X_{l,j}^{max}},$$

where $\dot{-}$ means:
$w \dot{-} u = \begin{cases} w - u, & \text{if } w \geq u \\ 0, & \text{if } w < u \end{cases}.$
The degree of uncertainty $\pi(i_g)$ is

$$\pi(i_g) = \frac{X_{l,j}^{max} - x_{l,j}^{G_g}(i_g) - \left(X_{l,j}^{min} \dot{-} x_{l,j}^{G_g}(i_g)\right)}{X_{l,j}^{max}}.$$

4.2 Determine the Interval Valued Intuitionistic Fuzzy Evaluation for the Group for Types of Movement by Periods

Here the interval valued intuitionistic fuzzy evaluation for types of movement (forward flexion, abduction, medial rotation, lateral rotation) in the form $\langle Q, P\rangle$ is proposed, Q, $P \subset [0, 1]$ are intervals, i.e., $Q = [\inf Q, \sup Q]$, $P = [\inf P, \sup P]$ and

$$\sup Q + \sup P \leq 1.$$

When we have some patient groups, treated in different ways we can obtain aggregated evaluation for them on the basis of the intuitionistic fuzzy evaluations of the separate patients in each group, but we can obtain interval valued intuitionistic fuzzy evaluation of the group, too.

In the first case, for the s-th group ($s = 1, 2, \ldots, w$) we obtain that the has the form

$$\langle \mu_s, \nu_s \rangle = \left\langle \frac{\sum_{i=1}^{n_s} \mu_{s,i}}{n_s}, \frac{\sum_{i=1}^{n_s} \nu_{s,i}}{n_s} \right\rangle$$

where n_s is the number of the patients in s-th group, and $\langle \mu_{s,i}, \nu_{s,i}\rangle$ is the final evaluation of the i-th patient in the s-th group.

When we like to obtain interval valued intuitionistic fuzzy evaluation of the s-th group, we construct numbers

$$Q_{s,inf} = inf\{\mu_{s,i} \,|\, i = 1, 2, \ldots, n_s\},$$

$$Q_{s,sup} = sup\{\mu_{s,i} \,|\, i = 1, 2, \ldots, n_s\},$$

$$P_{s,inf} = inf\{\nu_{s,i} \,|\, i = 1, 2, \ldots, n_s\},$$

$$P_{s,sup} = sup\{\nu_{s,i} \,|\, i = 1, 2, \ldots, n_s\}.$$

Now, we can construct the interval valued intuitionistic fuzzy evaluation

$$\langle Q_s, P_s \rangle = \big\langle [Q_{s,inf}, Q_{s,sup}], [P_{s,inf}, P_{s,sup}] \big\rangle.$$

When we like to compare two groups (a-th and b-th) on the bases of their results, we can use the following formulas (see [2]):

$$\langle Q_a, P_a \rangle \le \langle Q_b, P_b \rangle \text{ if and only if}$$

$$Q_{a,inf} \le Q_{b,inf}, Q_{a,sup} \le Q_{b,sup}, P_{a,inf} \ge P_{b,inf}, P_{a,sup} \ge P_{b,sup}.$$

In [2] other formulas for evaluation are discussed, too.

5 Conclusions

In the current paper the operative treatment of PHFx via angular-stable locking plate and PMMA augmentation was discussed. Our experience so far let us believe, that this method is a suitable solution due to ensured osteosynthesis stability, prospect for early mobilization, following good functional results and last, while complication rate does not go above complication level, stated in specialized medical literature.

A procedure that gives the possibility for the algorithmization of the method of forming the interval valued intuitionistic fuzzy evaluation to compare the achieved functional result from operative treatment of PHFx via angular-stable locking plate and PMMA augmentation in elderly patients to results of a similar group of patients, where bone autograft augmentation has been used by us. The suggested evaluation methodology and procedures are intended to make the patient's evaluations as objective as possible.

Acknowledgements. The authors are grateful for the support provided by the Bulgarian Ministry of Education and Science under the National Research Programme "Information and Communication Technologies for a Digital Single Market in Science, Education and Security" approved by DCM # 577 / 17.08.2018".

References

1. Atanassov, K.: Intuitionistic Fuzzy Sets. Springer, Heidelberg (1999)
2. Atanassov, K.: On Intuitionistic Fuzzy Sets Theory. Springer, Berlin (2012)
3. Klug, A., Wincheringer, D., Harth, J., Schmidt-Horlohé, K., Hoffmann, R., Gramlich, Y.: Complications after surgical treatment of proximal humerus fractures in the elderly—an analysis of complication patterns and risk factors for reverse shoulder arthroplasty and angular-stable plating. J. Shoulder Elbow Surg. **28**(9), 1674–1684 (2019)
4. Rusimov, L., et al.: Does supplemental intramedullary grafting increase stability of plated proximal humerus fractures? J. Orthop. Trauma **33**(4), 196–202 (2019)
5. Sproul, R., Iyengar, J., Devcic, Z., Feeley, B.: A systematic review of locking plate fixation of proximal humerus fractures. Injury **42**(4), 408–413 (2011)

6. Spross, C., Meester, J., Mazzucchelli, R.A., Puskás, G.J., Zdravkovic, V., Jost, B.: Evidence-based algorithm to treat patients with proximal humerus fractures—a prospective study with early clinical and overall performance results. J. Shoulder Elbow Surg. **28**(6), 1022–1032 (2019)
7. Thanasas, C., Kontakis, G., Angoules, A., Limb, D., Giannoudis, P.: Treatment of proximal humerus fractures with locking plates: a systematic review. J. Shoulder Elbow Surg. **18**(6), 837–844 (2009)
8. Zadeh, L.A.: Fuzzy sets. Inf. Control **8**, 333–353 (1965)

Opportunity to Detect Hazardous Materials in Water Using Intercriteria Analysis

Husein Yemendzhiev, Ralica Koleva, Valentin Nenov, and Vania Georgieva(✉)

Assen Zlatarov University, 1 Prof. Yakimov str., 8010 Burgas, Bulgaria
{hyemendzhiev,vnenov,vanya_georgieva}@btu.bg

Abstract. The detection of hazardous materials in water is crucial part of the water management and monitoring. However, analyzing the entire diversity toxic compounds, potentially presented in the water is not an easy, nor economical task and sometimes it is impossible, as well. This group of chemicals varies from heavy metals ions to complex organic molecules and bio-toxins. Finding a general indicator for the toxicity presented in water and aquatic environments could be an important step toward the development of early warning systems to support decision making in water monitoring.

In this study, an integrated approach combining information form 4 different sensors is used to evaluate the water status in terms of the toxic compounds presented. Bio-electrochemical reactor is used as a sensor to evaluate the specific biological response to the toxic compounds and pH, ORP and conductivity electrodes are used to find additional correlations between the water physical parameters and the inhibitors presented. To represent the two major groups of toxic chemicals, namely the heavy metals and the aromatic hydrocarbons, Cu^{2+} and phenol are used in different concentration in the tested models. The data obtained showed a good level of correlation and response of all the sensors and electrode used in terms of both presence and concentration of the tested pollutants. The application of the fuzzy logic approach to analyzing water quality provides an indicator of overall water toxicity.

Keywords: Toxicity evaluation · Hazardous materials · Water · Intuitionist fuzzy logic

1 Introduction

Various classification methods have been used for estimating the changing status and usability of natural waters. However, a discrepancy frequently arises from the lack of a clear distinction between each water type. Owing to inherent imprecision, some conventional methodologies in describing water quality conditions hide significant difficulties. In recent years, the fuzzy logic based methods have demonstrated to be appropriated to address uncertainty and subjectivity in environmental issues [1].

The "fuzzy" analysis is based on using approximations in the calculations rather than precise values to give a broad and potentially more useful response. Moreover, it allows disparate parameters to be combined in a meaningful way even though their values may

S. S. Sotirov et al. (Eds.): BioInfoMed 2020, LNNS 374, pp. 285–295, 2022.
https://doi.org/10.1007/978-3-030-96638-6_31

not be related. Just as apples and oranges are different but all represent the quality of fruitiness, so biochemical oxygen demand and chemical concentrations, for instance, may represent a particular aspect of water quality and so can be combined through fuzzy analysis.

The application of the fuzzy logic approach to analyzing water quality helps in providing an accurate indicator of overall status of the water. As a common example, the river ecosystems and the quality of the water depend generally on its alkalinity/acidity, dissolved oxygen (DO), biochemical oxygen demand (BOD), suspended solids, and the presence of chemicals including chlorides, phosphates, nitrates and sodium. However, in case of toxic compounds presence, the responsecould be different. In cases like this, Fuzzy logic designed as environmental indices can solve wrong data interpretation regarding a complex sets of multidimensional data.

In this study, we aimed the two major groups of toxic chemicals, namely heavy metals and organic compounds such as aromatic hydrocarbons. Heavy metals are highly poisonous even at very small concentration since they are not metabolized by the body and accumulate in the tissues. Among the toxic metals pollutions, copper contamination has the highest numbers of cases reported worldwide suggesting that the effects of copper-contaminated water upon crop cultivation and copper content in food must be investigated. Copper is an essential element for plant nutrients, but it becomes toxic at high concentrations. Copper is required for constituting enzymes catalyzing redox reactions, and is involved in photosynthetic functions. In addition, excess copper induces high levels of reactive oxygen species (ROS) and affects the photosystem in photosynthesis which subsequently reduces the yield or quality of crops. It should be mentioned that, regarding copper contained in natural water, a large range of water quality parameters such as total hardness, alkalinity, pH and dissolved organic carbon can modify its bioavailability and toxicity. As an example, the copper toxicity is higher in water with lower hardness and Dissolved Organic Carbone concentrations [2].

Regarding the Phenol as a toxic aromatic hydrocarbon compound, it is a common industrial substance involved in the production of chemicals such as xylenes, oils,plastics and several pharmaceuticals. The effluents which contains high amounts of phenol and phenolic compound are quite resistant to biological treatment due to the present of stable benzene ring. It can be accumulated and transported in the tissues before being eliminated through metabolic pathways. Whith this reason, pollution caused by phenol or phenolic compound exist in environment either at atmosphere, hydrosphere or terrestrial area became a great environment concern [3].

Nowadays environmental protection and water quality management has become an important issue for the society and industry. Many people are concerned about the quality of the environmental resources due to the complex nature of the information availble. The application of sophisticated systems in wastewater treatment needs adequate evaluation of the toxicity of the pollution. Recent advances in control engineering suggest that hybrid control strategies, integrating ideas and paradigms existing in different soft computing techniques, such as fuzzy logic may provide improved control of water quality [4]. In this sense, Fuzzy logic is a powerful methodology for solving problems with many applications in control and information processing. The use of a fuzzy controllers has significantly influenced the approach to control problems. Fuzzy set instruments are quite simple conceptually as they consist of an input stage, a processing stage, and an output stage.

The aim of the study is to find signal profiles and patterns specific for two groups of pollutants and to apply mathematical model able to perform toxicity level assessment and prediction of the type of the inhibitory influence. The copper and phenol are used as representatives of the two major groups of toxic compounds (heavy metals and organic chemicals). The toxicity status of the pollutants was evaluated through the signal profiles obtained from three type of sensors - bio-electrochemical device, pH and conductivity electrodes.

2 Experimental

2.1 Bio-Electrochemical System Construction and Operation

In this study, Microbial Fuel Cell type reacotr (MFC) was used as a bio-electrochemical sensor. The reactor was assembled as a cylindrical plexiglass body consists of two identical chambers - anodic and cathodic with total volume of 50 dm^3 each. The two compartments were separated by Nafion® 424 perfluorinted proton exchange membrane with 2 cm diameter. The electrodes were made of carbon cloth and stainless steel mesh as a current collector. The anode and cathode were with circular shape, diameter of 45 mm and distance between them 20 mm. To close the circuit, the electrodes were connected with external electric circuit loaded by 1 kΩ resistor. The anodic compartment of the MFC reactor which reached the operational state was fed by the tested water-toxic compound models. The cathode compartment was loaded 2% ferricyanide as a terminal electron acceptor.

2.2 Microorganisms and Cultivation Conditions

The electrochemically active microorganisms (electrogenes) were isolated from bottom sediments of "Poda" Protected Site located on the outfall of Lake Uzungeren, south of Burgas and the cultivated and enriched according to the procedure described by Mersinkova et al. 2020 [5]. The culture was suspended in fresh nutrient medium (LB without glucose to avoid fermentative metabolism) to a concentration of 10^7 CFU/dm^3. It was then loaded in the anode chamber of the MFC for biofilm formation and cell immobilization.

2.3 Toxicity Detection

As toxic agents two pollutants – phenol and copper ions in solution were applied. Two water systems was modeled, namely natural surface water and wastewater which differs in the organic matter content. The tested concentrations of the pollutants were as follows: 0; 0,5; 1; 2; 3; 4; 5; 20; 40 mg/dm^3 for phenol and 0; 5; 20; 40 mg/dm^3 for copper ions. A response signal for each model systems was profiled by the specific response of the bio-electrochemical sensor, pH, redox potential and conductivity electrodes. Conductivity was recorded by a HQ30d flexi conductometer (Hach Lange), the pH and redox potential by FiveEasy Plus pH meter (Mettler Toledo).

In order to evaluate the bio-toxicity the effect of the toxic compounds on the electrochemicaly active bacteria on the anode surface of the MFC was evaluated based on the fluctuations of the electric signal observed due to inhibition of the anode respiration systems of the microbial culture. All data obtained were summarized and prepared for modeling.

All experiments and chemical analyses were performed in triple and mean values are presented in the figures and tables.

2.4 Analysis: InterCriteria

This section provides some details on the InterCriteria analysis (IcrAData software) used to investigate the data sets. Full mathematical detail and justification for the procedure can be found in [6–8].

The water pollutants are defined as n objects O_i (where $i = 1, 2,.., n$) in a set O;

$$\{O_1, O_2, \ldots, O_n\} \in O.$$

Let $C(O_i)$ (where $i = 1, 2,.., n$) be a value assigned to object O_i given criteria C, such that $C(O)$ is a set:

$$C(O) = \{C(O_1), C(O_2), \ldots, C(O_n)\}$$

An Index Matrix (IM) with real elements [9] is composed, such that the objects O_i are the columns index set, and the criteria C – the rows index set of the matrix. The elements of the matrix are Intuitionistic Fuzzy Pairs (IFP), defined such that they show the degree of consonance (agreement) between two criteria. Mathematically, IFP is defined as an ordered number pair $<a, b>$ (for finding a relation the order does not matter), where $a, b \geq 0$ and $a + b \leq 0$.

For each criterion, a vector of all internal comparison $V(C)$ between criteria, conforming to the relations $R, \overline{R}$ and $\tilde{R}$, such that $V(C) = \{C_{1,n}, C_{2,n}, \ldots, C_{n,n}\}$ (no repeated elements (i, j) and (j, i) are treated as equivalent indices). Each element of the vector denotes

$$C_{i,j} = < C(O_i), C(O_j) > .$$

The degree of consonance between two criteria is defined by the number of matching elements in the internal comparison vectors (normalised), whereas the degree of dissonance is defined by the number of elements with opposing signs in those vectors (normalised). When the degrees of consonance and dissonance do not sum up to 1, their difference is the uncertainty of the obtained result.

3 Results and Discussions

For the purposes of this study, a lab scale installation has been prepared including the sensors and electrodes used. The installation was fed with the tested model solutions representing fresh water and wastewater with two different toxic components in different concentrations. Samples without added copper ions and phenol were used as a control. The data obtained (pH, redox potential, conductivity and bio-electrochemical signal) during these set of experiments are presented in Figs. 1, 2, 3 and 4.

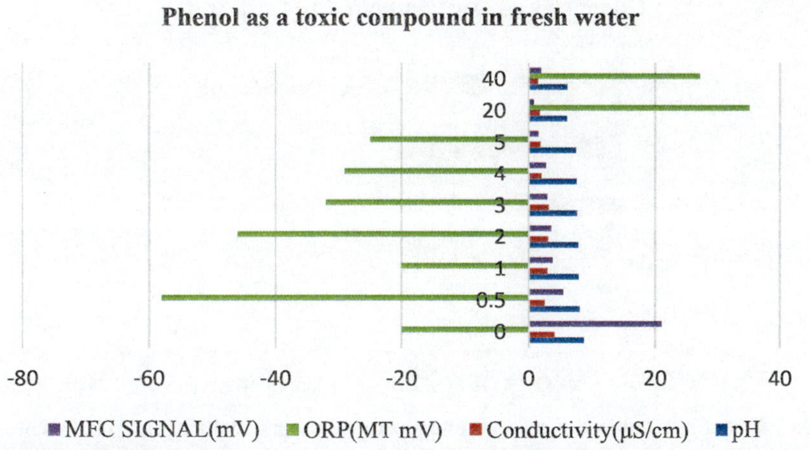

Fig. 1. Effect of different concentrations of phenol (from 0 to 40 mg/dm^3) in freshwater on the signal obtained from MFC, pH, redox potential and conductivity probes

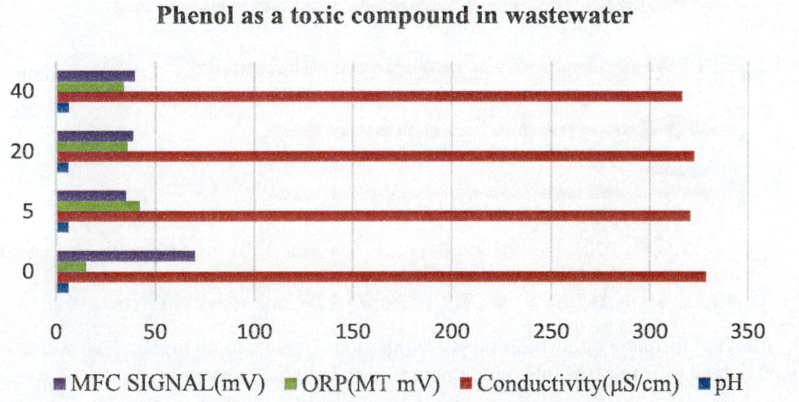

Fig. 2. Effect of different concentrations of phenol (from 0 to 40 mg/dm^3) in wastewater on the signal obtained from MFC, pH, redox potential and conductivity probes

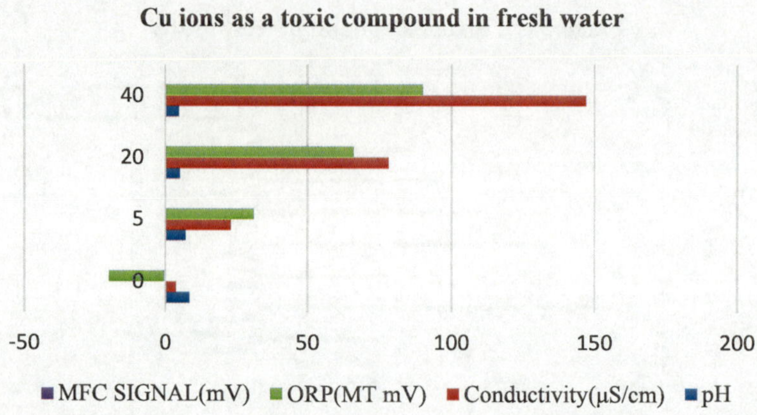

Fig. 3. Effect of different concentrations of copper ions (from 0 to 40 mg/dm^3) in freshwater on the signal obtained from MFC, pH, redox potential and conductivity probes

From this initial data, even in their raw format it is obvious that there is a certain amount of correlation between the status of the tested water and the parameter values observed. Even more, in dependence with the type of the toxic compound applied we observed some characteristic patterns of the signals. For instance, presence of phenol and copper ions in the water differs by their influence on the conductivity.

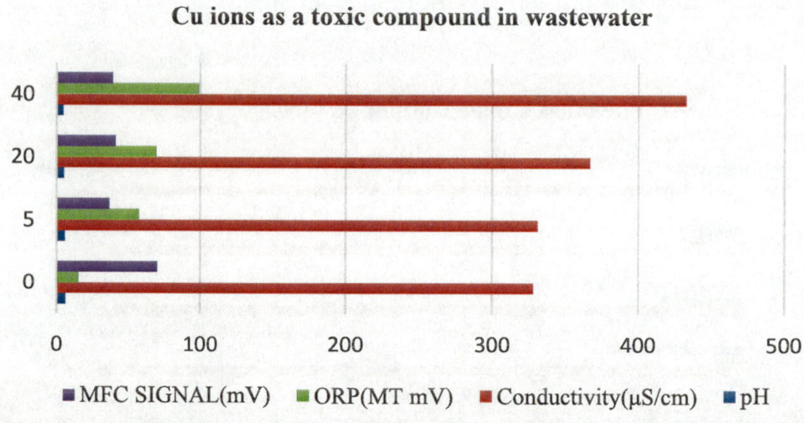

Fig. 4. Effect of different concentrations of copper ions (from 0 to 40 mg/dm^3) in wastewater on the signal obtained from MFC, pH, redox potential and conductivity probes

Both influences negatively the bio-electrochemical sensor by their inhibitory effect on the electrogenic bacteria but they have different grade of influence on pH. At this stage, our initial hypothesis that organic toxic compounds and heavy metals could be differentiated by their specific signal profiles or patterns was partially confirmed. Beside assessing toxicity levels (mostly through the bioelectrochemical sensor), we can predict the nature of the toxic compound presented.

Using IM input, InterCriteria analysis has been performed. The IFPs have been calculated for each physically meaningful pairing of the sixteen criteria (C_1, C_2,..., C_{16}). Following the scale by [9] for the defined consonance and dissonance intervals (Table 1), the interaction for all the explored criteria pairs are analysed (Table 2, 3, 4, 5).

Table 1. Definition of consonance and dissonance between each pair of criteria

$\mu_{C,C'}$	Meaning
[0.00–0.05]	Strong negative consonance
(0.05–0.15]	Negative consonance
(0.15–0.25]	Weak negative consonance
(0.25–0.33]	Weak dissonance
(0.33–0.43]	Dissonance
(0.43–0.57]	Strong dissonance
(0.57–0.67]	Dissonance
(0.67–0.75]	Weak dissonance
(0.75–0.85]	Weak positive consonance
(0.85–0.95]	Positive consonance
(0.95–1.00]	Strong positive consonance

4 Applying ICA Approach

We will use few index matrixes that contains 16 rows for different type of chemical parameters.

After the applying the ICA method we obtain index matrix (shown in the following tables) with intuitionistic fuzzy pairs that represents an intuitionistic fuzzy evaluation of the relations between every pair of criteria. The tables contain the coefficients μ.

The descriptions of different parameters are:

- C_1 – pH (bio-electrochemical reactor, fresh water, inhibitor: phenol)
- C_2 – Conductivity (Hach μS/cm) – (bio-electrochemical reactor, fresh water, inhibitor: phenol)
- C_3 – ORP (MT mV) – (bio-electrochemical reactor, fresh water, inhibitor: phenol)
- C_4 – BEC (mV, mA) – (bio-electrochemical reactor, fresh water, inhibitor: phenol)

- C_5 – pH (bio-electrochemical reactor, wastewater), inhibitor: phenol)
- C_6 – Conductivity (Hach µS/cm) – (bio-electrochemical reactor, wastewater, inhibitor: phenol)
- C_7 – ORP (MT mV) – (bio-electrochemical reactor, wastewater, inhibitor: phenol)
- C_8 – BEC (mV, mA) – (bio-electrochemical reactor, wastewater, inhibitor: phenol)
- C_9 – pH (bio-electrochemical reactor, phenol, inhibitor: fresh water)
- C_{10} – Conductivity (Hach µS/cm) – (bio-electrochemical reactor, phenol, inhibitor: fresh water)
- C_{11} – ORP (MT mV) – (bio-electrochemical reactor, phenol, inhibitor: fresh water)
- C_{12} – BEC (mV, mA) – (bio-electrochemical reactor, phenol, inhibitor: fresh water)
- C_{13} – pH (bio-electrochemical reactor, phenol, inhibitor: wastewater)
- C_{14} – Conductivity (Hach µS/cm) – (bio-electrochemical reactor, phenol, inhibitor: wastewater)
- C_{15} – ORP (MT mV) – (bio-electrochemical reactor, phenol, inhibitor: wastewater)
- C_{16} – BEC (mV, mA) – (bio-electrochemical reactor, phenol, inhibitor: wastewater)

After applying the ICA approach on the experimental data obtained for the studied parameters we could conlude:

- For the first table (Table 2) of calculated IFP for criteria in setting: inhibitor phenol in distilled water 1 pair of the criteria (C_1, C_2) is in weak positive consonance. The relation between this parameter is very strong;
- Next strong relation is between C_5 and C_8. According to the values of the Table 1 they have weak positive consonance. The similar behavior has also C_9 and C_{10}. They are in weak positive consonance;
- Next relation that is not very strong is for the pair C_9, C_{12} with positive consonance.
- The other notable connection is C_{10}, C_{12} (weak positive consonance)
- The parameters that are approximately independent with no practical relations are C_6-C_8, C_{10}-C_{11}, C_{13}-C_{14}, C_{14}-C_{15};
- The parameters that are in negative consonance are C_1-C_3, C_3-C_4, C_9-C_{11}, C_{13}-C_{15}.

Parameters that are in positive consonance demonstrate common behavior of reversed correlation. When only one of the parameters is changed, the consonance coefficients are near to 0.5, which is an evidence for lack of connection between individual parameters. Negative consonance indicates a discrepancy in the behavior of the parameters. In these cases, values of μ close to 0 are obtained.

Table 2. Calculated IFP for criteria in setting "inhibitor phenol in distilled water"

	C_1	C_2	C_3	C_4
C_1	1	0.805556	0.25	0
C_2	0.805556	1	0.333333	0
C_3	0.25	0.333333	1	0.027778
C_4	0	0	0.027778	1

Table 3. Calculated IFP for criteria in setting "inhibitor phenol in wastewater"

	C_5	C_6	C_7	C_8
C_5	1	0.5	0.166667	0.833333
C_6	0.5	1	0.333333	0.666667
C_7	0.166667	0.333333	1	0
C_8	0.833333	0.666667	0	1

	C_5	C_6	C_7	C_8
C_5	0	0.5	0.833333	0.166667
C_6	0.5	0	0.666667	0.333333
C_7	0.833333	0.666667	0	1
C_8	0.166667	0.333333	1	0

Table 4. Calculated IFP for criteria in setting "inhibitor fresh water in phenol"

	C_9	C_{10}	C_{11}	C_{12}
C_9	1	0.805556	0.25	0.972222
C_{10}	0.805556	1	0.333333	0.777778
C_{11}	0.25	0.333333	1	0.277778
C_{12}	0.972222	0.777778	0.277778	1

	C_9	C_{10}	C_{11}	C_{12}
C_9	0	0.194444	0.722222	0.027778
C_{10}	0.194444	0	0.638889	0.222222
C_{11}	0.722222	0.638889	0	0.694444
C_{12}	0.027778	0.222222	0.694444	0

Table 5. Calculated IFP for criteria in setting "inhibitor wastewater in phenol".

	C_{13}	C_{14}	C_{15}	C_{16}
C_{13}	1	0.472222	0.166667	0.25
C_{14}	0.472222	1	0.583333	0.5
C_{15}	0.166667	0.583333	1	0.5
C_{16}	0.25	0.5	0.5	1

	C_{13}	C_{14}	C_{15}	C_{16}
C_{13}	0	0.527778	0.694444	0.75
C_{14}	0.527778	0	0.277778	0.5
C_{15}	0.694444	0.277778	0	0.361111
C_{16}	0.75	0.5	0.361111	0

5 Conclusion

In conclusion, this study reveals the potential contribution of the approach applied in an important niche related to the water safety, environmental and human health. Application of InterCriteria analysis could be the first step toward sophistication of the water quality systems towards the development of integrated approach including data collection and processing, deep learning and AI. Such a system will be able to evaluate the levels of toxicity in water and to make predictions about the source and nature of the toxic compounds presented.

Acknowledgements. The work supported by Project BG05M2OP001-1.002-0019: "Clean technologies for sustainable environment–water, waste, energy for circular economy" (Clean&Circle), for development of a Centre of Competence and European Regional Development Fund through the Operational Program "Science and Education for Smart Growth" under contract UNITe No BG05M2OP001-1.001-0004 (2018–2023) is acknowledged.

References

1. Abdullah, P., Waseem, S., Bai, V.R., Mohsin, I.U.: Development of new water quality model using fuzzy logic system for Malaysia. Open Environ. Sci. **2**, 101–106 (2008)
2. Gao, Y., Feng, J., Zhu, J., & Zhu, L.: Predicting copper toxicity in zebrafish larvae under complex water chemistry conditions by using a toxicokinetic–toxicodynamic model. J. Hazard. Mater. **400** (2020)
3. Zlateva, P., Gerginova, M., Atanasov, B., Alexieva, Z.: Fuzzy model for toxicity evaluation of phenol derivatives degraded by yeast trichosporon cutaneum. Comptes rendus de l'Académie bulgare des Sciences **59**(12), 1253–1258 (2006)

4. Qu, R., Wang, X., Liu, Z., Yan, Z., Wang, Z.: Development of a model to predict the effect of water chemistry onthe acute toxicity of cadmium to Photobacterium phosphoreum. J. Hazard. Mater. **262**, 288–296 (2013)
5. Mersinkova, Y., Yemendzhiev, H.: Identification and characterization of natural habitats of electrochemically active bacteria. J. Adv. Biol. Biotechnol. **23**(1), 19–25 (2020)
6. Atanassov, K.: Intuitionistic fuzzy sets. Fuzzy Sets Syst. **20**(1), 87–96 (1986)
7. Atanassov, K.: Intuitionistic Fuzzy Sets. Springer, Heidelberg (1999)
8. Atanassov, K., Mavrov, D., Atanassova, V.: Intercriteria decision making: a new approach for multicriteria decision making, based on index matrices and intuitionistic fuzzy sets. Issues IFSs GNs **11**, 1–8 (2014)
9. Atanassov, K.: Index Matrices: Towards an Augmented Matrix Calculus. Springer, Heidelberg (2014)

Biomedical Approaches and Applications

A Survey on the Application of Mobile Communication Devices in Remote Cardiac Monitoring Systems

Ivo Iliev[1], Ivan Kanev[1], and Vessela Krasteva[2](✉)

[1] Faculty of Electronic Engineering and Technologies, Technical University of Sofia, 8 Kliment Ohridski blvd., 1000 Sofia, Bulgaria
izi@tu-sofia.bg, ikanev@icloud.com

[2] Institute of Biophysics and Biomedical Engineering, Bulgarian Academy of Sciences, Acad. G. Bonchev str. bl. 105, 1113 Sofia, Bulgaria
vessika@biomed.bas.bg

Abstract. The rapid improvements in wireless and cellular communication set the new mobile mHealth era of clinical care. Positive effects of early diagnosis and treatment, patient mobility, long-term cost-effectiveness and safety in the spectrum of cardiac morbidity and severity are largely expected. This study aims to answer the main question: "What is the current state of existing developments and applications based on mobile communication devices (GSM, smart watches, smartphones) for telemetric monitoring of cardiac patients via the electrocardiogram (ECG) and additional body sensors", presenting a systematic literature review (SLR) in the five largest databases for peer-reviewed literature on English papers published (2010–2020). SLR identified a number of 138 publications by the first filter on titles, abstracts and keywords: ("GSM"-or-"smartphone"-or-"smartwatch") and (("patient"-and-"telemetry") or ("patient"-and-"remote"-and-"monitoring") or ("ECG"-and-"telemetry") or ("ECG"-and-"remote"-and-"monitoring")): Scopus (50,36.2%), PubMed (24,17.4%), Springer (13,9.4%), Science Direct (24,17.4%), Web-of-Science (27,19.6%). SLR selected 117 unique records, which were subjected to analytical screening of titles and abstracts. SLR excluded 93 articles (79.5%) non-matching the specific question. A total of 24 papers were subjected to qualitative synthesis of the full-text. They present different technologies for remote patient monitoring (mostly using standard communication modules in smartphones GSM/GPRS/SMS/GPS, rarely other interfaces or mobile ad-hoc networks); sensors (ECG, photopletismogram, ballistocardiogram, termometer, accelerometer, piezo-sensors); continuous measurements (heart-rate, respiratory-rate, non-invasive blood pressure, heart-rhythm, oxygen saturation, energy expenditure, location information); alarm event detection (cardiac arrhythmia, hypertonia, fall); real-life applications (ambulatory monitoring of cardiac patients, cardiac rehabilitation, mobile ECG-detection system in ambulances, remote follow-up of cardiac implantable electronic devices, fetus heart-rhythm monitoring, etc.). Benefits and technological challenges behind mHealth are discussed, together with special aspects for patient privacy protection in client side when using mHealth.

S. S. Sotirov et al. (Eds.): BioInfoMed 2020, LNNS 374, pp. 299–311, 2022.
https://doi.org/10.1007/978-3-030-96638-6_32

Keywords: Electrocardiogram · GSM · Smartphone · mHealth · Remote patient monitoring · Telemetry · Cardiac activity · IoT · Systematic literature review

1 Introduction

Remote patient monitoring using Internet of Things (IoT) technologies provides a global connectivity able to transfer and store large amounts of physiological data in remote cloud-based healthcare systems [1–4]. The rapid improvements in wireless and cellular communications (from widely available Global System for Mobile (GSM) services with 3G, 4G to the new 5G Gateways) set the new mobile mHealth era of clinical care when physicians can follow their patients outside the traditional in-office setting and patients can be actively engaged in their care by simply using their smart mobile devices connected to different body sensor networks [5, 6]. Recent innovations made major advances in smartphone adapters for maintaining a portable, easy-to-use, cost-effective, and integrated environment for treatment, diagnosis, and point-of-care applications [7]. They trace the novel route towards high-quality medical services, supporting medical data communication, sharing, real-time monitoring, on-site diagnosis, rapid testing and health assistance of patients in rural areas, chronic or high-risk patients, elderly or disabled people who cannot benefit from ordinary clinical examinations but all need timely and accurate diagnosis and treatment by home healthcare applications [7–11]. Thus mobile phones have become an increasingly important platform for faster delivery of health interventions with reduced cost and enhanced health coverage [12]. Current IoT connectivity solutions face challenges related to the support of massive number of devices, device density, standardization, unobtrusive sensing, energy-efficiency, privacy and security [1–3, 13].

In the context of timely diagnosis and management of high risk cardiac patients with symptoms such as syncope, dizziness, chest pain, palpitations, or shortness of breath, which may correlate with intermittent cardiac arrhythmias, the ambulatory electrocardiogram (ECG) telemetry has been employed with consensus guidelines by the International Society for Holter and Noninvasive Electrocardiology (ISHNE) and the Heart Rhythm Society (HRS) [14]. The typical concept realization of a mobile ECG telemonitoring system is shown in Fig. 1. According to the vast literature review in [14], mobile cardiac telemetry (MCT) devices are often from 1 to 3-lead ECG recorders embedded as a patient module either in a patch, necklace pendant, or a chest belt carrier connected to a few chest ECG electrodes (Fig. 1). The MCT device includes standard analogue circuits – instrumentation amplifiers, filters, auxiliary circuits for reduction of the common-mode interference, analogue-to-digital converters, microcontroller for digital ECG signal processing and remote data transmitter. The transmitter can be based on low-consumption interfaces for close communication (e.g. Bluetooth Low Energy, XBee, SimplisiTI, LoRa), mobile networks of the telecommunication operators (GSM, GPRS, UMTS, 3G, 4G, 5G) or networks and protocols of the internet service providers (LAN, WLAN, TCP/IP). Worn continuously, MCT devices are capable of real-time ECG streaming, and often combine the functionality of traditional Holter event and loop recorders, for example, programmed to autodetect and autosend events at certain time (e.g., once every 10 min) to the remote reading center. The end receiver could be a special data server with an authorized access or a virtualized data cloud, a central data collection station in a hospital, a PC with expert analysis application, or even the smartphone of the attending physician (Fig. 1).

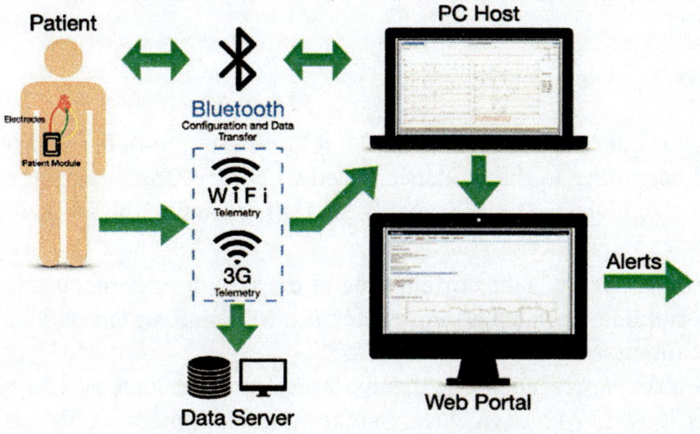

Fig. 1. A conceptual framework of a mobile cardiac telemetry system.

The benefit of using MCT is an immediate alarm generation upon an event detection without patient interaction or manual activation. Some shortcomings are mainly related to the continued engagement of qualified staff for the review and evaluation of telemetry data. Because of the non-standardized requirements for telemedicine and because there may be an incomplete exam of the patient due to the nature of the remote technology, there may be a higher chance of misdiagnosis. Another limitation of electrode-wire MCTs concerns the need of daily electrode changes, and thus, patient acceptance is reduced for long-term monitoring applications [14].

The possibility for event data transfer at unregulated time slots implies an uninterrupted connectivity of the patient module to the network, but causing some problems in the absence of good network coverage. On the other hand, the continuous connectivity leads to greater consumption and limits the duration of continuous monitoring. However, in such a conversion, the problems arise in establishing the connection and maintenance of the communication channel. Even at home, if the distance between the source and the receiver is large or there are partitions (reinforced concrete walls), the communication might not be established [15]. GSM services can overcome common communication problems, such as easy connection establishment, standardized communication and encryption protocols, reliable maintenance of the communication channel, found problematic in about 38% of remote cardiac monitoring events [16].

The more people (patients, consumers and physicians) are actively engaged with the mHealth, the better is the chance for shifting the classic medical paradigm from diagnostics and treatment towards predictive, preventive, personalized and participatory (P4) medicine [17]. Integration of mHealth adapters and applications in the high-tech smart devices (i.e. smartphones, smartwatches) to which most people are addicted can be their motivation for active participation in the personalized P4 medicine, which is greatly associated with improved quality of life.

The aim of this paper is to overview the scientific literature on the state-of-the-art applications of mobile communication devices in remote cardiac monitoring systems, with focus on the integration strategies of different body sensors, measurements of vital parameters, communication interfaces and areas of application.

2 Methods

The Systematic Literature Review (SLR) followed a pre-defined protocol, which was designed according to the evidence-based Cochrane Collaboration guidelines for performing SLRs of interventions in healthcare [18], including the following definitions:

1. *Main question*: "What is the current state of existing developments and applications based on mobile communication devices (GSM, smart watches, smartphones) for telemetric monitoring of cardiac patients?"
2. *Goal*: Search of papers indexed in the five largest abstract databases for peer-reviewed literature (Scopus, Web of Science, Science Direct, Spinger, PubMed).
3. *Inclusion criteria*:

   a. Content of titles, abstracts and keywords, including the following filters:

      - *Filter 1*: TITLE-ABS-KEY {("GSM" OR "smartphone" OR "smartwatch")}.
      - *Filter 2*: TITLE-ABS-KEY {(("patient" AND "telemetry") OR ("patient" AND "remote" AND "monitoring") OR ("ECG" AND "telemetry") OR ("ECG" AND "remote" AND "monitoring"))}.

   b. Up-to-date studies with a year of publication within the last 10 years (2010–2020), included by *Filter 3*: {PUBYEAR > 2009 AND PUBYEAR < 2021}.

4. *Exclusion criteria*:

   a. Papers that are not in English, excluded by the following filter:

      - *Filter 4*: {LIMIT-TO (LANGUAGE, "English")}

   b. Duplicated publications in the search lists of databases were automatically identified and excluded by a Python script, storing only unique records.

5. *Specific question*: "Does the paper present remote out-of-hospital ECG monitoring of patients and transmitting data via mobile communication devices?"
6. *Analysis plan*: The first step included analytical screening of all collected papers by two reviewers who comprehensively read the titles and abstracts, excluding papers non-matching the specific question. SLR papers were selected by consent between reviewers, then subjected to the second step of qualitative synthesis of the full-text.

3 Results

SLR was performed on June 17th 2020. SLR results are presented in the workflow (Fig. 2). A total of 138 publications were identified by Filters 1–4: Scopus (50, 36.2%), PubMed (24, 17.4%), Springer (13, 9.4%), Science Direct (24, 17.4%), Web-of-Science (27, 19.6%). The Python script for automatic scanning of record lists removed 21 duplicated publications. The collected 117 unique papers were further subjected to analytical screening of titles and abstracts that excluded 93 articles (79.5%) non-matching the specific question. Excluded articles covered the following topics:

- Clinical trials in a hospital setting; Monitoring of critical patients in ICUs;
- ECG telemetry in animals;
- Management of various communication architectures;
- Stress systems and clinical trials under special conditions;
- Processing and recording of ECG signals with large number of electrodes;

SLR finally selected 24 articles [19–42] for qualitative synthesis of the full-text.

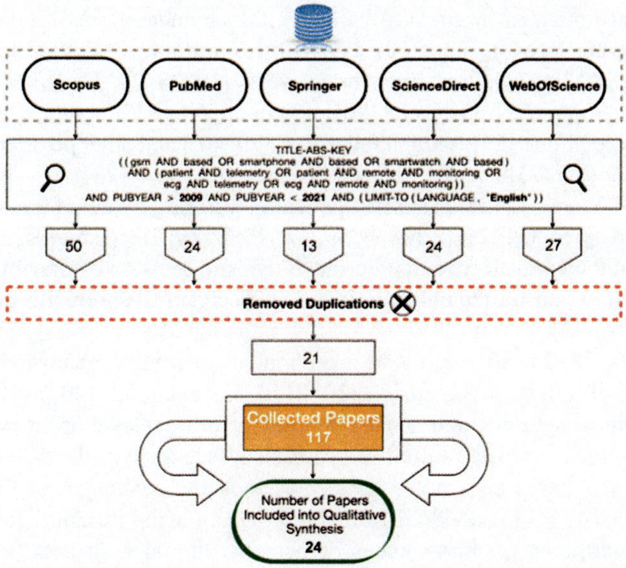

Fig. 2. SLR workflow: number of papers collected during SLR.

4 Discussion

The first focus of SLR qualitative synthesis was to analyze the type of the sensors, acquired signals, computed parameters and transmitted telemetric data.

All 24 SLR articles [19–42] presented telemonitoring of the heart activity, mainly by acquisition of reduced ECG lead systems (1 to 3 ECG channels) [19–28, 30, 32–36, 39–42] sampled at low rate (up to 300 Hz) in order to limit the transmitted data size. The long-term ECG monitoring in almost all telemonitoring systems relied on the standard adhesive wet ECG electrodes, except a few studies using manual contact dry metal Ag/AgCl electrodes [24, 27, 39, 41], capacitive ECG sensing electrodes [34], subcutaneous (implantable) electrodes [36], or some applications with very small and light wireless sensor nodes [30] or sensors embedded in a chest belt [40]. In one special application for monitoring of fetus heart rate, up to 8 ECG leads were acquired via 9 textile electrodes embedded in an elastic T-shirt [31].

Eight SLR articles (33%) showed mobile devices connected to a number of accessories with other sensors, as it was relatively easy to capture ECG simultaneously with other signals: photopletismogram (PPG) from a fingertip LED sensor [19, 26], piezo-sensors on a chest belt [19, 40], pressure sensor with piezoresistive transducer in an inflatable cuff [20, 21], 3-axis accelerometer [20, 26, 40], GPS [23, 26, 40], thermistor temperature body sensor [26, 40], ballistocardiogram measured by Wheatstone bridge based on resistance strain transducer [29], heart sounds by a digital stethoscope at the heart apex [32], GSM-embedded orientation sensors for user authentication [37].

Twelve SLR articles (50%) took benefit from remotely sent additional information related to a real-time measurement of various vital parameters, such as: heart rate [19–23, 26, 29, 34, 36, 39, 40], pulse rate [19, 26], respiration rate [19, 40], non-invasive blood pressure [20, 21, 33], heart arrhythmia events [19–23, 29, 33, 36] or ischemia ST-depression [23, 36], pulse transit time [26], skin temperature [26, 40], oxygen saturation [26], energy expenditure [26], sudden fall events [20, 40]) and other position information (GPS-based location [23, 26, 40], elapsed distance and speed [23]).

Fourteen SLR articles (58%) applied patient monitoring by direct transfer of signals in data packets, i.e. ECG [22–28, 30–34, 39, 41], PPG [26], heart sounds [32] for remote visualization of their waveform and precised diagnostic decision by the physician. Other five articles (21%) transmitted only continuously discrete measurements of vital parameters [19–21, 26, 40]. Seven (29%) of articles transmitted alarm messages (SMS alert [20, 21, 24, 29, 33, 34, 40]) on event detection (heart rhythm abnormalities [20, 21, 24, 29, 33, 34, 40], hyper/hypotension [20, 21, 33], sudden fall [20, 40]) thus gaining immediate medical attention and hence reducing the critical level of the patient.

The second focus of SLR qualitative synthesis was to analyze the type of the remote communication. Almost all articles (23, 96%) took the advantage of the broadband mobile connectivity (GSM/SMS/GPRS/3G/4G/5G) by using standard communication modules in smartphones, smartwatches or personal digital assistants (PDA) [17–24, 26–42], while a few used WiFi access points to the web-cloud [25, 40]. According to the provided data from one mobile operator, technical service 3G/3.5G communication including authorization with server required 5–6 s, of which greater part (app. 80%–90%) was for the authorization [33]. As the sent information was very small in size (<13 Kb), the significant difference in throughput between various types of connectivity (GPRS/EDGE – 30–250 kb/s and HSDPA – 0.8–6.4 Mb/s) resulted in nearly no difference for traffic delay. The delay was estimated at least 10 times larger for the further service of sending e-mail (pdf type report with size 120–138 Kb) from the web-server to the physician's smartphone or laptop, and that was mainly pending on the type of the operating system or hardware [33]. One experimental study showed that WiFi had a lower packet loss than 3G (1.52% vs. 6.51%), lower power consumption on data transmission (15% vs. 21%) and idle (3% vs. 6%), as well as lower delay (0.083 s vs. 0.157 s – both were quite small delays and did not cause serious influence on the system's real-time performance) [40].

Ten SLR articles (42%) transmitted ECG data by integrating the recording device as an element of a smart sensor network, i.e. IoT applications by connecting ECG monitoring node through WiFi (HTTP and MQTT protocols) to the IoT cloud [25]), or by connecting ECG monitoring node via a wireless interface (Bluetooth [26, 28, 31-33, 39, 40 or ZigBee [30]) to a GSM module and optionally then via 3G, WiFi to a remote healthcare server [26, 30, 31, 33, 39, 40]. One study reported statistics on the Bluetooth usage [33], which must respect short distance (<10 m) between the recorder and the mobile phone to secure Bluetooth wireless transfer with low miscommunication rate (0.7%). In two studies, the ECG monitoring node was directly connected to the standard audio GSM interface by transforming the ECG signal into sound modulation (a technology embedded in AliveCor™HeartMonitor) [27, 41].

The third focus of SLR qualitative analysis was on the software applications.

Four SLR articles (17%) proposed software applications mainly based on the biomedical toolkit in LabVIEW, embedded in the pre-processing part of the patient-connected module. Special graphic user interfaces (GUI) and algorithms were designed for continuous signal processing and online measurements of vital parameters, such as soft threshold based peak detector [19], noise removal in ECG by Wavelet transform [20]; heart rate variability (HRV) analysis [21] by fast Fourier transform (FFT) [20]; fall-detector by analysis of three-axis accelerometer vibrations [20]; blood pressure measurement by analysis of cuff pressure [20, 21]; LabVIEW development platform for ECG visualization and configuration of the sampling rate [30].

Ten SLR articles (42%) presented details on GSM and cloud-server applications: [22] proposed a remote ECG server service program purposed for continuous receiving and processing of ECG signal during real-time patient monitoring based on Linux platform with embedded VC++; [23] used server-side software to display incoming signals and system status, i.e. sweep-display of the ECG signal, self-scaling displays of heart rate and walking speed, elapsed time and distance indicators, and a map or satellite image display of the patient's location and path for real-time monitoring by a qualified exercise scientist; [25] used web browser GUI application to manage continuous data transfer and ECG visualization in IoT cloud, as well as storing ECG data for further analysis, implemented on the basis of three servers (HTTP, MQTT and storage server); [26] presented a server application for the display of vital parameters (real-time mode), ECG/PPG waveforms (in an emergency or for special care cases) and geographic map location of the monitored patient. For increased transmission efficiency, data compression and a simple error correction algorithm were implemented; [28] showed a PDA application in C# for the Windows Mobile.net platform to acquire, store, upload, download and display ECG recordings on GUI. Multiple measurement files from different sessions and patients were stored on the PDA using Microsoft SQL Server Compact 3.5, considered as a secure and searchable database for desktop and mobile devices; In

[31] the server site contained: a synchronization engine storing ECG signals sent by remote stations, a set of signal processing modules for measurement of the fetal heart rate (including suppression of noise and mother ECG while enhancing the fetus ECG component, and selection of the most informative track for the measurement) and a web application for analyzing and visualization of the results; [32] presented a mobile application running FFT-based noise suppression algorithms on heart sound ambient noises and ECG, as well as storage of hystorical data. It provided automatic data interpretation and guidance to physician; [33] presented mobile application for HRV analysis, which detected arrhythmia irregularities. Then it initiated sending of ECG packets or.pdf type ECG file from the loop recorder via GSM 3G to a remote web-server and the doctor's e-mail; [39] built two smartphone applications as a part of the telemedicine platform: (i) the patient's applications to help patients to record their ECG signal and exchange messages and medical information and symptoms, with health centers and physicians, (ii) the expert application, which allows an expert to record and monitor ECG signals in real time, as well as to review and analyse recordings, using algorithms running on the cloud server. Additional services were implemented to allow medical experts to exchange ECG recordings and medical knowledge for consultations with other experts. A library for real time ECG visualization and basic ECG processing was implemented, including noise filtering, ECG delineation, clustering, and classification; [40] implemented a GUI smartphone applications for real-time visualization of physiological curves of vital parameters (heart rate, respiratory rate, skin temperature, posture, patient activity), with embedded filtering algorithms to manage data gaps. A disadvantage was the ability to report only patient's status, without professional analysis and instructions.

The forth focus of SLR qualitative analysis was on the experimental setting, real-life applications of mobile cardiac tele-monitoring technologies and feedback results.

Eight articles (33%) were laboratory studies on a test bench wearable health monitoring system connected to a GSM module and a desktop PC [19–21, 28–30, 32, 34].

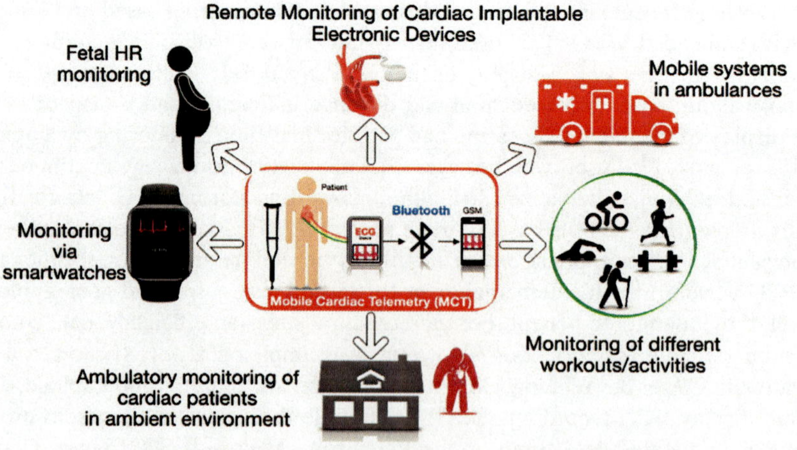

Fig. 3. Real life applications of mobile cardiac telemetry monitoring.

Fourteen articles (58%) presented real-life applications (Fig. 3), including: mobile ECG-detection system in ambulances with ECG transmitted to a remote medical center [22], cardiomobile system for remotely monitoring exercise in walking-based cardiac rehabilitation in 134 cardiac patients near their homes and physicians located in the hospital [23], remote follow-up monitoring of patients with pacemakers, as a complementary option of telemetry for early diagnosis and detection of functional problems in cardiac implantable electronic devices [24, 35, 36]; fetus heart-rhythm monitoring [31]; ambulatory monitoring of a healthy volunteer, showing no distortions of the transmitted ECG signal in both stationary and moving states [25]; ambulatory monitoring of five subjects (three linked by Wi-Fi networks, and two connected through 3G networks), showing the simultaneous observation of their vital parameters in real time [26]; ambulatory mobile monitoring and management of patients with heart failure in their ambient environment, showing the proper way of connecting the ECG recorder to acquire diagnostically meaningful 3 ECG leads (I, II and "quasi-V3/V4") [27]; ambulatory mobile monitoring of 73 cardiac patients and 32 arterial hypertension patients with ECG loop recorder and a smartwatch device [33]; remotely managed implantable loop recorders for telemonitoring of patients with specific symptoms and cardiac complaints, such as unexplained syncope, cryptogenic stroke, palpitations, chest pain, or monitoring recurrence of atrial fibrillation following ablation [36]; ambulatory monitoring of 391 patients (40 volunteers and 351 cardiac patients) in supine, sitting, and standing positions. Issues on the device placement vs. signal quality were discussed, showing that 96% of ECGs had the diagnostic potential for arrhythmia detection [39]; use of portable terminal for monitoring of walking, running, sitting, and sleeping volunteers with estimated measurement accuracy of 96.73% (heart rate), 94.85% (skin temperature), 98.15% (respiration rate) and 99.8% (posture) [40]; hand-held 1-lead ECG acquisition from 381 subjects (athletes and cardiac patients), providing high quality smartphone ECG signals with estimated sensitivities for automated detection of QRS delay (72.4%), atrial fibrillation (94.4%), AV block (77.3%) and specificities (>94.5%) [41].

Only four articles (17%) presented a feedback from the end users, reporting the estimated time taken to set up and take off the equipment, experienced technical and non-technical problems, provided qualitative statements about their experience of the program with a key measure of the quality of life [23]; Pacemaker patients who were enrolled and compliant with remote monitoring experienced significantly improved survival during 3 years of follow-up. The magnitude of benefit was similar to that observed in patients undergoing defibrillator implantation or cardiac resynchronization therapy [24]; According to quality of experience survey results in 73 cardiac patients and their attending physicians, the arrhythmia mobile monitoring was assessed as highly satisfactory, convenient, acceptably priced and easy to use service. Besides mobile monitoring of 32 arterial hypertension outpatients was also assessed as satisfactory and revealed elevated blood-pressure at nighttime, however reported as not cost-efficient [33]. According to the user query in [40], some volunteers were not very enthusiastic about the idea of continuously wearing a monitoring device. The reason could be the discomfort caused by the used chest belt with embedded sensors.

Seven SLR articles (29%) were reviews and addressed issues on the clinical usefulness of smartphone-based mobile ECG technologies and guidelines [42]. It was underlined that remote monitoring by wireless and cellular transmission has rapidly become the standard of care for patients with implantable electronic devices [24, 35, 36]. The improvements in patient outcomes were shown as shorter time-to-diagnosis and treatment, shorter and reduced hospitalizations, improved survival, patient satisfaction and improved quality of life. The benefits for the Healthcare system concerned improved accessibility for remote patients, fewer emergency room visits, fewer in-office visits (medical staff costs), less costs with shorter hospitalization, less travel costs for patients [35]. One of the remaining challenges is to ensure that remote monitoring systems are properly set up and used by patients and also that following facilities adopt changes in workflow patterns to efficiently gather and respond to transmitted data without adding to their follow-up burden [36]. Other problems related to the infrastructure needed to enroll, monitor, and manage patients [24], data transfer and storage, personal identification and patient privacy protection when using mobile communication devices in mHealth [37]. It was highlighted the motivation benefits related to the smartphone usability, effect of security on smartphones, sensor authentication usability, sensor security and application activities [37]. The problems on the triage and prioritization of large scale telemedicine patients with big data analysis has also been reviewed [38]. It suggested that analytic hierarchy process should be recommended to establish weights for evaluation criteria based on the experts' judgments, and "technique for order performance by similarity to ideal solution" should be recommended to provide an overall ranking of patients for different chronic diseases (heart, diabetes, blood pressure). Moreover, increasing the sources (such as, video, audio, image, medical sensors, and GPS) for triage for the prioritization and designation of appropriate emergency levels to patients remains an issue [38].

With the number of available mobile ECG technologies in the market [42], positive effects of early diagnosis and treatment, patient mobility, long-term cost-effectiveness and safety in the spectrum of cardiac morbidity and severity are largely expected.

Acknowledgements. This study was supported by the Bulgarian National Science Fund under Grant Ref. No. КП-06-H37/9 "Audio transformation (sonification) of the electrocardiogram – a new approach for remote monitoring of the heart activity".

References

1. Ahad, A., Tahir, M., Sheikh, M.A., Ahmed, K.I., Mughees, A., Numani, A.: Technologies trend towards 5G network for smart health-care using IoT: a review. Sensors **20**, 4047 (2020)
2. Wu, F., Wu, T., Yuce, M.R.: An internet-of-things (IoT) network system for connected safety and health monitoring applications. Sensors **19**(1), 21 (2019)
3. Yin, Y., Zeng, Y., Chen, X., Fan, Y.: The internet of things in healthcare: an overview. J. Ind. Inf. Integr. **1**, 3–13 (2016)
4. Jagadeeswari, V., Subramaniyaswamy, V., Logesh, R., Vijayakumar, V.: A study on medical Internet of things and big data in personalized healthcare system. Health Inf. Sci. Syst. **6**(1), 1–20 (2018). https://doi.org/10.1007/s13755-018-0049-x

5. Appelboom, G., et al.: Smart wearable body sensors for patient self-assessment and monitoring. Arch. Public Health **72**(1), 28 (2014)
6. Leu, F.Y., Ko, C.K., You, I., Choo, K.K.R., Ho, C.L.: A smartphone-based wearable sensors for monitoring real-time physiological data. Comput. Electr. Eng. **65**, 376–392 (2018)
7. Alawsi, T., Al-Bawi, Z.: A review of smartphone point-of-care adapter design. Eng. Rep. **1**, e12039 (2019)
8. Durán-Vega, L.A., et al.: An IoT system for remote health monitoring in elderly adults through a wearable device and mobile application. Geriatrics **4**(2), 34 (2019)
9. Pandor, A., et al.: Home telemonitoring or structured telephone support programmes after recent discharge in patients with heart failure: systematic review and economic evaluation. Health Technol. Assess. **17**(32), 1–207 (2013)
10. Scanaill, C.N., Carew, S., Barralon, P., Noury, N., Lyons, D., Lyons, G.M.: A review of approaches to mobility telemonitoring of the elderly in their living environment. Ann. Biomed. Eng. **34**, 547–563 (2006)
11. Nancarrow, S., Banbury, A., Buckley, J.: Evaluation of a national broadband network-enabled telehealth trial for older people with chronic disease. Aust. Health Rev. **40**(6), 641–648 (2016)
12. Klasnja, P., Pratt, W.: Healthcare in the pocket: mapping the space of mobile-phone health interventions. J. Biomed. Inform. **45**(1), 184–198 (2012)
13. Akbar, S., Coiera, E., Magrabi, F.: Safety concerns with consumer-facing mobile health applications and their consequences: a scoping review. J. Am. Med. Inform. Assoc. **27**(2), 330–340 (2020)
14. Steinberg, J.S., et al.: ISHNE-HRS expert consensus statement on ambulatory ECG and external cardiac monitoring/telemetry. Heart Rhythm **14**(7), e55–e96 (2017)
15. Din, Z.U., Bernold, L.E.: Experimental study of signal behavior for wireless communication in construction. Constr. Innov. **17**(4), 475–491 (2017)
16. Connecting Remote Cardiac Monitoring Issues with Care Areas. Pa. Patient Saf. Advis. **6**(3), 79–83 (2009)
17. Flores, M., Glusman, G., Brogaard, K., Price, N.D., Hood, L.: P4 medicine: how systems medicine will transform the healthcare sector and society. Pers. Med. **10**, 565–576 (2013)
18. Higgins, J.P.T., et al.: Cochrane Handbook for Systematic Reviews of Interventions, 2nd edn. John Wiley & Sons, Chichester UK (2019)
19. Jegan, R., Nimi, W.S., Ramson, S.R.: Sensors based biomedical framework to monitor patient's vital parameters. In: 5th International Conference on Devices, Circuits and Systems, pp. 96–100. IEEE, Coimbatore; India (2020)
20. Tewary, S., et al.: A novel approach towards designing a wearable smart health monitoring system measuring the vital parameters and emergency situations in real-time and providing the necessary medical care through telemedicine. In: 2016 IEEE Students' Conference on Electrical, Electronics and Computer Science, pp. 1–8. IEEE, Bhopal, India (2016)
21. Rajan, S., Sukanesh, R., Vijayprasath, S.: Design and development of mobile based smart tele-health care system for remote patients. Eur. J. Sci. Res. **70**(1), 148–158 (2012)
22. Yu, X., Wu, K., Hou, Z.: Design and implementation of ECG wireless transmission system based on ARM9. In: 2010 International Conference on Computer, Mechatronics, Control and Electronic Engineering, pp 211–213. IEEE, Changchun, China (2010)
23. Worringham, C., Rojek, A., Stewart, I.: Development and feasibility of a smartphone, ECG and GPS based system for remotely monitoring exercise in cardiac rehabilitation. PLoS One **6**(2), e14669 (2011)
24. Mittal, S.: Smartphone-based electrocardiographic and cardiac implantable electronic device monitoring. Cardiol. Rev. **25**(1), 12–16 (2017)
25. Yang, Z., Zhou, Q., Lei, L., Zheng, K., Xiang, W.: An IoT-cloud based wearable ECG monitoring system for smart healthcare. J. Med. Syst. **40**, 286 (2016)

26. Lee, Y.G., Jeong, W.S., Yoon, G.: Smartphone-based mobile health monitoring. Telemed. J. e-Health **18**(8), 585–590 (2012)
27. Madias, J.E.: A proposal for monitoring patients with heart failure via smart phone technology-based electrocardiograms. J. Electrocalriol. **49**(5), 699–706 (2016)
28. Istepanian, R.S.H., et al.: The potential of machine-to-machine communications for developing the next generation of mobile device monitoring systems. In: Roa Romero, L. (eds.) XIII Mediterranean Conference on Medical and Biological Engineering and Computing 2013. IFMBE Proceedings, vol. 41, pp. 1903–1906. Springer, Cham (2014). https://doi.org/10.1007/978-3-319-00846-2_469
29. Chunwu, W., Xu, W., Zhe, L., Kexin, Z., Dan, Y.: The ballistocardiogram signal monitoring system based on the GSM network. In: Long, M. (eds.) World Congress on Medical Physics and Biomedical Engineering 26–31 May 2012, Beijing, China. IFMBE Proceedings, vol. 39, pp. 1346–1349. Springer, Berlin, Heidelberg (2013). https://doi.org/10.1007/978-3-642-29305-4_353
30. Valchinov, E.S., Pallikarakis, N.E.: A wearable wireless ECG sensor: a design with a minimal number of parts. In: Bamidis, P.D., Pallikarakis, N. (eds.) XII Mediterranean Conference on Medical and Biological Engineering and Computing 2010. IFMBE Proceedings, vol. 29, pp. 288–291. Springer, Berlin, Heidelberg (2010)
31. Lanzola, G., Secci, I., Scarpellini, S., Fanelli, A., Magenes, G., Signorini, M.G.: A Mobile Remote Monitoring Service for Measuring Fetal Heart Rate. In: Roa Romero, L. (eds.) XIII Mediterranean Conference on Medical and Biological Engineering and Computing 2013. IFMBE Proceedings, vol. 41, pp. 1435–1438. Springer, Cham (2014). https://doi.org/10.1007/978-3-319-00846-2_355
32. Dutta, M., Dutta, S., Sikdar, S., Dutta, D., Sharma, G., Sharma, A.: Reliable, real-time, low cost cardiac health monitoring system for affordable patient care. In: Gupta, S., Bag, S., Ganguly, K., Sarkar, I., Biswas, P. (eds.) Advancements of Medical Electronics. LNB. pp. 281–289. Springer, New Delhi (2015). https://doi.org/10.1007/978-81-322-2256-9_26
33. Kirtava, Z., Gegenava, T., Gegenava, M., Simonia, G., Andronikashvili, I.: e-Health/m-Health applications for cardiac patients telemonitoring. In: Braidot, A., Hadad, A. (eds.) VI Latin American Congress on Biomedical Engineering CLAIB 2014, Paraná, Argentina 29, 30 & 31 October 2014. IFMBE Proceedings, vol. 49, pp. 825–828. Springer, Cham (2015). https://doi.org/10.1007/978-3-319-13117-7_210
34. Harikumar, R., Shivappriya, S.N.: Intelligent telemetry system for ECG monitoring. Indian Heart J. **66**(S2), S143 (2014)
35. Cheung, C.C., Deyell, M.W.: Remote monitoring of cardiac implantable electronic devices. Can. J Cardiol **34**(7), 941–944 (2018)
36. Burri, H., Varma, N.: Remote device management in patients with cardiac complaints. Card. Electrophysiol. Clin. **5**(3), 337–347 (2013)
37. Shuwandy, M.L., Zaidan, B.B., Zaidan, A.A., Albahri, A.S.: Sensor-based mHealth authentication for real-time remote healthcare monitoring system: a multilayer systematic review. J. Med. Syst. **43**(2), 33 (2019)
38. Kalid, N., Zaidan, A.A., Zaidan, B.B., Salman, O.H., Hashim, M., Muzammil, H.: Based real time remote health monitoring systems: a review on patients prioritization and related Big Data using body sensors information and communication technology. J. Med. Syst. **42**(2), 30 (2018)
39. Marouf, M., Vukomanovic, G., Saranovac, L., Bozic, M.: Multi-purpose ECG telemetry system. BioMed. Eng. OnLine **16**(1), 80 (2017)
40. Zhang, Y., Liu, H., Su, X., Jiang, P., Wei, D.: Remote mobile health monitoring system based on smart phone and browser/server structure. J. Healthcare Eng. **6**(4), 717–738 (2015)

41. Haberman, Z.C., et al.: Wireless smartphone ECG enables large-scale screening in diverse populations. J. Cardiovasc. Electrophysiol. **26**(5), 520–526 (2015)
42. Guzik, P., Malik, M.: ECG by mobile technologies. J. Electrocardiol. **49**(6), 894–901 (2016)

Check for updates

High-Risk Cardiac Patients' Follow-Up via Portable Telemonitoring Personal Analyzer: Applicability, Reliability and Accuracy

Ivo Iliev[1], Irena Jekova[2](✉), Serafim Tabakov[1], Krasimira Koshtikova[3], Nikolay Runev[3], and Emil Manov[3]

[1] Department of Electronics, Faculty of Electronic Engineering and Technologies, Technical University of Sofia, 8 Kliment Ohridski Blvd., 1000 Sofia, Bulgaria
izi@tu-sofia.bg
[2] Institute of Biophysics and Biomedical Engineering, Bulgarian Academy of Sciences, Acad. Georgy Bonchev Str., 1113 Sofia, Bulgaria
irena@biomed.bas.bg
[3] Department of Propedeutics of Internal Medicine, Clinic of Cardiology, University Hospital "Alexandrovska" – Sofia, Saint Georgy Sofiisky Str., 1431 Sofia, Bulgaria
{nrunev,doctor_emil_manov}@abv.bg

Abstract. Patients' telemetry deals with remote follow-up of vital parameters and clinical conditions without continuous direct contact between physician and patient. This study presents a telemetry system for real-time monitoring of cardiac patients and investigates its accuracy, reliability and applicability. The system includes personal analyzer (PA) for continuous acquisition and analysis of single-channel electrocardiogram (ECG), central unit and a console for the cardiologist. PA transfers to the central unit 10 s ECG segments on regular basis (5/10/15 min), on demand or when an alarm based on rhythm disorders and/or heartbeat abnormalities is generated.

The ECG telemetry system is tested on 85 hospitalized patients with cardiovascular diseases. In parallel, the patients' ECGs have been recorded by a Holter device Signa-Lyzer SD. An experienced cardiologist has compared the ECG segments that have provoked an alarm to the respective ECG segments in the Holter recording in order to judge about the correctness (accuracy) of the alarm, as well as has checked the Holter ECG recording for events that are missed by the PA (reliability).

Full correspondence between the PA alarms and the rhythms recorded via the Holter has been observed (100% accuracy). Considering the witnessed Holter events, PA reliably warns for 80.6% of the rhythm disorders. In the remaining 19.4% an alarm is not generated mainly due to the fact that the patients have been out of the telemetry range.

The observed results show that the tested PA could support timely diagnosis, prompt reaction and adequate therapeutic treatment in potentially life-threatening clinical conditions.

Keywords: Wireless electrocardiogram monitoring · Portable ECG module · Telemetry · Cardiac diseases

S. S. Sotirov et al. (Eds.): BioInfoMed 2020, LNNS 374, pp. 312–321, 2022.
https://doi.org/10.1007/978-3-030-96638-6_33

1 Introduction

Cardiovascular diseases including heart attacks, strokes, and other circulatory disturbances are one of the leading causes of premature deaths in the world, since the cardiology is a field in which every second counts. According to the Centre for Disease Control and Prevention, one of every four deaths in the United Stated is due to a heart disease [1]. The continuous monitoring of high risk cardiac patients is usually performed via 12-lead electrocardiographic (ECG) monitoring systems that are stationary and therefore restrict the patients' movement. On the other hand, tracking of cardiac patients could also be performed via an ECG holter device with the ability to record continuously ECG signals during normal daily activities for 1 to 7 days in 3 to 12- lead configurations. The drawback of this "classical" ECG holter is the lack of real-time analysis of the recoded signals, which could sometimes appear to be critical.

The telemedicine provides expert medical assistance to unqualified, remote personnel or to the patients themselves by means of telecommunication methods and applications [2]. In this respect, the developments in the telecommunication infrastructure are a precondition for the development in telemetry cardiac devices for events recording, which are based on ECG analysis as one of the most useful non-invasive tools for the diagnosis of cardiac diseases [3]. The event recording devices can be generally divided into loop and post-event recorders [4]. The loop recording approach relies on continuous long-term ECG acquisition during which the events are triggered either manually by the patient or automatically by embedded ECG analysis algorithms [5–7]. The post-event monitoring relies on patient-activated ECG recording, which is provoked by the appearance of some symptoms [4, 8–11].

Wireless hospital ECG telemetry systems are using continuous ward monitoring of cardiac arrhythmias usually via loop ECG recording devices. They implement distant ECG acquisition, analysis and transmission of information and alerts for rhythm disturbances to a central monitoring station in a quasi-real time [12]. Such ECG telemetry systems are typically used to evaluate symptoms such as chest pain, dizziness, syncope, palpitations, shortness of breath, transient loss of consciousness, etc., which may be correlated with intermittent cardiac arrhythmias [5–11, 13, 14], as well as to follow up asymptomatic atrial fibrillation in patients with cryptogenic stroke [13–15].

Although in the near past there were few published criteria for in-patient telemetry [16] and some hospitals have developed and implemented their own guidelines to identify patients who require telemetry [17], nowadays the norms are summarized in the United Healthcare Oxford Clinical Policy (2018) [18]. It states that telemetry tracking is mandatory for patient with an indication of "silent" atrial fibrillation; patients with cryptogenic stroke with potential origin atrial fibrillation; and patients after surgical intervention (ablation). Specific aspects of ambulatory ECG telemetry, such as duration; electrode types; construction of the device, data transmission; data processing algorithms; and data protection, have been extensively studied by the International Society for Holter and Noninvasive Electrocardiology [13].

The aim of this study is to investigate the applicability, reliability and accuracy of a telemonitoring system for real-time remote monitoring of cardiac patients with rhythm and conduction disorders.

2 Method

The portable telemonitoring personal analyzer (PA) under analysis is part of a telemetry system including also a central unit that receives the information from PA via hybrid network architecture (i.e. wireless bluetooth and wired RS485 interfaces), a console for the cardiologist and a specially developed software PatRis (Patients at risk).

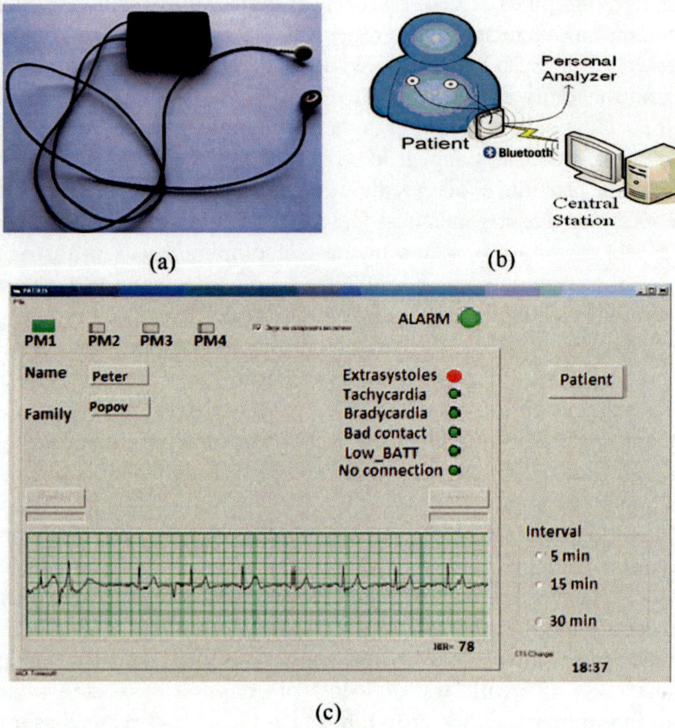

Fig. 1. Portable telemonitoring personal analyzer for acquisition and analysis of the patient's heart activity: (a) PA device; (b) Application of the PA device; (c) Main window of the software for communication between PA and the central unit (PatRis).

PA is a small size unit, which is continuously worn by the patient (Fig. 1) and supports the following functionalities:

- Continuous single-channel ECG acquisition at 250 Hz sampling rate via standard self-adhesive electrodes placed on the chest (lead I position);
- Real-time heartbeat detection based on a method described in [19];
- Real-time classification of normal and ectopic beats based on a method described in [20];
- Heart rate calculation;
- Alarms for:

  - Rhythms with ectopic beats;
  - Tachycardia;
  - Bradycardia;
  - Bad electrode contact;
  - Low battery;
  - No connection.

- Transfer of 10 s ECG segments controlled by the PatRis software on 3 reasons:

  - Over a predefined time interval (5, 15, 30 min);
  - When atypical heart activity is detected (i.e. rhythm disorders and/or heartbeat abnormalities);
  - "On-demand", after a request by the cardiologist.

In order to evaluate the accuracy and the reliability of the PA module, the patients' ECGs were recorded in parallel by a Holter device Signa-Lyzer SD (Signa Cor, Bulgaria), which provides:

- Continuous recording of 4 ECG leads (up to 48 h);
- Fast view of the ECG recording at windows with length 5, 10, 30, 60 min;
- Fast and accurate diagnosis of rhythm disorders;
- Analysis of the heart rate variability in the time domain;
- Automatic search of epochs with certain pathologies;
- Measurement of time and amplitude features.

An experienced cardiologist performed the following analyses of the ECG data that had been transferred by the estimated PA device with alarm option and the ECG data collected by the Signa-Lyzer SD Holter:

- Compared the ECG segments that had provoked an alarm to the respective ECG segments in the Holter recording and judged about the correctness (accuracy) of the alarm;
- Checked the Holter ECG recording for pathological events that are missed by the PA (reliability).

3 Results

3.1 Patients' Population

The telemetry ECG monitoring system was tested in the Cardiology unit of the Department of Propedeutics of Internal Medicine "Prof. Kirkovich" in University Hospital "Alexandrovska" – Sofia. To estimate the applicability of PA for high-risk cardiac patients follow-up the study included patients with ischemic heart disease with or without previous myocardial infarction (MI), as well as patients at high-risk with arterial hypertention, congestive heart failure (CHF), valvular heart disease, cardiomyopathy, myopericarditis, accompanying thyroid pathology, electrolyte imbalance.

The research involved 85 patients with cardiovascular diseases with the following demographic characteristics:

- Gender: 31 males (36.5%) and 54 females (63.5%);
- Age:
    - Mean (min-max) age: 64.8 (23–91) years old;
    - ≤60 years: 40 patients (46.5%); >60 years: 45 patients (52.3%);
- Hospital stay between 3 and 12 days, 5.1 days mean value.

The clinical characteristics of the patients' population are presented in Table 1.

Table 1. Clinical characteristics of the 85 patients involved in the study

Clinical characteristic	Total		Male		Female	
	Nb	%	Nb	%	Nb	%
Ischemic heart disease with MI	5	5.8	3	9.7	2	3.7
Ischemic heart disease without MI	23	26.7	9	29.0	14	25.9
Arterial hypertension						
Stage 1	2	2.3	1	3.2	1	1.9
Stage 2	56	65.1	21	67.7	35	64.8
Stage 3	10	11.6	4	12.9	6	11.1
Chronic congestive heart failure						
I functional class	2	2.3	0	0	2	3.7
II functional class	4	4.7	2	6.5	2	3.7
III functional class	50	58.1	17	54.8	33	61.1
IV functional class	2	2.3	0	0	2	3.7
Valvular heart disease	38	44.2	8	25.8	30	55.6
Cardiomyopathy	6	7.1	2	6.5	4	7.4
Pericarditis/Myopericarditis	13	15.1	2	6.5	11	20.4
Patients with pacemakers	6	7.1	3	9.7	3	5.6
Thyroid (examined in 46 patients)						
Normal function	38	44.2	14	42.2	24	44.4
Hypothyroidism	5	5.8	1	3.2	4	7.4
Hyperthyroidism ara>	3	3.5	2	6.6	1	1.9
Not examined	39	45.9	14	42.2	25	46.3

(continued)

Table 1. (*continued*)

Clinical characteristic	Total		Male		Female	
	Nb	%	Nb	%	Nb	%
Anemia	9	10.5	4	12.9	5	9.3
Electrolyte imbalance	5	5.8	2	6.5	3	5.6
Antiplatelet treatment	41	47.7	18	58.1	23	42.6
Anticoagulant treatment	36	41.9	11	35.5	25	46.3
Diabetes mellitus type 2	24	27.9	11	35.5	13	24.1
Ejection fraction: mean: 61%, *range*: (25–78)%						

3.2 Performance Results

Full correspondence between the PA alarms and the rhythms recorded via the Holter has been observed (i.e. Accuracy = 100%). Considering the observed Holter events, PA reliably warns for 80.6% of the rhythm disorders. In the remaining 19.4% an alarm has not been generated mainly due to the fact that the patients have been out of the telemetry range, which covers the cardiology ward (e.g. for imaging diagnostics and/or additional laboratory tests). Detailed results for the detection accuracy and reliability of PA are presented in Table 2, where the presence of a certain cardiac disorder is considered according to the observations over the Holter ECG recording.

Table 2. Accuracy and reliability of the tested portable personal analyzer for different rhythm and conduction disorders visible in the Holter ECG recordings

Rhythm disorder	Presence	Detection accuracy	Detection reliability
Supraventricular tachycardia	32.9% (28/85)	100% (28/28)	71.4% (20/28)
Supraventricular tachyarrhythmia	8.2% (7/85)	100% (7/7)	85.7% (6/7)
Bradycardia	14.1% (12/85)	100% (12/12)	75.0% (9/12)
Atrial ectopic beats	50.6% (43/85)	100% (43/43)	76.7% (33/43)
Ventricular ectopic beats	58.8% (50/85)	100% (50/50)	86.0% (43/50)
Atrial flutter	40.0% (34 /85)	100% (34/34)	91.2% (31/34)
Atrial fibrillation	5.9% (5/85)	100% (5/5)	80.0% (4/5)
Ventricular tachycardia	8.2% (7/85)	100% (7/7)	57.1% (4/7)
Conduction disturbances			
SA block	3.5% (3/85)	100% (3/3)	33.3% (1/3)
AV block	9.4% (8/85)	100% (8/8)	50.0% (4/8)
Bundle branch block	5.9% (5/85)	100% (5/5)	60.0% (3/5)

4 Discussion

The presented portable telemonitoring personal analyzer has been applied for continuous real-time follow-up of 85 high-risk cardiac patients. This device has prevented the need for restriction of the patients movement that would inevitably be posed by the application of a standard ECG monitoring system and at the same time has provided a real-time feedback that could not be achieved via a standard Holter ECG device.

The clinical characteristics of the patients' population (presented in Table 1) show higher prevalence of patients (mostly women) that correspond to the New York Heart Association (NYHA) III functional classification, i.e. chronic heart failure (CHF), with preserved ejection fraction (>50%). These patients feature with limited physical activity, i.e. they feel comfortable at rest, however, less than ordinary activities cause fatigue, palpitation, or dyspnea, which are mainly due to:

- Valvular heart disease,
- Ischaemic heart disease without MI, but with risk factors: arterial hypertension stage 2 and diabetes mellitus type 2 (with a frequency, comparable to that in the general population)
- Inflammatory diseases of the myocardium and pericardium
- Cardiomyopathies

Interestingly, electrolyte disturbances have been found in only 6% of the patients in the studied high-risk population with various rhythm disorders. On the other side, in 46 patients the level of thyroid hormons was evaluated and 17% of them revealed dysfunction: hypo- or hyperthyreoidism. Seven percent of our patients have undergone permanent pacemaker implantation.

From the technical point of view, PA has been designed to alarm for cardiac problems based on heartbeat (normal or ectopic) and heart rate abnormalities (tachycardia, bradycardia). Therefore, there is not full correspondence between the observed rhythms in the Holter ECG recording and the PA alarm, i.e. rhythms like supraventricular tachyarrhythmia, atrial flutter/fibrillation are normally detected by PA either as tachycardia, or as rhythms with more than 2 extrasystoles. Despite this, the detection of such heartbeats or rhythm problems would provoke an alarm and would drive the cardiologist attention. In this respect, Table 2 shows the potential of PA to warn for a cardiac problem without any claims that PA could support the correct classification of the listed rhythm and conduction disorders. The portable personal analyzer showed 100% accuracy for detection of rhythm and conduction disorders as compared to standard Holter ECG. This means that it has not generated any false alarms, which is a precondition to be well accepted by the cardiologists. PA reliably detects the rhythm disorders observed in the patients' population (71.4%–91.2%) except the ventricular tachycardia (57.1%) which is observed in 8.2% of the population (i.e. 7 patients). The last could be explained by the fact that the 3 not recognized VTs have not been sustained and at the moment of their occurrences the patients could be out of the cardiology unit (e.g. for additional tests in other departments – image diagnostics, urology, pulmonology, etc.), which do not fall within the scope of wireless communication of the telemetry system.

Unfortunately, the current version of the PA does not support reliable detection of conduction disorders (SA, AV, Bundle branch blocks), since usually they do not represent heart rate abnormalities and QRS complexes with different waveforms compared to the patient's sustained rhythm within the analyzed 10 s epoch. To cover the automatic detection of such conduction disorders, additional ECG analyses should be implemented in the PA module in future.

Currently, the advantage of PA compared to an ordinary Holter device is that when a rhythm disorder is detected within 10 s, an alarm is triggered and the ECG signal is transferred from PA to the central unit. More sustained conduction disturbances could be followed by periodical (5/15/30 min) or on-demand transfer of 10 s ECG segment.

From the medical point of view, the PA would allow timely diagnostic reaction and adequate therapeutic approach in potentially life-threatening clinical conditions:

- Proarrhythmic effects of antiarrhythmic drugs, especially when combined with antidepressants, antihistamines or macrolide antibiotics (risk of QT interval prolongation);
- Malignant ventricular dysrhythmias;
- Episodes of paroxysmal atrial fibrillation, needed anticoagulation for prevention of stroke and systemic embolism.

Moreover, the data, obtained with PA, might draw physicians' attention to hitherto undiagnosed thyroid gland dysfunction.

5 Conclusion

Our single-center study showed that the proposed portable telemonitoring personal analyzer with an excellent accuracy and high reliability (in comparison to standard Holter ECG) could detect different rhythm disorders. This device might be a useful tool for proper diagnostics and timely management of life-threatening arrhythmias in high-risk stabilized patients with congestive heart failure, valvular diseases, coronary syndromes or cardiomyopathies, taken out of the intensive care unit to the general cardiology ward. Moreover, this PA could be upgraded for more reliable detection of conduction disorders in clinical practice.

Acknowledgement. This work was supported by the Bulgarian National Science Fund under Grant Ref. No. КП-06-H37/9 "Audio transformation (sonification) of the electrocardiogram – a new approach for remote monitoring of the heart activity" and the European Regional Development Fund within the Operational Programme "Science and Education for Smart Growth 2014–2020" under the Project CoE "National center of mechatronics and clean technologies" BG05M2OP001-1.001-0008.

References

1. Komalasari, R., Nurjanah, N., Yoche, M.: Quality of life of people with cardiovascular disease: a descriptive study. Asian/Pacific Island Nurs. J. **4**(2), 92–96 (2019). https://doi.org/10.31372/20190402.1045

2. Kvedar, J., Coye, M., Everett, W.: Connected health: a review of technologies and strategies to improve patient care with telemedicine and telehealth. Health Aff (Millwood) **33**, 194–199 (2014)
3. Global ECG Telemetry Devices Market - Growth, Trends, and Forecast (2019 - 2024), November (2019). https://www.researchandmarkets.com/r/fwhusd
4. Marouf, M., Vukomanovic, G., Saranovac, L., Bozic, M.: Multipurpose ECG telemetry system. Biomed. Eng. Online **2017**, 1–20 (2017). https://doi.org/10.1186/s12938-017-0371-6
5. Linzer, M., Pritchett, E., Pontinen, M., McCarthy, E., Divine, G.: Incremental diagnostic yield of loop electrocardiographic recorders in unexplained syncope. Am. J. Cardiol. **66**(2), 214–219 (1990)
6. Zimetbaum, P., Kim, K., Josephson, M., Goldberger, A., Cohen, D.: Diagnostic yield and optimal duration of continuous-loop event monitoring for the diagnosis of palpitations: a cost-effectiveness analysis. Ann. Intern. Med. **128**(11), 890–895 (1998)
7. Kohno, R., Abe, H., Benditt, D.: Ambulatory electrocardiogram monitoring devices for evaluating transient loss of consciousness or other related symptoms. J Arrhythm. **33**, 583–589 (2017)
8. Wu, C., et al.: Utility of patient-activated cardiac event recorders in the detection of cardiac arrhythmias. J. Interv. Card. Electrophysiol. **8**(2), 117–120 (2003). https://doi.org/10.1023/A:1023604816368
9. Kaleschke, G., et al.: Prospective, multicenter validation of a simple, patient-operated electrocardiographic system for the detection of arrhythmias and electrocardiographic changes. Europace **11**(10), 1362–1368 (2009)
10. Park, M., de Asmundis, C., Chierchia, G., Sarkozy, A., Benatar, A., Brugada, P.: First experience of monitoring with cardiac event recorder electrocardiography omron system in childhood population for sporadic, potentially arrhythmia related symptoms. Europace **13**(9), 1335–1339 (2011)
11. Attanasio, P., et al.: Use of a patient-activated event recording system in patients with tachycardic palpitations: how long to follow up? Ann Noninvasive Electrocardiol. **20**, 566–569 (2015)
12. Turakhia, M., et al.: Latency of ECG displays of hospital telemetry systems: a science advisory from the american heart association. Circulation **126**(13), 1665–1669 (2012). https://doi.org/10.1161/CIR.0b013e31826ae459
13. Steinberg, J., Varma, N., Cygankiewicz, I., et al.: 2017 ISHNE-HRS expert consensus statement on ambulatory ECG and external cardiac monitoring/telemetry. Heart Rhythm **14**(7), e55–e96 (2017). https://doi.org/10.1016/j.hrthm.2017.03.038
14. Bocchiardo, M., Asteggiano, R.: ECG portable devices: example of e-Health strength and threats. e-J. Cardiol. Pract. **18**(25) (2020). https://www.escardio.org/Journals/E-Journal-of-Cardiology-Practice/Volume-18/ecg-portable-devices-example-of-e-health-strength-and-threats
15. Ramkumar, S., Nerlekar, N., D'Souza, D., Pol, D., Kalman, J., Marwick, T.: Atrial fibrillation detection using single lead portable electrocardiographic monitoring: a systematic review and meta-analysis. BMJ Open. **8**, e024178 (2018)
16. Emergency Cardiac Care Committee members: Recommended guidelines for in-hospital cardiac monitoring of adults for detection of arrhythmia. J. Am. Coll. Cardiol. **18**(6), 1431–1433 (1991)
17. Dhillon, S., et al.: Effectiveness of telemetry guidelines in predicting clinically significant arrhythmias in hospitalized patients. Cardiol. Res. **3**(1), 16–22 (2012)
18. United Healthcare Oxford Clinical Policy (2018). https://www.uhcprovider.com/en/policiesprotocols/commercial-policies/oxford-policies.html

19. Tabakov, S., Iliev, I., Krasteva, V.: Online digital filter and QRS detector applicable in low resource ECG monitoring systems. Ann. Biomed. Eng. **36**(11), 1805–1815 (2008). https://doi.org/10.1007/s10439-008-9553-5
20. Iliev, I., Krasteva, V., Tabakov, S.: Real-time detection of pathological cardiac events in the electrocardiogram. Physiol. Meas. **28**, 259–276 (2007)

Web-Based Software Tool for Electrocardiogram Annotation

Todor Stoyanov(✉)

Institute of Biophysics and Biomedical Engineering, Bulgarian Academy of Sciences, Acad. G. Bonchev str. bl. 105, 1113 Sofia, Bulgaria

Abstract. The manual annotation of large multilead ECG databases is a challenge, especially in the context for providing meaningful visualization, easy tools for annotation and simultaneous access from multiple experts via extended Internet connectivity. The aim of this work is to present the development platform and capabilities of internet-based software tool for the purpose of user-friendly manual annotation and delineation of heartbeats in 12-lead ECG databases. The annotation software consists of: (i) server-based application, written in Python under DJango framework with SQLite database manager; and (ii) web-based front-end application, created in Node Package Manager environment under React JavaScript framework. The server-based part contains procedures for managing the ECG records (i.e. receiving and saving ECG signals), saving manually annotated data, and generation of average beats. The front-end application contains: (i) graphical user interface for visualization of 12-lead ECG signals and management of the user input/output commands; (ii) annotation module, which provides tools to mark and correct the positions and type of QRS complexes; (iii) average beat module, which provides an option to switch between leads, provides graphical markers for annotation of fiducial points and time-intervals in the averaged heartbeat waveform and sets the average beat type; and (iv) rhythm type module, which shows the rhythm type and provides a possibility to change it. The software is opened for further developments. The presented annotation tool could be potentially used for annotation of large ECG databases without limitations on the number of ECG leads and number of annotations per ECG recording. This is in line with the new trends for accumulation of large ECG databases from multiple sources, which are the thoughtful platform for development of machine learning and especially deep learning algorithms for ECG signal processing and arrhythmia classification.

Keywords: Electrocardiogram (ECG) · Annotation software · Heartbeat types · Arrhythmia types · DJango framework · React JavaScript framework

1 Introduction

1.1 Background

The electrocardiogram (ECG) is a time domain representation of the body surface bioelectrical potentials originating from the electric activity of the heart. Usually it is used for non-invasive diagnosis of cardiac diseases.

S. S. Sotirov et al. (Eds.): BioInfoMed 2020, LNNS 374, pp. 322–331, 2022.
https://doi.org/10.1007/978-3-030-96638-6_34

The ECG analysis relies on automatic algorithms for signal quality estimation [1, 2], heartbeat detection [3, 4] and classification [5, 6], heart rate and morphological analysis for assessment of rhythm and conduction disorders [7–13], detection of acute ST changes [14, 15], prediction of defibrillation shock success [16], etc. To perform a reliable training and independent validation and testing of such methods the designers should have a large amount of ECG data that is correctly annotated in terms of noise content in the ECG recording, QRS positions, P-QRS-T segment delineation, heartbeat types, presence of arrhythmia or conduction disorder and its type, effect of the applied defibrillation shock, etc.

Recently, the need for specific and/or large enough ECG databases has been highlighted in the sense of pace pulses detection [17] and neural network training and evaluation [18–21], respectively. On the other hand, nowadays, many ECGs are being daily recorded worldwide for diagnostic purposes and they just need to be properly annotated, so that their use is not confined only to the particular patient diagnostic process.

1.2 Related Studies

Recently, a number of ECG annotation tools have been presented, including:

- LightWAVE [22] – an open source web-based software for observation and annotation of ECGs and other physiologic waveforms that is created as part of PhysioNet [23]. A modified version of the LightWAVE client allows automated and semi-automated annotation of heartbeats from ECG time series [24].
- WaveformECG [25] – an open source web-based platform supporting multi-lead ECG visualization, annotation of fiducial points and time intervals in ECG waveforms, diagnose annotation.
- EcgEditor [26] – a PC-based Java application for viewing, annotating and analyzing ECG measurements stored in the Unisens data format. It supports multichannel ECG visualization and measurements, manual annotation of artifacts, as well as triggers and involves plugin-algorithms for QRS detection.
- ECG_Viewer [27] – a PC-based Matlab GUI for reviewing, processing, and annotating ECG data files. It performs ECG filtering, beat detection using template matching, and inter-beat interval filtering.
- BSS_ECG [28] – a PhysioNet-compliant ECG signal visualizer with support for QRS annotations.
- LabelECG [29] – a web-based tool for ECG viewing and distributed ECG annotation. After login it supports selection of a dataset, visualization and annotation of ECG recordings, revision of the annotated records by the same annotator. This application has been used for annotation of the ECG recordings in the First China ECG Intelligent Competition, which has been dedicated to the development of algorithms for classification of normal ECGs, atrial fibrillation, early repolarization and T wave change, etc., from 12-lead resting ECGs with various time lengths [30].
- Hermes (TM) system [31] – web-based service, available through Physionet (www.physionet.org) that facilitates simple point-and-click user interaction and ECG data annotation.

- ViewECG Web [32] – an adaptable online viewer tool for ECG traces based on the HTML 5 standard that enables the end-user to interact with and annotate the rendered digital ECG.
- NOTOCORD [33] – a software tool that provides automated (real-time) and semi-automated (offline) ECG analyzes for detection of ECG fiducial points, a module for QT measurements and a solution for automated arrhythmia detection. The detected ECG characteristic points P, Q, R, S, T; arrhythmias and QT durations can be edited and validated.

1.3 Aim of the Study

The aim of this work is to present the development platform and capabilities of internet-based software tool for the purpose of user-friendly manual annotation and delineation of heartbeats in multilead ECG databases.

2 Method

The designed software supports annotation of single channel and standard 12-lead ECG. Each user can log in via a standard web browser on a PC or tablet and annotate a single or multiple ECG recordings during one session. After annotation, the user can save the data on the server until the next session.

Block-diagram of the designed web-based annotation tool is presented in Fig. 1. It involves a server part and a web-based front-end application.

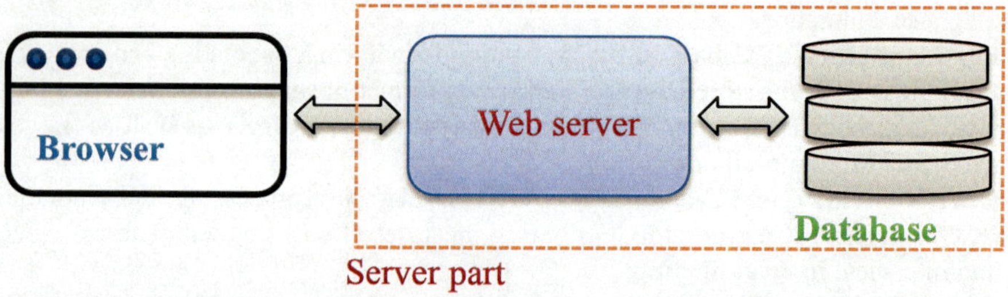

Fig. 1. Block-diagram of the web-based software tool for electrocardiogram annotation

2.1 Server Part

The server-based component is executed in Anaconda v.4.4.0 environment [34] and is written in Python programming language.

For the design of the web server part and the database a DJango v.3.0 framework [35] is created in Anaconda environment. DJango is a Python-based free and open-source web framework that follows the model-template-views (MTV) architectural pattern and provides tools for easy creation of complex database-driven websites. The framework emphasizes reusability and "pluggability" of components, less code and rapid development [35]. DJango also provides an optional administrative create, read, update and delete interface. In this work, the potential of DJango to store the ECG signals, annotation data, time delineation data and other is used. The server part uses a SQLite database, which is a relational database management system (RDBMS). SQLite is not a client-server database engine but embedded database software for local/client storage in application software such as web browsers. It is the most widely deployed database engine, which is used by several popular browsers, operating systems, and embedded systems [36].

The server part software implements procedures for:

- ECG records management, i.e. receiving and saving ECG signals;
- ECG pre-processing for filtering power line interference, baseline wandering and high-frequency noise. Each filter can be enabled/disabled.
- Annotation data management;
- Generation of average heartbeats for the annotated beat types in the ECG recording. For each lead the annotated beats are averaged within a window of (−150 +250) ms around the annotated QRS positions.

2.2 Web Based Front-End Application

The front-end application is created in Node Package Manager v.14.15.3 (NPM) environment [37] and uses JavaScript programming language. To build the web application various functional options of several available frameworks have been compared, including: Angular v.8 [38], Vue.js v.2.6 [39] and React.js v.16.13.0 [40]. React JavaScript framework has been chosen as the most popular and easy to use.

This front-end application implements:

- Login page, where each annotator can enter his/her ID and password;
- List page, which shows available ECG recordings;
- Annotation module that includes:
  - 'Info dialogue' – this item shows the current record's information as sampling rate (Hz), amplitude resolution (mV/LSB), record time duration (s);
  - 'Zoom' – a dropdown list for zoom in/out of the visual ECG trace. The dropdown list contains radio buttons for vertical zoom (2.5 mm/mV, 5 mm/mV, 10 mm/mV, 20 mm/mV, 40 mm/mV) and horizontal zoom (6.5 mm/s 12.5 mm/s, 25 mm/s, 50 mm/s 100 mm/s);
  - 'Show beats' – the user can highlight all beats of a selected type;

- 'All beats to…' – a dropdown list, which could be used by the user to change all beat types into a chosen new beat type;
- 'New beats' – a dropdown list, which could be used to set the type of new beats that need to be annotated.

This module provides the opportunity for manual annotation of the positions of all heartbeats. The beats are initially annotated as the beat type selected in the 'New beats' drop down list and could be further re-annotated by right-button click on the beat marker and selecting the beat type from the context menu. The embedded beat types are normal beat (N), supraventricular ectopic beat (S), ventricular ectopic beat (V), paced beat (P), questionable beat (Q), other beat type (O).

• Average beat module, which facilitates the annotation of fiducial points within the PQRST segment, such as P, QRS, QT and PQ begin/end, as well as the calculation of intervals that have diagnostic value. This module includes:

- 'Info dialogue' – this item shows the current record's information as sampling rate (Hz), amplitude resolution (mV/LSB), record time duration (s);
- 'Zoom' – a dropdown list for zoom in/out of the visual ECG trace. The dropdown list contains radio buttons for vertical zoom (80 mm/mV, 120 mm/mV, 160 mm/mV, 200 mm/mV) and horizontal zoom (200 mm/s 400 mm/s);
- 'Show lead' – a dropdown list, which allows the user to highlight a specific lead (the lead is bolded and stands infront all other leads, so that it is distinguished for delineation of the average beat);
- 'P, T waves On/Off' – two checkboxes that can be used to show/hide vertical markers for delineation of the PQ or/and TQ intervals;
- 'Select average beats' – a dropdown list, which provides the interface to the user to choose the average beat type that need to be delineated.

• Rhythm type module, which facilitates the annotation of the rhythm type. This module includes:

- 'Info dialogue' – this item shows the current record's information as sampling rate (Hz), amplitude resolution (mV/LSB), record time duration (s);
- 'Zoom' – a dropdown list for zoom in/out of the visual ECG trace. The dropdown list contains radio buttons for vertical zoom (2.5 mm/mV, 5 mm/mV, 10 mm/mV, 20 mm/mV, 40 mm/mV) and horizontal zoom (6.5 mm/s 12.5 mm/s, 25 mm/s, 50 mm/s 100 mm/s);
- 'Select rhythm' – a dropdown list, which could be used to select the rhythm type of the visualized ECG episode for which individual and average beats have been annotated. The available rhythm types are sinus rhythm, AV block (II, III degree), atrial flutter/fibrillation, ventricular fibrillation, bigeminy, trigeminy.

3 Results

The designed web based front-end application could be run under Windows 7 OS or higher, Internet Explorer 11, Google Chrome 69 or Mozilla Firefox browser 60.

The design of different modules of the web-based front-end ECG annotation application is illustrated in Figs. 2, 3, 4 and 5.

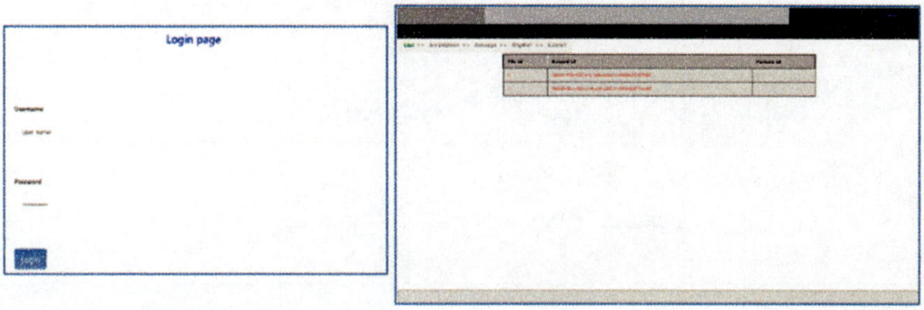

(a) (b)

Fig. 2. Visual interface of the developed web based annotation tool: (a) Login page and (b) List page of the ECG database

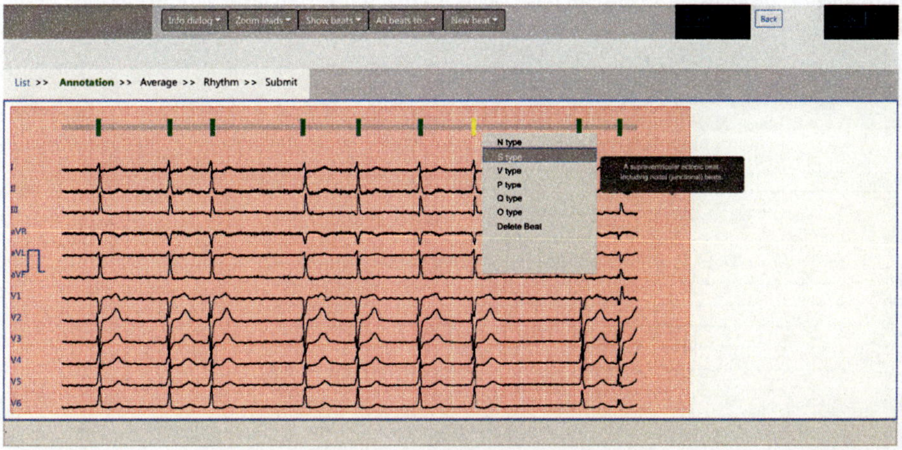

Fig. 3. Annotation module – main layout. This is the first page of the group of modules that handle the annotation of an ECG recording. It implements the manual annotation of the heartbeat positions (shown with vertical lines above lead I) and the heartbeat types, which could be re-annotated by right button click on the QRS mark and selection of the heartbeat type from a context menu list. Each heartbeat type is represented in different color (green for N beats and yellow for S beats in the example).

The designed software is open for further developments and improvements.

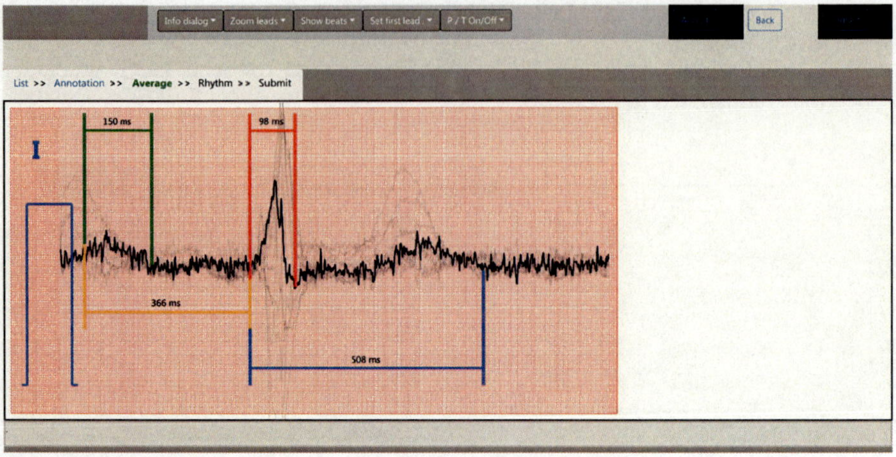

Fig. 4. Average beats module – main view with delineation markers. The averaged beats are generated by the server and displayed on the browser. The vertical delineation lines for P (green lines), QRS (red lines), QT (blue lines) and PQ (orange lines) begin/end are presented (as sent by the server). They could be corrected by mouse dragging. In this example, lead I is selected to be highlighted and stands up against all other leads, so that it is distinguished for delineation of the average beat.

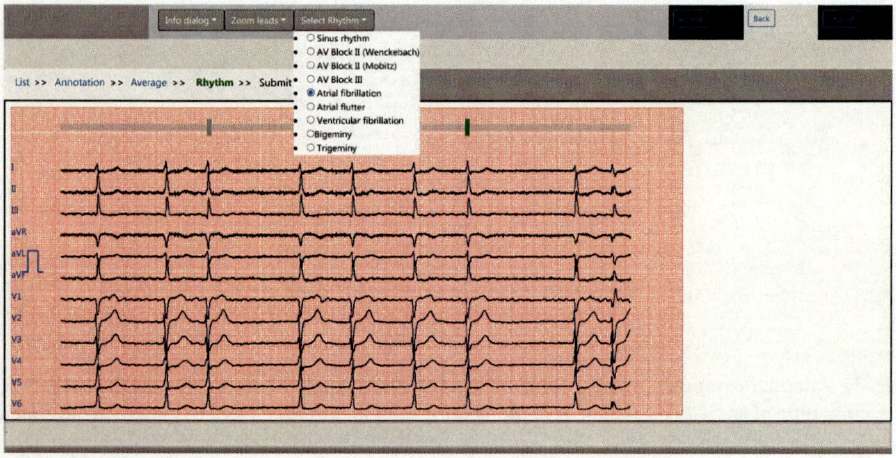

Fig. 5. Rhythm type module – main view with "Select Rhythm" drop down list. The initial rhythm type is sent by the server but could be manually changed by re-selection of the rhythm type in the drop down list as shown in the figure.

4 Discussion and Conclusions

This paper presents a web-based software tool for ECG annotation, which is designed in Anaconda environment (the server part) and NPM environment (the web-based front-end part). The Anaconda environment is selected due to the following advantages:

- Facility to add packages of different types and functionality in order to expand the potential of the server part.
- Facility to create different types of frameworks (e.g. DJango). Thus the user can use frameworks in different Python versions.
- Easy creation of a web server with added database support using the python scripting language.

The NPM environment provides similar advantages as Anaconda – i.e. flexibility and ability to add different packages, such as JSON package, which supports the work with JSON files and data; Axios package for data exchange between the web-based front end application and the server; Route package, which supports the navigation between pages, etc.

React.js frameworks (working in the NPM environment) is chosen for the web-based front-end application. This library is written in Java scripts, and has the following advantages:

- Existence of built-in functions for each stage from creation to removal of each component.
- Possibility to create separate classes (i.e. for each class there is a separate component). These classes can be built into each other but cannot be inherited.
- Easy building of components and web pages by means of 'props' variables and 'composition'.

The presented ECG annotation software is a complex tool that supports multiple annotations per record and has the capability to annotate large databases containing ECG recordings in JSON file format. It is open for addition of new features and further improvements. Considering that nowadays, many ECGs are being daily recorded worldwide for diagnostic purposes, such annotation tool could be useful for the accumulation of large annotated ECG databases that would be of great benefit for development, testing and evaluation of new algorithms for ECG signal analysis. Another application could be the training and examination of students in cardiology for correct heartbeat and arrhythmia annotation.

Future work is related with the implementation of algorithms for automatic beat detection, heartbeat classification and arrhythmia recognition that would facilitate the initial beat and rhythm annotation and would reduce the efforts for manual annotation.

Acknowledgement. This work was supported by the Bulgarian National Science Fund under Grant Ref. No. КП-06-H42/3 "Computer aided diagnosis of cardiac arrhythmias based on machine learning and deep neural networks".

References

1. Jekova, I., et al.: A real-time quality monitoring system for optimal recording of 12-lead resting ECG. Biomed. Signal Process. Control. **34**, 126–133 (2017)
2. Jekova, I., Krasteva, V., Christov, I., Abächerli, R.: Threshold-based system for noise detection in multilead ECG recordings. Physiol. Meas. **33**, 1463–1477 (2012)
3. Tabakov, S., Iliev, I., Krasteva, V.: Online digital filter and QRS detector applicable in low resource ECG monitoring systems. Ann. Biomed. Eng. **36**(11), 1805–1815 (2008)
4. Christov, I.: Real time electrocardiogram QRS detection using combined adaptive threshold. BioMed. Eng. OnLine. **3**, Article number 28 (2004)
5. Iliev, I., Krasteva, V., Tabakov, S.: Real-time detection of pathological cardiac events in the electrocardiogram. Physiol. Meas. **28**, 259–276 (2007)
6. Krasteva, V., Jekova, I., Leber, R., Schmid, R., Abächerli, R.: Superiority of classification tree versus cluster, fuzzy and discriminant models in a heartbeat classification system. PLoS One **10**(10), e0140123 (2015)
7. Jekova, I., Dushanova, J., Popivanov, D.: Method for ventricular fibrillation detection in the external electrocardiogram using nonlinear prediction. Physiol. Meas. **23**, 337–345 (2002)
8. Jekova, I., Mitev, P.: Detection of ventricular fibrillation and tachycardia from the surface ECG by a set of parameters acquired from four methods. Physiol. Meas. **23**, 629–634 (2002)
9. Krasteva, V., Jekova, I., Leber, R., Schmid, R., Abächerli, R.: Validation of arrhythmia detection library on bedside monitor data for triggering alarms in intensive care. Comput. Cardiol. **42**, 737–740 (2015)
10. Krasteva, V., Jekova, I., Dotsinsky, I., Didon, J.P.: Shock advisory system for heart rhythm analysis during cardiopulmonary resuscitation using a single ECG input of automated external defibrillators. Ann. Biomed. Eng. **38**, 1326–1336 (2010)
11. Krasteva, V., Jekova, I., Leber, R., Schmid, R., Abächerli, R.: Real-time arrhythmia detection with supplementary ECG quality and pulse wave monitoring for reduction of false alarms in ICU. Physiol. Meas. **37**, 1273–1297 (2016)
12. Christov, I., Krasteva, V., Simova, I., Neycheva, T., Schmid, R.: Ranking of the most reliable beat morphology and heart rate variability features for the detection of atrial fibrillation in short single-lead ECG. Physiological Measurement. **39**(6), 094005 (2018)
13. Jekova, I., Bortolan, G., Stoyanov, T., Dotsinsky, I.: Multi-type arrhythmia classification: assessment of the potential of time and frequency domain features and different classifiers. Int. J. Bioautom. **24**(2), 153–172 (2020)
14. Matveev, M., Krasteva, V., Naydenov, S., Donova, T.: Possibilities of signal-averaged orthogonal and vector electrocardiography for locating and size evaluation of acute myocardial infarction with ST-elevation. Anatol. J. Cardiol. **7**(1), 193–197 (2007)
15. Matveev, M., Naydenov, S., Krasteva, V., Donova, T., Christov, I.: Assessment of the infarct size in high-resolution electrocardiograms. Comput. Cardiol. **33**, 461–464 (2006)
16. Jekova, I., Mougeolle, F., Valance, A.: Defibrillation shock success estimation by a set of six parameters derived from the electrocardiogram. Physiol. Meas. **25**, 1179–1188 (2004)
17. Jekova, I., Iliev, I., Tabakov, S.: Application of Stockwell transform and Shannon energy for pace pulses detection in a single lead ECG corrupted by EMG artifacts. Appl. Sci. **10**(21), 7505 (2020)
18. Krittanawong, C., Zhang, H., Wang, Z., Aydar, M., Kitai, T.: Artificial intelligence in precision cardiovascular medicine. J. Am. Coll. Cardiol. **69**(21), 2657–2664 (2017)
19. Hannun, A., et al.: Cardiologist-level arrhythmia detection and classification in ambulatory electrocardiograms using a deep neural network. Nat. Med. **25**(1), 65 (2019)
20. Attia, Z.I., Kapa, S., Lopez-Jimenez, F., et al.: Screening for cardiac contractile dysfunction using an artificial intelligence–enabled electrocardiogram. Nat. Med. **25**(1), 70 (2019)

21. Krasteva, V., Ménétré, S., Didon, J.P., Jekova, I.: Fully convolutional deep neural networks with optimized hyperparameters for detection of shockable and Non-shockable rhythms. Sensors. **20**(10), s20102875 (2020)
22. Moody, G.B.: Lightwave: Waveform and annotation viewing and editing in a web browser. Comput. Cardiol. **40**, 17–20 (2013)
23. Goldberger, A., Amaral, L., Glass, L., et al.: PhysioBank, PhysioToolkit, and PhysioNet: Components of a new research resource for complex physiologic signals. Circulation **101**(23), e215–e220 (2000)
24. Citi, L., Olariu, C., Barbieri, R.: A LightWAVE client for semi-automated annotation of heart beats from ECG time series. Comput. Cardiol. **42**, 605–608 (2015)
25. Winslow, R.L., Granite, S., Jurado, C.: WaveformECG: a platform for visualizing, annotating, and analyzing ECG data. Comput. Sci. Eng. **18**(5), 36–46 (2016)
26. EcgEditor. https://github.com/Unisens/EcgEditor. Accessed 27 Oct 2020
27. ECG_Viewer. https://github.com/jramshur/ECG_Viewer. Accessed 27 Oct 2020
28. BSS_ECG, https://github.com/AdnanHidic/bss_ecg. Accessed 27 Oct 2020
29. Ding, Z., Qiu, S., Guo, Y.: LabelECG: a web-based tool for distributed electrocardiogram annotation. In: Liao, H., et al. (eds.) MLMECH 2019/CVII-STENT 2019. LNCS, vol. 11794, pp. 104–111. Springer, Cham (2019). https://doi.org/10.1007/978-3-030-33327-0_13
30. The First China ECG Intelligent Competition. http://mdi.ids.tsinghua.edu.cn/#/. Accessed 27 Oct 2020
31. Oefinger, M.B., Mark, R.G.: A web-based tool for cisualization and collaborative annotation of physiological databases. Comput. Cardiol. **32**, 163–165 (2005)
32. VIEWECG WEB. https://www.amps-llc.com/resting-ecgs/viewECG%20Web/ Accessed 18 Dec 2020
33. NOTOCORD. http://www.notocord.com/solutions/ecg. Accessed 18 Dec 2020
34. Anaconda. https://www.anaconda.com/products/individual. Accessed 18 Dec 2020
35. Django makes it easier to build better Web apps more quickly and with less code. https://www.djangoproject.com/. Accessed 27 Oct 2020
36. Most Widely Deployed SQL Database Estimates. https://sqlite.org/mostdeployed.html. Accessed 27 Oct 2020
37. Node Package Manager (NPM) environment. https://nodejs.org/en/. Accessed 18 Dec 2020
38. Angular. https://angular.io/. Accessed 18 Dec 2020
39. Vue.js Frame works. https://vuejs.org/. Accessed 18 Dec 2020
40. React.js. https://reactjs.org/. Accessed 18 Dec 2020

Differences in Ischemia Mechanism in Coronary Artery Disease and Cardiac Syndrome X

Mikhail Matveev[1(✉)], Olympia Roeva[2], Mitko Petrov[2], and Svetlin Tsonev[3]

[1] Department of Analysis and Processing of Biomedical Signals and Data, Institute of Biophysics and Biomedical Engineering, Bulgarian Academy of Sciences, Acad. G. Bonchev Str., Bl. 105, 1113 Sofia, Bulgaria
mgm@biomed.bas.bg

[2] Department of Bioinformatics and Mathematical Modelling, Institute of Biophysics and Biomedical Engineering, Bulgarian Academy of Sciences, Acad. G. Bonchev Str., Bl. 105, 1113 Sofia, Bulgaria
olympia@biomed.bas.bg, mpetrov@biomed.bg

[3] Department of Cardiology, Medical University – Sofia, 1 St. Georgi Sofiisky Str., 1431 Sofia, Bulgaria

Abstract. The Cardiac Syndrome X (CSX) includes patients with the triad of angina pectoris: a positive exercise-strain test and angiographically smooth coronary arteries. The most manifested clinical symptom is chest pain. The aim of the study is to explore the differences in ischemia mechanism in both groups – CSX and patients with Coronary Artery Disease (CAD), based on specific changes in Autonomic Cardiac Control (ACC). A novel approach for multicriteria decision making – InterCriteria Analysis (ICrA) is used to reveal relationships between 13 standard indicators for Heart Rate Variability (HRV) – in resting state (RS) and by sympathetic stimulation (handgrip test; HT) or parasympathetic stimulation (Valsalva manoeuvre; VM). Through these indicators the activity of ACC in two groups (patients with CSX or CAD), each one containing 40 patients, is evaluated. In the considered here three cases all pairs of the 13 indicators, without exception, do not show any dependencies. These results give reason to assume, that in the two groups there is a different response of the sympathetic autonomic system under conditions of stress with matching parasympathetic tone.

Keywords: Cardiac Syndrome X · Coronary artery disease · InterCriteria analysis · Heart rate variability · Autonomic cardiac control

1 Introduction

The concept of coronary X-syndrome includes patients, mostly women at a mean age of 53 years fulfilling the following 3 criteria: chest pain, positive exercise-strain test, normal coronary angiogram [1, 2]. The chest pain is a leading symptom of these patients, usually stronger and more prolonged than the typical angina pectoris as well as more difficult to control by the standard antiischaemic therapy [1]. As a result, patients are

S. S. Sotirov et al. (Eds.): BioInfoMed 2020, LNNS 374, pp. 332–341, 2022.
https://doi.org/10.1007/978-3-030-96638-6_35

re-hospitalized repeatedly, new unnecessary angiograms are being done, and quality of life of these patients worsens.

The idea of X-syndrome was introduced first by Kemp [2] explaining the unclear pathogenesis and therapeutic approach of these patients.

Statistical data show that not less than 33% women presenting with typical angina chest pain have clear coronary vessels, proved by selective coronary angiogram [3].

One of the most discussed ideas for the origin of this syndrome is myocardial ischemia [4]. An argument against the supposed pathogenetic mechanisms is the lack of disturbed left-ventricle function of patients with X-syndrome compared to those with coronary artery disease.

The aim of the study is to explore the differences in ischemia mechanism in both groups – Cardiac Syndrome X (CSX) and patients with Coronary Artery Disease (CAD), based on specific changes in Autonomic Cardiac Control (ACC). A novel approach for decision support, named InterCriteria Analysis (ICrA) [5] is applied to estimate the relationship between 13 standard indicators for Heart Rate Variability (HRV) used for evaluation of the activity of ACC in two groups of patients – patients with CSX and patients with CAD.

The ICrA is based on the apparatus of the index matrices (IM) [6–8] and on the intuitionistic fuzzy sets (IFS) [9, 10]. The approach has been successfully applied to decision making in various fields as biology/medicine [11], algorithms performance [12], mathematics [13], etc.

2 Material and Methods

2.1 Groups of Patients

Group 1*:* 40 patients with criteria for CSX.
Group 2*:* 40 patients, corresponding to the age of the first group, with proven invasive or multi-slice computed tomography CAD.

Patients with CSX and CAD were stratified by age, respectively: mean age 57.8 ± 9.60 years and 63.23 ± 10.12 years.

The following inclusion and exclusion criteria were used to form the two patient's groups:

- Female patients with a history of typical chest pain during exercise or at rest.
- Positive stress test (pain and ECG negative dynamics – ST depression in at least 2 adjacent leads), pharmacological stress test, outpatient ECG monitoring.
- Lack of hemodynamically significant coronary heart disease – "clean" coronary arteries by selective coronary angiography or multi-slice computer tomography.

Exclusion from the Experimental Group of the Study
Excluded from the present study were patients with evidence of Prinz metal angina, stroke, grade III hypertension, valvular disease, congenital heart malformations, congenital and acquired myocardial infarctions, arrhythmias or conduction heart disease (including left femoral block), pregnant women, severe motor or mental impairment due to various causes, neoplastic disease, liver failure, acute or chronic renal failure, chronic obstructive pulmonary disease.

2.2 Methods

Heart Rate Variability and Evaluation of the Autonomic Cardiac Control
A lot of knowledge has been accumulated, suggesting a clear link between the state of the vegetative nervous system and the cardiovascular pathology. These facts are an incentive for studies connected with determination of quantitative indices for evaluating the state of the vegetative nervous system (VNS). The HRV appears to be the most promising method in this respect [14].

The justified interest in HRV triggered the appearance of numerous studies based on different methodological platforms for analysis of the variability. The "*Heart Rate Variability. Standards of Measurement, Physiological Interpretation, and Clinical Use. Task Force of the European Society of Cardiology and the North American Society of Pacing and Electrophysiology*" [15] occupies a special place among them.

As the sequences of R-R intervals (Fig. 1) represent a time series from a mathematical and statistical point of view, their analysis is performed using time-domain and frequency-domain indices:

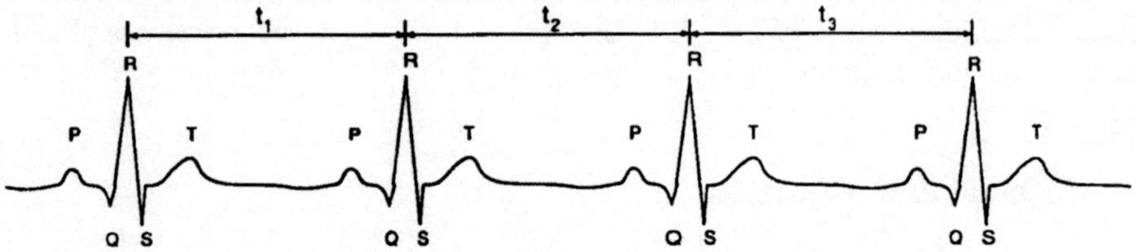

Fig. 1. R-R intervals

Two sets of indices are used in the ***time-domain*** to describe HRV:

A. From the RR-Tachogram

RRA (average value) $= \frac{1}{n}(t_1 + t_2 + \ \dots \ + t_n)$;

SDRR (standard deviation) $=$

$\sqrt{\left((t_1 - RRA)^2 + (t_2 - RRA)^2 + \ \dots \ + (t_n - RRA)^2\right)/n}$;

MDRR (mean (absolute) deviation) $=$

$\frac{1}{n}(|t_1 - RRA| + |t_2 - RRA| + \dots + |t_n - RRA|)$;

MED (median) – 50% of all values of R-R intervals are above and 50% below this value. In a symmetrical distribution this value equals average value. If the two values of R-R intervals do not correspond to each other, the distribution of the R-R intervals is not symmetrical.

B. From the dRR-tachogram.

dRRA (average value) $= \frac{1}{n-1}(dt_1 + dt_2 + \ \dots \ + dt_{n-1})$; where $dt_i = t_i - t_{i+1}$, $i = 1, n-1$;

dSDRR (standard deviation) $=$

$\sqrt{\left((dt_1 - dRRA)^2 + (dt_2 - dRRA)^2 + \ \dots \ + (dt_{n-1} - dRRA)^2\right)/(n-1)}$;

dMDRR (mean (absolute) deviation) $=$

$\frac{1}{n-1}(|dt_1 - dRRA| + |dt_2 - dRRA| + ... + |dt_{n-1} - dRRA|)$;

NN50 – number of pairs of adjacent RR (NN) intervals differing by more than 50 ms in the entire recording;

PNN50 – *NN50* divided by the total number of all NN intervals (indicated in %) or number of NN intervals differing by more than 50 ms per minute;

RMS (root mean-square deviation of the differences between adjacent RR intervals) $= \sqrt{(1/(n-1)) \sum_{i=1}^{n-1} dt_i^2}$.

In the ***frequency-domain*** the indices result from the spectral analysis of the sequences of R-R intervals through the Fourier transformation.

After frequency analysis (with FFT), the distribution of the R-R intervals is presented within the frequency-domain. The spectral high-frequency limit is 1/(2*RRA*). With short recordings (below 10 min), the analysed frequency ranges are differentiated as very low frequency (*VLF*), low frequency (*LF*) and high frequency (*HF*) (in ms^2), and as *LF/HF* ratio.

With a certain difference in the cited limits of these frequency ranges, we used in our studies for *VLF* 0.016–0.05 Hz; for *LF* 0.05–0.15 Hz; for *HF* 0.15–0.35 Hz. In the frequency band of up to 0.1 Hz (predominantly *LF*) the spectrum of the R-R intervals is mediated by fluctuations in the activity of the sympathetic nerve; the frequency band *HF* is mediated almost exclusively by the activity of the vagal nerve; the physiological interventions provoke a reciprocal change of the sympathetic and vagal response, i.e., that the autonomic balance is a ratio of neural mechanisms with opposite responses [16].

The HRV method is used to reveal relationships between the 13 standard indicators (*RRA, SDRR, MDRR, MED, dRRA, dSDRR, dMDRR* (*dMDNN*), *PNN50, RMS, VLF, LF, HF, LF/HF*) for HRV – in resting state (RS) and by sympathetic stimulation (handgrip test; HT) or parasympathetic stimulation (Valsalva manoeuvre; VM), through which we evaluated the activity of ACC in two groups of patients – with CSX and with CAD.

InterCriteria Analysis Background

An Intuitionistic Fuzzy Pair (IFP) [17] as an estimation of the degrees of "agreement" and "disagreement" between two criteria applied to different objects is obtained. An IFP is an ordered pair of real non-negative numbers $\langle \alpha, \beta \rangle$ such that $\alpha + \beta \leq 1$.

Consider an IM [6–8] whose index sets consist of the criteria (for rows) and objects (for columns). The elements of this IM are further assumed to be real numbers. An IM with index sets consisting of the criteria (for rows and for columns) with elements IFPs corresponding to the degrees of "agreement" and "disagreement" between the respective criteria is then constructed.

Let O denotes the set of all objects $O_1, O_2, \ldots, O_n$ being evaluated, and $C(O)$ be the set of values assigned to the objects by a given criterion C, i.e.,

$$O \stackrel{\text{def}}{=} \{O_1, O_2, \ldots, O_n\}; \;\; C(O) \stackrel{\text{def}}{=} \{C(O_1), C(O_2), \ldots, C(O_n)\}$$

Then, let $C^*(O) \stackrel{\text{def}}{=} \{\langle x, y \rangle | \, x \neq y \; \& \; \langle x, y \rangle \in C(O) \times C(O)\}$.

In order to find the "agreement" between two criteria, the vector of all internal comparisons of each criteria, which fulfil exactly one of the following three relations:

R, $\overline{R}$ and $\tilde{R}$, is constructed. In other words, it is required that for a fixed criterion C and any ordered pair $\langle x, y\rangle \in C^*(O)$ it is true:

$$\langle x, y\rangle \in R \Leftrightarrow \langle y, x\rangle \in \overline{R}, \tag{1}$$

$$\langle x, y\rangle \in \tilde{R} \Leftrightarrow \langle x, y\rangle \notin (R \cup \overline{R}), \tag{2}$$

$$R \cup \overline{R} \cup \tilde{R} = C^*(O). \tag{3}$$

Only a subset of $C(O) \times C(O)$ needs to be considered for the effective calculation of the vector of internal comparisons (further denoted by $V(C)$) since from Eqs. (1)–(3) it follows that if the relation between x and y is known, then so is the relation between y and x. Thus of interest are only the lexicographically ordered pairs $\langle x, y\rangle$. Denote for brevity $C_{i,j} = \langle C(O_i), C(O_j)\rangle$. Then, for a fixed criterion C, the vector with $n(n-1)/2$ elements is obtained:

$$V(C) = \{C_{1,2}, C_{1,3}, \ldots, C_{1,n}, C_{2,3}, C_{2,4}, \ldots, C_{2,n}, C_{3,4}, \ldots, C_{3,n}, \ldots, C_{n-1,n}\}.$$

Let $V(C)$ be replaced by $\hat{V}(C) = \langle \hat{V}_1(C), \hat{V}_2(C), \ldots, \hat{V}_{\frac{n(n-1)}{2}}(C)\rangle$, where for the k-th component $(1 \le k \le \frac{n(n-1)}{2})$:

$$\widehat{V_k}(C) = \begin{cases} 1, \text{ iff } V_k(C) \in R \\ -1, \text{ iff } V_k(C) \in \overline{R} \\ 0, \text{ otherwise} \end{cases}.$$

When comparing two criteria, the degree of "agreement" $\mu_{C,C'}$ is determined as the number of matching components of the respective vectors, divided by the length of the vector for normalization purposes. The degree of "disagreement" $\nu_{C,C'}$ is the number of components of opposing signs in the two vectors, again normalized by the length. Ordered pair $\langle \mu_{C,C'}, \nu_{C,C'}\rangle$ is an IFP. In the most of the obtained pairs, the sum $\mu_{C,C'} + \nu_{C,C'}$ is equal to 1. However, there may be some pairs, for which this sum is less than 1. The difference $\pi_{C,C'} = 1 - \mu_{C,C'} - \nu_{C,C'}$ is considered as a degree of "uncertainty".

Let $\alpha, \beta \in [0, 1]$ be given, so that $\alpha + \beta \le 1$. The criteria C and C' are in:

- $\langle \alpha, \beta\rangle$ Positive consonance (PC), if $(\mu_{C,C'} > \alpha)$ and $(\nu_{C,C'} < \beta)$;
- $\langle \alpha, \beta\rangle$ – Negative consonance (NC), if $(\mu_{C,C'} < \beta)$ and $(\nu_{C,C'} > \alpha)$;
- $\langle \alpha, \beta\rangle$ – Dissonance (D), otherwise.

The scale for defining the corresponding dissonance and consonance values, proposed in [18], has been used in the analysis of the results obtained here.

3 Results and Discussion

The input IMs for ICrA are constructed in the following form for the three cases (RS, HT and VM):

Case (RS, HT or VM)	$Patient_1$	$Patient_2$	...	$Patient_{40}$
C_1^{CAD}	$C_{1,1}^{CAD}$	$C_{1,2}^{CAD}$	...	$C_{1,40}^{CAD}$
C_2^{CAD}	$C_{2,1}^{CAD}$	$C_{2,2}^{CAD}$	...	$C_{2,40}^{CAD}$
$\vdots$	$\vdots$	$\vdots$	$\ddots$	$\vdots$
C_{13}^{CAD}	$C_{13,1}^{CAD}$	$C_{13,2}^{CAD}$	...	$C_{13,40}^{CAD}$
C_1^{CSX}	$C_{1,1}^{CSX}$	$C_{1,2}^{CSX}$	...	$C_{1,40}^{CSX}$
C_2^{CSX}	$C_{2,1}^{CSX}$	$C_{2,2}^{CSX}$	...	$C_{2,40}^{CSX}$
$\vdots$	$\vdots$	$\vdots$	$\ddots$	$\vdots$
C_{13}^{CSX}	$C_{13,1}^{CSX}$	$C_{13,2}^{CSX}$	...	$C_{13,40}^{CSX}$

where with C are denoted the 13 indicators as C = [*RRA, SDDR, MDRR, MED, dRRA, dSDRR, dMDRR, PNN50, RMS, VLF, LF, HF, LF/HF*] and $C_{i,j}$ are the values of each indicator i ($i = 1, 2, \ldots, 13$) for each patient j ($j = 1, 2, \ldots, 40$).

A cross-platform software implementing ICrA, called ICrAData [19], has been used. The obtained results, degree of "agreement" $\mu_{C,C'}$ and degree of "disagreement" $\nu_{C,C'}$ between considered 13 indicators for the three cases are presented as IMs in the form:

$\mu_{C,C'}$	C_1^{CSX}	...	C_{13}^{CSX}
C_1^{CAD}	μ_{C_1,C_1}	...	μ_{C_1,C_m}
$\vdots$	$\vdots$	$\ddots$	$\vdots$
C_{13}^{CAD}	μ_{C_m,C_1}	...	μ_{C_m,C_m}

and

$\nu_{C,C'}$	C_1^{CSX}	...	C_m^{CSX}
C_1^{CAD}	ν_{C_1,C_1}	...	ν_{C_1,C_m}
$\vdots$	$\vdots$	$\ddots$	$\vdots$
C_{13}^{CAD}	ν_{C_m,C_1}	...	ν_{C_m,C_m}

The presented results include only the relations between the indicators $C_1^{CAD}, \ldots, C_{13}^{CAD}$ and $C_1^{CSX}, \ldots, C_{13}^{CSX}$. The relations between indicators $C_1^{CAD}, \ldots, C_{13}^{CAD}$ and between indicators $C_1^{CSX}, \ldots, C_{13}^{CSX}$ are not subject to consideration.

3.1 Resting State

ICrA shows that no dependencies (correlations) have been established between the pair of the 13 indicators considered for patients with CSX and with CAD. The obtained $\mu_{C,C'}$

and $\nu_{C,C'}$ values are presented in IMRSμ and IMRSν, respectively. The results show that all criteria pairs are in dissonance, i.e. there is no relations between the 13 indicators.

3.2 Sympathetic Contour ACC (Handgrip Test; HT)

In this case, again, in general no dependencies were found between the pairs of the 13 indicators. Obtained results are presented in IMHTμ and IMHTν.

IMRSμ =

$\mu_{C,C'}$	RRA^{CSX}	SDRR^{CSX}	MDRR^{CSX}	MED^{CSX}	dRRA^{CSX}	dSDRR^{CSX}	dMDRR^{CSX}	PNN50^{CSX}	RMS^{CSX}	VLF^{CSX}	LF^{CSX}	HF^{CSX}	LF/HF^{CSX}
RRA^{CAD}	0.58	0.48	0.50	0.58	0.49	0.46	0.46	0.43	0.48	0.50	0.46	0.49	0.49
SDRR^{CAD}	0.55	0.54	0.54	0.54	0.52	0.54	0.52	0.49	0.53	0.50	0.51	0.51	0.47
MDRR^{CAD}	0.55	0.53	0.53	0.54	0.51	0.52	0.51	0.48	0.52	0.49	0.49	0.48	0.48
MED^{CAD}	0.58	0.48	0.50	0.58	0.49	0.45	0.46	0.43	0.48	0.50	0.46	0.49	0.49
dRRA^{CAD}	0.56	0.44	0.45	0.56	0.44	0.42	0.42	0.39	0.44	0.41	0.39	0.39	0.50
dSDRR^{CAD}	0.51	0.43	0.44	0.51	0.42	0.43	0.41	0.38	0.43	0.42	0.40	0.39	0.51
dMDRR^{CAD}	0.51	0.43	0.44	0.51	0.42	0.42	0.40	0.38	0.42	0.40	0.39	0.38	0.48
PNN50^{CAD}	0.33	0.27	0.29	0.33	0.27	0.27	0.26	0.27	0.27	0.24	0.22	0.23	0.29
RMS^{CAD}	0.54	0.44	0.45	0.54	0.44	0.43	0.42	0.39	0.44	0.42	0.40	0.40	0.50
VLF^{CAD}	0.53	0.57	0.56	0.53	0.54	0.57	0.55	0.54	0.56	0.50	0.55	0.56	0.46
LF^{CAD}	0.52	0.57	0.56	0.51	0.54	0.59	0.56	0.55	0.58	0.50	0.51	0.55	0.46
HF^{CAD}	0.55	0.48	0.49	0.54	0.47	0.49	0.47	0.45	0.48	0.45	0.44	0.43	0.51
LF/HF^{CAD}	0.45	0.51	0.49	0.45	0.52	0.55	0.54	0.53	0.54	0.53	0.56	0.59	0.42

IMRSν =

$\nu_{C,C'}$	RRA^{CSX}	SDRR^{CSX}	MDRR^{CSX}	MED^{CSX}	dRRA^{CSX}	dSDRR^{CSX}	dMDRR^{CSX}	PNN50^{CSX}	RMS^{CSX}	VLF^{CSX}	LF^{CSX}	HF^{CSX}	LF/HF^{CSX}
RRA^{CAD}	0.42	0.49	0.47	0.42	0.48	0.52	0.50	0.49	0.50	0.49	0.53	0.51	0.51
SDRR^{CAD}	0.41	0.40	0.40	0.42	0.43	0.41	0.41	0.40	0.42	0.46	0.45	0.45	0.49
MDRR^{CAD}	0.40	0.41	0.41	0.41	0.43	0.42	0.42	0.41	0.42	0.46	0.46	0.47	0.48
MED^{CAD}	0.42	0.49	0.48	0.42	0.48	0.53	0.51	0.50	0.50	0.49	0.53	0.51	0.51
dRRA^{CAD}	0.37	0.48	0.46	0.37	0.47	0.49	0.48	0.48	0.48	0.52	0.54	0.54	0.43
dSDRR^{CAD}	0.42	0.48	0.47	0.42	0.49	0.49	0.49	0.48	0.48	0.51	0.53	0.54	0.42
dMDRR^{CAD}	0.37	0.44	0.43	0.37	0.45	0.45	0.45	0.44	0.44	0.47	0.49	0.50	0.40
PNN50^{CAD}	0.19	0.25	0.25	0.20	0.27	0.25	0.26	0.25	0.26	0.28	0.31	0.30	0.24
RMS^{CAD}	0.40	0.48	0.47	0.40	0.48	0.49	0.49	0.49	0.48	0.52	0.54	0.54	0.44
VLF^{CAD}	0.46	0.40	0.41	0.46	0.43	0.41	0.41	0.38	0.41	0.49	0.44	0.43	0.53
LF^{CAD}	0.46	0.39	0.40	0.47	0.43	0.37	0.39	0.36	0.39	0.48	0.47	0.44	0.53
HF^{CAD}	0.43	0.48	0.47	0.44	0.49	0.47	0.48	0.46	0.48	0.53	0.54	0.55	0.47
LF/HF^{CAD}	0.55	0.46	0.48	0.54	0.45	0.43	0.42	0.40	0.43	0.47	0.43	0.41	0.58

IMHTμ =

$\mu_{C,C'}$	RRA^{CSX}	$SDRR^{CSX}$	$MDRR^{CSX}$	MED^{CSX}	$dRRA^{CSX}$	$dSDRR^{CSX}$	$dMDRR^{CSX}$	$PNN50^{CSX}$	RMS^{CSX}	VLF^{CSX}	LF^{CSX}	HF^{CSX}	LF/HF^{CSX}
RRA^{CAD}	0.56	0.47	0.49	0.56	0.44	0.40	0.39	0.35	0.42	0.54	0.47	0.43	0.56
$SDRR^{CAD}$	0.55	0.51	0.52	0.55	0.50	0.49	0.49	0.45	0.49	0.56	0.56	0.49	0.54
$MDRR^{CAD}$	0.56	0.51	0.52	0.55	0.50	0.50	0.49	0.45	0.49	0.55	0.56	0.49	0.55
MED^{CAD}	0.56	0.46	0.49	0.57	0.44	0.40	0.38	0.35	0.42	0.54	0.47	0.43	0.57
$dRRA^{CAD}$	0.57	0.42	0.42	0.57	0.42	0.39	0.38	0.34	0.40	0.48	0.40	0.37	0.58
$dSDRR^{CAD}$	0.55	0.43	0.42	0.55	0.42	0.39	0.38	0.35	0.40	0.50	0.40	0.38	0.56
$dMDRR^{CAD}$	0.55	0.42	0.42	0.55	0.42	0.39	0.38	0.35	0.40	0.49	0.38	0.36	0.55
$PNN50^{CAD}$	0.30	0.21	0.21	0.31	0.23	0.21	0.21	0.22	0.21	0.26	0.18	0.19	0.25
RMS^{CAD}	0.57	0.43	0.43	0.57	0.43	0.41	0.39	0.35	0.41	0.51	0.41	0.38	0.58
VLF^{CAD}	0.51	0.49	0.49	0.51	0.47	0.48	0.48	0.43	0.47	0.54	0.52	0.48	0.55
LF^{CAD}	0.53	0.49	0.47	0.53	0.50	0.51	0.50	0.46	0.51	0.49	0.44	0.47	0.48
HF^{CAD}	0.55	0.46	0.45	0.55	0.45	0.44	0.43	0.40	0.45	0.51	0.43	0.41	0.52
LF/HF^{CAD}	0.41	0.48	0.48	0.39	0.45	0.47	0.44	0.40	0.47	0.46	0.51	0.49	0.52

IMHTν =

$\nu_{C,C'}$	RRA^{CSX}	$SDRR^{CSX}$	$MDRR^{CSX}$	MED^{CSX}	$dRRA^{CSX}$	$dSDRR^{CSX}$	$dMDRR^{CSX}$	$PNN50^{CSX}$	RMS^{CSX}	VLF^{CSX}	LF^{CSX}	HF^{CSX}	LF/HF^{CSX}
RRA^{CAD}	0.43	0.51	0.48	0.42	0.52	0.57	0.56	0.51	0.56	0.45	0.52	0.56	0.43
$SDRR^{CAD}$	0.41	0.44	0.43	0.41	0.43	0.45	0.43	0.40	0.46	0.39	0.41	0.47	0.42
$MDRR^{CAD}$	0.41	0.44	0.43	0.41	0.44	0.45	0.43	0.39	0.46	0.41	0.41	0.47	0.42
MED^{CAD}	0.43	0.52	0.48	0.42	0.52	0.58	0.57	0.52	0.56	0.45	0.52	0.56	0.42
$dRRA^{CAD}$	0.37	0.51	0.50	0.37	0.49	0.53	0.52	0.48	0.53	0.45	0.54	0.57	0.37
$dSDRR^{CAD}$	0.37	0.49	0.48	0.37	0.48	0.51	0.50	0.46	0.51	0.42	0.52	0.54	0.36
$dMDRR^{CAD}$	0.36	0.48	0.47	0.35	0.46	0.51	0.49	0.45	0.50	0.42	0.52	0.54	0.36
$PNN50^{CAD}$	0.15	0.25	0.23	0.15	0.22	0.25	0.24	0.22	0.24	0.19	0.28	0.27	0.20
RMS^{CAD}	0.37	0.51	0.50	0.37	0.49	0.53	0.52	0.48	0.53	0.44	0.54	0.56	0.37
VLF^{CAD}	0.47	0.48	0.47	0.47	0.48	0.48	0.46	0.42	0.50	0.44	0.46	0.50	0.43
LF^{CAD}	0.43	0.47	0.47	0.43	0.44	0.43	0.43	0.39	0.44	0.47	0.52	0.49	0.48
HF^{CAD}	0.37	0.45	0.45	0.37	0.44	0.46	0.44	0.40	0.46	0.40	0.49	0.51	0.41
LF/HF^{CAD}	0.59	0.51	0.50	0.59	0.52	0.51	0.51	0.46	0.51	0.53	0.49	0.49	0.48

3.3 ACC Parasympathetic Stimulation (Valsalva Manoeuver; VM)

In this case, all pairs of the 13 indicators, without exception, do not show any dependencies (correlations). All considered pairs are in dissonance. Due to the similarity with the above results, the results in this case are not presented.

The lack of significant positive or negative relations observed here show that the CSX and CAD patients show different behaviour both in parasympathetic stimulation and in the response of the sympathetic autonomic system.

4 Conclusion

In this paper a novel approach ICrA is applied to reveal relationships between 13 standard indicators for Heart Rate Variability, respectively in resting state and by sympathetic stimulation (handgrip test) or parasympathetic stimulation (Valsalva manoeuvre). Based on these indicators the activity of ACC in two groups of 40 patients – with CSX and with CAD is evaluated.

The obtained results show that at rest and at stimulation of the sympathetic part of the ACC there are differences in the balance of the autonomic system between the groups with CSX and CAD. In parasympathetic stimulation there is no difference in the direction of change of vagal tone. Obviously, the sympathetic proportion in patients with CSX has a greater potential for increased activity than in patients with CAD. The parasympathetic part in CSX patients was less active than in CAD patients. These results give reason to assume, that the two considered groups show a different response of the sympathetic autonomic system under conditions of stress with matching parasympathetic tone.

Acknowledgements. Work presented here is partially supported by the Bulgarian National Scientific Fund under Grant KP-06-N22/1 "Theoretical Research and Applications of InterCriteria Analysis".

References

1. Kaski, J.C.: Pathophysiology and management of patients with chest pain and normal coronary arteriograms (cardiac syndrome X). Circulation **109**(5), 568–572 (2004)
2. Kemp, H.G.: Left ventricular function in patients with the angina syndrome and normal coronary arteriograms. Am. J. Cardiol. **32**, 375–376 (1973)
3. Kemp Jr., H.G., Vokonas, P.S., Cohn, P.F., Gorlin, R.: The angina syndrome associated with normal coronary arteriograms. Report of a six year experience. Am. J. Med. **54**, 73542 (1973)
4. Rosano, G.M.C., et al.: Failure to demonstrate myocardial ischaemia in patients with angina and normal coronary arteries. Evaluation by continuous coronary sinus pH monitoring and lactate metabolism. Eur. Heart J. **17**(8), 1175–1180 (1996)
5. Atanassov, K., Mavrov, D., Atanassova, V.: Intercriteria decision making: a new approach for multicriteria decision making, based on index matrices and intuitionistic fuzzy sets. Issues Intuition. Fuzzy Sets Gener. Nets **11**, 1–8 (2014)
6. Atanassov, K.: Generalized index matrices. Compt. Rendus l'Acad. Bulgare Sci. **40**(11), 15–18 (1987)

7. Atanassov, K.: On index matrices, part 1: standard cases. Adv. Stud. Contemp. Math. **20**(2), 291–302 (2010)
8. Atanassov, K.: Index Matrices: Towards an Augmented Matrix Calculus. Springer, Cham (2014)
9. Atanassov, K.: On Intuitionistic Fuzzy Sets Theory. Springer, Berlin (2012). https://doi.org/10.1007/978-3-642-29127-2
10. Atanassov, K.: Intuitionistic fuzzy sets. VII ITKR session, Sofia, 20–23 June 1983. Reprinted: Int. J. Bioautom. **20**(S1), S1–S6 (2016)
11. Todinova, S., et al.: Blood plasma thermograms dataset analysis by means of InterCriteria and correlation analyses for the case of colorectal cancer. Int. J. Bioautom. **20**(1), 115–124 (2016)
12. Sotirov, S., et al.: A hybrid approach for modular neural network design using intercriteria analysis and intuitionistic fuzzy logic. Complexity **2018**, Art. ID 3927951 (2018)
13. Atanassov, K., Vassilev, P.: On the intuitionistic fuzzy sets of n-th type. In: Gawęda, A., Kacprzyk, J., Rutkowski, L., Yen, G. (eds.) Advances in Data Analysis with Computational Intelligence Methods. Studies in Computational Intelligence, vol. 738, pp. 265–274. Springer, Cham (2018). https://doi.org/10.1007/978-3-319-67946-4_10
14. Matveev, M., Prokopova, R., Nachev, C.: Normal and Abnormal Circadian Characteristics in Autonomic Cardiac Control: New Opportunities for Cardiac Risk Prevention. Nova Science Publishers, Inc., New York (2006)
15. Task Force of the European Society of Cardiology and the North American Society of Pacing and Electrophysiology. Heart rate variability. Standards of measurement, physiological interpretation and clinical use. Circulation **93**, 1043–1065 (1996)
16. Malliani, A., Pagani, M., Montano, N., Mela, G.S.: Sympathovagal balance: a reappraisal. Circulation **98**, 2640–2643 (1998)
17. Atanassov, K., Szmidt, E., Kacprzyk, J.: On intuitionistic fuzzy pairs. Notes Intuition. Fuzzy Sets **19**(3), 1–13 (2013)
18. Atanassov, K., Atanassova, V., Gluhchev, G.: InterCriteria analysis: ideas and problems. Notes Intuition. Fuzzy Sets **21**(1), 81–88 (2015)
19. Ikonomov, N., Vassilev, P., Roeva, O.: ICrAData – software for intercriteria analysis. Int. J. Bioautom. **22**(1), 1–10 (2018)

A FEM Study of Mechanical Behavior of S-Shaped-Stent

Mihail Mihalev[1(✉)], Lilia Vladimirova-Mihaleva[2], Chavdar Hardalov[3], and Docho Tsankov[4]

[1] Faculty of Mechanical Engineering, Department of Precision Engineering and Measurement, Technical University of Sofia, 8 Kl. Ochridski Blvd, 1000 Sofia, Bulgaria
mmihalev@tu-sofia.bg

[2] Faculty of Physics, Department of Optics, Sofia University "St. Klimen Ochridski", 5. J. Bourchier Blvd, Sofia, Bulgaria

[3] Faculty of Applied Mathematics and Informatics, Department of Physics, Technical University of Sofia, 8 Kl. Ochridski Blvd, 1000 Sofia, Bulgaria

[4] Faculty of Automatics, Department of Electrical Motion Automation Systems, Technical University of Sofia, 8 Kl. Ochridski Blvd, 1000 Sofia, Bulgaria

Abstract. The paper presents results of simulations of mechanical behavior of S-shaped stents of different lengths.

The mechanical behavior of the stent strongly depends on the stent strut topology and the material properties. The presented stent strut topology differs from the known devices, usually used in the angioplasty. The aim of the study is to suggest a different stent topology and design and to model, simulate and compare its mechanical behavior and parameters with other clinically used stents.

The suggested stent design approach is based on the Escher tessellation, consisting in periodic tiling of a specific shape pattern without gapes and/or overlaps. As a result, the array of multiplied slot patterns forms the desired stent struts and the S-formed bridges. The method enables generation of variety of new and unique stent geometries.

The following mechanical properties of the bare metal stent (medical stainless-steel Grade 316L) are estimated: recoil, foreshortening, dogboning as well as critical stresses and they are compared with commercial stents. All graphical modelling and computer simulations are performed using SolidWorks and COMSOL Multiphysics.

Unlike the traditional stents the stents of suggested design show unexpected negative values of dogboning. All other parameters are comparable.

The results of the modelling and simulation open room for further design, new stent properties and new design algorithms.

Keywords: Solid mechanics · Coronary stents · FEM · Comsol

1 Introduction

The first coronary stent was suggested in 1990 by Julio Palmaz and Richard Schatz, approved in 1994 by US Food and Drug Agency (FDA). The original form of the stent

S. S. Sotirov et al. (Eds.): BioInfoMed 2020, LNNS 374, pp. 342–353, 2022.
https://doi.org/10.1007/978-3-030-96638-6_36

mesh was patented and it is known as "Palmaz-Schatz" stent (PS). Nowadays many types of stent grafts are known. One of the most used type is the slotted tube stent, fabricated from different metals. Usually, bare metal stent (BMS) grafts are made from a stainless steel of Grade 316L tube with customized outer and inner diameter. A laser (usually solid-state Nd^{3+}:YAG or fiber) is used for forming the geometry of the slots [1]. Then, after isotropic etching, filleted struts with desired cross-section dimensions are built.

The mechanical behavior of the stent strongly depends on both geometry (length, shape, strut thickness) and material properties.

The finite element analysis (FEA) based on finite element method (FEM) is established as a standard tool for study of mechanical parameters of stents. Many published works about behavior of the stents can be found: [1–5]. Most of the researchers prefer to investigate the mechanical parameters of stents based on PS stent [6] due to their high symmetry class which facilitates the calculations [7].

The experience in stent applications led FDA to standardize the characteristics of the stents [8].

In [9] the authors show that one most important stent parameter – dogboning – can be reduced by asymmetric stent geometry. The authors again emphasize on PS stent. Chua et al. in [10], Qui et al. in [11] introduced the idea of non-linear dependance of the elastic recoil and dogboning on geometrical parameters of the stent, especially on its length.

The presented investigations open room for a stent geometry, different from PS pattern. It is based on the Escher tessellation [12, 13] consisting in periodic tiling of a specific shape without any gapes and/or overlaps. As a result, the array of multiplied slot patterns forms the desired stent struts and the S-formed bridges. The method enables generation of variety of new and unique stent geometries. Following the Escher tessellation approach, new mechanical behavior of the stents is expected.

2 Materials and Methods

2.1 Geometry of PS and ISMA Stents

The Palmas-Schatz stent (see Fig. 1) consists of struts, parallel to the tube axis, connected by bridges, thus forming rectangular slots - repeated unit cells (RUC) with high rotational and translational symmetry. PS stent obeys two independent symmetry types: rotational of order 12 (1/12 of the circumference) and translation symmetry of order 2 [12, 14]. These symmetries enable the modelling of only one small unit (Fig. 2), which can be multiplied by rotation, translation and/or mirroring, forming the whole stent (Fig. 1), thus facilitating the calculations by minimization of the number of elements and Dots of Freedom (DOF).

The company ISMA EOOD, Sofia, Bulgaria, suggested a stent design, based on Escher tessellation, called by us ISMA stent [15]. Figure 3 represents a view of the whole stent. Its slot shape differs significantly from the diamond slot shape of the PS stent as it can be seen comparing Fig. 1 and Fig. 3. The RUC of ISMA stent consists of struts, which edges are neither parallel nor perpendicular to the rotational tube axis (Fig. 4). The struts and RUC's are obtained by translation and rotation of the slot shape in circumferential and longitudinal directions by Escher Tessellation. Every strut consists of circular arcs with common centers and radii r and $R = r + \Delta$, where Δ is the size of the strut, indicated in Fig. 4.

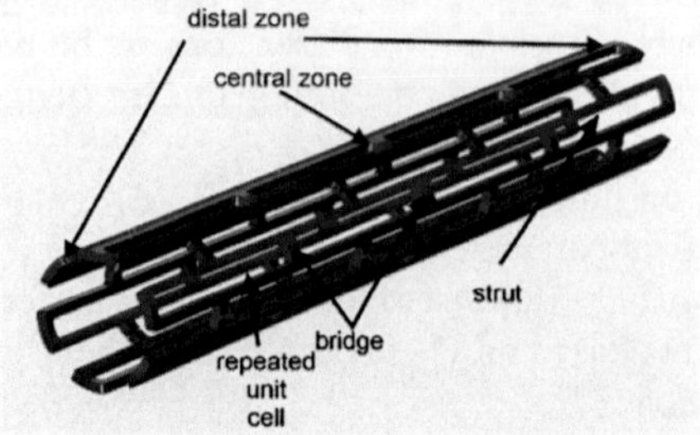

Fig. 1. Design of standard PS stent

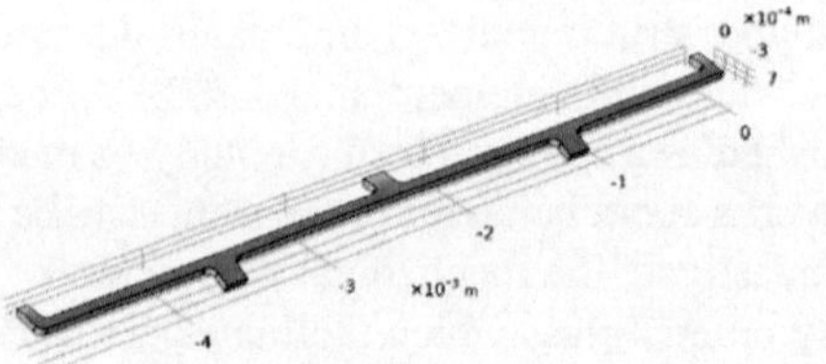

Fig. 2. Repeated unit of PS stent

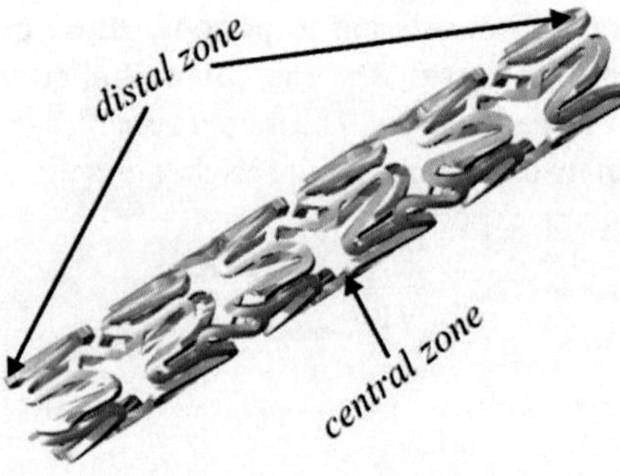

Fig. 3. View of ISMA stent

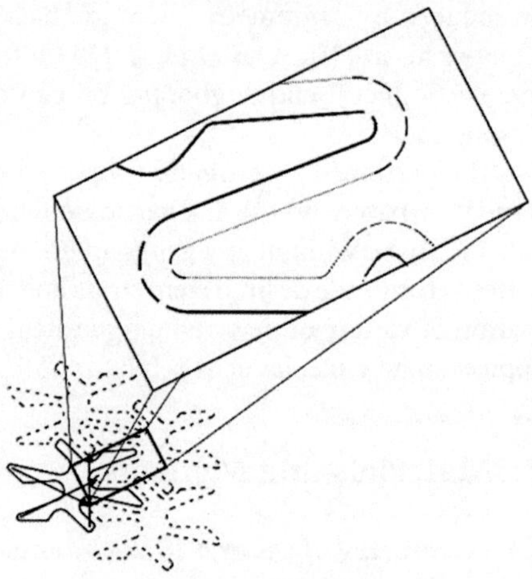

Fig. 4. Tessellation of ISMA stent

The ISMA stent geometry is more complex – rotation of RUC by the order 6 (1/6 of circumference), followed by translation by 2/3 RUC along the tube axis. The filleting of the strut shape, especially in the distal zone of ISMA stent, leads to more tight behavior of the mesh compared with PS. The lower symmetry class and more complexity of the geometry can be a serious problem for numerical calculations. On other side, this different stent design seems more suitable for the purposes of angioplasty stenting. The tessellation method for ISMA stent is shown in Fig. 4, depicted by dash lines. The slot shape is shifted so that center of the small arc coincides with the center of the arc with larger radius. This procedure resembles Escher tessellation and can be used for achieving other stent designs.

2.2 3D Modelling

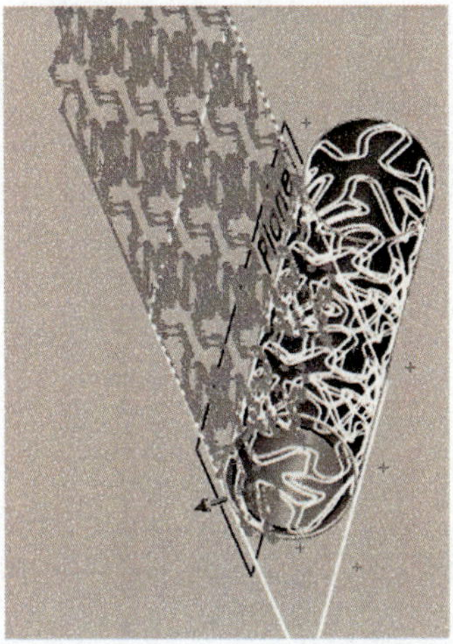

Fig. 5. ISMA stent wrapping with planar drawing screenshot from SolidWorks)

As shown in [14], the whole PS stent can be assembled in the COMSOL Multiphysics® environment from a small unit by translation and rotation. In the case of ISMA stent this approach is not applicable because of the lower symmetry. A 3D model (Fig. 5) of the ISMA stent is developed in DS SolidWorks and imported into COMSOL Multiphysics® in full scale as a solid [15].

A tube with prescribed stent lengths (10 and 16 mm) and inner and outer diameters (1.35 and 1.55 mm) is extruded. The method for building 3D solid model of the stent resembles the movement of the cutting laser beam. Planar unfolded drawing of the stent geometry is placed as a DXF sketch on a plain, tangential to outer cylindrical surface of the tube (see plain sketch Fig. 5). The SolidWorks feature "Wrap" is used to fold the geometry around the cylindrical surface (see the wrapped sketch around the cylindrical tube in Fig. 5). The option "Deboss" imitates the laser cutting of the slots, removing the material from the tube, thus building the stent struts [16]. A versatility of the "Deboss" option of "Wrap" feature consist in the possibility to obtain the whole stent in only one step without any additional processing on the 3D model.

2.3 FEM Simulation

Once the solid model of the ISMA stent is prepared, it is imported into COMSOL Multiphysics®. A piecewise triangle function *r0* is defined, which sets the stent diameter evolution as a function of one dimensionless parameter *param* from 0 to 2. *Param* is used to calculate implicitly the balloon pressure to reach the assigned diameter.

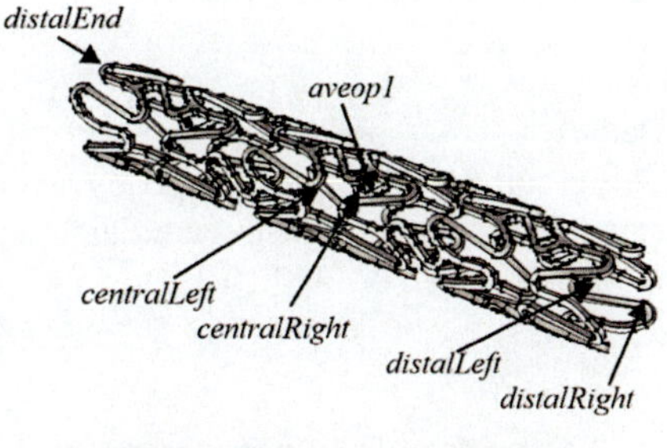

Fig. 6. Points for calculations

The COMSOL "Average" operator (*aveop1*, see Fig. 6), defined over a strut edge, is used to evaluate the real inflation of the stent at each *param* step. This in contribution with the piecewise function *r0* is used to solve the needed pressure *p* for achieving the assigned stent diameter. The corresponding equation is introduced as a global equation option of the Solid Mechanics feature of the model. The pressure *p* is obtained by solving the equation:

$$aveop1(r) - r0(param) = 0 \tag{1}$$

Another COMSOL "Integration" operators (*distalRight, distalLeft, centralRight, centralLef, and distalEnd*, see Fig. 5) are used for calculations of the real stent diameter $2r$ at distal and central zones, indicated in Fig. 1 and Fig. 3, as well as to calculate the stent length changes during expansion of the catheter balloon.

Structural mechanics module (Solid mechanics) of COMSOL Multiphysics is implemented. Stationary solution is chosen [17]. The stent inflation is simulated by applying a radial pressure p on the inner surface or the struts as a boundary condition. The needed pressure is obtained by solution of Eq. 1.

2.4 Stent Parameters

Stent parameters are standardized by the FDA in [8].

Dogboning

The dogboning DB is the ratio of the difference between stent distal and central diameter to the center diameter:

$$DB = \frac{d_{distal} - d_{central}}{d_{central}} \cdot 100\% \tag{2}$$

Foreshortening

The foreshortening FS is the ratio of the difference between stent length at expanded state l_{expanded} and the stent length in crimped state l_0 to its expanded state length:

$$FS = \frac{l_{expanded} - l_0}{l_0} \cdot 100\% \tag{3}$$

Recoil

The recoil R is the percentage by which the diameter of a stent decreases from its expanded diameter to its relaxed diameter:

$$R = \frac{d_{expanded} - d_{relaxed}}{d_{relaxed}} \cdot 100\% \tag{4}$$

The values of the parameters are calculated at characteristic points on the mesh, selected as settings in the COMSOL integration operator.

Material

The austenitic stainless-steel Grade 316L is the most used metallic biomaterial in orthopedics applications, especially in the manufacture of articulated prostheses and as structural elements in fracture fixation, since it has high mechanical strength and it is biologically inactive [18]. BMS are fabricated also from stainless-steel 316L. Elastoplastic nearly incompressible material is assumed. As plasticity model "Large plastic strains" is chosen. Linear isotropic model describes the hardening of the material after reaching the initial yield stress.

As material stainless-steel Grade 316L with following material properties is used, obtained from the material library, embedded in COMSOL:

Young modulus: $E = 200$ GPa
Yield stress: $\sigma_{\text{yield}} = 170$ MPa
Poison's ratio: $\nu = 0{,}27$
Isotropic tangent modulus: $E_t = 2$ GPa
Ultimate tensile strength: 627 MPa

Mesh
A swept mesh replaces the standard mesh. The mesh type is very suitable for thin geometries. The swept mesh extends the mesh on source region into the target sides, creating prismatic elements which can be handled more effectively with far less elements [19]. Free triangle mesh is applied over the outer surface elements of the struts. Additionally, a swept mesh with distribution operation is applied on all other strut surfaces to mesh complex solid and surface regions.

Solver
The mechanical deformation at every pressure should be calculated. COMSOL Multiphysics affords an automation opportunity, called Parametric solver, which collaborates with the default stationary solver. Parametric solver is added to the default stationary solver. A dimensionless parameter *param* with a step of 0.01 in the range between 0 and 2 estimates the desired stent inflated diameter between initial and end value, thus imitating the time evolution of the stent (Eq. 1). The direct default solver calculates at each step of *param* the mechanical characteristics as prescribed in the parametric solver node.

3 Results and Discussion

A comparative analysis of mechanical properties of PS and ISMA stent for stent lengths of both 10 and 16 mm is performed. All other stent design parameters (initial inner stent diameter – 1.3 mm, strut cross-section 100×100 μm and material – stainless steel grade 316L) are also chosen the same.

The authors in [18] classify the lesions as short (<10 mm) and long (>10 mm) lesions. Lesion length is an important predictor of in-stent restenosis in vascular angioplasty stenting. In accordance with [20] two lengths of 10 and 16 mm, corresponding to short and long lesions, are simulated.

The simulation results are comparably presented for both PS and ISMA stents in Fig. 7, 8, 9, 10, 11, 12, 13, 14, 15, 16 and 17 as a function of the applied pressure, which is implicitly calculated for each value of the parameter *param*.

3.1 Diameters

Figure 7, 8, 9 and 10 illustrate the evolution of the distal and central diameters of both stents as a function of the calculated pressure *p* for the corresponding parameter *param*. The diameter is evaluated as a mean distance, measured between two mutually opposite points in the distal and central zones.

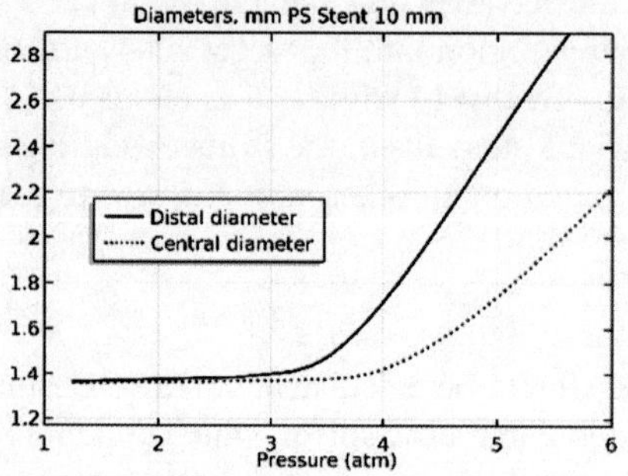

Fig. 7. Stent diameter vs. pressure *p* PS stent, length of 10 mm

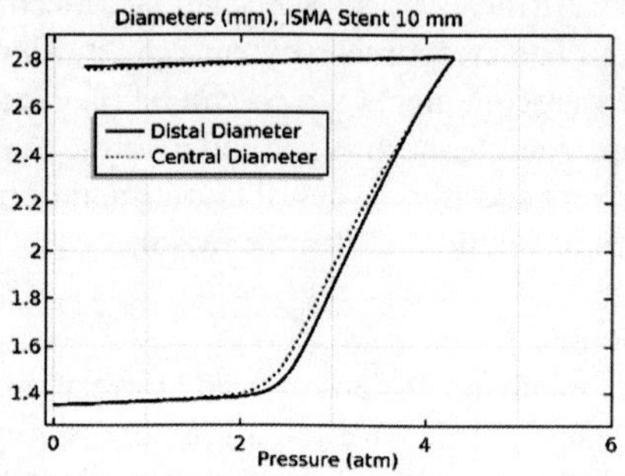

Fig. 8. Stent diameter vs. pressure *p*, ISMA stent of 10 mm

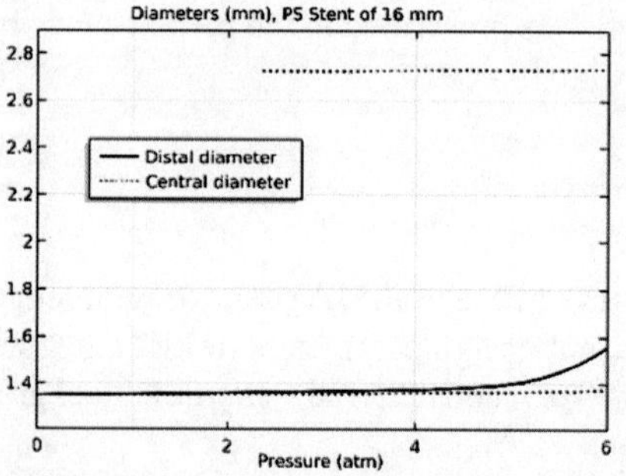

Fig. 9. Stent diameter vs. pressure *p*, PS stent of 16 mm

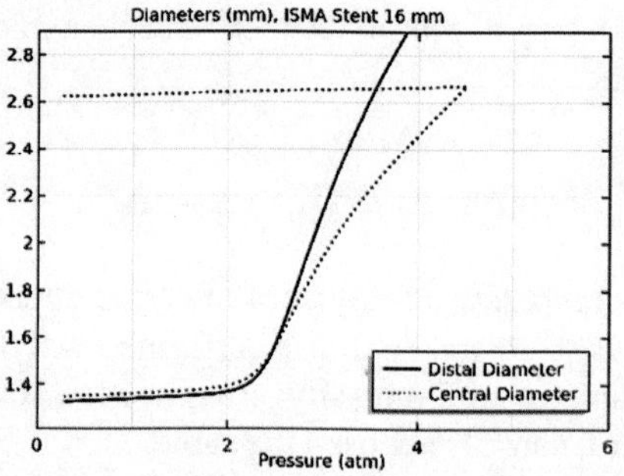

Fig. 10. Stent diameter vs. pressure *p*, ISMA stent of 16 mm

It can be clearly seen that a desired stent diameter is obtained with an ISMA stent at lower pressure than in a PS stent for both stent lengths.

With the larger stent length, the needed pressure rises and this effect is more expressed in case of PS stent. From the figures it can be seen that for a distal diameter of 2.8 mm appr. 4.5 atm are needed, if ISMA stents are inflated, and appr. 5.5 atm for 10 mm PS stent. The PS stent with 16 mm lengths reaches at 6 atm a diameter of only 1.6 mm.

3.2 Dogboning

Figure 11, 12, 13 and 14 reveal the dogboning of both stent types as a function of applied pressure and stent length. The PS stents show definitely positive dogboning, i.e. the distal diameter is larger than the central one. On the contrary, the ISMA stent diameter is larger in the central zones (negative values of dogboning called also barrel-shape [5]) for moderate lengths and show positive values of the dogboning for large stents. Implanting a stent with negative dobgoning enables a larger vessel lumen exactly at the point of narrowing (lesion) thus restoring the normal blood flow. Maxima of dogboning can be recognized in both stent types. The position of these maxima (positive or negative) moves into the higher pressures when the stent length increases. This change of the sign of dogboning of the ISMA stent could be explained by the presumption that with the increasing of stent length the struts in the central zones become more "tight", while the distal ends are getting loose.

Generally, ISMA stent shows significantly lower maximal dogboning versus PS stent (4% vs. 30% at 10 mm and 14% vs. 15% at 16 mm stent at lower diameter of PS stent, see Fig. 7, 8, 9 and 10). The observed "hysteresis" can be explained by the recoil of the stent which is evaluated in the next subsections.

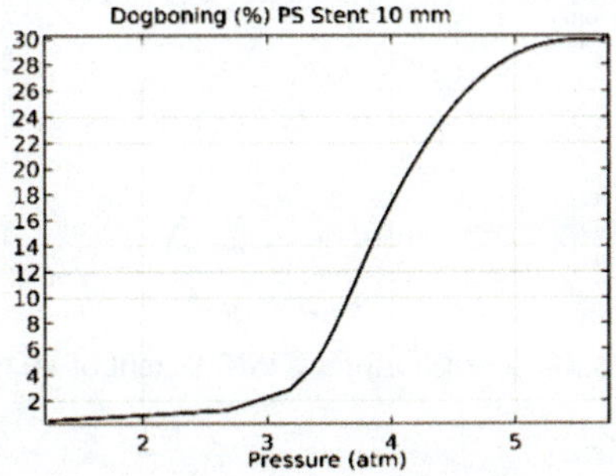

Fig. 11. Dogboning in % vs. pressure *p* for PS stent of 10 mm

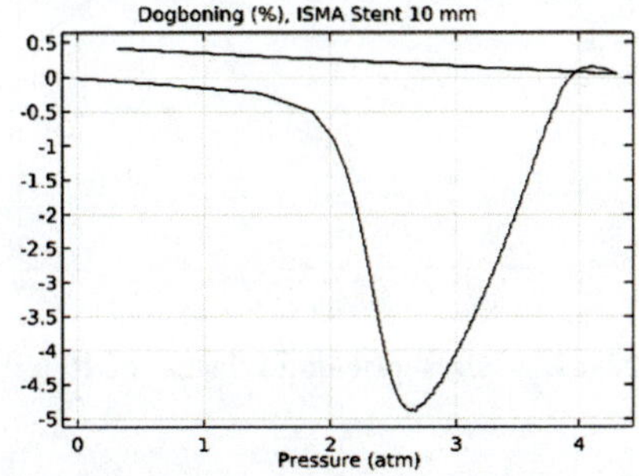

Fig. 12. Dogboning in % vs. pressure *p* for ISMA stent of 10 mm

3.3 Foreshortening

Both PS and ISMA stents show similar foreshortening as depicted in Fig. 15, 16, 17 and 18. The foreshortening increases rapidly and nonlinearly when the balloon pressure is nearly half of the pressure for achieving the desired diameter (2.8 mm in this example). The recoil of both types of stent is significantly higher for longer stents.

3.4 Recoil

Because we expect changes of the recoil with the stent length, the recoil is measured for both distal and central zones. The recoil is calculated for maximal balloon pressure for both PS and ISMA stent. The results are presented in Table 1.

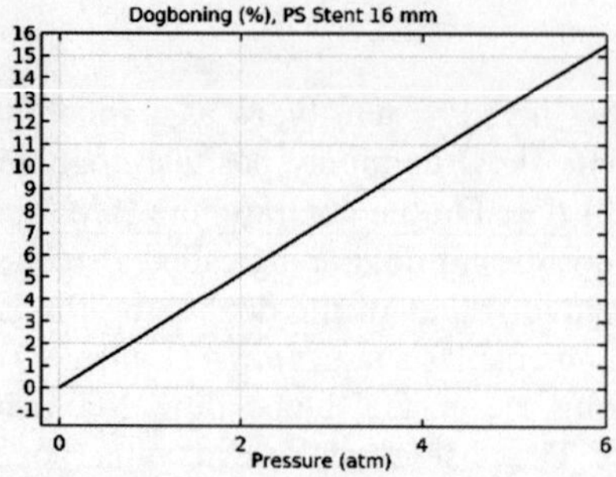

Fig. 13. Dogboning in % vs. pressure p for PS stent of 16 mm

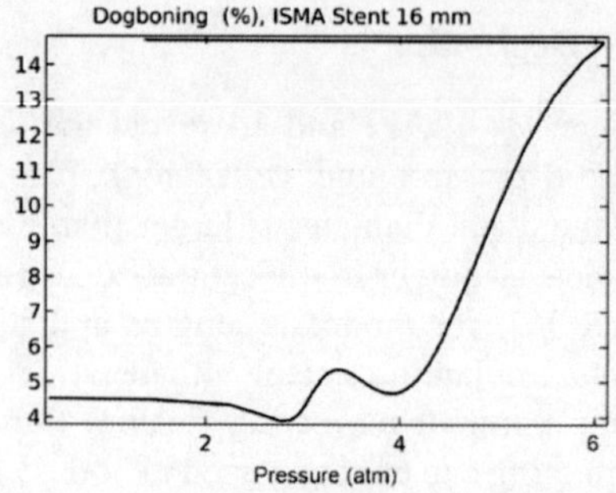

Fig. 14. Dogboning in % vs. pressure p for ISMA stent of 16 mm

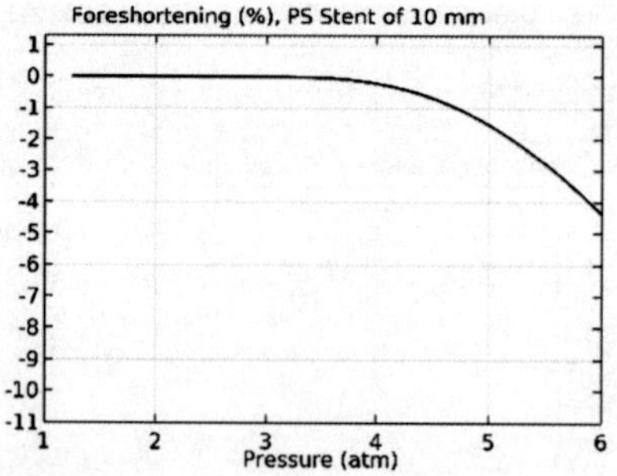

Fig. 15. Foreshortening PS stent of 10 mm

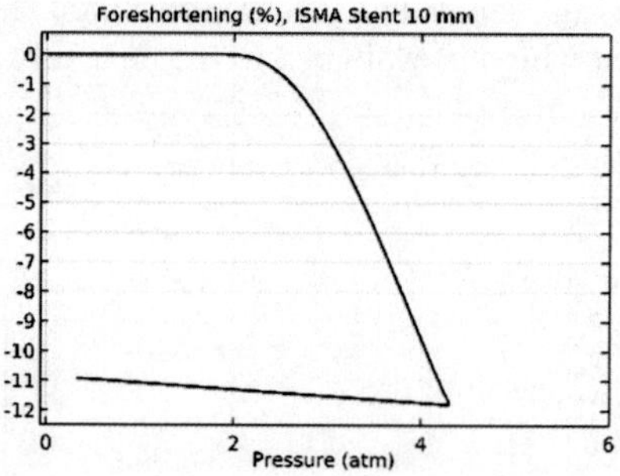

Fig. 16. Foreshortening ISMA stent of 10 mm

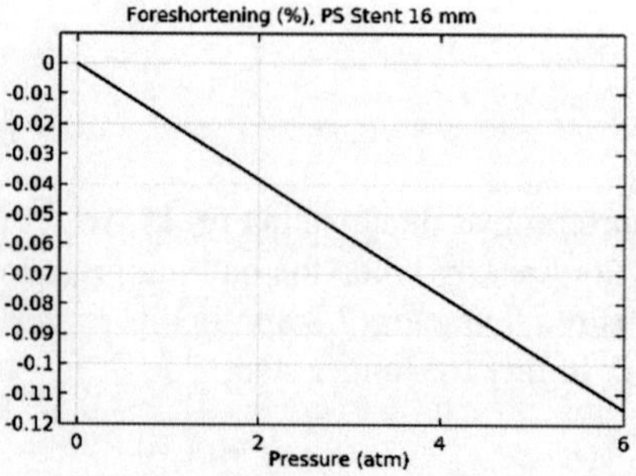

Fig. 17. Foreshortening PS stent of 16 mm

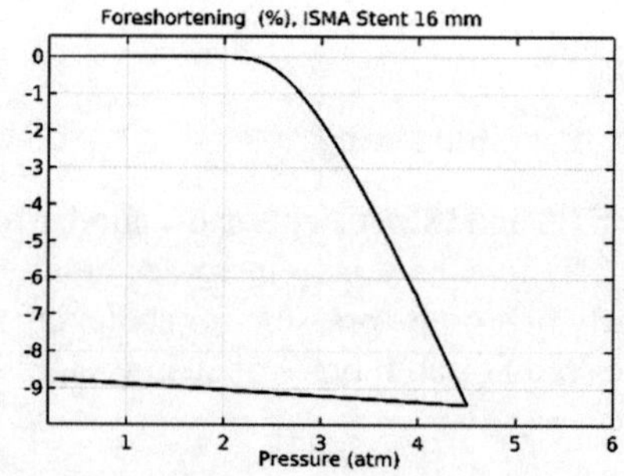

Fig. 18. Foreshortening ISMA stent of 16 mm

In fact, the stent recoil varies with the length, as it can be seen from Table 1. The PS stent reveals significantly lower recoil in the central and distal zones than the ISMA stent. The obtained results are comparable with the reported recoils in reference [5] for commercially available stents.

Table 1. Central and distal recoil for both PS and ISMA stents

Stent length	Recoil (%)			
	PS		ISMA	
	Central	Distal	Central	Distal
10 mm	0.92	0.85	1.32	1.63
16 mm	0.53	0.16	1.39	1.35

3.5 Von Mises Stress

Von Mises stresses, calculated for balloon pressure at diameters of 2.8 mm for both PS and ISMA stent, are presented in Table 2. Clearly defined critical stress is required by the standards [8]. As a measurement the von Mises stress is chosen. The maximal von Mises stress at all mesh points of the stents is reflected in the table.

Table 2. Maximal von Mises stress in PS and ISMA stent

Stent length	Von Mises stresses (MPa)	
	PS	ISMA
10 mm	478	421
16 mm	581	442

As it can be clearly seen, von Mises stresses in both types of stents are much lower than the UTS (627 MPa) at all stent mesh points, which is a criterium for possible stent failure.

PS and ISMA stents are of different design type – diamond and S-shaped, respectively. The geometry of ISMA is obtained using Escher tessellation. The results enable an evaluation in what extent the different design affects the mechanical behavior and parameters (dogboning, recoil, foreshortening) of the stent after inflation.

The analysis is performed by using of FEA at identical initial conditions (material, initial diameter, thickness of the tube and stent length). From point of view of applied inflation pressure, similar results can be obtained at lower pressure in the case of ISMA stent. This enables a more gentle treatment of stenosis and is an indication of long durability of the device. The barrel-shaped (negative) dogboning in the case of ISMA stent is an additional important result leading to positive health effect targeting better restoring of the vessel's lumen. The other studied standard parameters are comparable.

The design method, based on Escher tessellation, opens field for future optimization of the stent parameters changing the geometry in a versatile way. In the future an algorithm for computer automation of the whole design process can be also developed.

4 Conclusions

A stent design, based on Escher tessellation, is presented. This stent shape leads to mechanical parameters which meet the requirements of the FDA requirements [8]. The investigation is performed by comparative analysis with PS stent. Most of the mechanical parameters are comparable, although the proposed ISMA stent demonstrates unexpected barrel-shaped dogboning.

This type of design, based on Escher tessellation, enables variety of strut forms with other perspective mechanical properties.

Acknowledgements. The used software COMSOL Multiphysics 5.5 and the license have been delivered under Project BG05M2OP001-1.001-0008 "National Centre for Mechatronics and Clean Technologies", funded by Operational Program "Science and Education for Smart Growth" 2014–2020.

The authors gratefully appreciate the support of the Management of the Project, as well as of the Operational Program "Science and Education for Smart Growth" 2014–2020.

The supplying of technical and constructive documentation of the stent by ISMA EOOD is highly acknowledged.

References

1. Muhammadm, N., Al Bakri Abdullah, M., Saleh, M., Li, L.: Laser cutting of coronary stents: progress and development in laser based stent cutting technology. In: Key Engineering Materials, vol. 660, pp. 345–350 (2015)
2. Reddy, J.: An Introduction to Finite Element Method, 3rd edn. McGraw-Hill Education, New York (2005)
3. Chandrupatla, T.: Finite Element Analysis for Engineering and Technology. Universities Press, India (2004)
4. Kolston, P.: Finite elememt modelling: a new tool for the biologist. Phil. Trans. R. Soc. Lond. A **358**, 611–631 (2000)
5. Umer, M., Ali, M., Mubashar, A., Mir, M.: Computational modeling of balloon-expandable in coronary artery using the finite element method. Res. Rep. Clin. Cardiol. **10**, 43–56 (2019)
6. Jung, T., Kim, J.Y.: Finite element structural analysis of self-expandable stent deployment in a curved stenotic artery. J. Mech. Sci. Technol. **30**(7), 3143–3149 (2016). https://doi.org/10.1007/s12206-016-0624-5
7. Schievano, S., et al.: Finite element analysis of stent deployment: understanding stent fracture in percutaneous pulmonary valve implantation. J. Interv. Cardiol. **20**(6), 546–554 (2007)

8. Non-Clinical Engineering Tests and Recommended Labeling for Intravascular Stents and Associated Delivery Systems - Guidance for Industry and FDA Staff, U.S. Department of Health and Human Services Food and Drug Administration Center for Devices and Radiological Health (2020). https://www.fda.gov/regulatory-information/search-fda-guidance-documents/non-clinical-engineering-tests-and-recommended-labeling-intravascular-stents-and-associated-delivery
9. De Beule, M., Van Impe, R., Verhegghe, B., Segers, P., Verdonck, P.: Finite element analysis and stent design: reduction of dogboning. Technol. Health Care **14**, 233–241 (2006)
10. David Chua, S., MacDonald, B., Hashmi, M.: Effects of varying slotted tube (stent) geometry on its expansion behavior using finite element method. J. Mater. Process. Technol. **155–156**, 1764–1771 (2004)
11. Li, H., Qiu, T., Zhu, B., Wu, J., Wang, X.: Design optimization of coronary stent based on finite element models. Comput. Simul. Cardiovasc. Syst. **2013**, 1–10 (2013)
12. Lang, R.J.: Twists, Tilings, and Tessellations. CRC Press Taylor & Francis Group (2018)
13. Mihalev, M.: Escher tessellation for design of slotted tube vascular stent. Int. Sci. J. Ind. 4.0 **3**(1), 30–33 (2018)
14. Gomes, I.V., Puga, H., Alves, J.L.: Shape and functional optimization of biodegradable magnesium stents for manufacturing by ultrasonic-microcasting technique. Int. J. Interact. Des. Manuf. (IJIDeM), 1059–1069 (2017)
15. Mihalev, M., Hardalov, Ch., Vladimirova–Mihaleva, L., Stefanov, P., Georgiev, V., Dzherekarov, H.: Nummerical modelling of a real coronary stent. In: AIP Conference Proceedings, vol. 2075, pp. 170007-1–170007-6 (2012)
16. Lombard, M.: SolidWorks® 2009 Bible. Wiley Publishing Inc., Indianapolis (2009)
17. COMSOL Structural Mechanics Module User's Guide. www.doc.comsol.com/5.5/doc/com.comsol.help.sme/StructuralMechanicsModuleUsersGuide.pdf&usg=AOvVaw1fBb_ZoIYU-Y3VR8mdAbSx
18. Silva, G., Baldissera, M.R., de Sousa Trichês, E., Cardoso, K.R.: Preparation and characterization of stainless steel 316L/HA biocomposite. Mat. Res. **16**(2), 304–309 (2013)
19. COMSOL Multiphysics Reference Manual. doc.comsol.com/5.5/doc/com.comsol.help.comsol
20. Lin, Y.H., et al.: The impact of lesion length on angiographic restenosis after vertebral artery origin stenting. Eur. J. Vasc. Endovasc. Surg. **32**, 379–385 (2006)

A Method for Fast-Acting Analysis of Quality of Unknown Olive Oils Using Fluorescence Spectroscopy

Vanya Slavova[1], Lilia Vladimirova-Mihaleva[2], and Mihail Mihalev[3](✉)

[1] Department of Plant Breeding, Maritsa Vegetable Crops Research Institute, 32, Brezovsko shosse Str., 4003 Plovdiv, Bulgaria
vania_plachkova@abv.bg

[2] Faculty of Physics, Sofia University "Str. Kliment Ohridski", 5 James Bourchier Blvd., 1164 Sofia, Bulgaria
vladimirova@phys.uni-sofia.bg

[3] Faculty of Mechanical Engineering, Technical University Sofia, 8 Kliment Ochridski Blvd., Sofia, Bulgaria
mmihalev@tu-sofia.bg

Abstract. Fluorescence spectroscopy is a widely used method from biology to agriculture and from the natural to the technical sciences. It is convenient and easy to be applied on samples of plant and animal origin.

Since the fluorescence signal is often very weak and is emitted in all directions, in order to avoid saturation of the detector, the useful fluorescence signal is measured in a direction 45° to the excitation radiation. For the purposes of the study four laser wavelengths - 370 nm, 395 nm, 450 nm and 425 nm, are chosen to excite the samples.

For first time several types of olive oils from three regions of Europe - Greece, Italy and Spain, are object of investigation. This study examines different types - extra virgin, virgin, olive pomace oil and olive oils (for characterization of the olive oil types COI/OT/NC no. 1 Resolution No. RES-2/91-IV/04 and COI/T.15/NC no 3–25 (revised June 2003) standards are used). Data are obtained for extra virgin olive oil, virgin olive oil, olive pomace oils and cold-pressed olive oil which originate from the model oil was chosen from different producers. Analysis of the data shows that extra virgin olive oils have a pronounced fluorescence peak of 680 nm followed by 720 nm while the olive pomace oil and cold-pressed olive oils are shifted to the left i.e., about 600 nm with symmetrically arranged additional peaks.

Fluorescence spectra for unknown olive oils were obtained. The data show that the method using fluorescence spectroscopy, developed by the authors, allows the classification of the type of studied olive oil based on its characteristic fluorescent spectrum.

Keywords: Olive oil · Fluorescence spectroscopy

S. S. Sotirov et al. (Eds.): BioInfoMed 2020, LNNS 374, pp. 354–362, 2022.
https://doi.org/10.1007/978-3-030-96638-6_37

1 Introduction

Vegetable and essential oils are a type of fatty fluid. The quality of fatty fluids used in the food and cosmetics industries is an important issue because they are often mixed with some cheap vegetable fats such as soy, sunflower, or rapeseed oil. The standard chemical methods used to make oils are slow, expensive, and need special equipment. For this reason, methods for characterizing fatty fluids by spectral methods are increasingly being improved [1, 2].

Fluorescence spectroscopy is widely used to assess the authenticity of olive oil, as well as to set up the origin of other vegetable oils [3]. Its main advantages are that no preliminary preparation of the sample is required, no chemical reagents are required. Fluorescence spectroscopy is used to detect impurities of hazelnut oil in olive oil [4]. Several methods have been known in recent years to analyze the differences between olive oils using fluorescent methods [2–5]. Attempts have been made to detect impurities of sunflower oil in model olive oil systems [6, 7]. Particularly promising is the fluorescent technique of excitation-matrix matrices, which are the subject of research in the configuration of successfully tested experimental installations, allowing to find very accurately the origin and composition of mixtures of oils as excitation sources are in the ultraviolet and visible ranges [8].

In addition to the difficulty of distinguishing one oil from another, consumers face another difficulty, such as the impurities used by producers when producing particular vegetable oil for the food industry.

The purpose of this study is to set up an algorithm by which it will be possible to easily and quickly distinguishing olive oils from each other (quality and origin) by fluorescence spectroscopy as well as to detect unwanted impurities in their composition because in the retail chains the addition of rapeseed oil or another type of oils is more and more often observed.

Known analytical techniques for authenticating vegetable oils apply chromatographic methods such as gas and high-performance liquid chromatography.

These technologies are time consuming, expensive, require and produce hazardous chemical waste. Fluorescence spectroscopy can be used as theirs alternative, as it is non-destructive, does not require chemical reagents, and is applied in real time.

In our work, fluorescence spectroscopy is applied to detect unknown quality olive oils. The model oil was chosen to be olive oil. Different types of olive oils have been studied - *extra virgin olive oil, virgin olive oil, olive pomace oil* and *cold-pressed olive oil* from three different regions in Europe.

2 Materials and Methods

2.1 Materials

Olive oils from three regions in Europe were studied - Greece, Italy and Spain. The analyzed oils types are: *extra virgin, olive oils olive pomace oils* and *cold-pressed olive oil.*

In the article, only the regions in which the olive oils are produced are pronounced, without tradenames to avoid their promotion. The obtained results are average value of five samples of fluorescent spectra of olive oils.

2.2 Methods

The most suitable laser radiation is in the UV region of the light spectrum in the range $245 \div 255$ nm, which should be with low power ~3 mW. The radiation passes through forming optics, which allows an emission angle of the source signal of 7°. The forming optics is a hemispherical lens made of N-BAK2 glass.

The radiation is led from the source to the spectrometer through the optics forming unit by means of a quartz fiber-optic light guide. The secondary radiation from the illuminated sample (visible spectrum) - illuminated by the exposed UV radiation is coupled to the detector unit by means of light guide optics. The quartz multimode fiber has a step index of refractive index and numerical aperture 0.22. The detector has a photosensitive area of 1.9968×1.9968 mm. Its sensitivity is in the range from 200 nm to 1100 nm.

The forming optics after the laser emitting unit is determined mainly due to their refractive, dispersion, and thermo-optical properties, as well as due to their transparency in the UV range $240 \div 280$ nm. To bring the radiation to the spectrometer, we consider as most appropriate to use a multimode optical fiber (with a core diameter of 125 μm). The optical part of the laser, LEDs and lenses are in accordance with the characteristics of the radiation - UV as well as the detecting device. The detecting device has a minimum sensitivity of 1.3 μW in the spectral range $200 \div 1100$ nm.

2.3 Experimental Set Up

It is preferable to use a laser diode (LD) as a source in the circuit, as its spectral width is very small. The LED's used in the experiment have a relatively wide spectral radiation width of about 30–40 nm and the angular distribution of its radiation is in a large angular range of $\pm 30°$.

As the fluorescence is often very weak and is emitted spherically in all directions, in order to saturate the detector, the useful fluorescence signal is measured in a direction that is under 45° to the excitation radiation (Fig. 1).

It was chosen to work with four laser wavelengths 370 nm, 395 nm, 450 nm and 425 nm. The LED irradiates the sample and its emission wavelength is transmitted through the optical fiber to the spectrometer. The sensitivity of the spectrometer is in the range from 200 nm to 1100 nm. Its resolution is about $\delta\lambda = 3$ nm.

The experimental setup is compact enough and can be adjusted on an area of 38 cm in length and 42 cm in width. The developed installation will be applied for the practical development of algorithms for fast-acting analysis of unknown olive oils.

Quantitative biological and chemical analyses were not performed in this work.

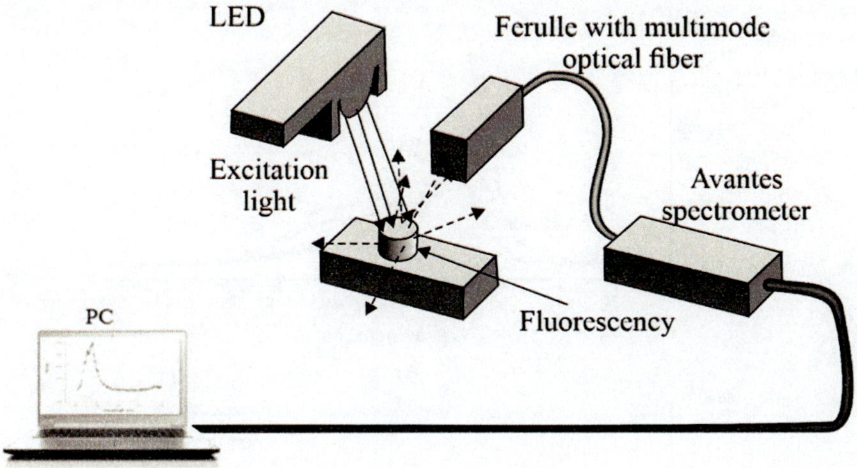

Fig. 1. Scheme of experiments

3 Results and Discussion

Figure 2 presents fluorescence spectra of *extra virgin olive oils* from three different regions of Europe - Greece, Italy and Spain. They were chosen because of the specific technologies used in the cultivation of olives and the processing of olive oil.

The experiments are performed with different excitation wavelengths, the most suitable is the spectral line at 395 nm. It can be seen that the samples of extra virgin olive oil are characterized by fluorescent peaks at 680 and 720 nm.

The difference in their intensity in our opinion is related to the environment and soil characteristics.

The spectra obtained by fluorescence spectroscopy show that virgin olive oils are the most sensitive to radiation with a wavelength of 395 nm, as Extra Virgin olive oils. Also, the fluorescence intensity is many times lower than that of Extra Virgin oils. Figure 3 shows that we have two characteristic peaks with a wavelength of 500 and 675 nm. From Fig. 2 it can be clearly seen that the position of the most intensive peak of extra virgin olive oil is independent on the geographic origin of the olive oil. In our opinion this peak can be retrieved as characteristic spectral line for Extra virgin olive oil.

In the study of *Olive pomace Oils* (Fig. 4) we observe a different spectrum compared to the spectra of *Extra Virgin* and *Virgin Olive Oils*. The spectra are also stretched with three characteristic peaks at 400, 500 and 670 nm. There is an expansion and shift to the blue-green part of the spectrum.

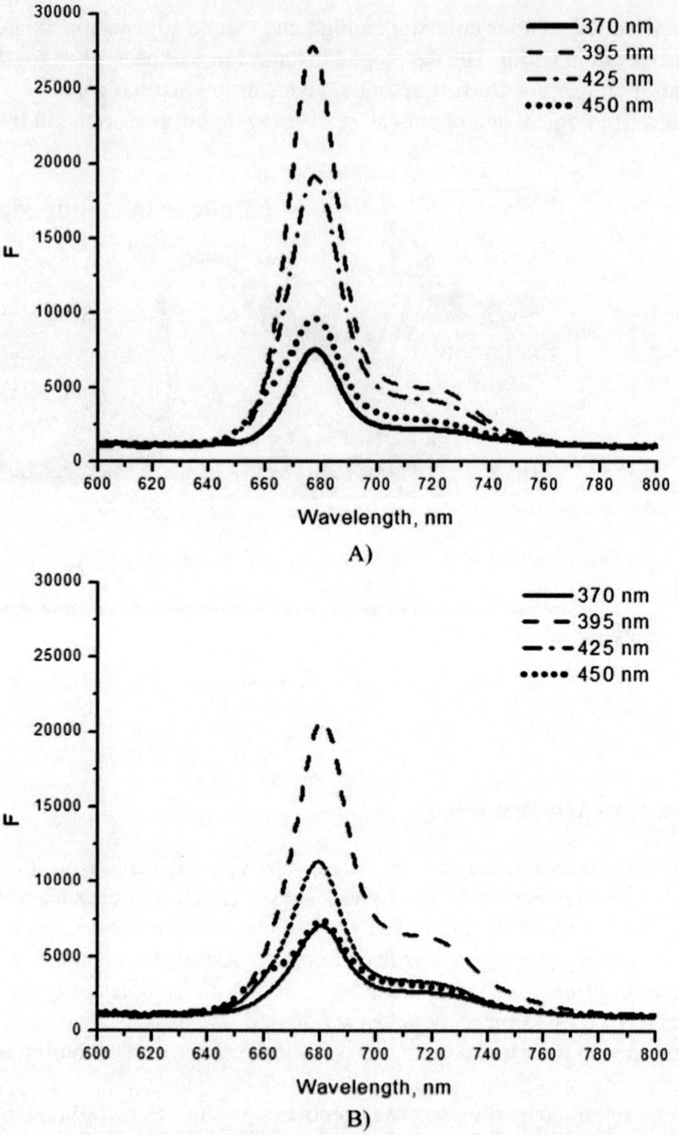

Fig. 2. Study of Extra Virgin Olive Oils from three different regions: A) Greece, B) Italy and C) Spain

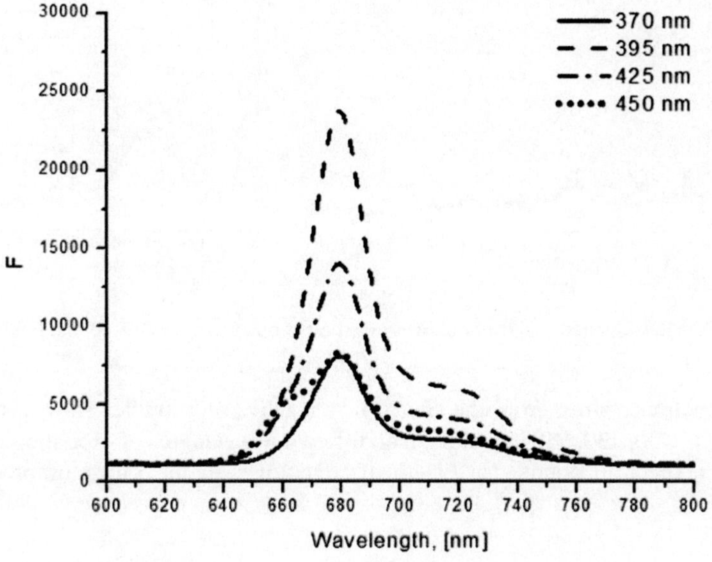

C)

Fig. 2. continued

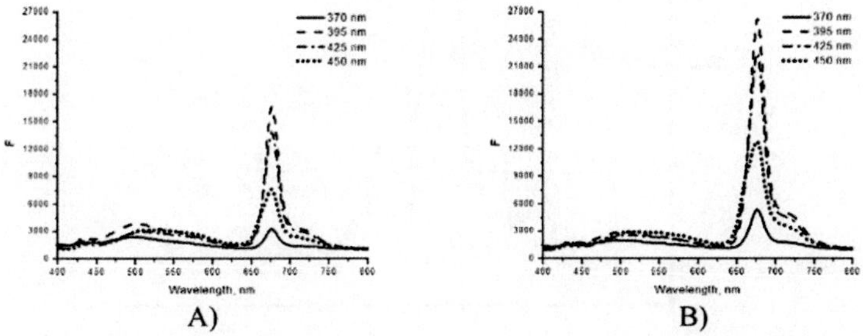

Fig. 3. Results for experiments with *Virgin Olive Oil* A) Italy and B) Spain

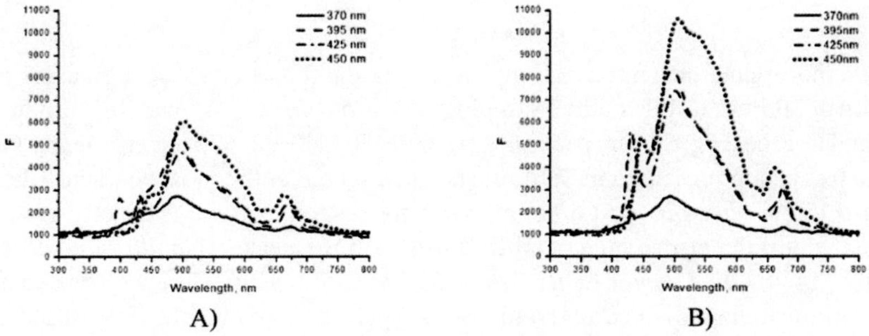

Fig. 4. Fluorescence spectra of Olive pomace oil from A) Italy and B) Spain

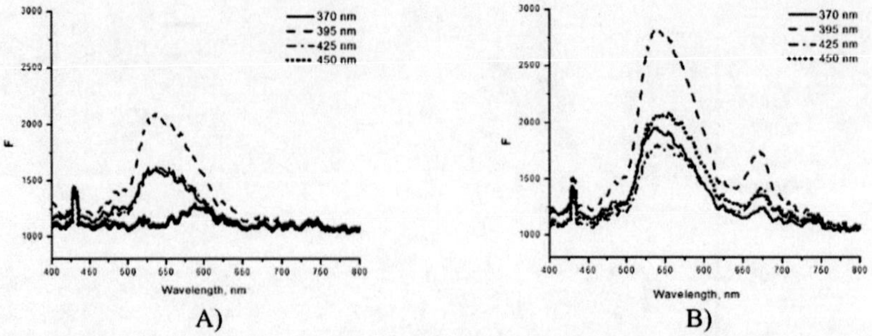

Fig. 5. Fluorescence spectra of cold-pressed olive oils from A) Greece and B) Spain

The spectra obtained from the *Cold-Pressed Olive Oils* studies show pronounced peaks at 330, 400, 430, 540 and 670 nm at different wavelengths of excitation radiation. (Fig. 5). The spectrum obtained at 395 nm of excitation radiation is the most pronounced.

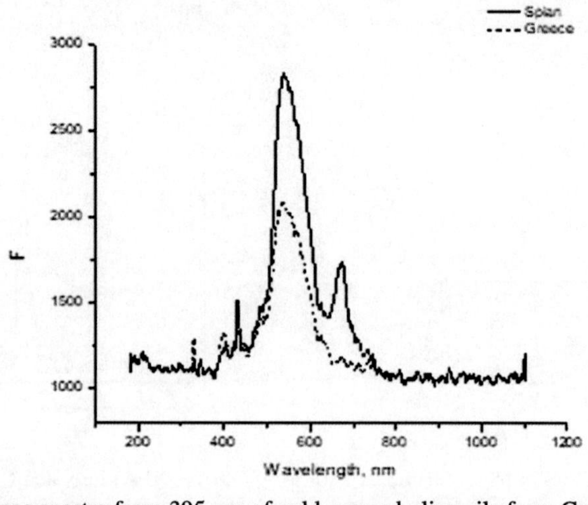

Fig. 6. Fluorescence spectra from 395 nm of cold-pressed olive oils from Greece (dashed line) and Spain (continuous line)

For a more accurate determination, we present the fluorescence spectra at excitation radiation at 395 nm of the studied samples from *Cold-Pressed Olive Oils*, produced in Spain and Greece (Fig. 6). The peaks of 330, 400 430, 540 and 670 nm are clearly visible.

The results are presented in Table 1. It can be seen that the most characteristic peak for *Extra Virgin olive oils* is at 680 nm, while for rust olive oils it is exactly at 676 nm, which is within the error range. And to distinguish the peak of *dahlia olive oil* at 500 nm can be used. *Olive Pomace Oils* also peaked at 675 $\pm$ 3 nm. Here we observe a peak at 397 nm for Italian olive oil and 440 nm for Spanish olive oil. The flow of 500 nm is observed in both olive oils. That is, for more unrefined olive oils, a distinction can be made between the region.

For Cold-pressed Olive Oils, three peaks of 429, 535 and 671 nm were observed. They are observed in the studied olive oils from Greece and Spain.

Table 1. The fluorescent response of the studied olive oil for different irradiation wave length (nm)

Country	Greece				Italy				Spain			
Type olive oils	Irradiation wave length (nm)											
	370	395	425	450	370	395	425	450	370	395	425	450
Peak (nm)												
Extra Virgin Olive Oils	677	677	677	677	680	680	679	680	679	679	679	679
Virgin Olive Oils	–	–	–	–	676	676	676	675	676	676	676	675
Olive Pomace Oils	–	–	–	–	398	398	396	397	433	441	441	447
					501	492	496	502	496	496	500	504
					675	671	671	662	676	675	674	662
Cold-pressed Olive Oils	429	429	429	429	–	–	–	–	429	429	429	429
	535	536	536	536					540	539	539	539
	671	671	671	671					671	671	671	671

4 Conclusions

This paper examines different types of olive oil from three regions of Europe - Greece, Italy and Spain. Fluorescent spectra were obtained for *Extra Virgin, Virgin, Pomace* and *Cold-Pressed Olive Oils* for excitation radiation at 370, 395, 425 and 490 nm.

The most suitable for future research is the excitation radiation with a wavelength of 395 nm. At this wavelength, the olive oils have the most accurate spectra and the peaks characterizing the different species are most pronounced.

It has been shown that, despite the different geographical location, different types of olive oil (for example: *Extra Virgin*) show similar fluorescent characteristics.

With fluorescent spectroscopy we can easily ascertain the difference between *Virgin olive oil* and *Extra Virgin olive oil*. Both have well expressed different characteristic peaks, so that *Extra Virgin olive oils* can be reliably distinguished from other olive oil types, as example.

Generally, olive oil types can be qualitatively estimated by their fluorescent characteristics. Any other peaks in the spectrum of an unknown olive oil will be an indication of presence of impurities. These impurities can be quantitatively estimated by their fluorescent characteristic peaks, as example for rapeseed and sunflower oils in [9–15].

Therefore, fluorescence spectroscopy as a fast and accurate method can be used to determine the type of unknown olive oil.

The specific algorithm of fluorescent spectroscopy which was developed is characterized by precision, fast response and easy feasibility. Fluorescence spectroscopy is used for the analysis of olive oils. It is shown that fluorescence spectroscopy is an efficient and fast method for analysis of purity and composition-native olive oil and juice, which are an integral part of a healthy lifestyle.

Acknowledgements. The researchers are grateful to the management of the Center of competence (*Clean and Circle*), project BG05M2OP001-1.002-0019 - "Clean technologies for a sustainable environment - water, waste, energy for a circular economy" for the support.

References

1. Phakawat, T., Soottawat, B.: Essential oils, extraction, bioactivities, and their uses for food preservation. J. Food Sci. **79**, 1231–1249 (2014)
2. Mignani, A., et al.: Visible and near-infrared absorption spectroscopy by an integrating sphere and optical fibers for quantifying and discriminating the adulteration of extra virgin olive oil from Tuscany. Anal. Bioanal. Chem. **399**(3), 1315–1324 (2010)
3. Lia, F., Mangion, M., Farrugia, C.: Application of elemental analysis via energy dispersive x-ray fluorescence (ED-XRF) for the authentication of maltese extra virgin olive oil. Agriculture **10**, 71 (2010)
4. Arlorio, M., et al.: Olive oil adulterated with hazelnut oils: simula-tion to identify possible risks to allergic consumers. Food Addit. Contam. Part A Chem. Anal. Control Expo Risk Assess **27**(1), 11–18 (2010)
5. Cheikhousman, R., et al.: Fluorescence spectroscopy for monitoring deterioration of extra virgin olive oil during heating. Anal. Bioanal. Chem. **382**, 1438–1443 (2005)
6. Monfreda, M., Gobbi, L., Grippa, A.: Blends of olive oil and sunflower oil: characterisation and olive oil quantification using fatty acid composition and chemometric tools. Food Chem. **134**(4), 2283–2290 (2012)
7. Negroni, M., D'Agostina, A., Arnoldi, A.: Effects of olive, canola, and sunflower oils on the formation of volatiles from the Maillard reaction of lysine with xylose and glucose. J. Agric. Food Chem. **49**(1), 439–445 (2021)
8. Boskou, B.: Olive Oil – Constituents, Quality, Health Properties and Bioconversions, 1st edn., 63–88. Intechopen (2012) https://www.intechopen.com/books/olive-oil-constituents-quality-health-properties-and-bioconversions
9. Sikorska, E., et al.: Front-face fluorescence spectroscopy and chemometrics for quality control of cold-pressed rapeseed oil during storage. Foods **8**, 665 (2019)
10. Mas, P.A., Boureveresse, D.J.-R.: Birlouez-Aragon, I: fluorescence spectroscopy for monitoring rapeseed oil upon heating. Czech J. Food Sci. **22**, 127–129 (2020)
11. Nikolova, K., et al.: Physico-chemical properties of sunflower oil enriched with ω-3 fatty acids. Bul. Chem. Commun. **46**(3), 473–478 (2014)
12. Poulli, K., Mousdisa, G., Georgiou, C.: Rapid synchronous fluorescence method for virgin olive oil adulteration assessment. Food Chem. **105**(1), 369–375 (2007)
13. Kozak, W., Samotyja, U.: The use of oxygen content determination method based on fluorescence quenching for rapeseed oil shelf-life assessment. Food Control **33**(1), 162–165 (2013)
14. Sikorska, E., Gliszczyńska-Świgło, A., Khmelinskii, I., Sikorski, M.: Synchronous fluorescence spectroscopy of edible vegetable oils. Quantification of tocopherols. J. Agric. Food Chem. **53**(18), 6988–6994 (2005)
15. Mu, T., et al.: Fluorescence polarization technique: a new method for vegetable oils classification. Anal. Methods **7**, 5175–5179 (2015)

Resveratrol Stiffens 1-palmitoyl-2-oleoyl-sn-glycero-3-phosphocholine Bilayers

Victoria Vitkova[1](✉), Rusina Hazarosova[2], Krassimira Antonova[1], Denitsa Mitkova[1], Vesela Yordanova[2], Albena Momchilova[2], and Galya Staneva[2]

[1] Georgi Nadjakov Institute of Solid State Physics, Bulgarian Academy of Sciences, 72 Tsarigradsko Chaussee, blvd., 1784 Sofia, Bulgaria
victoria@issp.bas.bg

[2] Institute of Biophysics and Biomedical Engineering, Bulgarian Academy of Sciences, Acad. G. Bonchev Str., bl. 21, 1113 Sofia, Bulgaria

Abstract. In response to exogenous stress some plant species synthesize the phytoalexin resveratrol (3, 5, 4′-trihydroxy-*trans*-stilbene). Found in the skins and seeds of red grapes, red wines, peanuts and other nutrients this polyphenolic compound has been recognized as beneficial in the prevention of oxidative damage in the human organism. The mechanism by which resveratrol exerts its pleiotropic effects is still unclear. Here we study its influence on the structure, mechanics and electrical properties of biomimetic lipid systems composed of the monounsaturated lipid 1-palmitoyl-2-oleoyl-sn-glycero-3-phosphocholine (POPC). The degree of hydration of lipid molecules in the bilayers is assessed by means of Laurdan fluorescence spectroscopy of large unilamellar vesicles. In POPC vesicles, we report enhanced lipid ordering at all concentrations of the polyphenol studied. The increased degree of lipid order in the POPC-containing matrix supports the hypothesis that resveratrol can be considered as a "filler", located parallel to the lipid molecules. Thermal shape fluctuation analysis of nearly spherical giant unilamellar vesicles is applied to probe the membrane mechanics in the presence of resveratrol. Higher bending elasticity modulus of POPC bilayers is measured at increasing the polyphenol concentration. At 200 μmol/L of resveratrol the membrane bending rigidity is reported to increase by nearly 20% compared to its value in bidistilled water. This finding is coherent with the results from fluorescence spectroscopy testifying to the denser packing of POPC molecules induced by resveratrol. The reported results provide knowledge about the phytoalexin's effect on the structural organization of membrane lipids as well as on the bilayer mechanical properties. Revealing the molecular basis of resveratrol-membrane interactions helps developing future applications of the polyphenol in medicinal chemistry and pharmacology.

Keywords: Resveratrol · Lipid bilayers · Vesicles

S. S. Sotirov et al. (Eds.): BioInfoMed 2020, LNNS 374, pp. 363–371, 2022.
https://doi.org/10.1007/978-3-030-96638-6_38

1 Introduction

The natural polyphenolic compound of plant origin, resveratrol (3, 5, 4′-trihydroxy-*trans*-stilbene), has been extensively studied [1] in view of its use as a dietary supplement and a therapeutic agent. The potential cardioprotective effect of resveratrol supplementation has been probed in recent clinical trials [2–4]. Resveratrol supplementation has been reported to be effective in improving glycemic control and possibly providing a potential adjuvant for the treatment and management of diabetes [5]. Data reported in the literature indicates that resveratrol's influence on the blood pressure is complex. While the phytoalexin significantly decreases the systolic level at higher doses, it has no significant effects on diastolic levels of blood pressure [3]. Recent results suggest that resveratrol supplementation considerably reduces total cholesterol and triacylglycerol concentrations in individuals with dyslipidemia [4]. With regard to pharmacological and biomedical applications, in order to deepen the comprehension of resveratrol beneficial effects on cardiovascular function and metabolism, profound studies of the molecular mechanisms of its action at cellular level are required.

In the present *in-vitro* study, we address the interaction of the polyphenolic compound with biological membranes by probing the effect of the phytoalexin on the structure, mechanical and electrical properties of biocompatible model systems. Modeling basic physical properties of biomembranes, lipid vesicles are studied at controlled experimental parameters (composition, viscosity, ionic strength, etc.) [6, 7]. Molecular packing, bending elasticity and electrical capacitance of one-component phospholipid bilayers are measured to quantify the effect of resveratrol on the fluidity, deformability and dielectric properties of membranes.

2 Materials and Methods

2.1 Materials

The synthetic monounsaturated phospholipid 1-palmitoyl-2-oleoyl-sn-glycero-3-phosphocholine (POPC, Avanti Polar Lipids Inc. AL, USA) is used. Resveratrol (trans-3, 5, 4′-trihydrostilbene) is purchased from Santa Cruz Biotechnology (Dallas, TX).

When measuring the electrical capacitance of membranes, in order to achieve the desired conductivity ratio of the aqueous phase inside and outside the vesicles, the conductivity of the suspending aqueous phase is adjusted by the addition of 0.1 mmol/L of sodium chloride (Sigma-Aldrich, Germany). The probe 6-dodecanoyl-N, N-dimethyl-2-naphthylamine (Laurdan, Sigma Aldrich, Germany) is used in fluorescence spectroscopy measurements.

All substances used for the preparation of vesicle suspensions are used as purchased without further purification. Aqueous solutions are prepared with water bidistilled in a quartz distiller.

2.2 Vesicle Preparation

Giant unilamellar vesicles (GUVs) are produced by electroformation [8] following previously established protocols in a chamber with two indium-tin oxide (ITO)-coated glass plates, separated by a spacer made from polydimethylsiloxane (PDMS, Dow Corning, Germany) as described in [9]. Lipid depositions were made on the ITO-side of each of the electrodes by spreading a small quantity (~25 μL) of lipid solution of 1 g/L POPC in chloroform-methanol mixture (9:1 volume parts, analytical grade). After the complete drying of the lipid under vacuum, the 4 mL-internal volume of the electroformation chamber is completely filled with bidistilled water. Alternating voltage with amplitude of 2 V peak-to-peak and frequency of 10 Hz is subsequently applied to the chamber. In several hours a wealth of unilamellar vesicles is obtained. Single, detached quasispherical vesicles without membrane defects are subjected to fluctuation analysis or electrodeformation experiments.

Large unilamellar vesicles (LUVs) are formed by the extrusion method [10]. The lipid is dissolved in chloroform/methanol (1:1 v/v) and subsequently mixed with Laurdan in the initial organic solution at a 1:200 probe:lipid molar ratio. After removing the solvent under a stream of oxygen-free dry nitrogen, the residues are kept under vacuum overnight and afterwards filtered (0.45 μm) double-distilled water is added at room temperature (22 °C) to achieve 1 mM of Laurdan-labelled lipid. Heating at 60 °C for 5 min, vortexing for 5 min are followed by sonication for 1 min, and cooling in ice for 5 min. The multilamellar vesicles obtained at this stage are then extruded with a LiposoFast small-volume extruder equipped with polycarbonate filters (Avestin, Ottawa, Canada) as follows: 11 extrusions through 800 nm, followed by 21 extrusions through 100 nm filters. The final lipid concentration in cuvette is 200 μmol/L.

All GUV and LUV samples are measured the same day.

2.3 Laurdan Fluorescence Spectroscopy

Vesicles are Laurdan labeled (1/200 Laurdan/lipid) and formed at 1 mM of POPC and different concentrations of resveratrol (50 μM (lipid:Resv = 20:1) 100 μM (10:1), 200 μM (5:1) and 500 μM (2:1)). They are prepared by evaporation of the solvents from the lipid solution up to dryness by using a stream of nitrogen; the lipid film is then left under vacuum overnight to remove all traces of the organic solvent. The dried lipid film was dispersed with 10 mM Hepes buffer (150 mM NaCl, pH7.4) and the mixture was vortexed and sonicated to yield multilamellar vesicles (MLVs). Lipid suspensions are extruded through polycarbonate filters with a diameter pore of 800 and 100 nm to form large unilamellar vesicles (LUVs) [10].

The lipid packing in LUVs in the presence of resveratrol is probed by fluorescence spectroscopy of Laurdan fluorophore [11, 12], sensitive to the polarity of its environment at the glycerol backbone of lipid molecules. The Laurdan dipole moment increases upon excitation leading to the rearrangement of the surrounding water molecules. The shift of the emission maximum can be quantified by the Generalized Polarization (*GP*) parameter as follows:

$$GP = (I_{440} - I_{490})/(I_{440} + I_{490}), \tag{1}$$

where I_{440} and I_{490} are the emission intensities at 440 and 490 nm. Theoretically the values for the *GP* function go from -1.0 (being least ordered) to $+1.0$ (being most ordered) [12]. Steady-state fluorescence measurements are performed by means of a FP-8300 spectrofluorimeter (Jasco) equipped with a Xenon arc lamp. Excitation wavelength for Laurdan is 355 nm. Heating temperature scans are performed from 27° to 57 °C. The emission spectra are recorded from 390 to 600 nm in triplicate and the calculated *GP* values are averaged.

2.4 Thermal Shape Fluctuation Analysis

According to the theoretical model describing the mechanical properties of lipid bilayers [13, 14] the elastic energy per unit area, F_c, is given by the expression:

$$F_c = k_c(c_1 + c_2 - c_0)^2/2 + k_G c_1 c_2 \tag{2}$$

where k_c represents the measured quantity, namely the bending modulus, c_1 and c_2 denote the principle curvatures, c_0 is the spontaneous curvature and k_G stands for the Gaussian curvature modulus of the bilayer. The contribution of the latter in the case of GUVs' closed membranes vanishes. Thermal shape fluctuation analysis (TSFA) of giant lipid vesicles is applied to assess the bending modulus, k_c, of lipid bilayers in the presence of up to 200 μM of resveratrol.

In the present study the visualization of the membrane fluctuations is carried out by a phase-contrast inverted microscope (Axiovert 100, Zeiss, Germany) equipped with a two-channel thermocontrol (±0.1 °C) system operating both with the microscope plate and the oil-immersed objective (100×, numerical aperture 1.25). The image-recording is performed by means of a CCD camera (C3582, Hamamatsu Photonics, Japan) and a frame grabber (DT3155, Data Translation, USA, 768 × 576 8-bit pixels, pixel size: 0.106 μm/pix) mounted in a PCI-slot of a Linux-operating station for digitization and further analysis [15–18]. The stroboscopic illumination of the observed vesicles, synchronized with the camera, permitted the fast modes of fluctuations to be recorded and processed [15]. All measurements of the bending elasticity are carried out at (25 ± 0.1)°C. Only visibly fluctuating quasispherical vesicles larger than 10 μm in diameter are chosen for TSFA. Several hundred images are captured once per second and subjected to TSFA for calculation of the membrane bending constant and tension of every vesicle studied [16–18]. The value of the bending modulus for every resveratrol concentration reported is obtained over an ensemble of not less than ten GUVs.

2.5 Electrodeformation of GUVs

The specific electrical capacitance, C_m, of POPC membranes at 50 μM of resveratrol is evaluated from the frequency-dependent deformation of GUVs in alternating electric field [19]. The chamber for electrodeformation measurements consists of two parallel glass slides, separated by a 0.5 mm-thick inert spacer (Sigma-Aldrich Inc., USA). AC electric field from an arbitrary waveform generator (33120 A, HP/Agilent, CA, USA) is applied to a pair of two rectangular parallel ITO-electrodes deposited on the lower inner surface of the chamber at 1 mm apart. The measurement is performed by varying the frequency of the imposed uniform field in the range of 0.2–3 kHz, the applied root mean square voltage not exceeding 7 V, thus providing in our experiment AC fields below the critical values for membrane electroporation [20]. The vesicle electrodeformation is observed and recorded using an inverted phase-contrast microscope (IM3, Optika, Italy) equipped with a dry objective (×40, 0.65 numerical aperture) and Axiocam ERc 5s camera 5 MP (Zeiss, Germany) connected to a computer for the digitization of the recorded signal with 0.055 μm/pixel.

The ratio between the conductivity of the bulk solution enclosed by the vesicle membrane and the conductivity of the suspending medium is $\Lambda = 0.82$, corresponding to a more conductive suspending solution. Data analysis is performed following the original approach of Salipante et al. [21] as described in [22].

3 Results and Discussion

In order to investigate the effect of resveratrol on the structure of phospholipid bilayers, the packing of lipid molecules is probed via fluorescence spectroscopy of the fluorophore Laurdan. Its generalized polarization (see Eq. 1) is measured as a function of the temperature in pure monounsaturated liposomes as shown in Fig. 1. Heating the sample leads to the gradual decrease of Laurdan *GP* in the temperature range from 27° to 57 °C studied. In the whole temperature range, the lipid vesicles exhibited liquid-disordered (L_d) phase condition defined from the negative *GP* values.

Laurdan *GP* values of POPC bilayers grow with increasing the content of resveratrol. The addition of 50 μM of resveratrol to POPC vesicles does not lead to any changes in the *GP* values (Fig. 1). It can be translated as POPC molecular ordering at the lipid-water interface where Laurdan molecule is located, unaltered by the presence of resveratrol. Laurdan *GP* values of POPC bilayers grow with increasing the phytoalexin content from 100 to 500 μM, which means that the polyphenol increases the stiffness of the bilayer. The polyphenol has the most pronounced hardening effect at 500 μM. Resveratrol shows high affinity to POPC vesicles, probably due to the great membrane fluidity of this lipid, promoting the interaction and penetration of the fluorescent label deeply in the bilayer. Hence, the hypothesis arise that the phytoalexin assumes vertical position in the nonpolar region of the membrane, thus playing a role of "filler", ordering and protecting the lipid bilayer.

Our results are in agreement to DPH and Laurdan fluorescence spectroscopy data by Han et al. who reported evidences for the protective effect of resveratrol on DOPC vesicles treated with hydroxyl radicals [23]. DOPC unsaturated lipid membranes are characterized by high membrane fluidity, which means that reactive oxygen species (ROS) would easily permeate the membrane. The preliminary treatment by resveratrol however, would make DOPC vesicles less prone to ROS action by rendering the membrane more ordered and dehydrated.

Since the polyphenol compound resveratrol has been shown to modulate the membrane biophysical properties it might be suggested that one of the roles of polyphenols is to tune the lipid order, fluidity and organization in the membranes.

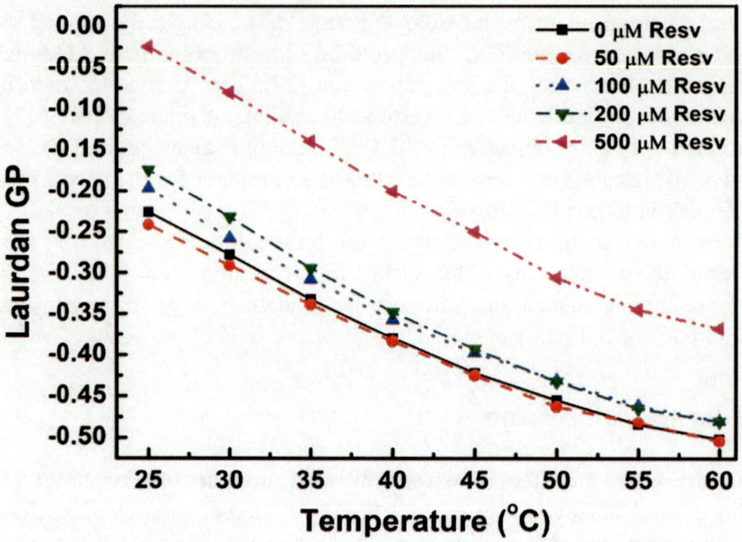

Fig. 1. Laurdan generalized polarization (*GP*) as a function of temperature in an absent and a presence of 50, 100, 200 and 500 μM resveratrol-containing POPC vesicles. *GP* values are proportional to the lipid order in the membrane. The temperature increase causes a decrease of *GP* in treated and untreated liposomes. The addition of resveratrol up to 500 μM leads to *GP* increase in POPC membranes, i.e. resveratrol exerts ordering effect on POPC vesicles.

The specific electrical capacitance of POPC bilayers is measured to decrease twofold in the presence of 50 μmol/L of resveratrol (see Table 1).

Table 1. Specific capacitance C_m of POPC membranes obtained from GUV electrodeformation in the presence of 50 μM of resveratrol.

Resveratrol content, μmol/L	Specific capacitance, C_m, μF/cm^2
0	0.52 ± 0.03
50	0.26 ± 0.02

The lower specific electrical capacitance of resveratrol-containing membranes compared to pure POPC matrix, accompanied by rigidification of the bilayer measured, suggests membrane thickening effect of the polyphenol interacting with the hydrophobic core of the bilayer.

The increased degree of lipid order in POPC bilayers measured by Laurdan fluorescence spectroscopy corresponds to the results from the bending elasticity measurements, providing evidences about the rigidifying effect of the polyphenol (see Fig. 2). Resveratrol-treated POPC bilayers are characterized by higher bending constant compared to the control pure POPC membrane. Higher bending modulus of resveratrol-treated membranes: nearly 20%-increase at 200 μmol/L of resveratrol compared to its

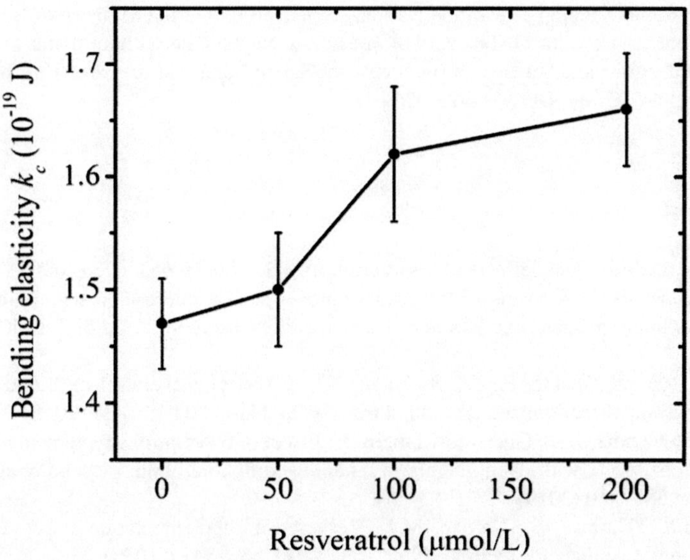

Fig. 2. Bending constant k_C of POPC membranes in the presence of up to 200 μmol/L of resveratrol at pH 5.6 and (25 ± 0.1)°C. The bending rigidity increases with increasing resveratrol total concentration.

value in bidistilled water. Resveratrol conformation parallel to the lipid molecules in the bilayer is thus hypothesized.

4 Conclusion

The effect of resveratrol (trans-3, 5, 4′-trihydrostilbene) on the mechanics, structure and dielectric properties of phospholipid bilayers is probed.

The obtained results show that the polyphenol alters the membrane mechanical properties in a way that the bending elasticity of POPC bilayers increases with resveratrol concentration. Nearly 20%-growth of the bending modulus is reported at 200 μmol/L of the phytoalexin compared to its value in bidistilled water. The higher bending modulus combined with the increased lipid packing of resveratrol-treated membranes measured by Laurdan fluorescence spectroscopy suggests resveratrol conformation parallel to the lipid molecules in the bilayer. The lower specific electrical capacitance of resveratrol-containing membranes compared to pure POPC matrix, accompanied by rigidification of the bilayer measured, testifies to the membrane thickening effect of resveratrol interacting with the hydrophobic core of the bilayer.

The knowledge about the effect of the phytoalexin on the structural organization of membrane lipids as well as on the bilayer mechanical and electrical properties contributes to reveal the molecular basis of resveratrol-membrane interactions fostering the development of future applications of the polyphenol in medicinal chemistry and pharmacology.

Acknowledgments. Funding from the National Science Fund at the Ministry of Education and Science of Bulgaria (Grant DN08-7/2016) is acknowledged. The authors thank the Bulgarian Ministry of Education and Science for support by the Scientific Infrastructure on Cell Technologies in Biomedicine (SICTB) (DO1-154/28/08/2018).

References

1. Frémont, L.: Biological effects of resveratrol. Life Sci. **66**(8), 663–673 (2000)
2. Agarwal, B., et al.: Resveratrol for primary prevention of atherosclerosis: clinical trial evidence for improved gene expression in vascular endothelium. Int. J. Cardiol. **166**(1), 246–248 (2013)
3. Liu, Y., Ma, W., Zhang, P., He, S., Huang, D.: Effect of resveratrol on blood pressure: a meta-analysis of randomized controlled trials. Clin. Nutr. **34**(1), 27–34 (2015)
4. Simental-Mendía, L.E., Guerrero-Romero, F.: Effect of resveratrol supplementation on lipid profile in subjects with dyslipidemia: a randomized double-blind, placebo-controlled trial. Nutrition **58**, 7–10 (2019)
5. Bhatt, J.K., Thomas, S., Nanjan, M.J.: Resveratrol supplementation improves glycemic control in type 2 diabetes mellitus. Nutr. Res. **32**(7), 537–541 (2012)
6. Luisi, P.L., Walde, P.: Giant Vesicles. Wiley, Chichester (2000)
7. Dimova, R.: Giant vesicles: a biomimetic tool for membrane characterization. In: Iglič, A. (ed.) Advances in Planar Lipid Bilayers and Liposomes. Elsevier, Burlington (2012)
8. Angelova, M.I., Soleau, S., Meleard, P., Faucon, J.-F., Bothorel, P.: Preparation of giant vesicles by external AC fields. Kinetics and application. Prog. Colloid. Polym. Sci. **89**, 127–131 (1992)
9. Vitkova, V., Antonova, K., Popkirov, G., Mitov, M.D., Ermakov, Y.A., Bivas, I.: Electrical resistivity of the liquid phase of vesicular suspensions prepared by different methods. J. Phys.: Conf. Ser. **253**, 012059 (2010)
10. Vitkova, V., et al.: Elasticity and phase behaviour of biomimetic membrane systems containing archaeal lipids. Colloids Surf. A Physicochem. Eng. Asp. **601**, 124974 (2020)
11. Parasassi, T., Gratton, E.: Membrane lipid domains and dynamics as detected by Laurdan fluorescence. J. Fluoresc. **5**(1), 59–69 (1995)
12. Parasassi, T., De Stasio, G., D'Ubaldo, A., Gratton, E.: Phase fluctuation in phospholipid membranes revealed by Laurdan fluorescence. Biophys. J. **57**(6), 1179–1186 (1990)
13. Helfrich, W.: Elastic properties of lipid bilayers: theory and possible experiments. Z. Naturforsch. **28c**, 693–703 (1973)
14. Evans, E.A.: A new material concept for the red cell membrane. Biophys. J. **13**(9), 926–940 (1973)
15. Genova, J., Vitkova, V., Aladgem, L., Meleard, P., Mitov, M.D.: Using stroboscopic illumination to improve the precision of the bending modulus measured by the analysis of thermally induced shape fluctuations of giant vesicles. Bulg. J. Phys. **31**, 68–75 (2004)
16. Mitov, M.D., Faucon, J.F., Méléard, P., Bothorel, P.: Thermal fluctuations of membranes. In: Gokel, G.W. (ed.) Advances in Supramolecular Chemistry, pp. 93–139. JAI Press Inc., Greenwich (1992)
17. Vitkova, V., Misbah, C.: Dynamics of lipid vesicles – from thermal fluctuations to rheology. In: Iglic, A. (ed.) Advances in Planar Lipid Bilayers and Liposomes, pp. 258–292. Academic Press, Burlington (2011)
18. Genova, J., Vitkova, V., Bivas, I.: Registration and analysis of the shape fluctuations of nearly spherical lipid vesicles. Phys. Rev. E **88**, 022707 (2013)

19. Vlahovska, P.M., Gracia, R.S., Aranda-Espinoza, S., Dimova, R.: Electrohydrodynamic model of vesicle deformation in alternating electric fields. Biophys. J. **96**(12), 4789–4803 (2009)
20. Needham, D., Hochmuth, R.M.: Electro-mechanical permeabilization of lipid vesicles. Role of membrane tension and compressibility. Biophys. J. **55**(5), 1001–1009 (1989)
21. Salipante, P.F., Knorr, R.L., Dimova, R., Vlahovska, P.M.: Electrodeformation method for measuring the capacitance of bilayer membranes. Soft Matter **8**, 3810–3816 (2012)
22. Vitkova, V., Mitkova, D., Antonova, K., Popkirov, G., Dimova, R.: Sucrose solutions alter the electric capacitance and dielectric permittivity of lipid bilayers. Colloids Surf. A Physicochem. Eng. Asp. **557**, 51–57 (2018)
23. Han, J., Suga, K., Hayashi, K., Okamoto, Y., Umakoshi, H.: Multi-level characterization of the membrane properties of resveratrol-incorporated liposomes. J. Phys. Chem. B **121**(16), 4091–4098 (2017)

Biomedical Imaging and Image Processing

Gabor Features for the Classification and Evaluation of Chromogenic In-Situ Hybridization Images

Stoyan Pavlov[1,3](✉), Galina Momcheva[2,3], Pavlina Burlakova[2], Simeon Atanasov[2], Dimo Stoyanov[1,3], Martin Ivanov[1,3], and Anton Tonchev[1]

[1] Department of Anatomy and Cell Biology, Faculty of Medicine, Medical University "Prof. Dr Paraskev Stoyanov", Prof. Marin Drinov Str. 55, 9000 Varna, Bulgaria

[2] Department of Computer Science, Varna Free University "Chernorizets Hrabar", Yanko Slavchev str. 84, Chayka resort, 9007 Varna, Bulgaria

[3] Research Group for Advanced Computational Bioimaging, Research Institute of Medical University "Prof. Dr Paraskev Stoyanov", Prof. Marin Drinov Str. 55, 9000 Varna, Bulgaria

Abstract. High-throughput chromogenic in-situ hybridization (CISH) is a bright-field microscopic technique that reveals the spatial distribution of gene expression in animal cells and tissues by an easily detectable coloured precipitate. The "golden standard" for the grading of CISH-stained tissues involves qualitative scoring by a domain expert. This method is biased, suffers from low reproducibility, and lowers the efficiency of high-throughput experiments. A few quantitative image analysis approaches resolve these issues, but the proposed methods are sensitive to experimental conditions or require expert adjustment of multiple parameters. The idea of our research team is to extract textural information from CISH-images that will be used to generate a feature space for semantic segmentation and functional analysis of gene expression. In our current work, we explore the idea by unsupervised classification based on features generated via Gabor energy filters. The tissue was divided into overlapping 150 μm tiles and processed with a Gabor filter bank (5 wavelengths, 16 directions, bandwidth 1.4). The results for the 16 directions at each wavelength were combined by a maximum superposition into a single image, and the mean grey value, standard deviation and entropy were measured. After appropriate dimensionality reduction, the tiles were classified by a fuzzy C-means algorithm. Four experts without prior knowledge of the classification results evaluated the strength and pattern of gene expression of a set of randomly selected tiles, and independently each class in the original whole-slide images. A comparison between the class-scale and tile-scale evaluations was used to assess the usefulness of the selected features.

Keywords: Chromogenic in-situ hybridization · Texture analysis · Machine learning · Feature extraction

S. S. Sotirov et al. (Eds.): BioInfoMed 2020, LNNS 374, pp. 375–383, 2022.
https://doi.org/10.1007/978-3-030-96638-6_39

1 Introduction

The high-throughput chromogenic in-situ hybridization (CISH) is a method to detect known mRNA sequences in the cells of a specific tissue or organ by complementary hybridization with nucleotide probes tagged with enzyme-linked haptens [1]. The hybridization is revealed by an easily detectable coloured precipitate that can be documented using a standard bright-field microscopic imaging system. CISH is primarily used to localize the specific mRNA fragments of the sought gene product and the cell that produces it in fixed tissues [2, 3]. The positive signals correspond to cells that actively produce (aka express) the product of a gene under investigation.

This paper presents a proof of concept for the usability of spatial features extracted through Gabor filtering for the segmentation or the analysis of CISH whole slide images in high-throughput imaging experiments. The paper is organized as follows: In Sect. 2, we explain the rationale and motifs behind the proposed approach. Section 3 discusses the selected texture features. In Sect. 4, we outline the experiment and its results. Section 5 discusses the evaluation by domain experts. In Sect. 6, we conclude with some remarks on further developments.

2 Rationale

2.1 CISH Evaluation

Currently, the gold standard for gene expression grading in CISH stained tissue slides is the assessment by an expert. This approach involves visual inspection of the cell expression and manual labelling (positive or negative) or grading depending on the amount of the observed precipitate. The usual visual scoring is based on cellular gene expression strength (“negative”, “low”, “moderate”, or “strong”), and patterns of expression (“ubiquitous”, “regional” or “scattered”) [4, 5]. This method is highly biased and suffers from low reproducibility as it strongly depends on the conditions (Fig. 1) and the expertise of the annotator. Furthermore, it is slow and is not particularly effective in high-throughput experiments, when large amounts of image data are generated at high speeds. There are several attempts to design automated workflows for an unbiased evaluation of gene expression. Celldetect is an open-source algorithm for automatic localization and grading of cellular gene expression in whole-slide CISH images [6]. The algorithm performs intensity-based thresholding, classification, and labelling at a reduced spatial resolution (approximately a single cell per pixel). While this approach significantly reduces bias and improves reproducibility, it still suffers from the high variability between images and staining batches inherent to the method. The good reproducibility of Celldetect relies heavily on the human operator to select the proper parameters to account for the intensity variability between images and batches. Another reliable quantitative workflow is developed by Allen Brain Institute and is implemented in the annotation of their Mouse Brain ISH Atlas [7, 8]. The workflow evaluates gene expression in high-resolution images with normalized brightness based on the grey value of the corresponding pixels in combination with advanced filtering to account for different patterns of expression [9, 10]. Both approaches are very efficient but are sensitive to brightness fluctuations, and noise

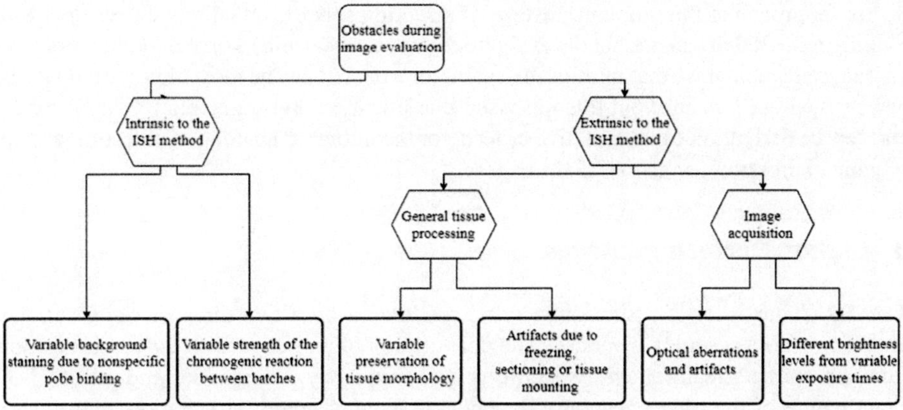

Fig. 1. Most common sources of variability in CISH-stained microscopic images.

introduced by the ISH procedure or the image acquisition and require stringent control of experimental conditions.

A robust and automated workflow can facilitate reproducibility between experiments and between labs and can ensure the acquisition of comparable data from imaging CISH experiments.

During the manual evaluation, experienced annotators can recognize similar levels of gene expression despite significant differences in the overall brightness between images. This process relies on an implicit evaluation of brightness and colour distribution within the locality of the observed tissue. The most prominent property of an image that is related to the local fluctuations of intensity and colour is the texture. Recent research showed that second-order textural features could be used successfully to classify and localize gene expression patterns to specific cerebellar cortical layers [11] or to identify mRNA-enriched sites in the hippocampal region of the brain [12].

2.2 Search for Spatial Features

The texture is a repeating pattern or a function of spatial variation of the brightness intensity of the pixels. Texture analysis plays an essential role in face recognition, surface defect detection, pattern recognition, and medical image analysis by the extraction of meaningful information from digital images. Texture feature extraction refers to the process of computing characteristics of an image, which numerically describes textural image properties and may include scalar numbers, discrete histograms, or empirical distributions [13, 14].

The spatial distribution of brightness variations in this type of images is highly variable and inhomogeneous. That is why a particular standard method for feature extraction can not satisfy all requirements of the analysis of CISH-stained images. The idea of our research team is to use different approaches in the analysis of CISH images that will be used for structural segmentation and functional analysis of gene expression.

As the level of expression is very difficult to quantify in CISH images due to its nonlinear relationship with the amount of precipitated dye, we decided to apply an inverse approach to the problem. Instead of selecting specific quantitative measures and looking for a fit that can predict the evaluator's grades, we decided first to find features for an image segmentation that mimics the evaluator's decision. The most important features can be then used as an input to supervised classification and segmentation algorithms, and for the derivation of quantitative criteria for the unbiased evaluation and comparison of gene expression levels in CISH images.

3 Gabor Derived Features

The 2D Gabor filter is a sinusoidal wave modulated by a two-dimensional Gaussian envelope with a complex response. The convolution of an image with the spatial Gabor function extracts features with specific spatial frequency, orientation, and phase. The exact nature of the filter's response depends on its wavelength, orientation, phase, and the spread of the Gaussian envelope in the x- and y-direction [15].

The Gabor filter returns an image of the details with the specified spatial frequency and highest response in the direction of the filter. The scale of the filter (i.e., the size of the emphasized details) depends on its wavelength and bandwidth. The bandwidth, in turn, determines the ratio between the wavelength and the standard deviation of the Gaussian envelope. As the spatial selectivity of simple cells in the primate visual cortex has a median bandwidth of 1.4 (which corresponds to a ratio between wavelength and sigma approximately 0.42) [16, 17], we chose this value for the filter. We extracted the local features from Gabor Energy (root sum square of the real and imaginary output). The filter bank includes five wavelengths (2, 4, 8, 16, and 32 px) oriented in 16 directions each. The outputs from all directions at a given scale were combined in one omnidirectional image using maximum superposition. The final output contains all maximal responses for each direction of the filter at the given scale. The features for classification were measured in these multidirectional outputs of the Gabor energy filter. We used three measurements – Mean grey value, Standard Deviation, and Shannon entropy, which results in 15 features in total.

4 Implementation and Results

4.1 Image Filtering and Feature Extraction

The described features were measured in overlapping neighbourhoods with size 150 × 150 μm from the original whole slide images. First, the tissue portion of each image was masked, and only the pixels with coordinates within the tissue mask were included in the scanning process. The images were scanned at full scale (0.5 μm/px) with square neighbourhoods with size 150 μm and horizontal and vertical steps 100 μm. The overlap ensured that objects located at the border between regions would be captured adequately by the algorithm. Each tile was processed separately with the described Gabor filter bank, and fifteen features were extracted. The process was programmed in Python Programming Language (Python Software Foundation, https://www.python.org/) and libraries in the SciPy ecosystem [18–21]. For image navigation, we used the OpenSlide library for visualization of microscopic whole slide images [22].

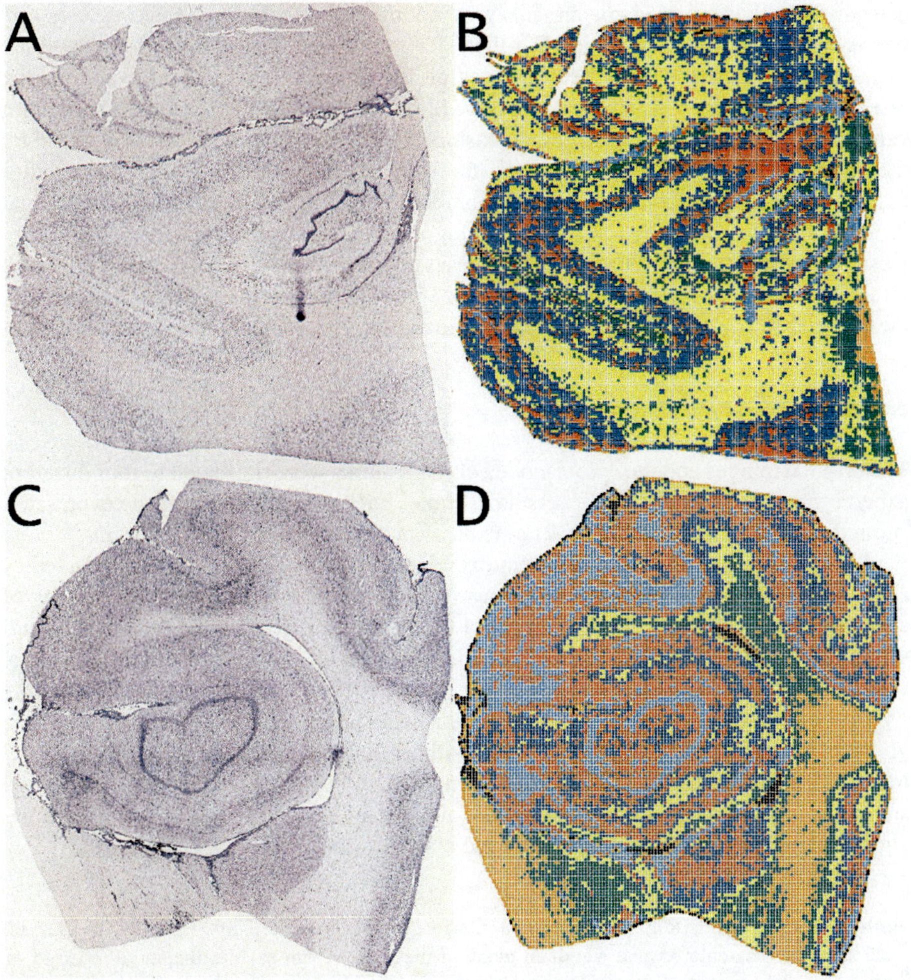

Fig. 2. Result of the Fuzzy C-Means clustering. Original whole slide images (A, C) stained with the CISH-technique for two different gene products and colour-coded spatial maps (B, E) of the clustering based on Gabor energy filter derived features.

4.2 Fuzzy C-means Clustering

To reduce the sources of measurement noise, improve accuracy in clustering, and reduce the computational cost, the dimensionality of the feature matrix was reduced by Principal Component Analysis (PCA) [23]. As the annotator is evaluating the CISH images mainly on two properties - strength and pattern (density) of expression, it seems appropriate to choose the first two principal components for further analyses. Additionally, a higher number of dimensions influences negatively the fuzzy C means algorithm (FCM), which we chose for the unsupervised clustering [24]. Our choice of the fuzzy counterpart of the more popular k-means algorithm was driven mainly by the notion that groups and categories in living organisms are fuzzy by nature and an approach that captures this fuzziness will be more beneficial. For the selection of a cluster number, we looped the FCM between two and ten cluster centres on the two components and evaluated the results with the fuzzy partition coefficient [25]. While the FPC reaches a maximum value at two clusters, we chose seven clusters. This decision is once again dictated by the nature of the processed image data: the standard protocol for evaluation includes at least six separate groupings, and this number of classes allows us to capture gradual differences in gene expression while keeping the FPC at acceptable values. The fuzzy clustering of the reduced data produced classes that, on inspection, separated the images into regions with different expression levels (Fig. 2). Even for the unprepared observer, the contrast distribution in the unprocessed images corresponds to the cluster mapping.

5 Expert Evaluation

The original images and the colour-coded classifications were presented to four different experts for evaluation of the clusters as a whole. Each evaluator graded the expression strength (0–3) and the density (0–2) of expressing cells in each class (Table 1).

Since the evaluators had a different experience, the final grades were calculated as a weighted mean of the individual grades rounded to an integer score. The grades of evaluators one and two received weight 1.0, evaluator 3 grades had weight 2.0, and the most experienced grader received weight 3. Finally, we compared the tile-scale evaluations to the class-scale evaluations.

In the next stage, a set of 137 randomly selected tiles were evaluated by the same group of experts using the same criteria without knowledge about how the Fuzzy C Means algorithm classified the corresponding tile. In this way, the evaluators assigned strength and density values to selected tiles at the same scale (150 μm x 150 μm) as the algorithm performed its measurements and classification.

Table 1. Expert evaluation of the expression strength (**Str**) and density (**Den**) of the classes. Final scores are calculated as a weighted mean of the scores from each evaluator rounded to an integer (weights are given in parentheses after the evaluator's designation). For comparison with the tile-scale evaluations, we provide the median tile scores per class of 137 randomly selected tiles graded similarly.

Class	Evaluator 1 (weight = 1)		Evaluator 2 (weight = 1)		Evaluator 3 (weight = 2)		Evaluator 4 (weight = 3)		Weighted grade		Median tile score	
	Str	Den	Str	Den	Str	Den	Str	Den	Str	Den	Str	Den
Class 0	0.0	0.0	0.0	0.0	0.0	0.0	0.0	0.0	**0**	**0**	**0**	**0**
Class 1	0.0	0.0	1.0	1.0	0.0	1.0	1.0	1.0	**1**	**1**	**1**	**1**

(continued)

Table 1. *(continued)*

Class	Evaluator 1 (weight = 1)		Evaluator 2 (weight = 1)		Evaluator 3 (weight = 2)		Evaluator 4 (weight = 3)		Weighted grade		Median tile score	
	Str	Den	Str	Den	Str	Den	Str	Den	Str	Den	Str	Den
Class 2	3.0	2.0	3.0	2.0	3.0	2.0	3.0	2.0	**3**	**2**	**3**	**2**
Class 3	1.0	1.0	3.0	1.0	1.0	1.0	1.0	1.0	**1**	**1**	**2**	**1**
Class 4	1.0	1.0	1.0	1.0	2.0	1.0	2.0	2.0	**2**	**1**	**2**	**1**
Class 5	1.0	1.0	3.0	2.0	2.0	2.0	2.0	2.0	**2**	**2**	**2**	**1**
Class 6	3.0	2.0	3.0	2.0	3.0	1.0	2.0	2.0	**3**	**2**	**3**	**2**

As we assumed by design that the clusters defined by the described algorithm must include tiles of similar expression level and density, we analyzed the agreement between the tile-scale scores and the corresponding class scores in a confusion matrix for the 137 randomly selected tiles (Table 2). While the median of the individual tile scores per class corresponds very well to the class grades (Table 1), the overall accuracy of the prediction of the tile's score from its class score is low. One explanation for this is that the evaluators looking at the individual tiles are assessing the expression strength and density with details at a smaller scale.

Table 2. Agreement between expert evaluations of the individual tiles and their corresponding classes assigned by the algorithm. The class grades were used as the conditional "ground truth".

Property	Levels	Precision	Recall	F1-score	N	Accuracy	Cohen's kappa
Expression strength	No expression	0.92	0.44	0.6	27	–	–
	Low	0.38	0.29	0.33	35	–	–
	Moderate	0.33	0.58	0.42	38	–	–
	High	0.65	0.54	0.59	37	–	–
	Overall	0.55	0.47	0.48	137	0.47	0.28
Density of expressing cells	No expression	1	0.45	0.62	29	–	–
	Sparse	0.58	0.55	0.56	64	–	–
	Dense	0.47	0.68	0.56	44	–	–
	Overall	0.68	0.56	0.58	137	0.57	0.31

Additionally, there are lots of "misclassifications" between neighbouring levels of the properties that reduce the calculated accuracy. Another reason is the presence of artefactual staining in some of the randomly selected tiles. Those tiles were accurately evaluated as nonexpressing by the human raters. The clustering algorithm, however, classifies the tiles based on the distribution of intensities in the image and does not discriminate between artefacts and valid staining.

6 Conclusion

We presented the use of a set of features derived from Gabor filtered images of CISH-stained tissues. The selected features allowed the clustering of the image regions into classes with generally good correspondence to the different expression patterns as described by expert evaluators in CISH images. The accuracy of prediction of the tile-scale evaluation by the class-scale grade was low, but it must be noted that this approach is not intended to (and does not) measure the ability of the algorithm to correctly discriminate between the different levels of expression strength and density of the expressing cells. The low accuracy stems mainly from the different scales at which the evaluators asses the randomly selected tiles without knowledge of their neighborhood and the classes in the whole image. As the median scores of the tiles in each class were in agreement with the overall class score assigned by the human raters, we may conclude that the proposed features have discriminating power similar to the human evaluation. Next step should be to analyze and select the most important of these features and combine them with other approaches into a feature space which can be used for efficient training of supervised segmentation algorithms and derivation of unbiased indicators for the quantitative comparison of gene expression patterns in CISH-stained tissues. An essential issue in the analysis of CISH-images is the distinction and removal of nonspecific staining. This problem can be addressed by the implementation of a similar carefully constructed feature space after differential analysis of positive and negative control images.

References

1. Corthell, J.T.: In situ hybridization. In: Basic Molecular Protocols in Neuroscience: Tips, Tricks, and Pitfalls, pp. 105–111. Elsevier (2014)
2. Jensen, E.: Technical review: *in situ* hybridization: AR insights. Anat. Rec. **297**, 1349–1353 (2014). https://doi.org/10.1002/ar.22944
3. McFadden, G.I.: In situ hybridization, chapter 12. In: Methods in Cell Biology, pp. 165–183. Elsevier (1995)
4. Eichele, G., Diez-Roux, G.: High-throughput analysis of gene expression on tissue sections by in situ hybridization. Methods **53**, 417–423 (2011). https://doi.org/10.1016/j.ymeth.2010.12.020
5. Visel, A., Thaller, C., Eichele, G.: GenePaint.org: an atlas of gene expression patterns in the mouse embryo. Nucleic Acids Res. **32**, D552–D556 (2004). https://doi.org/10.1093/nar/gkh029
6. Carson, J.P., Eichele, G., Chiu, W.: A method for automated detection of gene expression required for the establishment of a digital transcriptome-wide gene expression atlas. J. Microsc. **217**, 275–281 (2005). https://doi.org/10.1111/j.1365-2818.2005.01450.x

7. ISH Data: Allen Brain Atlas: Mouse Brain. https://mouse.brain-map.org/. Accessed 7 July 2020
8. Ng, L., Pathak, S.D., Kuan, C., et al.: Neuroinformatics for genome-wide 3D gene expression mapping in the mouse brain. IEEE/ACM Trans. Comput. Biol. Bioinform. **4**, 382–393 (2007). https://doi.org/10.1109/tcbb.2007.1035
9. Informatics Data Processing in: Documentation - Allen Mouse Brain Atlas. http://help.brain-map.org/download/attachments/2818169/InformaticsDataProcessing.pdf?version=1&modificationDate=1319667590884&api=v2. Accessed 7 July 2020
10. Lein, E.S., Hawrylycz, M.J., Ao, N., et al.: Genome-wide atlas of gene expression in the adult mouse brain. Nature **445**, 168–176 (2007). https://doi.org/10.1038/nature05453
11. Kirsch, L., Liscovitch, N., Chechik, G.: Localizing genes to cerebellar layers by classifying ISH images. PLoS Comput. Biol. **8**, e1002790 (2012). https://doi.org/10.1371/journal.pcbi.1002790
12. Ugolotti, R., Mesejo, P., Zongaro, S., et al.: Visual search of neuropil-enriched RNAs from brain in situ hybridization data through the image analysis pipeline hippo-ATESC. PLoS ONE **8**, e74481 (2013). https://doi.org/10.1371/journal.pone.0074481
13. Armi, L., Fekri-Ershad, S.: Texture image analysis and texture classification methods - a review. arXiv:190406554 (Cs) (2019)
14. Materka, A.: Texture analysis methodologies for magnetic resonance imaging. Dialogues Clin. Neurosci. **6**, 243–250 (2004)
15. Bianconi, F., Fernández, A.: Evaluation of the effects of Gabor filter parameters on texture classification. Pattern Recognit. **40**, 3325–3335 (2007). https://doi.org/10.1016/j.patcog.2007.04.023
16. De Valois, R.L., Albrecht, D.G., Thorell, L.G.: Spatial frequency selectivity of cells in macaque visual cortex. Vis. Res. **22**, 545–559 (1982). https://doi.org/10.1016/0042-6989(82)90113-4
17. Petkov, N., Kruizinga, P.: Computational models of visual neurons specialised in the detection of periodic and aperiodic oriented visual stimuli: bar and grating cells. Biol. Cybern. **76**, 83–96 (1997). https://doi.org/10.1007/s004220050323
18. Hunter, J.D.: Matplotlib: a 2D graphics environment. Comput. Sci. Eng. **9**, 90–95 (2007). https://doi.org/10.1109/MCSE.2007.55
19. Pedregosa, F., Varoquaux, G., Gramfort, A., et al.: Scikit-learn: machine learning in python. J. Mach. Learn. Res **12**, 2825–2830 (2011)
20. SciPy 1.0 Contributors, Virtanen, P., Gommers, R., et al.: SciPy 1.0: fundamental algorithms for scientific computing in Python. Nat. Methods **17**, 261–272 (2020). https://doi.org/10.1038/s41592-019-0686-2
21. van der Walt, S., Colbert, S.C., Varoquaux, G.: The NumPy array: a structure for efficient numerical computation. Comput. Sci. Eng. **13**, 22–30 (2011). https://doi.org/10.1109/MCSE.2011.37
22. Goode, A., Gilbert, B., Harkes, J., et al.: OpenSlide: a vendor-neutral software foundation for digital pathology. J. Pathol. Inform. **4**, 27 (2013). https://doi.org/10.4103/2153-3539.119005
23. Han, J., Kamber, M.: Data Mining: Concepts and Techniques. Elsevier, Burlington (2012)
24. Winkler, R., Klawonn, F., Kruse, R.: Fuzzy c-means in high dimensional spaces. Int. J. Fuzzy Syst. Appl. **1**, 1–16 (2011). https://doi.org/10.4018/ijfsa.2011010101
25. Kim, D.-W., Lee, K.H., Lee, D.: Fuzzy cluster validation index based on inter-cluster proximity. Pattern Recogn. Lett. **24**, 2561–2574 (2003). https://doi.org/10.1016/S0167-8655(03)00101-6

An Approach for Development of a Physical Breast Phantom for X-ray Imaging Using an Inkjet Printer: Preliminary Results

Tihomir Georgiev, Kristina Bliznakova(✉), Iliyan Kolev, Nikolay Dukov, and Zhivko Bliznakov

Medical University of Varna, 9002 Varna, Bulgaria
kristina.bliznakova@mu-varna.bg

Abstract. Anthropomorphic physical body phantoms turn out to be a valuable tool for evaluation, calibration and quality assurance of radiology equipment. Physical shape, composition and radiological equivalency are essential requirements for these phantoms. Nowadays, 3D printing technologies provide vast possibilities for production of anthropomorphic phantoms. Recently, simple printing on paper sheets was explored to produce realistic phantoms that are less expensive and do not require a specialized equipment. The current preliminary study evaluates the feasibility of an approach to create a physical anthropomorphic breast phantom using a conventional inkjet printer and plain office paper. Several x-ray contrast substances can be used as ink additives in order to achieve x-ray attenuation. In this study, we use a solution of potassium iodide (KI) in distilled water due to its low price, easy dissolution and easy insertion into the printer cartridge. A segmented slice from a patient breast MRI was printed on several sets of paper sheets. Each set was x-rayed at 40 kVp, 60 kVp and 80 kVp. Regions of Interest (ROIs) were assigned on the x-ray images to evaluate several parameters such as minimum, maximum, and average grey value. Subjective evaluation was visually performed, while objective evaluation was performed through a comparison of calculated values within a given set and for the ROI between the different sets. This preliminary study showed that the proposed approach is suitable for the development of realistic physical breast phantoms for x-ray imaging research. Further work is related to the creation of a complete "paper" phantom and its evaluation for use in mammography imaging.

Keywords: Anthropomorphic physical phantoms · Breast phantoms · X-ray imaging · Inkjet printed phantoms · Potassium iodide radiology · Mammography

1 Introduction

Anthropomorphic phantoms are valuable tool used to assess the efficiency of a given treatment, the development and optimization of a new medical technology, as well as assessment of new algorithms for image reconstruction [1]. There are several methods to create physical phantoms. The earliest ones used casting, some of them for example contained inclusions of liquid or solid iodine containing mixtures inside the phantoms [2].

S. S. Sotirov et al. (Eds.): BioInfoMed 2020, LNNS 374, pp. 384–389, 2022.
https://doi.org/10.1007/978-3-030-96638-6_40

The development of the 3D printing technologies allowed for 3D printed anthropomorphic models to be designed and created [3, 4]. Usually, 3D printing methods make use of either patient medical images or computer generated anthropomorphic models. Both of these methods allow for easier design based on model simulation and pre-processing. In addition, 3D models are easily shaped to be anthropomorphic in contrast to cast or solidified ones that often lack anthropomorphism.

Prints of potassium iodide (KI) on a plain paper have also been exploited to create anthropomorphic 3D phantoms. Jahnke *et al.* [5] proposed a method for physical phantoms based on patient specific CT image printed on paper using KI solution as ink. Ikejimba *et al.* [6] produced a 3D breast phantom based on a computational breast phantom built from mathematical equations, which after its modelling was printed slice by slice on a parchment paper using an inkjet printer with cartridges filled with iohexol. Another implementation used different types of parchment paper to mimic the radiological properties of the breast adipose tissue and a special ink with zinc acetate for the glandular tissue [7]. Furthermore, Pegues *et al.* [8] used two types of ink to mimic the properties of the liver tissue. In this paper, we present an approach for developing of a phantom printed with ordinary inkjet printer and KI and its evaluation in clinical conditions. This preliminary study is the initial step in achieving the final goal of our-long term study to compare different printing methods, including 3D printing [9] for manufacturing of anthropomorphic models for x-ray imaging studies.

2 Materials and Methods

2.1 Phantom Development

Calibration Phantom. In a previous study carried out by the team, calibration prints were simulated by using the in-house developed software *XRAYImagingSimulator* [10]. The calibration prints were then printed on a plain paper with an inkjet printer HP DeskJet 1050A and using a mixture of 600 mg of potassium iodide dissolved in 1 ml water added to 1 ml of standard ink. The study showed that the glandular tissue of the breast can be effectively represented by the proposed solution, while the adipose tissue was successfully mimicked by the 'clear' plain paper.

Breast Phantom. The computational 3D breast phantom is obtained based on segmentation procedure, applied on patient's image. Five sets of physical phantoms were printed using an image from the segmented patient set. A segmented image is shown on Fig. 1.

The image was printed on a plain office paper (80 g m^{-2}) using the commercially available inkjet printer HP Officejet 5510. The ink cartridge was filled with aqueous solution of potassium iodine (KI) 600 mg per 1 ml and mixed with standard ink in a ratio of 1:1. The segmented image was enlarged 3 times and printed in four sets – Nos 1, 2, 3 and 4. Each set consisted of ten sheets of paper, while all of the sets were printed with a single cartridge without refiling. Additionally, set No 5 was created by stacking sets 1, 2, 3 and 4. The thickness of the paper was measured and confirmed to be 0.1 mm. The thickness of sets 1 to 4 was measured 1 mm, while set 5 was 4 mm thick in total.

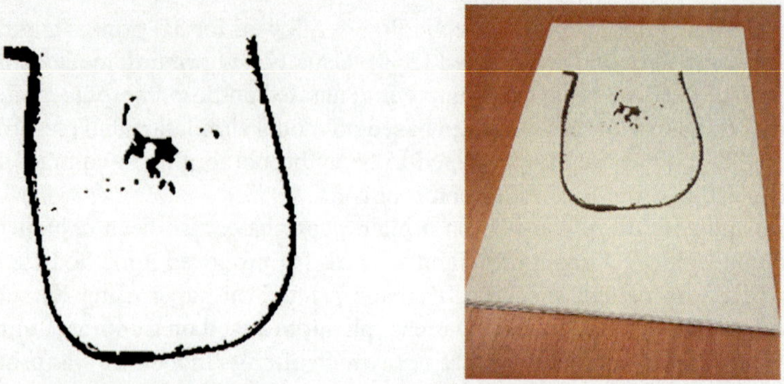

Fig. 1. Segmented image (left) and the printed phantom (right) (Set 5).

2.2 Experimental Work

X-ray image from each set of the resulting phantoms was acquired using Canon CXDI X-ray at St. Marina University Hospital in Varna with three different kVp settings: 40 kVp, 60 kVp, 80 kVp. Figure 2 displays the radiographic images of sets 1 and 5 obtained at the different voltage settings.

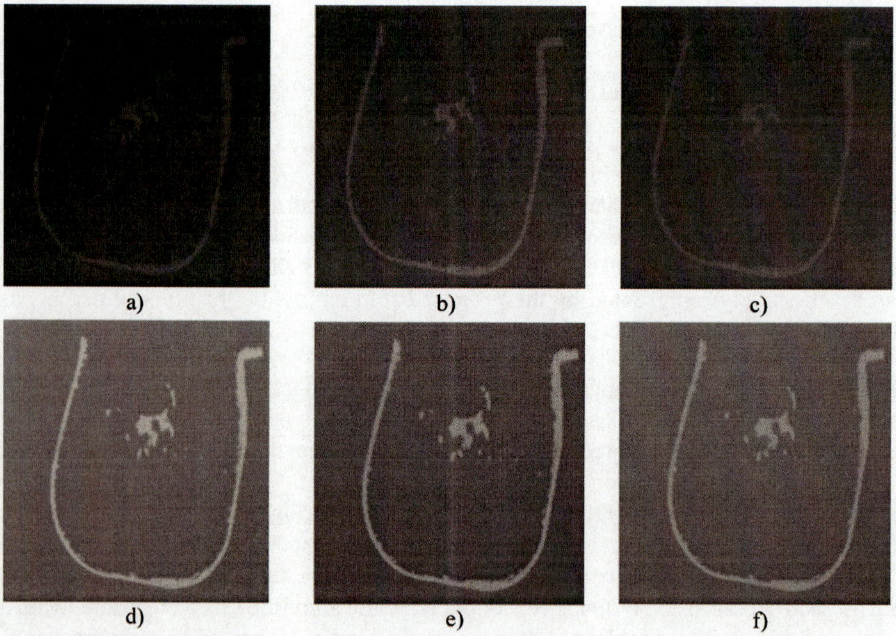

Fig. 2. Acquired images of (a) set 1 at 40 kVp, (b) set 1 at 60 kVp, (c) set 1 at 80 kVp, (d) set 5 at 40 kVp, (e) set 5 at 60 kVp, (f) set 5 at 80 kVp.

2.3 Processing of Data

All x-ray images were evaluated both subjectively and objectively in order to appropriately assess the quality of the created phantoms. Subjective evaluation was performed by visual observation of the images, obtained under different incident exposure settings.

For the objective evaluation, the x-ray images were processed using the open source image processing program ImageJ. The adopted procedure for the evaluation involves defining Regions of Interest (ROIs) on each image in order to compare pixel values inside the ROIs. These ROIs were positioned near the nipple (NN), in the middle of the image (MI) and on the upper right part of the image (radiological left) - opposite of the nipple (OP).

In order to reduce the influence of possible differences between the sheets (due to paper defects) as well as radiation influences during the imaging process, a dynamic range within each ROI was calculated as following:

$$\text{Dynamic range} = \frac{MAX\ PIXEL\ VALUE}{MIN\ PIXEL\ VALUE} \tag{1}$$

$$\text{LN (range)} = \ln\text{ (dynamic range)} \tag{2}$$

Finally, the average values and standard deviations for the dynamic ranges at a given voltage for sets 1 to 3 were calculated.

3 Results and Discussion

Fifteen x-ray images were acquired in total. Subjectively, the contrast of sets 1, 2, and 3 for the same voltage settings appears similar. The contrast of the x-ray image of phantom set 5, created by stacking all the other sets, was highest compared to the other sets as expected, while the contrast of set 4 was the lowest amongst the sets. In the latter case, a possible explanation is that set 4 was the last printed set and eventually less quantity of the potassium iodine (KI) was deposited on the sheets.

When comparing the dynamic ranges in terms of values for each set and ROI, as expected, the dynamic ranges for the ROIs of set 4 were less than the corresponding ranges of the remaining sets, while the dynamic ranges of the ROIs in set 5 were greater than those of the remaining sets.

Results for the dynamic ranges and natural logarithm of the dynamic ranges are listed in Table 1. Overall, the results indicate that the largest differences for dynamic ranges and therefore for the x-ray attenuations in different ROIs within the same image are depicted at 60 kVp for set 1 and at 40 kVp for the remaining sets. Additionally, dynamic ranges for the same region and the same voltage but on different images are close except for set 4, which was printed last and set 5 (Fig. 3). Another exception is the middle ROIs at 60 kVp. The larger difference for set 5 is explained with the larger number of stacked sheets, and thus thicker irradiated set. The smaller values for set 4 are due to the fact that this set was printed lastly, and eventually less KI on certain areas was deposited. The origin of other differences can be varying ink deposition during print, small differences in positioning of the paper sheets during x-ray imaging, external noise, etc.

Table 1. Dynamic ranges of different sets and voltages.

		40 kVp		60 kVp		80 kVp	
		Dynamic range	LN (range)	Dynamic range	LN (range)	Dynamic range	LN (range)
Set 1	OP	1.9168	0.6507	1.7968	0.5860	1.7112	0.5372
	MI	1.9086	0.6464	1.9547	0.6703	1.7483	0.5587
	NN	1.9058	0.6449	1.8839	0.6333	1.8264	0.6024
Set 2	OP	1.8387	0.6091	1.8529	0.6167	1.7498	0.5595
	MI	1.8092	0.5929	1.7845	0.5792	1.7788	0.5759
	NN	1.9407	0.6630	1.7470	0.5579	1.7208	0.5428
Set 3	OP	1.8750	0.6286	1.7748	0.5737	1.7498	0.5595
	MI	1.8534	0.6170	1.7527	0.5612	1.7301	0.5482
	NN	2.0338	0.7099	1.8187	0.5981	1.7415	0.5547
Set 4	OP	1.8012	0.5885	1.6459	0.4983	1.6282	0.4875
	MI	1.7989	0.5872	1.6350	0.4917	1.6188	0.4817
	NN	1.9124	0.6484	1.6728	0.5145	1.6482	0.4997
Set 5	OP	2.0869	0.7357	2.1667	0.7732	2.0304	0.7082
	MI	2.0718	0.7284	2.1188	0.7508	1.9766	0.6814
	NN	2.2203	0.7976	2.2530	0.8123	2.1008	0.7423

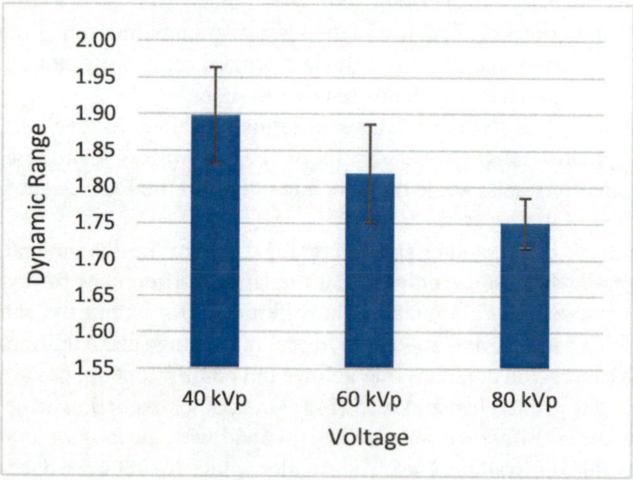

Fig. 3. Average values and standard deviation of the dynamic ranges for sets 1, 2 and 3.

4 Conclusion

Both subjective and objective evaluations of the initial phantoms demonstrate that the method can be used as a reliable technique for creating realistic 3D anthropomorphic phantoms. However, there are several issues that have to be addressed and will be examined during the next tests – what is the KI deposition, does it decrease after consecutive prints, is there a strict linear relation between real image prints and attenuation, does the printed images keep their quality or it deteriorates over time.

Acknowledgements. This work is supported by the Bulgarian National Science Fund under grant agreement DN 17/2.

References

1. Bliznakova, K., Buliev I., Bliznakov Z.: Anthropomorphic Phantoms in Image Quality and Patient Dose Optimization. A EUTEMPE Network Book. IOP Publishing (2018)
2. Leitner, R., Knogler, T., Homolka, P.: Development and production of a prototype iodine contrast phantom for CEDEM. Phys. Med. Biol. **58**, N25–N35 (2013)
3. Carton, A.-K., Bakoc, P., Ullberg, C., Helen, D., Maidment, A.D.A.: Development of a physical 3D anthropomorphic breast phantom. Med. Phys. **38**(2), 891–896 (2011)
4. Clark, M., Ghammraoui B., Badal, A.: Reproducing 2D breast mammography images with 3D printed phantoms. In: Medical Imaging 2016: Physics of Medical Imaging, Proceedings of SPIE, vol. 9783, p. 97830B. SPIE (2016)
5. Jahnke, P., et al.: Radiopaque three-dimensional printing: a method to create realistic CT phantoms. Radiology **282**(2), 569–575 (2017)
6. Ikejimba, L., et al.: Method for task-based evaluation of clinical FFDM and DBT systems using an anthropomorphic breast phantom. In: Proceedings of SPIE 10718, 14th International Workshop on Breast Imaging (IWBI 2018), p. 107180Q, 6 July 2018
7. Ikejimba, L., Salad, J., Makeev, A., Graff, C., Ghammraoui, B., et al.: Anthropomorphic breast phantoms for evaluation of FFDM/DBT and breast CT using inkjet printing. In: Proceedings of SPIE 10573, Medical Imaging 2018: Physics of Medical Imaging, p. 1057308, 9 March 2018
8. Pegues, H., Knudsen, J., Huayu, T, Gehm, M., Wiley, B., et al.: Using inkjet 3D printing to create contrast-enhances textures physical phantoms for CT. In: Proceedings of SPIE 10948, Medical Imaging 2019: Physics of Medical Imaging, p. 109484Z, 17 May 2019
9. Ivanov, D., et al.: Suitability of low density materials for 3D printing of physical breast phantoms. Phys. Med. Biol. **63**, 175020 (2018)
10. Bliznakova, K., et al.: Experimental validation of a radiographic simulation code using breast phantom for X-ray imaging. Comput. Biol. Med. **40**(2), 208–214 (2010)

Influence of Incident Mammography Spectra and Breast Phantom Characteristics on Mammography Features

Yanka Baneva and Kristina Bliznakova(✉)

Medical University of Varna, 9002 Varna, Bulgaria
kristina.bliznakova@mu-varna.bg

Abstract. The aim of this work is to evaluate the effect of incident mammography spectra and breast phantom characteristics on the extracted mammography features. For this purpose, four digital breast phantoms were generated by using a dedicated in-house developed software application. The phantoms differ in the content, the thickness and materials used in to simulate the gland, adipose and skin tissues. Another software tool, *XRAYImagingSimulator*, was used to generate x-ray images under typical mammography geometry. Eight mammography x-ray spectra were simulated. An image quality platform was used to calculate the following features: kurtosis, skewness, fractal analysis and the β factor from the power law spectra. The simulations showed that for the same breast thickness and both breast types (GAG, GGA), the GAG combination results in higher skewness values, which indicates that this breast phantom is less dense than the GGA phantom combination. Fractal values are higher for the denser breast GGA, while the kurtosis is higher for the GAG breast phantoms at mean energies of 16.37 keV and 18.28 keV. This simulation study concludes that the fractal dimension depends on the incident x-ray spectra, while the statistical features kurtosis and skewness remain almost constant. These features may be used as basic parameters in the design of software applications for breast density evaluation. The comparisons between the two basic phantoms (GGA and GAG) reveals which material combination might be suitable for the representation of a denser breast, i.e. GGA.

Keywords: Breast digital phantoms · Mammography · Breast imaging techniques · Simulations · Mammography features

1 Introduction

Breast cancer is a disease which is curable if is detected on time. For this, except for the screening programs where the cancer might be detected in an early stage, the mammograms need to provide the truth information about the presence or not of a cancerous cell. Density of the breast is important for the quality of the image parameter of the imaging system because dense tissue can hide the cancer [1]. The presence of dense breast tissue lowers the sensitivity of mammography [2]. Dose in mammography depends on the size and composition of the breast as well as the imaging device and exposure settings

S. S. Sotirov et al. (Eds.): BioInfoMed 2020, LNNS 374, pp. 390–399, 2022.
https://doi.org/10.1007/978-3-030-96638-6_41

selected materials and methods [3, 4]. For thicker and/or denser breasts, a higher-energy x-ray beam is needed to achieve adequate tissue penetration [5]. Therefore, optimization of all mammography units is essential for the early detection of the breast cancer. In this task, the availability of breast phantoms is a requirement. Clinical investigations might be expensive and time consuming, software phantoms instead are better alternative. Computer models mimic the anatomy and shape of the female breast and are used in simulation of x-ray imaging and generation of x-ray projection [6].

Goal of this study is to generate different in thickness and content breast phantoms and their synthetic mammograms by using dedicated x-ray software simulator and to evaluate the effect of incident mammography spectra and breast phantom characteristics on the extracted mammography features. Based on the derived results from the mammography projection images of modelled breast phantoms, a decision is taken for the composition of glandular and dense physical breast phantoms.

2 Materials and Methods

2.1 Phantoms

For the purpose of our work, we generated digital phantoms with different compositions, summarized in Table 1. For the model of the phantom a software application called *LCMFRGen* was used [7]. This application allows the creation of software versions of the physical phantom based on parameters like sizes of the container and the spheres, as well as the material characteristics [8–11]. The computational phantom consists of two main semi-cylinders and a slab object approximating the side wall of the physical model. Spheres with different diameters and compositions are used to represent the inhomogeneity in the breast (Fig. 1).

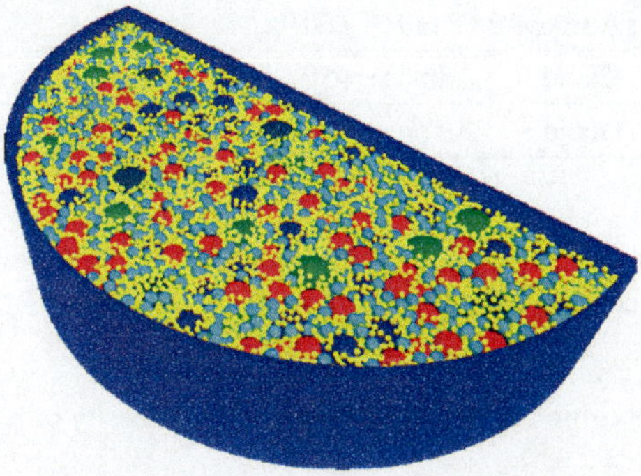

Fig. 1. Generated software phantom with *LCMFRGen* [7].

Four phantoms with different thiknesses and compositions were generated. The first phantom **Phantom 1** consist of a container and a container filling material made from gland, while the spheres material is adipose. Two phantoms are designed with the same material composition but different thiknesses: 50 mm and 60 mm. The algorithm for the sphere's generation implements the sphere packing concept for unequal non-overlapping spheres in a container space. The diameters of the spheres are: 7.94 mm, 6.35 mm, 4.76 mm, 3.175 mm, 1.59 mm and 0.79 mm. The number of the created ellipsoids is 37854 for the phantom with thickness 50 mm and 44926 for the phantom with container height of 60 mm. The container and the spheres of **Phantom 2** are simulated from gland tissue, while the container filling material is assigned adipose tissue, as shown in Table 1. Again, two computational phantoms are generated. The number of the created ellipsoids is 37534 for the phantom with thickness 50 mm and 45425 for the phantom with container height of 60 mm. The density by volume for **Phantom 1** is 36%, while for **Phantom 2** is 64%. Respectively, the density by mass for **Phantom 1** is 38%, while for **Phantom 2** is 65%. The densities of the generated breast models are calculated by using the following equations:

$$\textit{Breast Density by volume} = \frac{\text{Volume of gland, } cm^3}{\text{Volume of gland} + \text{Volume of adipose, } cm^3} * 100, \% \tag{1}$$

$$\textit{Breast Density by mass} = \frac{\text{Mass of gland, g}}{\text{Mass of gland} + \text{Mass of adipose, g}} * 100, \% \tag{2}$$

Table 1. Description of the computational breast phantoms, generated with the LUCMFRGen software tool used in the study.

Phantoms	Container	Spheres	Filling material	Container height	Number of spheres	Density by volume, %	Density by mass, %
Phantom1	Gland	Adipose	Gland	50	37854	36%	38%
	Gland	Adipose	Gland	60	44926	36%	38%
Phantom2	Gland	Gland	Adipose	50	37534	64%	65%
	Gland	Gland	Adipose	60	45425	64%	65%

The elemental composition of the materials used for modelling the parts of the phantoms are summarized in Table 2.

Table 2. Elemental composition as percentage weights and density of used breast substitutes in the study.

	H	C	N	O	Na	P	S	Cl	Density, g/cm^3
Adipose tissue	0.114	0.598	0.007	0.278	0.001	–	0.001	0.001	0.95
Gland	0.106	0.332	0.030	0.527	0.001	0.001	0.002	0.001	1.02

2.2 Generation of Projection Images

XRAYImagingSimulator - an in-house developed software tool is used for the creation of the planar x-ray projection images [12]. This software tool consists of two main modules: a) a module for phantom modelling, based on solid and voxel geometry, and b) a module for analytical x-ray image formation modelling, and allows simulation of 2D and 3D x-ray imaging techniques such as radiography, tomosynthesis, cone beam CT. In the present study the module for analytical x-ray image formation is used, as simulated mammography images are obtained for source-to-isocenter and source-to-detector distances are 600 mm, and 660 mm, respectively. The detector resolution is 10 pixels/mm, while the image size is 3500 × 3500 pixels.

For the generation of the images, eight mammography x-ray spectra are modelled and shown in Fig. 2: Mo 25, Mo/Rh 27, Mo 27, Mo/Rh 28, Mo 30, Rh/Rh 29, Rh 30, Rh/Rh 31. The entrance surface exposure was set to 3 mGy.

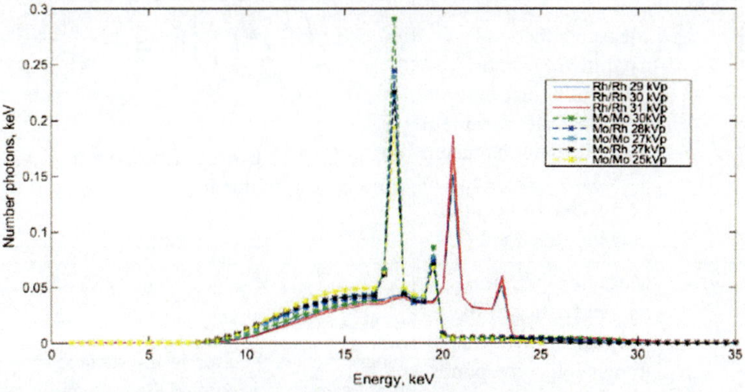

Fig. 2. Modelled x-ray mammography spectra used in the study.

Simulated x-ray spectra and their average energies are shown in Table 3:

Table 3. Simulated x-ray spectra and their mean energies.

Spectrum	Mo/Mo 25 kVp	Mo/Rh 27 kVp	Mo/Mo 27 kVp	Mo/Rh 28 kVp	Mo/Mo 30 kVp	Rh/Rh 29 kVp	Rh/ Rh 30 kVp	Rh/Rh 31 kVp
E, kV	15.95	16.2	16.37	16.41	16.93	18.28	18.52	18.75

2.3 Quality Platform

An in-house image quality software tool was used to calculate four basic features of the obtained mammography images from the phantoms: kurtosis, skewness, fractal analysis and the β factor from the power law spectra [13, 14]. The mathematical formulation of these descriptors is adapted from and shown in Table 4.

Table 4. Features extracted from projection images.

Extracted feature	Purpose	Formula
Kurtosis	Measures the relative tailedness (relatively to the normal distribution) or flatness of a distribution	$\mathrm{Kurt}[X] = \frac{\mu_4}{\sigma^4} = \frac{E[(X-\mu)^4]}{\left(E[(X-\mu)^2]\right)^2}$
		E is the expectation operator, where μ4 is the fourth moment about the mean (μ) and σ is the standard deviation
Skewness	Measures the asymmetry of the probability distribution of a real-valued random variable about its mean. The skewness value can be positive or negative, or even undefined	$\gamma_1 = \mathrm{E}\left[\left(\left(\frac{X-\mu}{\sigma}\right)^3\right)\right] = \frac{\mu_3}{\sigma^3} = \frac{E[(X-\mu)^3]}{\left(E[(X-\mu)^2]\right)^{\frac{3}{2}}} = \frac{\kappa_3}{\kappa_2^{3/2}}$
		where E is the expectation operator, μ3 is the third central moment
Fractal analysis	Measures the fractal characteristics of the images. A lower fractal dimension corresponds to smoother texture and opposite	$A(\varepsilon) = \sum_{x,y} \varepsilon^2 + \sum_{x,y} \varepsilon(\lvert i_\varepsilon(x,y) - i_\varepsilon(x, y+1)\rvert + \lvert i_\varepsilon(x, y) - i_\varepsilon(x+1, y)\rvert)$
		where A(ε) is the area of the surface for a given pixel size ε, ε^2 is sum of the area of each pixel
Power law spectral analysis	The power spectrum describes the signal and noise into frequency components of which the low frequency range can be characterized by a power law. The exponent of this power law gives then a measure of the texture or anatomical clutter	$F(k, l) = \frac{1}{MN}\sum_{a=0}^{M-1}\sum_{b=0}^{N-1} f(a, b)\exp\left[-2\pi i\left(\frac{ka}{M} + \frac{lb}{N}\right)\right]$, where k (0, 1, ..., M − 1) and l (0, 1, ...N − 1) are the relative spatial frequencies in the two directions and M × N is the image size of f(a, b)
		The total power spectrum per concentric ring is further transformed to $\log_2$ (total power spectrum) and plotted versus the octave number. The plot is almost linear and the data are fitted to a line obtained using linear regression analysis. The slope of the line m is related to exponential parameter β as: m = β − 2

The quality platform is divided in 3 main parts (Fig. 3). The first part is related to the choice of the image- DICOM files, RAW image files or picture image files. The second part is related to the calculations of the image descriptors. The third part is related to the region under investigation and the definition of smaller regions of interest, which are needed for the calculations, with or without overlapping, as shown in Fig. 3.

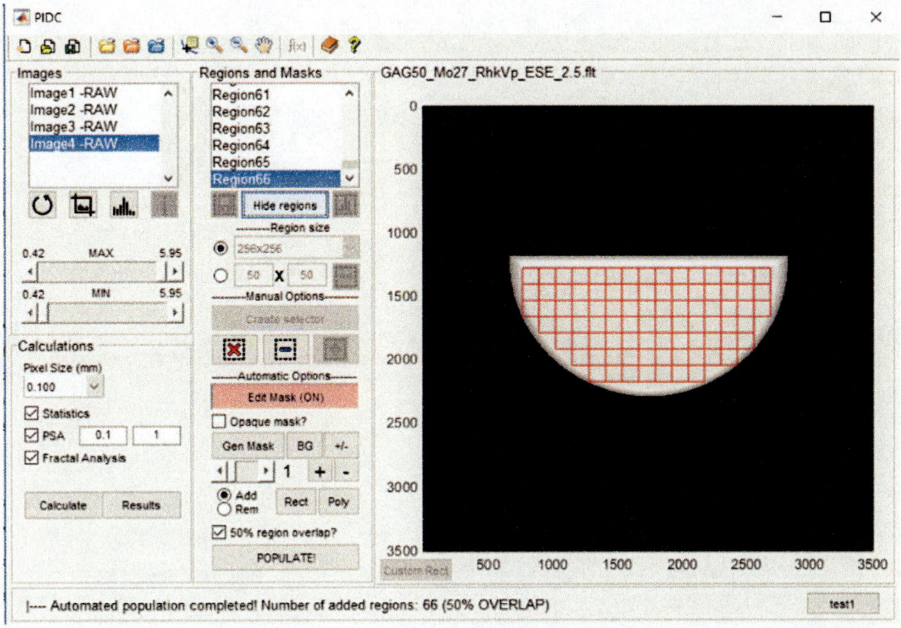

Fig. 3. A screenshot from the software tool for extracting of image features.

3 Results and Discussion

Simulated x-ray images of the generated projection images of the phantoms are shown in Fig. 4 for different spectra and phantoms.

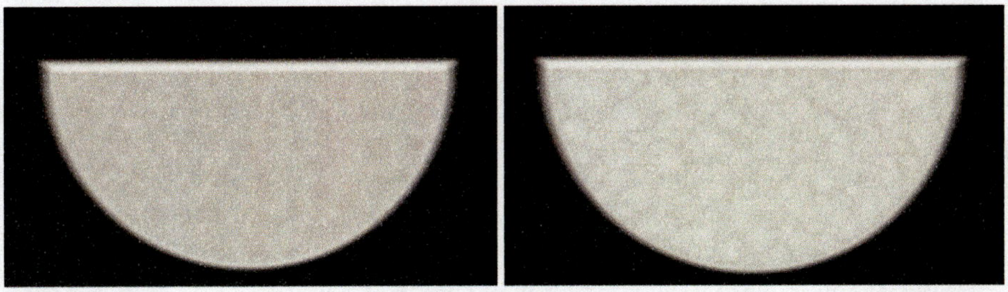

(a) (b)

Fig. 4. Simulated x-ray images of: (a) phantom GAG50 and (b) Phantom GGA50 with Mo25kVp

3.1 Influence of Phantom Materials

The influence of the materials used in breast phantoms on the extracted features is shown in Fig. 5.

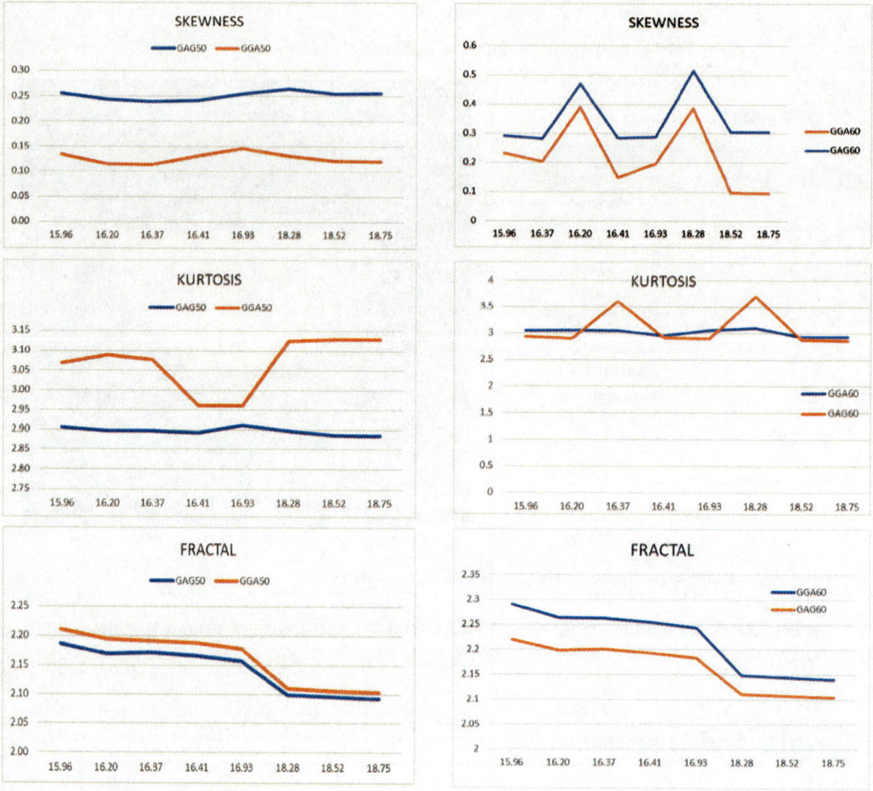

(a) (b)

Fig. 5. Comparison of the basic features of the obtained mammography images from: a) Phantoms GAG and GGA with thickness 50 mm, b) Phantoms GAG and GGA with thikness 60 mm. The values on the x-axes show the mean energy of the eight mammography spectra in keV.

The results from Fig. 5 show that for the same phantom thickness:

(a) *Skewness* has higher values for GAG (a gland for the container and the container filling material, adipose for the spheres) in comparison with GGA (gland for the container material and for the spheres, and adipose for the container filling material). This suggests that the GGA combination represents denser breast than GAG, because a dense breast would have lower skewness values [15].

(b) *Kurtosis* is higher for the GGA50 phantom (gland for the container material and for the spheres, and adipose for the container filling material) in comparison to

GAG50 (a gland for the container and the container filling material, adipose for the spheres).The value of the kurtosis is higher in the phantom combination GAG 60 than the resut from phantom combination GGA60 just for the energies 16.37 kV and 18.28 kV.

(c) *Fractal dimension* is higher for GGA the phantom with both thicknesses 50 mm and 60 mm in comparison to the phantom GAG with thicknesses 50 mm and 60 mm.

3.2 Influence of Phantom Thickness

Figure 6 summarizes the influence of phantom thickness on the extracted mammographic features for the GGA phantom and the different incident spectra.

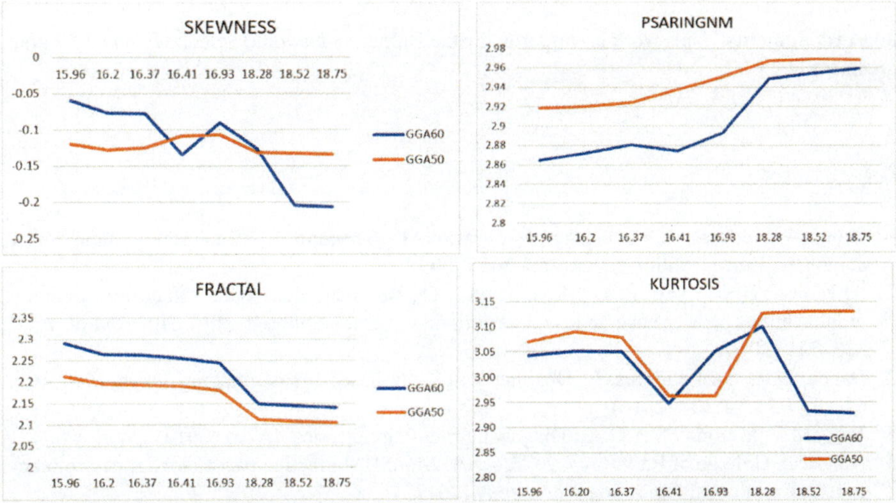

Fig. 6. The influence of the phantom thickness on the extracted features for GGA for the different incident spectra. The values on the x-axes show the mean energy of the eight mammography spectra in keV.

The skewness values are higher for the phantom with the higher thickness for lower mammographic incident energies, while the kurtosis values show to be opposite compared to the skewness.

The phantom with higher thickness shows higher value for the fractal dimensions. Similarly, the higher the phantom thickness is, higher β factors from the power law spectra are obtained.

For both phantom combination GAG and GGA, the fractal dimensions decreases with the increase of incident mammography energy, while the β factor from the power law spectra increases with the increase of the incident mammography energies.

4 Conclusions

Fractal dimension depends on the incident x-ray spectra, while the statistical features kurtosis and skewness remain almost constant. Increasing the mean energy of the mammography spectra results in lower fractal values and higher β factors from the power law spectra. The comparison between the two basic phantoms (GGA and GAG) reveals which material combination might be use for the representation of a denser breast, i.e. GGA, while GAG phantom represents fatty breast. The gene-mutation carriers tend to have denser breast tissue than do those in low-risk women. These features may be used as basic parameters in the design of software applications for breast density evaluation. Currently, the results from more than twenty radiographic features extracted from the images are under evaluation. They will be used in the design of deep learning system to classify mammographic breast density.

Acknowledgments. This work is supported by the Bulgarian National Science Fund under grant agreement DN 17/2.

References

1. Schulz-Wendtland, R., Fuchsjäger, M., Wacker, T., Hermann, K.-P.: Digital mammography: an update. Eur. J. Radiol. **72**(2), 258–265 (2009)
2. Al Mousa, D.S., Ryan, E.A., Mello-Thoms, C., Brennan, P.C.: What effect does mammographic breast density have on lesion detection in digital mammography? Clin. Radiol. **69**(4), 333–341 (2014)
3. Dance, D.R., Sechopoulos, I.: Dosimetry in x-ray-based breast imaging. Phys. Med. Biol. **61**(19), R271–R304 (2016)
4. Jaffe, M.J., Maidment, A.D.A.: Diagnostic radiology physics. IAEA (2012)
5. American College of Radiology: ACR – AAPM – SIIM practice parameter for determinants of image quality in digital mammography (2017)
6. Bliznakova, K.: The advent of anthropomorphic three-dimensional breast phantoms for X-ray imaging. Phys. Med. **79**, 145–161 (2020)
7. Bliznakova, K.: Development of breast software phantom dedicated for research and educational purposes. RAD Assoc. J. **2**, 14–16 (2017)
8. Cockmartin, L., Marshall, N., Bosmans, H.: Design and evaluation of a phantom with structured background for digital mammography and breast tomosynthesis. In: Maidment, A.D.A., Bakic, P.R., Gavenonis, S. (eds.) IWDM 2012. Lecture Notes in Computer Science, vol. 7361, pp. 642–649. Springer, Cham (2012). https://doi.org/10.1007/978-3-642-31271-7_83
9. Cockmartin, L., Bosmans, H., Bliznakova, K., et al.: Creation of realistic structured backgrounds using adipose compartment models in a test object for breast imaging performance analysis, in Radiological Society of North America (RSNA) 2016. In: Scientific Assembly and Annual Meeting (2016)
10. Cockmartin, L., Bosmans, H., Marshall, N.W.: Comparative power law analysis of structured breast phantom and patient images in digital mammography and breast tomosynthesis. Med. Phys. **40**(8), 081920 (2013)
11. Cockmartin, L., et al.: Design and application of a structured phantom for detection performance comparison between breast tomosynthesis and digital mammography. Phys. Med. Biol. **62**(3), 758–780 (2017)

12. Bliznakova, K., Speller, R., Horrocks, J., Liaparinos, P., Kolitsi, Z., Pallikarakis, N.: Experimental validation of a radiographic simulation code using breast phantom for X-ray imaging. Comput. Biol. Med. **40**(2), 208–214 (2010)
13. Marinov, S., et al.: Radiomics software for breast imaging optimization and simulation studies. Phys. Med. **89**, 114–128 (2021)
14. Baneva, Y., et al.: Evaluation of breast software model for X-ray 2D and 3D mammography imaging. Phys. Med. **32**, 255–256 (2016)
15. Li, H., Giger, M.L., Olopade, O.I., Margolis, A., Lan, L., Chinander, M.R.: Computerized texture analysis of mammographic parenchymal patterns of digitized mammograms. Acad. Radiol. **12**(7), 863–873 (2005)

Comparative Morphometric Image Analysis of LYVE-1 and Podoplanin in HIV Infected Preeclamptic Women

Vesselina Merhar[1](✉), Onankoy Onyangunga[2], Jagidesa Moodley[3], and Thajasvarie Naicker[2]

[1] Department of Biology, Medical Genetics and Microbiology, Faculty of Medicine, University Prof. Dr Asen Zlatarov, Burgas, Bulgaria

[2] Optics and Imaging Centre, Doris Duke Medical Research Institute, College of Health Sciences, University of KwaZulu-Natal, Durban, South Africa
naickera@ukzn.ac.za

[3] Women's Health and HIV Research Centre, College of Health Sciences, University of KwaZulu-Natal, Durban, South Africa
jmog@ukzn.ac.za

Abstract. *Introduction:* Preeclampsia is a serious medical complication and its exact pathogenesis is still unclear. The dysregulation of angiogenesis in preeclampsia is well known however, the role of lymphangiogenesis in the synergy of HIV infection and preeclampsia remains poorly understood. This study applied computer assisted morphometric image analysis to compare the immunohistochemical distribution of two lymphatic markers podoplanin and LYVE-1 in the placenta of women with HIV associated preeclampsia.

Materials and Methods: Placental tissue was obtained from normotensive and preeclamptic women stratified according to their HIV status. The preeclamptic group was divided into early ($<$ 34 weeks; EOPE) and late ($>$34 weeks; LOPE) onset preeclampsia. Immunostaining was performed using mouse anti-human LYVE-1 and D2-40 podoplanin antibodies and was morphometrically evaluated.

Results: LYVE-1 immunostaining was localized within endothelium of the arterial supply and venous drainage of both conducting and exchange villi as well as within medial cells of arteries. In contrast, podoplanin was immunolocalized in a reticular-like stromal complex within the exchange and conducting villi. In general, there was a decline in the expression of both lymphatic markers in preeclamptic versus normotensive women, being more prominent during EOPE than LOPE. Both LYVE-1 and podoplanin were upregulated in HIV^+ *vs.* HIV^- groups regardless of pregnancy and villi type.

Conclusion: This study provides a novel insight into the differential distribution of lymphatic markers, LYVE-1 versus podoplanin immunoexpression in the fetal circulation of HIV-infected preeclampsia. It highlights the benefits of computer assisted morphometric image analysis as a reliable tool for quantification of immunostaining in biological tissue.

Keywords: Lymphatic vessel endothelial hyaluronan receptor-1 (LYVE-1) · Morphometry · Podoplanin

S. S. Sotirov et al. (Eds.): BioInfoMed 2020, LNNS 374, pp. 400–408, 2022.
https://doi.org/10.1007/978-3-030-96638-6_42

1 Introduction

Preeclampsia (PE) is defined as new-onset hypertension presenting after 20 weeks of gestation with or without proteinuria and with multi organ dysfunction [1]. If left untreated it can develop to eclampsia, a life-threatening occurrence of seizures. To-date there is no known cure for PE however, resolution of the condition involves delivery of the placenta [2]. This suggests that the etiology of pre-eclampsia lies in the placenta and the placental bed.

It is well known that PE is initiated by defective trophoblast invasion within the placental bed [3]. The non- physiological transformation of spiral arteries within the myometrium is accompanied by hypofusion of the placenta [4]. In PE, the placenta becomes hypoxic (deprived of adequate oxygen supply) and starts to release anti-angiogenic factors into circulation thereby inducing systemic inflammation, oxidative stress, and endothelial cell dysfunction [5]. These mediators pre-empt the maternal syndrome of PE (presence of hypertension, proteinuria, liver dysfunction, oedema, eclampsia, *etc.*). The hypertension and proteinuria of PE emanate from endogenous anti-angiogenic soluble fms-like tyrosine kinase-1 (sFlt-1, sVEGFR-1) that neutralises vascular endothelial growth factor (VEGF) and placental growth factor (PlGF) [6–8]. Notably, sFlt1 and sEng are reduced in HIV positive compared to HIV negative pregnancy suggesting a counteraction of the immune hyper reactivity of PE. However anti-retroviral therapy is associated with immune reconstitution and would negate this effect [9–11].

Similar to angiogenesis, lymphangiogenesis is the formation of lymphatic vessels from pre-existing lymphatic vessels. It is a protagonist of fluid homeostasis and serves an immunological function by transporting white blood cells and antigen-presenting cells to the lymphoid organs where they elicit immune responses [12]. The lymphatic system has also been implicated in various pathological conditions, *e.g.*, lymphedema and hypertension [13]. These findings have enabled a better understanding of the molecular mechanisms governing lymphangiogenesis during normal and pathological conditions [14–16]. In view of this important role in the normal and pathological processes, the lymphatic system has become a subject of great interest in the recent years. Studies on lymphangiogenesis and distribution of the lymphatic system within different tissues, however, have been hampered by difficulties in their identification/recognition.

Data on lymphangiogenesis within the placenta are limited and conflicting [17–19] and there is a paucity of data on on its' role in the synergy of HIV associated pre-eclampsia. Recent investigations by our group have shown that LYVE-1 and podoplanin lymphatic markers are immunoexpressed in the placentas of HIV infected preeclamptic women [20, 21]. We applied computer assisted morphometric image analysis to quantify the immunohistochemical distribution of these markers. Morphometric image analysis is a quantitative description of geometric features of tissues, cells, cell organelles or biomarkers and brings an element of accuracy to the microscopic imaging itself. It has become one of the most widely used qualitative method in biomedical studies and pathology. This study compares the immunohistochemical distribution of two lymphatic markers podoplanin and LYVE-1 in the placenta of women with HIV associated preeclampsia.

2 Materials and Methods

2.1 Ethical Approval and Placenta Collection

Ethical approval for the study was obtained from the Biomedical Research Ethics Committee, University of KwaZulu-Natal (BE: 040/12). Written informed consent was obtained prior to enrollment in the study.

Placental specimens utilized in this study were obtained from pre-eclamptic and normotensive pregnancies that were further divided into two subgroups depending on their HIV status (*i.e.*, HIV-positive and HIV-negative). PE was further grouped into two sub-groups based on the gestational age: early (EOPE; <34 weeks) and late (LOPE; >34 weeks) onset pre-eclampsia. The normotensive group included women with previous caesarean section with no other medical condition requiring elective caesarean section. The study group consisted of normotensive HIV negative (N−, n = 30; n = number of participants); normotensive HIV positive (N+, n = 30); early onset HIV negative (EOPE−, n = 30); early onset pre-eclamptic HIV positive (EOPE+, n = 30); late onset pre-eclamptic HIV negative (LOPE−, n = 30) and late onset pre-eclamptic HIV positive (LOPE+, n = 30) groups.

2.2 Immunohistochemistry

Tissue preparation for microscopy and immunohistochemistry was performed using a procedure described previously [20]. Specific antibodies against lymphatic markers were a mouse anti-human lymphatic vascular endothelial hyaluronan receptor-1 antibody (LYVE-1) and a monoclonal mouse anti-human clone D2-40 (podoplanin). The secondary antibody was a goat secondary horseradish peroxidase antibody, followed by DAB (3, 3'-diaminobenzidine) as the chromogen. Specificity of labelling was assessed by examination of control sections in which the primary antibody was replaced with non-immune serum of the same IgG class or with a buffer control.

2.3 Computer-Assisted Morphometric Analysis of LYVE-1 and Podoplanin Immuno Expression

All specimens were viewed using Axioscope A1 microscope (Carl Zeiss, Germany). The images were captured with AxioCamMRc5 digital FireWire camera coupled with AxioVision software for image acquisition, processing and analysis (both from Carl Zeiss, Germany). At least 4 slides per tissue block were viewed and the microscopic images were obtained from at least four randomly selected fields of view per slide captured at 20x magnification. We ensured that the camera settings and scaling were the same throughout the whole viewing process by making a workflow customization. This allowed us to make repetitive viewings with the same accuracy and to compare the results. We used the same RGB threshold and intensity settings for all archival of images. Morphometry was performed using the automated software mode after calibration with the same settings applied to all images. The automated mode included: filtering of the images, including the removal of background noise, thresholding of the immunostaining and morphometric analyses LYVE-1 and podoplanin expression were determined as a percentage of immunostaining within the conducting and exchange placental villi as described previously [20]. Since the conducting and exchange villi significantly differ in size, we had to apply a different approach in measuring the amount of the immunostaining within them. We applied a single object analysis to the large conducting villi, where each villa was framed and the amount of the label (threshold) was expresses as percentage per frame area. The amount of immunostaining within the much smaller exchange villi was determent by two phase threshold and expressed as a total percentage within the frame area. After measurement, the data were transferred to MS-Excel spreadsheet for a further analysis. LYVE-1 and podoplanin expression was determined as a percentage of immunostaining (brown chromogen) within a frame area of placental villi.

2.4 Statistical Analysis

IBM SPSS statistics software version 23 was used to analyze the data. Results are presented as parametric or non-parametric. To determine the statistical difference between study groups, the Kruskal–Wallis test was used for non-parametric data. One-way ANOVA testing was used to examine the effect of HIV infection (HIV+ *vs.* HIV−), pregnancy type (normotensive *vs.* preeclamptic), and villus type (exchange *vs.* conducting) on PDPN immunostaining. A $p < 0.05$ value was considered statistically significant.

3 Results

3.1 Qualitative LYVE-1 and Podoplanin Immunoexpression Within the Placenta

In our study we found a significant difference between the immunolocalization of LYVE-1 and podoplanin within the placenta (Fig. 1). LYVE-1 was localized to endothelial cells, of both the arterial and venous supply of the conducting villi and endothelial cells of capillaries within the exchange villi (both intermediate and terminal) (Fig. 1 A & B).

In contrast, podoplanin localization was conspicuous and expressed in a reticular-like complex of the mesenchymal stroma around arteries of the placenta. There was an absence of immunostaining within vascular endothelial cells of both conducting and exchange (intermediate and terminal) villi (Fig. 1 C & D).

3.2 Morphometric Analysis of LYVE-1 and Podoplanin Immunostaining Within the Placental Villi

Percentage LYVE-1 and podoplanin (Mean ± SD) immunoexpression within conducting and exchange villi across study groups are shown in Tables 1 and 2.

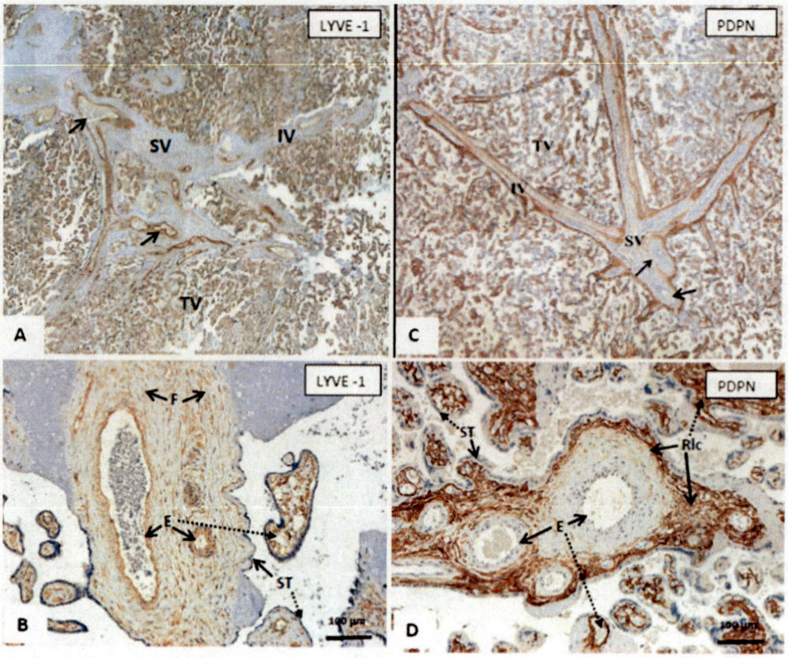

Fig. 1. LYVE-1 (A & B) and podoplanin (C & D) immunolocalization of normotensive placental villi. A & C are composite images of LYVE-1 and podoplanin (PDNP) captured at an initial magnification 5x. *Abbreviations:* SV – stem villi; IV – intermediate villi; TV – terminal villi; E - Endothelial cells; ST – Syncytiotrophoblast; F – fibroblast-like cells; Rlc - Reticular-like complex; black arrows on images A and C point out to arteries within the conducting villi.

Table 1. LYVE-1 and podoplanin (PDNP) immuno-reactivity based on pregnancy type. Results represent the percentage of immunostaining per frame area of conducting and exchange villi and are expressed as mean ± SD.

Label type	Villi type	N	EOPE	LOPE
LYVE-1	Conducting	14.94 ± 4.57	13.18 ± 5.91	18.50 ± 4.24
PDPN	Conducting	8.20 ± 3.58	6.81 ± 2.90	9.07 ± 3.66
LYVE-1	Exchange	16.01 ± 7.47	11.27 ± 5.88	14.71 ± 7.01
PDPN	Exchange	17.33 ± 3.77	15.15 ± 2.88	18.80 ± 4.03

Both lymphatic markers were down regulated on EOPE group compared with the LOPE and the normotensive group. It was notable that the podoplanin expression in the conducting villi was significantly lower compared with its expression in the exchange villi and the LYVE-1 expression in the same villi.

Considering the HIV status across the study population ($n = 180$), the LYVE-1 and PDPN expression within the exchange villi was significantly higher in HIV^{+} compared to

Table 2. LYVE-1 and podoplanin (PDNP) immuno-reactivity based on their HIV status. Results represent the percentage of immunostaining per frame area of conducting and exchange villi and are expressed as mean ± SD.

Label type	Villi type	HIV^-	HIV^+
LYVE-1	Conducting	14.81 ± 4.72	16.25 ± 4.29
PDPN	Conducting	8.40 ± 3.18	7.66 ± 3.57
LYVE-1	Exchange	11.89 ± 6.33	16.10 ± 7.21
PDPN	Exchange	15.06 ± 3.33	18.00 ± 3.23

the HIV^- group irrespective of their pregnancy type. There was no significant difference between the immunoexpression of both markers in the conducting villi of HIV positive and negative groups.

It should be noted that the amount of the podoplanin label in the conducting villi was significantly lower compared to the rest of the groups (Table 1 and 2).

4 Discussion

A precise calculation of the amount of immunostaining on microscopic images of tissue is challenging. There are many factors that might influence the accuracy of measurement, *e.g.*, tissue staining, correct parameters used during microscopy viewing (balance between brightness, contrast, saturation, sharpness *etc.*,) and a very important factor of the correct outlining of the regions of interest (ROIs) within which the actual measurements are performed. With the introduction of new automated technologies for specimen preparation, image viewing and analysis these difficulties may be overcome. In our study we used an automated stainer for simultaneous staining of batches of placental tissue under similar experimental conditions and then proceeded to utilize computer assisted morphometric image analysis for accurate measurements.

From a histomorphological aspect, villi ensure maximal contact area with maternal blood thus ensuring exchange of nutrients, oxygen and waste exchange. They emerge from the chorionic plate as large conducting (stem) villi with well-defined blood vessels and branch into increasingly smaller exchange (intermediate and terminal) villi. The terminal villi contain capillaries where direct exchange takes place [22]. In light of the different calibration settings for morphometric analyses of each type villi, single object analysis was used for calculating the immunostaining in conducting villi whilst a two-phase threshold analysis to define the percentage label within the exchange villi was conducted [20].

In our study, there was a decline in the expression of both lymphatic markers in preeclamptic versus normotensive women, irrespective of both HIV status and pregnancy type. A decline in sVEGFR-3 and sVEGFR-3 with a concomitant elevation of VEGF-C has been reported however, the ratio of sVEGFR-3/VEGF-C is further down-regulated in PE [4]. Clinically, in PE there is often a disproportionate accrual of interstitial fluid that may corroborate a dysfunctional lymphatic drainage that eventuates in oedema [23].

In our study, lymphatic markers were down regulated in EOPE group compared to the LOPE and normotensive groups. Abnormal decidual lymphangiogenesis is linked to a decline of decidual Tregs (T-regulatory cells) in severe PE [24]. In our study, vascular endothelial cells rather than lymphatic vessel endothelial cells were immunostained by LYVE-1 whilst PDPN immunostaining showed a reticular interstitial matrix. This made us conclude that there was an absence of lymphatic vessels within the placental villi. Several studies support our observation of an absence of lymphatic vessels within the placenta [18, 20, 24, 25]. However, lymphatic vessels were observed within the decidua [25]. Nonetheless we suggest that the role of the lymphatic system is taken up by macrophage like cells and that the reticular PDPN network serves to mop up the extra fluid and take it to into the vascular circulation.

Based on HIV status the immunostaining of LYVE-1 and PDPN within the exchange villi was significantly higher in HIV^+ compared to the HIV^- group irrespective of their pregnancy type. In contrast the immunostaining of both LYVE-1 and PDPN within the conducting villi were similar in HIV^+ and HIV^- groups. The HIV-1 accessory proteins tat (trans-activator of transcription) and p17 mimic VEGF, promoting angiogenesis. However anti-retroviral therapy such as HAART (High Active Retroviral Therapy) reconstitutes immune response, decreases the inflammatory milieu of PE whilst decreasing angiogenesis. Moreover, nucleoside reverse transcriptase inhibitors prompt mitochondrial oxidative stress and weakens receptor tyrosine kinase signalling thereby affecting EC proliferation and invasion and inhibiting lymphangiogenesis [26]. In HIV infection, HIV-1 matrix protein p17 directly generates a pro-lymphangiogenic microenvironment [27]. Moreover, HIV-1 gp120 prompts the release of both angiogenic, and lymphangiogenic elements from human lung mast cells hence may have a similar activation role in PE [28].

In conclusion, this study, shows a similar immunostaining expression of LYVE-1 and PDPN in the placenta of HIV+ compared to the HIV- group irrespective of their pregnancy type. This similarity may be attributed to HIV-1 proteins and the immune reconstitution effect of antiretroviral therapy. It highlights the benefits of computer assisted morphometric image analysis as a reliable tool for quantification of immunostaining in biological tissue.

References

1. Brown, M.A., Magee, L.A., Kenny, L.C., Karumanchi, S.A., McCarthy, F.P., Saito, S., et al.: Hypertensive disorders of pregnancy: ISSHP classification, diagnosis, and management recommendations for international practice. Hypertension **72**(1), 24–43 (2018)
2. Lambert, G., Brichant, J.F., Hartstein, G., Bonhomme, V., Dewandre, P.Y.: Pre-eclampsia: an update. Acta Anaesthesiologica Belgica **65**, 137–149 (2014)
3. Burton, G.J., Woods, A.W., Jauniaux, E., Kingdom, J.C.P.: Rheological and physiological consequences of conversion of the maternal spiral arteries for uteroplacental blood flow during human pregnancy. Placenta **30**(6), 473–482 (2009)
4. Gathiram, P., Moodley, J.: Pre-eclampsia: its pathogenesis and pathophysiology. Cardiovasc. J. Afr. **27**(2), 71–78 (2016)
5. Rana, S., Lemoine, E., Granger, J.P., Karumanchi, S.A.: Preeclampsia: pathophysiology, challenges, and perspectives. Circ. Res. **124**(7), 1094–1112 (2019)

6. Mutter, W.P., Karumanchi, S.A.: Molecular mechanisms of preeclampsia. Microvasc. Res. **75**(1), 1–8 (2008)
7. Maynard, S.E., Min, J.-Y., Merchan, J., et al.: Excess placental soluble FMS-like tyrosine kinase 1 (sFlt1) may contribute to endothelial dysfunction, hypertension, and proteinuria in preeclampsia. J. Clin. Invest. **111**(5), 649–658 (2003)
8. Lely, A.T., Salahuddin, S., Holwerda, K.M., et al.: Circulating lymphangiogenetic factors in preeclampsia. Hypertens. Pregnancy **32**(1), 42–49 (2013)
9. Govener, N., Moodley, J., Gathiram, P., Naicker, T.: Soluble fms-like tyrosine kinase-1 in HIV infected pre-eclamptic South African Black women. Placenta **35**, 618–624 (2014)
10. Govender, N., Naicker, T., Moodley, J.: Endoglin in HIV-associated preeclamptic placentae. Hypertens Pregnancy **34**, 342–354 (2015)
11. Maharaj, N.R., Phulukdaree, A., Nagiah, S., Ramkaran, P., Tiloke, C., Chuturgoon, A.A.: Pro-inflammatory cytokine levels in HIV infected and uninfected pregnant women with and without preeclampsia. PLoS ONE **12**(1), 1–9 (2017)
12. Alitalo, K., Carmeliet, P.: Molecular mechanism of lymphangiogenesis in health and disease. Cancer Cell **1**, 219–227 (2002)
13. Martel, C., et al.: Lymphatic vasculature mediates macrophage reverse cholesterol transport in mice. J. Clin. Invest. **123**, 1571–1579 (2013)
14. Tammela, T., Alitalo, K.: Lymphangiogenesis: molecular mechanisms and future promise. Cell **140**, 46–476 (2010)
15. Martinez-Corral, I., Makinen, T.: Regulation of lymphatic vascular morphogenesis: implications for pathological (tumor) lymphangiogenesis. Exp. Cell Res. **319**, 1618–1625 (2013)
16. Proulx, S.T., Detmar, M.: Molecular mechanism and imaging of lymphatic metastasis. Exp. Cell Res. **319**, 1611–1617 (2013)
17. Volchek, M., Girling, J.E., Lash, G.E., et al.: Lymphatics in the human endometrium disappear during decidualization. Human Reprod. **25**(10), 2455–2464 (2010)
18. Wang, Y., Sun, J., Gu, Y., et al.: D2–40/podoplanin expression in the human placenta. Placenta **32**, 27–32 (2011)
19. Castro, E., Tony, P.W., Galambos, C.: Neither normal nor diseased placentas contain lymphatic vessels. Placenta **4**, 310–316 (2011)
20. Onyangunga, O.A., Moodley, J., Merhar, V., Ofusori, D.A., Naicker, T.: Lymphatic vascular endothelial hyaluronan receptor-1 immunoexpression in placenta of HIV infected pre-eclamptic women. J. Reprod. Immunol. **117**, 81–88 (2016)
21. Onyangunga, O., Moodley, J., Odun-Ayo, F., Naicker, T.: Immunohistochemical localization of podoplanin in the placentas of HIV-positive preeclamptic women. Turk. J. Med. Scis. **48**, 916–924 (2018)
22. Ross, M.H., Paulina, W.: Histology. A text and atlas, 7th edn. Wolters Kluwer, Philadelphia (2015)
23. Shange, G.P., Moodley, J., Naicker, T.: Effect of vascular endothelial growth factors a, c, and d in HIV-associated pre-eclampsia. Hypertens. Pregnancy **36**, 196–203 (2017)
24. Junk, Y.J., Park, Y., Kim, H.S., Lee, H.J., et al.: Abnormal lymphatic vessel development is associated with decreased decidual regulatory T cells in severe preeclampsia. Am. J. Reprod. Immunol. **80**(1), e12970 (2018)
25. Liu, H., Li, Y., Zhang, J., Rao, M., Liang, H., Liu, G.: The defect of both angiogenesis and lymphangiogenesis is involved in preeclampsia. Placenta **36**(3), 279–286 (2015)
26. Song, L., Ding, S., Ge, Z., Zhu, H., et al.: Nucleoside/nucleotide reverse transcriptase inhibitors attenuate angiogenesis and lymphangiogenesis by impairing receptor tyrosine kinases signalling in endothelial cells. Br. J. Pharmacol. **175**, 1241–1259 (2018)

27. Caccuri, F., Rueckert, C., Giagulli, C., Schulze, K., et al.: HIV-1 matrix protein p17 promotes lymphangiogenesis and activates the endothelin-1/endothelin b receptor axis. Arterioscler. Thromb. Vasc. Biol. **34**, 846–856 (2014)
28. Marone, G., Rossi, F.W., Pecoraro, A., Pucino, V., et al.: HIV gp120 induces the release of proinflammatory, angiogenic, and lymphangiogenic factors from human lung mast cells. Vaccines **8**, 208 (2020)

Excitable Structures and Motor Activity

Effect of Changes in the Intracellular Resistivity of Skeletal Muscle Fibre on Intracellular and Extracellular Potentials

Vladimir G. Dimitrov and Alexander G. Dimitrov(✉)

Bulgarian Academy of Sciences, Institute of Biophysics and Biomedical Engineering, Acad. G. Bonchev Str., Bl. 105, Sofia 1113, Bulgaria
vgdi3@abv.bg, agd@biomed.bas.bg

Abstract. Intensive muscle contractions are accompanied by metabolic changes that could affect the ionic composition of the cytoplasm of muscle fibre and thus its resistivity. We aim to study the effects of intracellular resistivity on intracellular action potentials (IAPs) and extracellular potentials (EPs). The Hodgkin-Huxley type mathematical model was used to simulate the intracellular processes in a muscle fibre. The spatial potential profile of the IAP was used to calculate the EP. The results showed that changes in the intracellular resistivity had practically no effect on the amplitude and shape of IAP in the temporal domain. The main effect was on the velocity of IAP propagation and thus on the length of IAP profile in the spatial domain. Changes in the length of the IAP profile affected the amplitude of the EPs above the end-plate region and above the middle of the fibre semilength to a similar extent. Effects of changes in the velocity of IAP propagation on the duration of the main negative phase of EPs were similar for detection above the end-plate region and above the middle of the fibre semilength. The results will help in analysis of changes in the M-waves obtained during and after muscle activity.

Keywords: H-H type model · Amplitude of extracellular potential · Duration of extracellular potential · Propagation velocity · Electrode position

1 Introduction

The electrically evoked muscle action potential (M-wave) is often used in sports, clinical and scientific studies for a non-invasive estimating of changes in the functional state of skeletal muscles (Botter and Merletti 2016). M-wave is the sum of extracellular potentials (EPs) produced by individual fibres of the muscle. The EPs are generated by intracellular action potentials (IAPs) that are induced at the end-plates and propagate to the ends of the respective muscle fibres, in response to stimulation. The IAP and velocity of its propagation along the fibre can be changed during muscle activity (Hanson and Persson 1971; Hanson 1974). This could reflect changes in activity of channels, transporters, pumps, etc. in the sarcolemma and in the tubular system membrane. Intensive muscle contractions are also accompanied by metabolic changes that could affect the ionic composition of the cytoplasm of muscle fibres and the muscle fibre water content

S. S. Sotirov et al. (Eds.): BioInfoMed 2020, LNNS 374, pp. 411–420, 2022.
https://doi.org/10.1007/978-3-030-96638-6_43

(Sjogaard et al. 1985; Okada 2004; Hoffman 2009; Kay and Blaustein 2019), which could influence the fibre resistivity. One of the aims of this paper is to clarify the effects of changes in the intracellular resistivity of a muscle fibre on IAPs and EPs produced by the fibre.

Another problem that exists in experimental studies is the choice of the proper site for M-wave detection. Detection above the end-plate region is often used. However, it has long been thought (Cupido et al. 1996) that such detection is not sensitive to changes in the velocity of IAP propagation because the EPs at this site are produced before the impulses propagate away from the end-plate. Thus, another aim of the present study is to verify the sensitivity of EPs detected above the end-plate and above the middle of the fibre semilength to changes in the intracellular resistivity that might affect the velocity of IAP propagation along the fibre.

2 Method

The method used was mathematical modelling. We based on the Hodgkin-Huxley (1952) type model for human muscle fibre at normal temperature (Stephanova and Dimitrov 1982; Kleinpenning et al. 1990). We simulated changes in the cable properties of muscle fibre related to changes in the intracellular resistivity and studied their effects on IAPs and EPs. The initial cable parameters are represented in Table 1. In our simulations, the initial intracellular resistivity (r_i in Table 1) was changed proportionally to a coefficient k_r, i.e. $r_{i}\_new = k_r \cdot r_i$, where $k_r = 1/9$; 1/4; 1; 4 or 9.

The cylindrical fibre 80 mm long was represented through 801 segments of equal length. A rectangular stimulus 0.2 ms duration was applied to the midpoint of the fibre, which simulated the end-plate (Fig. 1).

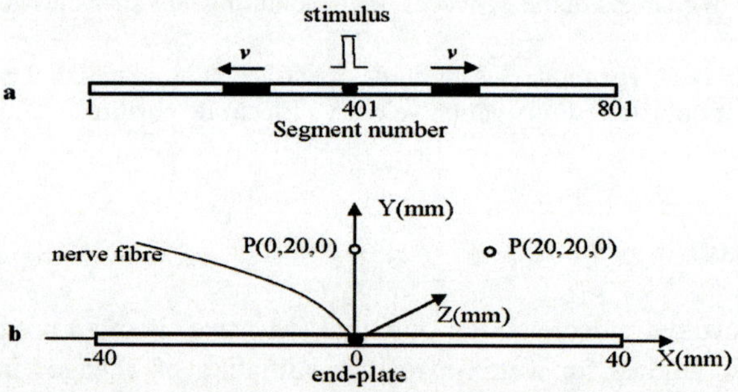

Fig. 1. Schematic representation of a muscle fibre used in simulations of (a) intracellular action potentials (IAPs) and (b) extracellular potentials (EPs).

Then, a discretized version of IAP along the fibre ($\varphi_{iap}(x)$) was calculated for each time point (row in the IAP matrix). IAP in the time domain at any point along the fibre was the column in the IAP matrix corresponding to that point. To demonstrate the IAP as a function of time for different cases of intracellular resistivity, we chose the midpoint of the fibre semilength, i.e. 601st segment of the fibre (Fig. 1a), which corresponded to x = 20 mm from the end-plate (Fig. 1b).

Table 1. Initial parameters of human muscle fibre used in simulations of intracellular action potentials

Model parameter	Parameter value
Temperature, T (°C)	37
Intracellular axial step size, Δx (cm)	0.01
Intracellular time step size, Δt (ms)	0.01
Fibre radius, a (μm)	27.5
Intracellular conductivity, σ_i ($\Omega^{-1}cm^{-1}$)	0.01
Intracellular resistivity, $r_i = 1/\sigma_i$ (Ω cm)	100
Intracellular resistance per unit length $R_i = r_i/(\pi a^2)$ (Ω cm^{-1})	$4.2091 \cdot 10^6$
Membrane capacitance, C (μF cm^{-2})	1
Temperature coefficient for conductance, Q_{10}	1.5
Temperature correction factor for conductance at T = 37°C, QT = $Q_{10}^{(T-2)/10}$	4.1335
Maximum sodium conductance per unit area at T = 2°C, g_{Na2} (Ω^{-1} cm^{-2})*	$70 \cdot 10^{-3}$
Maximum sodium conductance per unit area, T = 37°C, $g_{Na} = g_{Na2}QT(\Omega^{-1}$ cm$^{-2})$	$289.3 \cdot 10^{-3}$
Maximum potassium conductance per unit area at T = 2°C, g_{K2} (Ω^{-1} cm^{-2})*	$20 \cdot 10^{-3}$
Maximum potassium conductance per unit area, T = 37°C, $g_K = g_{K2}QT(\Omega^{-1}$ cm$^{-2})$	$82.67 \cdot 10^{-3}$
Maximum leak conductance per unit area at T = 37°C, gL (Ω^{-1} cm^{-2})	2
Temperature coefficient for rate constants, Q_{10}	$0.24 \cdot 10^{-3}$
Temperature correction factor for rate constants at T = 37°C, QT = $Q_{10}^{(T-2)/10}$	11.3
Membrane resting potential, E_r (mV)	−77
Sodium equilibrium potential, E_{Na} (mV)	30
Potassium equilibrium potential, E_K (mV)	−70
Leakage potential, E_L (mV)	−77.4
Initial fibre semilengths, L (mm)	40

* Adrian et al. (1970)

The spatial profile of IAP along the fibre ($\varphi_{iap}(x)$) for each time point (t) was then used to calculate EPs ($\varphi_e(P,t)$). The EP produced at the point of observation $P(x_o, y_o, z_o)$ in the infinite volume conductor (Dimitrov and Dimitrova 1998; Dimitrova and Dimitrov 2006) is:

$$\varphi_e(x_o, y_o, z_o, t) = -K \int_S dS \int_{-\infty}^{\infty} \frac{\partial \varphi_{iap}(x)}{\partial x} \cdot \frac{\partial (1/r)}{\partial x} \cdot dx = -K \cdot \pi a^2 \int_{-\infty}^{\infty} \frac{\partial \varphi_{iap}(x)}{\partial x} \cdot \frac{\partial (1/r)}{\partial x} \cdot dx \quad (1)$$

where $r = \sqrt{(x_o - x_i)^2 + (y_o - y_i)^2 + (z_o - z_i)^2}$ is the distance from the corresponding i-th fibre segment to the observation point $P(x_o, y_o, z_o)$; dS is an element of the cross-sectional area of the fibre. In the case of cylindrical fibre considered in the present study, the first integral is reduced to the area of the fibre cross-section, i.e. to πa^2 term, in which a - is the fibre radius.

To simulate detection with surface electrodes, we calculated EPs produced by a single muscle fibre at radial distance $y_o = 20$ mm from the fibre axis (Fig. 1b). The point monopolar electrode $P(x_o, y_o, 0)$ was located either above the end-plate ($x_o = 0$) or above the middle of the fibre semilength ($x_o = 20$ mm).

3 Results

3.1 Effect on the Size of the Stimulation Current

Intracellular resistivity affected the size of the stimulation current required to activate the fibre. The current sufficient for the fibre with the initial parameters was found to be subthreshold for the fibre with a lower resistivity. Therefore, the magnitude of the stimulation current used was changed in inverse proportion to the square root of the corresponding k_r. Such a current was also useful because it induced similar latency of the IAPs at the end-plate. This reduced errors in estimating the velocity of propagation of the IAPs in fibres with various resistivities.

3.2 Effect on Intracellular Action Potentials

Changes in the intracellular resistivity did not practically affect the shape, amplitude, and duration of the IAP in the temporal domain (Fig. 2). However, the IAPs reached the midpoint of the fibre semilength (601st segment, x = 20mm from the end-plate, Fig. 1) with different latencies (l), which reflected changes in the velocity of IAP propagation along the fibre with variation in intracellular resistivity. The velocity can be estimated using the x/l ratio. For the fibre with initial parameters (black solid line in Fig. 2), the initial velocity (v_i) was about $v_i = 4$ m/s. As follows from Fig. 2, the velocity was inversely proportional to the square root of the corresponding changes in the resistivity, k_r, i.e. $v_{new} = v_i/\sqrt{k_r} = v_i \cdot K_{vel}$, where $K_{vel} = 1/\sqrt{k_r}$ (Fig. 3).

Thus, in contrast to identical IAPs in the temporal domain (Fig. 2), changes in the propagation velocity (Fig. 3) would lead to proportional variations in the length of the spatial profile of IAP along the fibre (Fig. 4). As follows from Fig. 4, in the case of muscle fibres, two kinds of IAP spatial profiles can be identified. One of them is that formed at

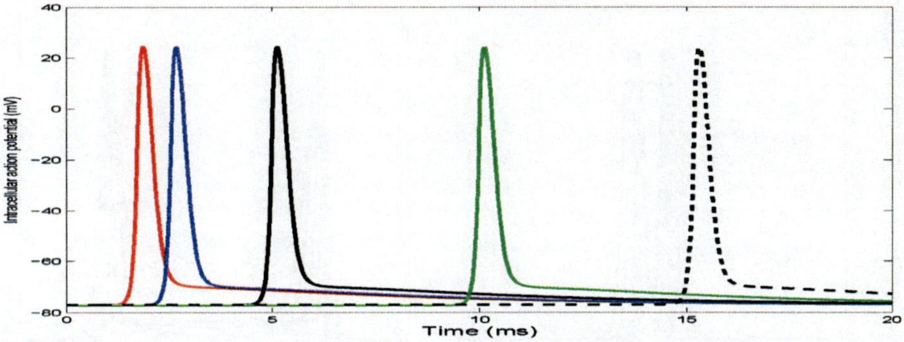

Fig. 2. Intracellular action potential in the time domain for fibres with different intracellular resistivity: $k_r = 1/9$ (red line); $k_r = 1/4$ (blue line); $k_r = 1$ (black bold line –the fibre with the initial resistivity); $k_r = 4$ (green line); $k_r = 9$ (black dashed line).

the end-plate region, where two waves of excitation arise (Fig. 4a). Another kind is that formed outside the end-plate region (Fig. 4b), where the two waves propagating to the fibre ends are already far from each other and from the end-plate. These peculiarities have to reflect on EPs, because generation of EPs is the spatial problem (Eq. 1).

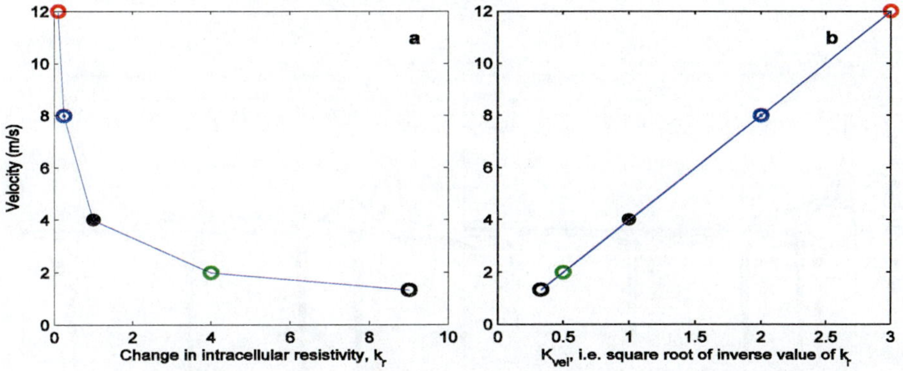

Fig. 3. Velocity of IAP propagation along the muscle fibre as (a) a function of changes in the intracellular resistivity k_r, and (b) a function of $K_{vel} = 1/\sqrt{k_r}$. Colors of points correspond to colors of lines in Fig. 2.

3.3 Effect on Extracellular Potentials

The amplitudes of the main phases of EPs produced by fibres with different intracellular resistivity differ (Fig. 5). Note that in Fig. 5 the negativity of EPs is shown as an upward deflection, as accepted in electrophysiology. The amplitude of EPs detected above the end-plate region (Fig. 5a) is higher and duration of their negative phase is shorter than those of EPs detected further away from the end-plate (Fig. 5b). Duration of the negative

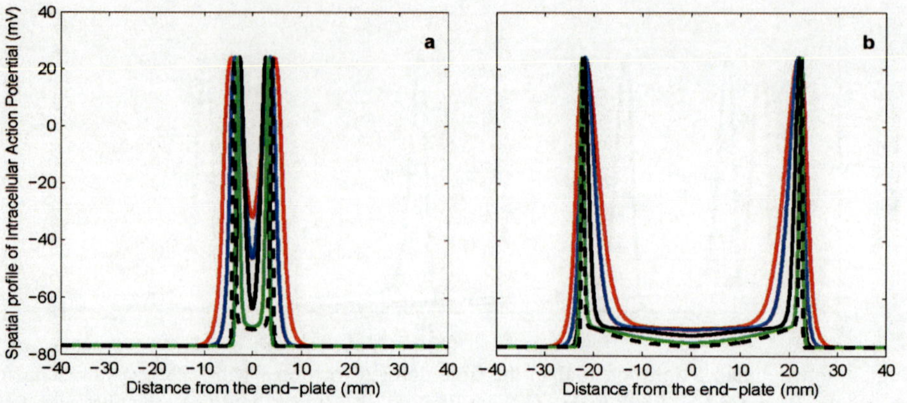

Fig. 4. Spatial profiles of IAPs in the end-plate region (a) and in the midpoint of the fibre semilength (b). The IAP spatial profiles are represented for the moments of time, at which the negative maxima of the respective EPs from Fig. 5 are produced.

phase of EP was defined by the time interval between the zero crossing points. The higher the velocity, the proportionally shorter is the negative phase (Fig. 6). Changes in the amplitude and duration of the negative phase with changes in intracellular resistivity are similar at both sites of the EP detection.

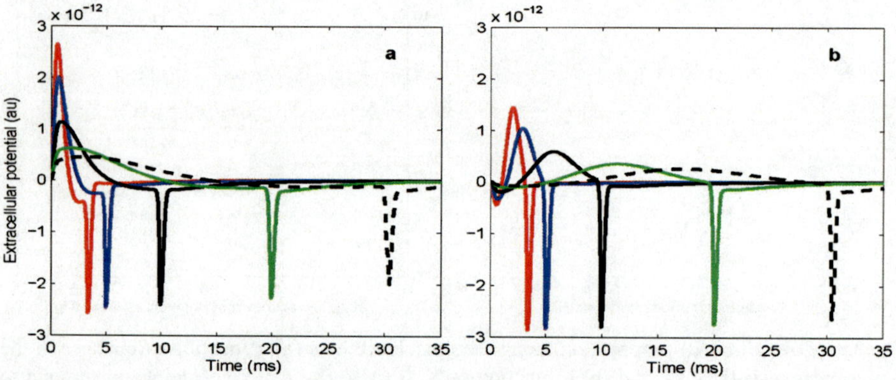

Fig. 5. Extracellular potentials (EPs) produced at a point electrode located at radial distance from the fibre axis y = 20 mm (a) above the end-plate and (b) above the midpoint of the fibre semilength. The colors of lines correspond to those in Fig. 2. Negativity of EPs is shown as an upward deflection as accepted in electrophysiology.

The fibre considered in the present paper is symmetrical in respect to the end-plate. Therefore, the excitation waves reached the fibre ends and extinguished there simultaneously creating a single terminal phase (sharp downward deflections in Figs. 5a and 5b), whose shape repeated that of IAP in the temporal domain. Hence, the velocity of

IAP propagation can be also estimated through the ratio of the fibre semilength (40 mm) and the time of appearance of the spike in the terminal phase of EPs.

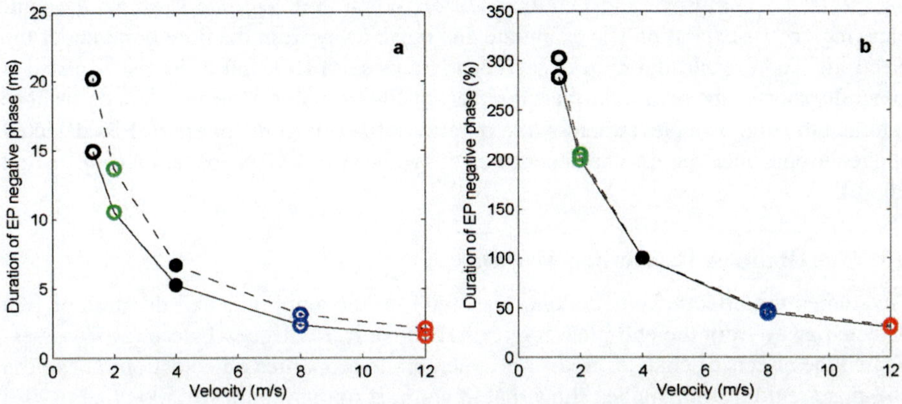

Fig. 6. Effect of propagation velocity on duration of the negative phase of EP detected above the end-plate region (solid line) and above the fibre semilength (dashed line); (a) duration is expressed in ms; (b) duration is expressed in percents in relation to the initial values. Colors of points correspond to colors of lines in Fig. 2.

4 Discussion

Despite intensive research, the mechanisms of changes even in the simple parameters that characterize M-wave shape and size, such as the amplitude and duration of the different phases, are still not well understood (Rodrigues-Falces et al. 2019).

4.1 Amplitude of the Negative Phase of EPs

Our results show that even IAPs identical in the time domain (Fig. 2) can induce EPs that differ considerably in their amplitude and shape (Fig. 5), if they have different spatial profile length because of their different propagation velocity. Figure 4 represents IAP spatial profiles formed in the end-plate area (Fig. 4a) and in the middle of the fibre semilength (Fig. 4b). They correspond to the moments of time when the maximum of the negative phase of the particular EP is generated (Fig. 5). The IAP profiles corresponding to various intracellular resistivity, differ in their spatial lengths (Fig. 4) and thus in the effects on amplitude of detected EPs (Dimitrova 1973; Dimitrova and Dimitrov 2002; 2006). Thus, EPs detected above the end plate region are also sensitive to parameters of propagation. Above the end-plate region, both excitation waves are located symmetrically in respect to the point of EP detection. This makes the amplitudes of the EPs detected above the end-plate greater (Fig. 5a) than the amplitudes of EPs detected above the middle of the fibre semilength (Fig. 5b).

4.2 Duration of the Negative Phase of EPs

Above the midpoint of the fibre semilength, the negative phase of EP is formed when the closest wave of excitation approaches and moves away from the fibre point under the electrode (Dimitrova and Dimitrov 2006). When the electrode is above the end-plate region, the excitation waves initiate and move away from the fibre point under the electrode. As a result, duration of the negative phase of EPs and of its rising phase is generally shorter above the end-plate region (Fig. 5a) than above the midpoint of the fibre semilength (Fig. 5b). Nevertheless, the duration of the negative phase of EPs detected above the end-plate region also depends on the velocity of IAP propagation (Fig. 5a and Fig. 6).

4.3 The Obtained Results and M-wave

We studied the effect of intracellular resistivity on the amplitude and duration of EPs detected away from the end-plate region and above it. Rodrigues-Falces (2016) investigated the effects of changes in the fibre membrane properties on detections above the end-plate region. Both studies show that in contrast to opinion of Cupido et al. (1996) the EPs detected above the end-plate are sensitive to changes in parameters of IAP propagation. Our results are for a single muscle fibre. M-wave is the sum of individual single fibre EPs. If the muscle were consisted of fibres with similar initial properties and similar metabolic changes because of muscle activity, then both sites of M-wave detection would be similarly sensitive to changes in the intracellular resistivity. In the case of scattered diameters of fibres and differences in metabolic variations, the place of M-wave detection could affect the M-wave changes. The individual EPs produced above the end-plate region are synchronised, like those represented in Fig. 5a, as they are produced almost simultaneously during electrical stimulation. Their summation would be much more efficient (in the sense of greater M-wave amplitude) than that above the middle of the muscle semilength. In the latter case, the individual waves of excitation approach the middle of the fibre semilenght already desynchronized (Fig. 2). Similar is valid for EPs (Fig. 5b). However, the desynchronization should increase the duration of the negative phase of the M-wave more considerably away from the end-plate than above the end-plate.

Thus, we can conclude that both surface detections above the end-plate region and away from it are appropriate for detecting changes in the velocity of IAP propagation along the fibres, although they have their specifics. If both detections are available in experimental studies, their combination could be useful for estimating the scatter of propagation velocities in fibres, which may be induced also by changes in intracellular resistivity.

The range of changes in the intracellular resistivity considered in the present study was from 1/9 to 9 i.e. 81 times. What are the actual variations in metabolites induced by muscle activity and how they could affect the specific value of intracellular resistivity remains to be clarified, but as follows from our studies, they may affect the velocity of IAP propagation. Changes in the intracellular resistivity affected also the stimulus strength necessary for activation of fibres. This might make some fibres unexcitable or might increase delay of the IAP arising at the end-plate. This in turn might reduce also the M-wave amplitude.

5 Conclusions

Probable changes in the intracellular resistivity induced by muscle activity might result in: (1) changes in the excitability of single muscle fibres and thus in the amplitude of the M-wave; (2) no effect on IAP amplitude, shape, and duration in the temporal domain if the stimulus is above threshold; (3) changes in the IAP propagation velocity that is inversely proportional to the square root of the corresponding changes in resistivity; (4) decrease of the duration of the negative phase of EPs when the intracellular resistivity decreases and vice versa; (5) changes in the length of the IAP spatial profile and enlargement of the EP amplitude when the intracellular resistivity is reduced and vice versa; (6) despite their peculiarities, the EPs and M-waves detected above the end-plate region and away from it are sensitive to changes in the intracellular resistivity and thus in velocity of IAP propagation along the fibre.

References

Adrian, R.H., Chandler, W.K., Hodgkin, A.L.: Voltage clamp experiments in striated muscle fibres. J. Physiol. **208**(3), 607–644 (1970)

Botter, A., Merletti, R.: EMG of electrically stimulated muscles. In: Merletti, R., Farina, D. (eds.) Surface Electromyography. Physiology, Engineering, and Applications. IEEE Press Series in Biomedical Engineering, pp. 311–332. Wiley, New Jersey (2016)

Cupido, C.M., Galea, V., McComas, A.J.: Potentiation and depression of the M wave in human biceps brachii. J Physiol. **491**(2), 541–550 (1996)

Dimitrov, G.V., Dimitrova, N.A.: Precise and fast calculation of the motor unit potentials detected by a point and rectangular plate electrode. Med. Eng. Phys. **20**(5), 374–381 (1998)

Dimitrova, N.: Influence of the length of the depolarized zone on the extracellular potential field of a single unmyelinated nerve fibre. Electromyogr. Clin. Neurophysiol. **13**(5), 547–558 (1973)

Dimitrova, N.A., Dimitrov, G.V.: Amplitude-related characteristics of motor unit and M-wave potentials during fatigue. A simulation study using literature data on intracellular potential changes found in vitro. J. Electromyogr. Kinesiol. **12**(5), 339–349 (2002)

Dimitrova, N.A., Dimitrov, G.V.: Electromyography (EMG) modeling. In: Metin, A. (ed.) Wiley Encyclopedia of Biomedical Engineering. Wiley, New Jersey (2006)

Hanson, J., Persson, A.: Changes in the action potential and contraction of isolated frog muscle after repetitive stimulation. Acta Physiol. Scand. **81**(3), 340–348 (1971)

Hanson, J.: Effects of repetitive stimulation on membrane potentials and twitch in human and rat intercostal muscle fibres. Acta Physiol. Scand. **92**(2), 238–248 (1974)

Hodgkin, A.L., Huxley, A.F.: A quantitative description of membrane current and its application to conduction and excitation in nerve. J Physiol. **117**(4), 500–544 (1952)

Hoffman, E.K., Lambert, I.H., Pedersen, S.F.: Physiology of cell volume regulation in vertebrates. Physiol. Rev. **89**(1), 193–277 (2009)

Kay, A.R., Blaustein, M.P.: Evolution of our understanding of cell volume regulation by the pump-leak mechanism. J. Gen. Physiol. **151**(4), 407–416 (2019)

Kleinpenning, P.H., Gootzen, T.H.J.M., Van Oosterom, A., Stegeman, D.F.: The equivalent source description representing the extinction of an action potential at a muscle fiber ending. Math. Biosci. **101**(1), 41–61 (1990)

Okada, Y.: Ion channels and transporters involved in cell volume regulation and sensor mechanisms. Cell Biochem. Biophys. **41**(2), 233–258 (2004)

Rodriguez-Falces, J.: The formation of extracellular potentials over the innervation zone: are these potentials affected by changes in fibre membrane properties? Med. Biol. Eng. Comput. **54**(12), 1845–1858 (2016). https://doi.org/10.1007/s11517-016-1487-8

Rodriguez-Falces, J., Vieira, T., Place, N., Botter, A.: Potentiation of the first and second phases of the M wave after maximal voluntary contractions in the biceps brachii muscle. Med. Biol. Eng. Comput. **57**(10), 2231–2244 (2019). https://doi.org/10.1007/s11517-019-02025-7

Sjogaard, G., Adams, R.P., Saltin, B.: Water and ion shifts in skeletal muscle of humans with intense dynamic knee extension. Am J Physiol. **248**(2 Pt2), R190–R196 (1985)

Stephanova, D.I., Dimitrov, G.V.: Mathematical modelling of ionic processes in human skeletal muscle fibres. Electromyogr. Clin. Neurophysiol. **22**(5), 329–347 (1982)

InterCriteria Decision Making Approach for Osteoarthritis Disease Analysis

Bistra Zaharieva, Lyubka Doukovska(✉), and Slaviiana Danailova

Intelligent Systems Department, Institute of Information and Communication Technologies, Bulgarian Academy of Sciences, Acad. Georgi Bonchev Str., bl. 2, Sofia 1113, Bulgaria
lyubka.doukovska@iict.bas.bg

Abstract. The InterCriteria decision making approach is based on two fundamental concepts: intuitionistic fuzzy sets and index matrices. This paper continues a series of research on the application of approach of InterCriteria Analysis (ICrA) on real medical data. The ICrA is applied on the results of physiotherapeutic treatment and kinesiotherapeutical program characteristics. The main goal is an improvement of general quality of patients' life through practices of specific methodology of kinesiotherapy. The knee performs the function of a medial motor link in the kinetic chain of the lower limb. The osteoarthritis is one of the most devastating chronic conditions that affect people around the world. The paper describes achieved results from using a multicriteria decision making approach of analyzing the treatment results of patients with Osteoarthritis disease in order to improve the process of taking decisions about the treatment course.

Keywords: Human health · Osteoarthritis disease · Kinesiotherapy · Intelligent systems · InterCriteria analysis

1 Introduction

One of the most common diseases of the locomotory system is the osteoarthritis of the knee joint. It affects mainly middle-aged people, prevailing female. The main reason why the knee is being worn out are old age, static abnormalities, overloading, traumas, inner joint injures and diseases. All these unfavorable factors lead to surgical intervention and inserting of artificial joints at a later stage. However, before this surgical interference the patient can undergo kinesiotherapy which could have quite favorable effect on their condition. In this way the endoprosthesis of the knee joint is delayed in time. For the purposes of this research we analyzed the results received from the observation of 10 patients [2]. The patients have been treated by specialized kinesiotherapy methodology aiming at increasing the volume of movement in the cervical partition and decreasing limited mobility. While taking decisions about the treatment process we have applied ICrA of the data received from the observation of the patient's state in order to find the subordination between the restorations of the separate movements in the knee joint.

The InterCriteria decision making approach is based on two fundamental concepts: intuitionistic fuzzy sets and index matrices. The detailed analysis of the functionality of

S. S. Sotirov et al. (Eds.): BioInfoMed 2020, LNNS 374, pp. 421–432, 2022.
https://doi.org/10.1007/978-3-030-96638-6_44

this decision making approach, as well as its specifics, are presented in the paper [3]. The theoretical basis on which its development this approach is presented in a series of publications [8–11]. In recent years, this approach been used successfully in the analysis of real medical data of patients with Behterev's disease, results are published in papers [13–15]. The paper describes achieved results from using a multicriteria decision making approach of analyzing the treatment results of patients with Osteoarthritis disease in order to improve the process of taking decisions about the treatment course.

2 Disease

Osteoarthritis of the knee is a slowly progressive, not primarily inflammatory degenerative disease (osteoarthritis) of the knee joint. Joint wear and tear occurs when the cartilage that acts as a buffer between the bones breaks down, leading to pain in the joint structures. Pain is usually felt when bending the knee, when climbing stairs and also after long periods sitting or lying down. Pain in the morning is a common complaint in the early stages of knee joint osteoarthritis. The joints feel stiff. The pain usually subsides after the joint has been able to warm up. A worn joint can easily become inflamed. Osteoarthritis that was initially non-inflammatory therefore becomes activated osteoarthritis with inflammatory phases. The aim is to keep the inflammatory phases as brief as possible by using medication.

The gonartrosis leads to medical, social and psychological problems, deriving primarily from the accompanying restricted movements of the affected lower limb and the tendency of invadilizing the people suffering from it. The main reasons why the knee joint is being worn out are old age, static abnormalities (e.g. overloading, traumas, inner joint injuries and diseases). Among the most widely-spread causes are the constant overloading, obesity, hard physical labor and hypokinesia [1].

All these unfavorable factors lead to surgical intervention and inserting of artificial joints at a later stage. However, before this surgical interference the patient can undergo kinesiotherapy which could have quite favorable effect on their condition. In this way the endoprosthesis of the knee joint is delayed in time.

3 Etiology

The knee joint is one of the largest joints which bears the weight of the body and is of essential importance for the correct posture and gait. It has got three sections- outer, inner and patellofemoral. The gonartrosis can be isolated only in one of these sections, but it can, as well, affect two or even all three of them. The one located in the inner part of the knee is exposed to greater mechanical load and is, therefore, most often affected [4].

Gonartrosis is a chronic degenerative illness of the knee joint, which is a part of the locomotory system. Changes begin in the joint cartilage and gradually spread to the joint capsule and the connections, involving as well the bones which form the knee joint.

4 Symptoms

Typical signs of osteoarthritis in the knee joint are:

- Knee pain when climbing stairs and when walking on uneven ground.
- Pain is particularly bad when starting to move after long periods of rest and then slowly subsides only to recur after prolonged periods of stress.
- Symptoms increase after carrying heavy loads.
- In the course, there may be a noticeable swelling.
- In the advanced stages, crunching and crackling noises can be heard when the knee is moved.
- The joint is particularly sensitive in damp and cold weather conditions.

The more advanced the osteoarthritis, the more pronounced the symptoms: The pain becomes more frequent and often occurs even when at rest. Knee mobility continues to decrease.

5 Kinesiotherapy

The aim of kinesiotherapy with gonartrosis is, as with all other degenerative joint illnesses, to rehabilitate the locomotory function of the lower limbs to the maximum possible degree. According to some sources [5] most exercises with patients having osteoarthritis are based on the biomechanical concept stating that muscles have an important protective function over the joints through sustaining normal alignment and absorption of stress and pressure. Patients with osteoarthritis also show muscular fiber atrophy which is more clearly expressed than the one typical of aging. In this way the pressure which cannot be taken by these weakened muscles, tendons and ligaments goes to the joint cartilage and the liable bone.

The goal of this research is to develop a specialized methodology of treatment, including exercises and techniques for joint mobilization and mobilization of the nervous system, and to investigate its efficiency with patients having primary one-sided gonartrosis.

Ten patients with clinical diagnosis of knee joint arthritis have been examined. All of them are female, with moderately expressed motor deficit and pain symptoms occurring while moving. The average age of the patients is 62,9 years. The length of their complaints is between 5 and 7 years. From this group 4 patients have more distinct symptoms in their right limb, while 6 of them- in their left limb. With all patients we applied a methodology of kinesiotherapy and registered the changes in their functional condition. All patients underwent 10 procedures.

6 Methods of Examination

Angulation is a method of measuring the volume of joint mobility. The results are registered using SFTR methodology where S is used to determine the sagittal plane, F is the frontal plane, T is the transversal plane and R stands for the rotary movements. The results are written with three numbers, the middle part showing the initial (zero) position of the joint. We used a universal two-shouldered octant to measure the physiologically active volume of movement in the knee joint and the results were reported according to the standard SFTR methodology [6].

The generic tool for estimation of the general health state (Social Functioning, SF 36) does not give a common result. It is divided into physical and mental components and estimates different aspects like vitality, physical function, pain, general condition, emotional state, social realization, etc. This instrument has got a wide application because it can be used regardless of the age or the pathology of the patients.

The generic tool provides information about the functional and general state from the patients' point of view [7]. The following indexes are being quantitatively estimated:

1. Flexion (FL) – the maximum flexion of the knee joint is about 150 degrees. When there are diseases like gonartrosis, this flexion decreases to 100–110°.
2. Physical Functioning (PF) – it describes the degree of limiting the performance of physical activities by the physical condition. Lower results on this scale indicate that the motor activity of the patient is considerably restricted by the state of their health.
3. Role-Physical Functioning (RPF) – it describes the influence of the physical state on the everyday role functioning. Lower results on this scale indicate that everyday activity is considerably restricted by the physical condition of the patient.
4. Role-Emotional Functioning (REF) – it estimates the degree to which the emotional state prevents the working process or other everyday activities. Lower results on this scale are interpreted as restriction in performing everyday work due to deteriorated emotional state.
5. General Health (GH) – estimates patients' health condition at the present moment and their treatment perspectives. The lower the result on this scale gets, the lower the esteem about the health condition is.
6. Vitality (VT) – it characterizes the feeling of energy and strength, or the opposite, the feeling of weakness. The lower result indicates that the patient gets easily tired and that their vital activity has lowered.
7. Social Functioning (SF) – it determines the degree of restriction of social activity by the physical or emotional state. Lower results indicate considerable restriction of social contacts, decrease of the level of communication due to deteriorated physical and emotional state.
8. Bodily Pain (BP) – it determines the ability to fulfill everyday activities, including work at home and outside. Lower results on this scale indicate that pain considerably restricts patient's activity.
9. Mental Health (MH) – it is characterised by the presence of depressive mood, anxiety and the total percent of positive emotions. Lower results indicate the presence of depressive, unrestful experiences and mental adversity.

7 InterCriteria Decision Making Approach

The presented multicriteria decision making approach is based on two fundamental concepts: intuitionistic fuzzy sets and index matrices. It is called 'InterCriteria decision making' approach [3].

Intuitionistic fuzzy sets defined by Atanassov [8–10] represent an extension of the concept of fuzzy sets, as defined by Zadeh [12], exhibiting function $\mu_A(x)$ defining the membership of an element x to the set A, evaluated in the [0; 1] interval. The difference between fuzzy sets and intuitionistic fuzzy sets (IFSs) is in the presence of a second function $\nu_A(x)$ defining the non-membership of the element x to the set A, where:

$$\begin{gathered} 0 < \mu_A(x) \leq 1, \\ 0 \leq \nu_A(x) \leq 1, \\ 0 \leq \mu_A(x) + \nu_A(x) \leq 1. \end{gathered} \tag{1}$$

The IFS itself is formally denoted by:

$$\mathrm{A} = \{\langle \mathrm{x}, \mu_A(x), \nu_A(x) \,|x \in \mathrm{E}\rangle\} \tag{2}$$

Comparison between elements of any two IFSs, say A and B, involves pairwise comparisons between their respective elements' degrees of membership and non-membership to both sets.

The second concept on which the proposed method relies is the concept of index matrix, a matrix which features two index sets. The theory behind the index matrices is described in [11]. Here we will start with the index matrix M with index sets with m rows $\{C_1, \ldots, C_m\}$ and n columns $\{O_1, \ldots, O_n\}$:

$$M = \begin{array}{c|ccccccc} & O_1 & \cdots & O_k & \cdots & O_l & \cdots & O_n \\ \hline C_1 & a_{C_1,O_1} & \cdots & a_{C_1,O_k} & \cdots & a_{C_1,O_l} & \cdots & a_{C_1,O_n} \\ \vdots & \vdots & \ddots & \vdots & \ddots & \vdots & \ddots & \vdots \\ C_i & a_{C_i,O_1} & \cdots & a_{C_i,O_k} & \cdots & a_{C_i,O_l} & \cdots & a_{C_i,O_n} \\ \vdots & \vdots & \ddots & \vdots & \ddots & \vdots & \ddots & \vdots \\ C_j & a_{C_j,O_1} & \cdots & a_{C_j,O_k} & \cdots & a_{C_j,O_l} & \cdots & a_{C_j,O_n} \\ \vdots & \vdots & \ddots & \vdots & \ddots & \vdots & \ddots & \vdots \\ C_m & a_{C_m,O_1} & \cdots & a_{C_m,O_j} & \cdots & a_{C_m,O_l} & \cdots & a_{C_m,O_n} \end{array}, \tag{3}$$

where for every p, q ($1 \leq p \leq m$, $1 \leq q \leq n$), C_p is a criterion (in our case, one of the twelve pillars), O_q in an evaluated object, aC_pO_q is the evaluation of the q-th object against the p-th criterion, and it is defined as a real number or another object that is comparable according to relation R with all the rest elements of the index matrix M, so that for each i, j, k it holds the relation $R(aC_kO_i, aC_kO_j)$. The relation R has dual relation $\overline{R}$, which is true in the cases when relation R is false, and vice versa.

For the needs of our decision making method, pairwise comparisons between every two different criteria are made along all evaluated objects. During the comparison, it is maintained one counter of the number of times when the relation R holds, and another counter for the dual relation.

Let $S^{\mu}_{k,l}$ be the number of cases in which the relations $R(aC_kO_i, aC_kO_j)$ and $R(aC_lO_i, aC_lO_j)$ are simultaneously satisfied.

Let also $S^{\nu}_{k,l}$ be the number of cases in which the relations $R(aC_kO_i, aC_kO_j)$ and its dual $\overline{R}(aC_lO_i, aC_lO_j)$ are simultaneously satisfied. As the total number of pairwise comparisons between the object is $n(n-1)/2$, it is seen that there hold the inequalities:

$$0 \leq S^{\mu}_{k,l} + S^{\nu}_{k,l} \leq \frac{n(n-1)}{2} \tag{4}$$

For every k, l, such that $1 \leq k \leq 1 \leq m$, and for $n \geq 2$ two numbers are defined:

$$\mu_{C_k,C_l} = 2\frac{S^{\mu}_{k,l}}{n(n-1)}, \quad \nu_{C_k,C_l} = 2\frac{S^{\nu}_{k,l}}{n(n-1)} \tag{5}$$

The pair constructed from these two numbers plays the role of the intuitionistic fuzzy evaluation of the relations that can be established between any two criteria C_k and C_l. In this way the index matrix M that relates evaluated objects with evaluating criteria can be transformed to another index matrix $M*$ that gives the relations among the criteria:

$$M^* = \begin{array}{c|ccc} & C_1 & \cdots & C_m \\ \hline C_1 & \langle \mu_{C_1,C_1}, \nu_{C_1,C_1} \rangle & \cdots & \langle \mu_{C_1,C_m}, \nu_{C_1,C_m} \rangle \\ \vdots & \vdots & \ddots & \vdots \\ C_m & \langle \mu_{C_m,C_1}, \nu_{C_m,C_1} \rangle & \cdots & \langle \mu_{C_m,C_m}, \nu_{C_m,C_m} \rangle \end{array}. \tag{6}$$

The final step of the algorithm is to determine the degrees of correlation between the criteria, depending on the user's choice of μ and ν. We call these correlations between the criteria: 'positive consonance', 'negative consonance' or 'dissonance'.

Let α, $\beta \in [0; 1]$ be given, so that $\alpha + \beta \leq 1$. We call that criteria C_k and C_l are in:

- (α, β) - positive consonance, if $\mu C_k, C_l > \alpha$ and $\nu C_k, C_l < \beta$;
- (α, β) - negative consonance, if $\mu C_k, C_l < \beta$ and $\nu C_k, C_l > \alpha$;
- (α, β) - dissonance, otherwise.

Obviously, the larger α and/or the smaller β, the fewer number of criteria may be simultaneously connected with the relation of (α, β) - positive consonance. For practical purposes, it carries the most information when either the positive or the negative consonance is as large as possible, while the cases of dissonance are less informative and can be skipped.

8 Results and Discussion

In this research we analyzed the results received from the observation of 10 patients. The patients have been treated by specialized kinesiotherapy methodology aiming at increasing the volume of movement in the cervical partition and decreasing limited mobility. As a result of kinesioterapeutical (KT) program the following summarized results have been presented in Table 1 for the beginning of the course and in Table 2 for the end of the course of treatment of the observed patients. In tabulation of results from the patients observation will be using the indices based on the generic tool for estimation of the general health state (Social Functioning, SF 36).

Table 1. Results before KT program

№	FL	PF	RPF	REF	VT	SF	BP	GH	MH
1	105	40	25	30	48	37	45	30	25
2	110	55	40	50	68	62	50.5	80	12.1
3	96	30.5	54	46.6	40	30	49.9	60	48
4	105	44.7	38	40	45	59	60	50	69
5	100	50	60	49	50	60	40	55	70
6	112	28	40	35	40	47	60	30	20
7	95	40	55	45	60	50	40	56	51
8	90	50.8	40.2	45	49	50	60	40	37
9	105	63.3	30	39	50	55	40	50	59
10	90	40	50	49	50	40	50	40	35

Table 2. Results after KT program

№	FL	PF	RPF	REF	VT	SF	BP	GH	MH
1	115	95	55	66.7	52	37.5	22.5	35	50
2	120	70	60	69.9	68	62.5	12.5	80	50
3	113	50	60	50	70	50	15	80	60
4	117	60	45.6	47	60	66	40	55	75
5	110	59	61	55	59	75	20	60	85.7
6	118	40	60	55	60	58	40	49	30
7	105	49	65	59	71	65	31	60	70
8	103	65	40	49	55	59	50	49	49
9	118	70	45	49	70	65	20	69	68
10	110	59	55	59	59	57	45	59	44

The obtained results show a change in the observed indicators, which leads to an improvement in the health status of the patients. This gives us reason to apply the InterCriteria decision making approach to analyze these indicators.

Table 3. Membership pairs of the intuitionistic fuzzy IC correlations at the beginning of the observation

μ	FL	PF	RPF	REF	VT	SF	BP	GH	MH
FL	1	0.422222	0.311111	0.311111	0.355556	0.555556	0.444444	0.466667	0.377778
PF	0.422222	1	0.377778	0.533333	0.644444	0.733333	0.333333	0.533333	0.577778
RPF	0.311111	0.377778	1	0.711111	0.511111	0.488889	0.288889	0.6	0.644444
REF	0.311111	0.533333	0.711111	1	0.666667	0.644444	0.4	0.711111	0.466667
VT	0.355556	0.644444	0.511111	0.666667	1	0.666667	0.266667	0.622222	0.466667
SF	0.555556	0.733333	0.488889	0.644444	0.666667	1	0.422222	0.644444	0.644444
BP	0.444444	0.333333	0.288889	0.4	0.266667	0.422222	1	0.311111	0.266667
GH	0.466667	0.533333	0.6	0.711111	0.622222	0.644444	0.311111	1	0.577778
MH	0.377778	0.577778	0.644444	0.466667	0.466667	0.644444	0.266667	0.577778	1

Table 4. Non-membership pairs of the intuitionistic fuzzy IC correlations at the beginning of the observation

ν	FL	PF	RPF	REF	VT	SF	BP	GH	MH
FL	0	0.422222	0.577778	0.555556	0.466667	0.333333	0.333333	0.466667	0.533333
PF	0.422222	0	0.533333	0.355556	0.2	0.177778	0.466667	0.333333	0.355556
RPF	0.577778	0.533333	0	0.222222	0.377778	0.466667	0.555556	0.311111	0.333333
REF	0.555556	0.355556	0.222222	0	0.244444	0.333333	0.422222	0.177778	0.488889
VT	0.466667	0.2	0.377778	0.244444	0	0.222222	0.555556	0.222222	0.444444
SF	0.333333	0.177778	0.466667	0.333333	0.222222	0	0.422222	0.266667	0.333333
BP	0.333333	0.466667	0.555556	0.422222	0.555556	0.422222	0	0.488889	0.6
GH	0.466667	0.333333	0.311111	0.177778	0.222222	0.266667	0.488889	0	0.355556
MH	0.533333	0.355556	0.333333	0.488889	0.444444	0.333333	0.6	0.355556	0

Table 5. Membership pairs of the intuitionistic fuzzy InterCriteria correlations at the end of the observation

μ	FL	PF	RPF	REF	VT	SF	BP	GH	MH
FL	1	0.444444	0.311111	0.288889	0.511111	0.422222	0.466667	0.355556	0.466667
PF	0.444444	1	0.222222	0.466667	0.333333	0.444444	0.4	0.4	0.488889
RPF	0.311111	0.222222	1	0.6	0.555556	0.511111	0.288889	0.577778	0.533333
REF	0.288889	0.466667	0.6	1	0.4	0.311111	0.333333	0.444444	0.355556
VT	0.511111	0.333333	0.555556	0.4	1	0.555556	0.311111	0.688889	0.555556
SF	0.422222	0.444444	0.511111	0.311111	0.555556	1	0.444444	0.533333	0.777778
BP	0.466667	0.4	0.288889	0.333333	0.311111	0.444444	1	0.155556	0.377778
GH	0.355556	0.4	0.577778	0.444444	0.688889	0.533333	0.155556	1	0.533333
MH	0.466667	0.488889	0.533333	0.355556	0.555556	0.777778	0.377778	0.533333	1

Table 6. Non-membership pairs of the intuitionistic fuzzy IC correlations at the end of the observation

ν	FL	PF	RPF	REF	VT	SF	BP	GH	MH
FL	0	0.511111	0.555556	0.6	0.422222	0.511111	0.444444	0.533333	0.466667
PF	0.511111	0	0.644444	0.422222	0.6	0.488889	0.511111	0.488889	0.444444
RPF	0.555556	0.644444	0	0.244444	0.288889	0.377778	0.577778	0.311111	0.355556
REF	0.6	0.422222	0.244444	0	0.466667	0.6	0.555556	0.422222	0.555556
VT	0.422222	0.6	0.288889	0.466667	0	0.355556	0.622222	0.177778	0.355556
SF	0.511111	0.488889	0.377778	0.6	0.355556	0	0.488889	0.377778	0.177778
BP	0.444444	0.511111	0.577778	0.555556	0.622222	0.488889	0	0.733333	0.555556
GH	0.533333	0.488889	0.311111	0.422222	0.177778	0.377778	0.733333	0	0.377778
MH	0.466667	0.444444	0.355556	0.555556	0.355556	0.177778	0.555556	0.377778	0

While taking decisions about the treatment process we have applied InterCriteria analysis of the data received from the observation of the patient's state. The results achieved with InterCriteria decision making approach are presented on the tables - Table 3, Table 4, Table 5 and Table 6.

In Fig. 1 are presented the geometric interpretations of the obtained results, showing the change in the distribution of the criteria at the beginning and end of the observation.

The Tables 3 and 4 present the results of the patients at the beginning of the observation. The results obtained from the analysis show correlation between two of the indicators – Role-Emotional Functioning (REF) and Role-Physical Functioning (RPF). The correlation described between these two factors is typical because one's health physical function in determines the emotional condition, to a large extend. When a person's

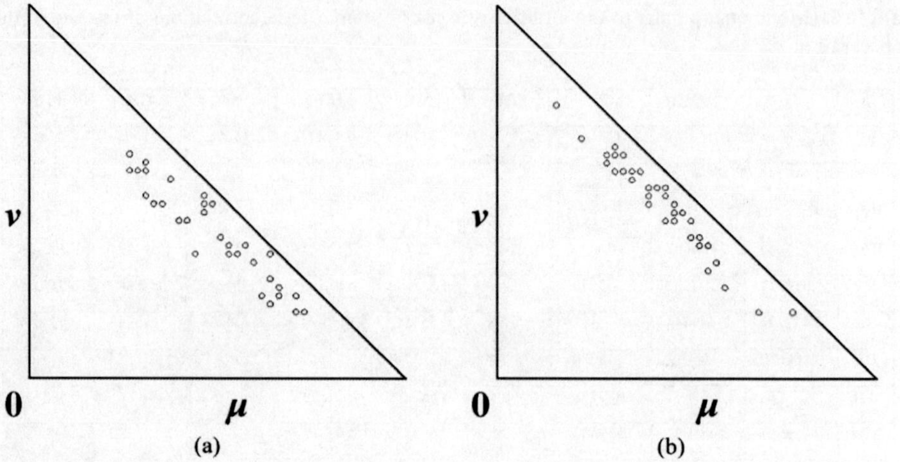

Fig. 1. Geometrical visualization of the InterCriteria correlations onto the intuitionistic fuzzy interpretational triangle at the beginning of the observation (a) and at the end of the observation (b)

health condition is satisfactory, both physically and emotionally, they feel confident and their social functioning is improved. On the contrary, if a person is sick this leads to negative feelings (both emotionally and physically) and prevents them from having a normal social life. The patient gets depressed, becomes introvert and shows no interest in going out.

The Tables 5 and 6 present the results of the patients at the end of the observation. The achieved results shows a decrease in the relationship between the two criteria was observed, but it still remained with a high degree of correlation. The results of the analysis show correlation between the pairs of criteria – General Health (GH) - Vitality (VT) and between the pairs of criteria – Mental Health (MH) - Social Functioning (SF). The assessment of patients' health condition at the present moment and their treatment perspectives is highly dependent on their feeling of energy and strength.

It is known that the criterion Mental Health (MH) determines the presence of depressive mood, anxiety and the total percent of positive emotions. The criterion Social Functioning (SF) determines the degree of restriction of social activity by the physical or emotional state. Both criteria are related to the patient's emotional state and have a strong impact on the recovery process and on the achievement of successful results of the course of the physiotherapeutic treatment and kinesiotherapeutical program.

The presented results of this study are closely linked to medical practice and once again confirm the claim of universality in the application of the InterCriteria decision making approach.

9 Conclusion

This paper describes the use of a multicriteria decision making approach to analyze the treatment outcomes of patients with osteoarthritis. The results obtained from the analysis of the criteria for assessing the health status of the observed patients show the presence of relationships between the criteria that are not strongly correlated. This fact gives us reason to believe that the criteria chosen for the observing of patients have a high degree of independence. This presupposes the existence of additional requirements to the choice of evaluation criteria in order to be able to choose a better specialized methodology for kinesitherapy. The conclusions, based on patient observations, confirm the already known results of previous studies showing that the muscles in the human body are closely related and the improvement of one movement will lead to the improvement of another, which also necessitates the appropriate choice of specific kinesitherapy methodology.

The results presented in this paper from the application of the InterCriteria decision making approach are a step towards improving the decision making process regarding the treatment course in order to improve the overall quality of life of patients with osteoarthritis.

Acknowledgment. The authors are thankful for the support provided by the Bulgarian National Science Fund under Grant Ref. № KP-06-N22/1.

References

1. Michael, W.P., et al.: The epidemiology, etiology, diagnosis, and treatment of osteoarthritis of the knee. Deutsches Ärtzeblatt International **107**(9), 152–162 (2010)
2. Otzetov, K.: Study of the effects of applying the specialized methodology of KT with patients having initial gonartrosis, PhD Thesis, Sofia (2014)
3. Atanassov, K., Mavrov, D., Atanassova, V.: InterCriteria decision making. A new approach for multicriteria decision making, based on index matrices and intuitionistic fuzzy sets. In: Proceedings of the 12th International Workshop on Intuitionistic Fuzzy Sets and Generalized Nets, Warsaw, Poland (2013)
4. Altman, R., Asch, E., Bloch, D., et al.: Development of criteria for the classification and reporting of osteoarthritis of the knee. Arthritis Rheum **29**, 1039–1049 (1986)
5. Van Baar, M.E., Assendelft, W.J., Dekker, J., et al.: Effectiveness of exercise therapy in patients with osteoarthritis of the hip or knee. Arthritis Rheum, **42**, 1361–9 16 (1999)
6. Dimitrova, E., Popov, H.: Instruction manual of exercises in functional diagnostics of the locomotory system, Sofia (2006)
7. Ware, J., Kosinski, M., Dewey, J.E., Gandek, B.: How to Score and Interpret Single–Item Health Status Measures: A Manual for Users of the SF-8 Health Survey. Lincoln RI: Quality Metric Incorporated (2001)
8. Atanassov, K.: Intuitionistic fuzzy sets, VII ITKR session 1983, deposed in centr. Sci.-Techn. Library of the Bulg. Acad. Sci. 1697/84 Reprinted in: Int. J. Bioautom. **20**, S1-S6 (2016). (in Bulgarian)
9. Atanassov, K.: Intuitionistic fuzzy sets. Fuzzy Sets Syst. **20**(1), 87–96 (1986)
10. Atanassov, K.: Intuitionistic Fuzzy Sets: Theory and Applications. Physica-Verlag, Heidelberg (1999)
11. Atanassov K., Generalized index matrices. Comptes rendus de l'Academie Bulgare des Sci. **40**(11), 15–18 (1987). ISSN: 1310–1331
12. Zadeh, L.A.: Fuzzy sets. Int. J. Inf. Control **8**, 333–353 (1965)
13. Zaharieva, B., Doukovska, L., Radeva, I., Ribagin, S.: Intercriteria analysis approach to behtetrev's disease analysis. Notes Intuit. Fuzzy Sets **23**(2), 119–127 (2017). Prof. Marin Drinov Publishing House, ISSN: 1310–4926

14. Zaharieva, B., Doukovska, L., Ribagin, S., Michalíková, A., Radeva, I.: Intercriteria analysis of behterev's kinesitherapy program. Notes Intuit. Fuzzy Sets **23**(3), 69–80 (2017). Prof. Marin Drinov Academic Publishing House, ISSN: 1310–4926
15. Zaharieva, B., Doukovska, L., Ribagin, S., Radeva, I.: Intercriteria analysis of data obtained from patients with behterev's disease. Int. J. Bioautom. **24**(1), 5–14 (2020). Prof. Marin Drinov Academic Publishing House, ISSN: 1314–1902, https://doi.org/10.7546/ijba.2020.24.1.000507

A Risk Assessment Study on Musculoskeletal Disorders in Computer Users Based on A Modified Nordic Musculoskeletal Questionnaire

Sonya Vachinska, Valentina Markova(✉), and Todor Ganchev

Technical University of Varna, 1 Studentska str., Varna, Bulgaria
{s_vachinska,via,tganchev}@tu-varna.bg

Abstract. We present a recent study on musculoskeletal disorders (MSDs) risk assessment in computer users. For this research, we developed a modification of the Nordic Musculoskeletal Questionnaire (NMQ), where respondents are asked to report only about pain and musculoskeletal discomfort, which they consider linked to the prolonged work with a computer. The modified NMQ also collects information about the respondents' opinion concerning the ergonomic aspects of their workplace. Based on the modified NMQ, we collected responses from 60 people who usually use PCs in their work routine. Based on the post-test discussions with the participants, we found out that the questionnaire brought an educational benefit because it made the participants aware of specific inappropriate habits and body postures. Specifically, the test provoked the participants to think about the trunk's proper posture, upper and lower limbs, the distance to the screen, the position with respect to the keyboard, mouse, etc., which was perceived as beneficial and useful. Thought-provoking outcomes of our study are the conclusions that there is an urgent need for proper training of the average computer user on the health compliant work positions and healthy habits. Furthermore, energetic efforts are required to popularize good practices for the safe use of computers and related standards. The results obtained to this end motivate the development and promotion of adequate proactive measures for computer users' long-term health preservation.

Keywords: Ergonomics · Musculoskeletal disorders · Computer workplace

1 Introduction

Musculoskeletal disorders (MSDs) are one of the most common work-related ailments in Europe [1] and accounted for more than one-third (35%) of all occupational injuries and illnesses from work in the United States [2]. They are the number-one type of lost-time work injury and are the leading cause of work disability in the developed economies [3]. Nowadays, MSDs affect millions of workers and cost employers billions of euros for treatment, hospitalization, reduced quality of production, or low productivity.

S. S. Sotirov et al. (Eds.): BioInfoMed 2020, LNNS 374, pp. 433–444, 2022.
https://doi.org/10.1007/978-3-030-96638-6_45

MSDs at work is a term that summarizes several injuries and disorders of the body in the muscles, bones, nerves, etc. [1, 5]. Many body areas can be affected by adults and adolescents too [4] - the back, neck, buttock, upper limbs - shoulders, elbows, wrists, hands, fingers, and lower limbs - hip, knees, ankles [1, 6–9]. MSDs result in back pain, neck pain or numbness, cramps, limitations of mobility at limbs. Aiming for a better quality of life and the workers' well-being, the respective governments and companies should strive to find the right approach for tackling MSDs [5, 10]. Maintaining a good worker's health and reducing day-offs from illness also speak about good business thinking and social responsibility from a company's side.

To this end, computer-daily office work has been associated with many risk factors for developing musculoskeletal symptoms in the neck, shoulders, and upper back, such as constrained postures for a long time, insufficient variation of body posture, and repetitive movements on upper limbs while using the keyboard and mouse [6, 7]. Repeatedly, temporal and postural factors have been associated with musculoskeletal symptoms or disorders in workers who use computers [15, 17, 18]. The risk factors associated with MSDs in the workplace are often referred to as ergonomics risk factors and human factors. The experts would contend that the contributory psychosocial factors are also part of the ergonomics remit. An experienced ergonomist will also develop safety workstations or solutions without risk [11, 12, 15, 23] because in some countries at least (in Bulgaria too), ergonomics has its roots origins in engineering.

To better understand the efficiency of various strategies, policies, and actions to address MSDs, the EU appointed a European-level survey. As a result, the appointed EU panel published a report with 25 different initiatives from 14 different countries [12]. This report analyzed the effect of various types of actions, ranging from awareness-raising campaigns to direct interventions and enforcement action. In the EU, it is up to each country to decide what measures to take to tackle the aging population, improving health, and ensuring good working capacity combined with long service. For example, France adopts a strong focus on preventing workplace risk and imposes specific actions on the design of equipment and workplaces. In 2015, Germany passed an act to strengthen health promotion and preventive health care, subsequently grown to the Joint German occupational safety and health strategy, which policy put the focus on the collaboration between different partners – government, employee and employer, insurance and health care organizations [12, 20]. According to the sixth wave (2015) of the European Working Conditions Survey (EWCS), workers reporting MSDs in the last 12 months for managers, technicians, and associate professionals (according to International Standard Classification of Occupations 2008 - ISCO-08) are 39% [13]. They use a computer or laptop a minimum of 5 days per week and during all-day work. The same survey summarizes problems as muscular pains in the shoulders, neck, or upper limbs reported by workers increase from 30% for age under 25 to 48% for persons at 55+ years old. The increase in musculoskeletal disorders for EU for six years is +5.9% (for the year 2007 is 54.2%, and for the year 2013 is 60.1%), and only for Bulgaria, it is +5.5% (for the year 2007 is 37.3% and for the year 2013 is 42.8%) according to Eurostat and Labor Force Surveys (LFSs) [19]. For the same period, MSD cases among workers in Luxembourg, Hungary, Germany, and the United Kingdom have decreased to 21.8%, 17.3%, 10.4%, and 10.0%, respectively. The countries with declining MSDs complaints [12, 19, 20] have introduced recurrent training in schools and companies and convinced employees and their leaders of the importance of maintaining human health because training costs are lower than treatment costs or health insurance.

Computer work in Bulgaria is usually sedentary work, mainly involving computer use-PCs or laptops for reading and typing. Thus, workers are usually required to sit for long hours in front of a computer [26], which leads to a 59% risk for MSDs according to research ESENER-3, 2019 [19]. Many individuals experience musculoskeletal discomforts, particularly at the buttock and low back regions during prolonged sitting [7]. In this posture, the abdominal muscles relax, the back bends in the lower back and shoulders, and the neck and head are often brought forward. A popular alternative to sitting on an office chair while performing computer work is to sit on an exercise ball [8] because it counteracts the relaxation of the abdominal muscles and keeps the back sent due to striving to keep balance on the ball. Some authors recommend changing the posture using sit-stand workstations (some of the time spent sitting followed by standing and so on) and thus reduce cardiometabolic risks associated with sedentary behavior [6]. Others recommend avoiding static standing posture due to lower limb strain (swelling, varicose veins) in lower limbs [9]. Analyzing all studies discussed so far, it is clear that there is a need for in-depth research on workers' knowledge about the risks to their health during the working day. With a view to the daily use of software products among university professors, programmers, engineers, and other professionals, we focused on studying the work habits of people who use computers daily.

The current study aimed to assess computer users' movement or inactivity and obtain information about MSDs based on a modified Nordic Musculoskeletal Questionnaire (NMQ) and self-assessment body posture visualization according to ISO 11226, ROSA and RULA methods [14–17]. We developed a modified NMQ, where the respondents are asked to report only about pain and musculoskeletal discomfort they consider linked to prolonged work with a computer. We hypothesized that people working on a computer for a long time are aware of the risk they expose their bodies and are sufficiently competent, educated, and familiar with maintaining their health and ability to work. We expected that people would have information on avoiding or reducing the risk of MSDs and what preventive measures to use. Specifically, the good practices for MSDs prevention include good time-task work management; neutral and correct working postures; correct hand movements - smaller amplitude and angles of unfolding/folding of large muscles; use of smaller muscle groups; avoiding prolonged inactivity of the body, including prolonged sitting; periodic stretching of immobilized muscles; training and increasing the tone of the spinal muscles, directly affecting the maintenance of a straight back and a healthy waist.

2 Method

2.1 Standards and Recommendations

The daily work with a computer is connected with the layout and organization of each specific workplace. There are legal regulations for the safety of computer equipment (series standards EN ISO 9241 - Ergonomics of Human System Interaction), while furniture and arrangement of the workplace are the employer or employee's choice or preference.

Their knowledge of ergonomics is based on the equipment and layout (arrangement of a screen, keyboard, mouse, printer, document stand, etc.) for each specific computer location. On the other hand, according to the European framework directive on safety and health at work (Directive 89/391 EEC), the employer is obliged to provide healthy and safe workplaces to his employees by guaranteeing them minimum safety and health requirements throughout Europe. Simultaneously, the Member States are allowed to maintain or establish more strict regulations, such as requirements for targeted training and periodic provision of information on risks at work [10]. The computer interfaces, such as a monitor, keyboard, and mouse, need to meet all safety requirements of the EN ISO 9241 as separate physical devices. However, workers' daily use can create a risk of MSDs if people do not have explicit, reasoned, and periodic instructions on correct - body posture, movement of upper limbs, or right equipment placement [23, 27].

Based on the analysis of good practices and specific needs, we developed a questionnaire, which helps us to assess the extent to which the respondents were aware of the issues connected to the MSDs, as well as to provoke some awareness about the importance of the prevention and knowledge about the severity of the impact of these diseases. The purposely-developed questionnaire has two main functions: (1) it collects information about computer habits, and (2) provides clear textual or visual information on how it is safe to work with a computer. These are useful to both the employee and his employer because the respondent is informed or thinks about behaving more safely in front of a computer (e.g., changing the posture for a certain period or doing short stretching exercises). On the other hand, the supervisor receives information on where and what needs to be improved to reduce the risk of MSDs.

2.2 Questionnaire Design

The questionnaire used in the present work builds on the Nordic Musculoskeletal Questionnaire (NMQ) [14]; however, we modified it due to this research's specific requirements. Specifically, the present study focuses on evaluating the levels of discomfort and pain resulting from prolonged periods of computer work characterized by limited and rare changes in body posture and poor from an ergonomic perspective organization of the workplace. Thus, building on the NMQ, we added questions related to the established standards for an ergonomic computer workplace [23], which respect to work safety and efficiency. Thus, in the modified NMQ [29, 30], respondents are asked to report only pain and musculoskeletal discomfort, which they consider linked to the prolonged work with a computer. An additional aspect during the modified questionnaire development was to make people aware of the ergonomic risk from MSDs via structured and multiple-choice variants. Specifically, there are two types of questions: general and specific - focusing on the back and limb position during computer work. The purpose of the general questions is surveying, while the specific ones permit a more profound analysis. Many of these questions contain pictures with different positions of the head, torso, and limbs. Each picture responds on a scale from 1 to 5 (1 is good, 5 is dangerous for the body). Finally, the participants can calculate the risk for any disorders due to incorrect body posture, inadequate work organization, and movements.

We used the Green-Orange-Red code, shown in Fig. 1, for risk assessment as an EAWS index [9, 20]. This code provides an intuitive visualization of the EWAS index, which depends on the behavior analysis and workspace compliance with the ergonomic requirements. Thus, it provides meaningful feedback to the respondents and illustrates. In this sense, the questionnaire serves as a self-assessment tool, which allows the respondent to personally identify the MSD risks, figure out the weaknesses in his working style, and consequently realize the reasons for them. Additionally, the respondents get informed about correct postures at work, the correct position of the monitor, mouse, or keyboard, level for adjusting the height of the chair or table/desk, ways of organizing breaks for stretching or resting the eyes, etc., to avoid muscle damage during prolonged monotonous work and to preserve human health.

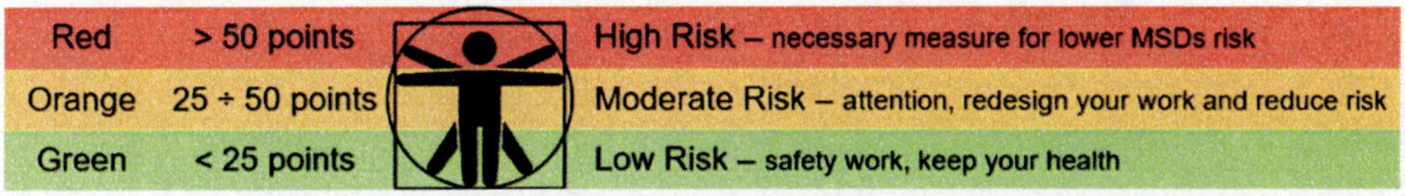

Fig. 1. Index for self-assessment risk code with recommendations.

3 Implementation and Results

In the current study, we targeted people above 20 years old whose professional scope implies PCs' use during most of their working time. In total, 60 participants were recruited to fill in the modified NMQ. Specifically, our study was structured in two parts: general questions (focusing on the substance of the problem) and specific questions, aiming to give a detailed picture of the effects of particular back and limb positions during the usage of computers as part of our lives. In total, 44 women (average age of 38.9 years) and 16 men (average age of 41.7 years) took part in the study. The analysis shows that their body mass index (BMI) was: BMI <25 for 32 people (normal: 4 men and 25 women, underweight: 3 women), 14 people were overweight, and 13 obese. Half of the respondents (50%) do not practice any physical exercises or sports, maintaining good muscle tonus and body sustainability; however, 26.7% do sports gum every day. The following Table 1 shows a summary of all statistical information of participants. The analyses ascertain that 80% of participators pertain to the orange code (moderate risk), and 20% are in the dangerous red code (high risk for development of MSD). Moreover, there was no single person whose responses pertain entirely to the green (safety) zone.

In Table 2 and Fig. 2, we present the reported discomfort or pain due to prolonged computer use according to the respondents. The most significant discomfort is in the neck (45.0%) regardless of the frequency of its occurrence, followed by that in the upper back (35.0%) and shoulders (25.0%). The highest frequency of total discomfort throughout the body (53.3%) is observed 1–2 times per week, while prolonged all-day pain affects areas such as the neck (3.3%), upper back (1.7%), and buttocks (1.7%).

As shown in Table 2, despite the relatively smaller number of people (30.0%) who report pain 3–4 times a week, there is again a predominance of discomfort in the neck (18.3%), upper back (11.7%), and shoulders (11.7%) of the respondents. Pain is the first

Table 1. Main characteristics of the population under study.

Gender	Number	Age (years)			Weight (kg)			Height (m)		
		Mean	Min	Max	Mean	Min	Max	Mean	Min	Max
Men	16	41.7	20	75	88.1	55	120	1.75	1.60	1.87
Women	44	38.9	20	66	65.8	50	105	1.65	1.53	1.74
Declined answer, (kg/m)	−1 men				−1 men			−1 men		
BMI men N (%)	Underweight < 18.50 0 (0)		Normal 18.5–24.99 4 (25)		Overweight 25–29.99 5 (31.3)			Obese ≥ 30 6 (37.5)		
BMI women, N (%)	Underweight 3 (6.8)		Normal 25 (56.8)		Overweight 9 (20.5)			Obese 7 (15.9)		
Regular sports activities, N (%)	Every day 16 (26.7)		1–2 times per week 14 (23.3)					Not sport activities 30 (50)		

Table 2. Participant's frequency report for discomfort or pain from prolonged computer work

Body map area N = 60	1–2 times per week	3–4 times per week	Every day (once)	Every day (several times)	Every day (all day long)	Total discomfort
Head, %	15.0	1.7	1.7	–	–	**18.3**
Neck, %	16.7	18.3	6.7	–	3.3	**45.0**
Upper back, %	13.3	11.7	6.7	1.7	1.7	**35.0**
Shoulder, %	6.7	11.7	6.7	–	–	**25.0**
Lower back, %	6.7	3.3	3.3	6.7	–	**20.0**
Elbows, %	5.0	1.7	1.7	–	–	**8.3**
Wrists/Hands, %	5.0	1.7	1.7	1.7	–	**10.0**
Buttock, %	5.0	3.3	5.0	–	1.7	**15.0**
Knee, %	5.0	–	1.7	–	–	**6.7**
Ankles/Feet, %	6.7	–	5.0	–	–	**11.7**
Total people with pain or discomfort, N (%)	**32** (53.3%)	**18** (30%)	**15** (25%)	**6** (10%)	**4** (6.7%)	No discomfort or pain: **10** (16.7%)

symptom of the body that alarms us that something goes wrong, and measures should be taken. Therefore, no pain should be ignored, no matter how often it occurs.

Only 10 participants in the survey (16.7%) reported that they do not have complaints about discomfort or pain, but half of them are under 33 and have predominantly normal BMI. Despite the lack of pain, they pertain to the moderate risk zone (orange) due to the reported sitting position (80%) while using a laptop (50%) for two or more hours with no interruptions (60%), and cumulative computer use of more than 6 h (70%) per day.

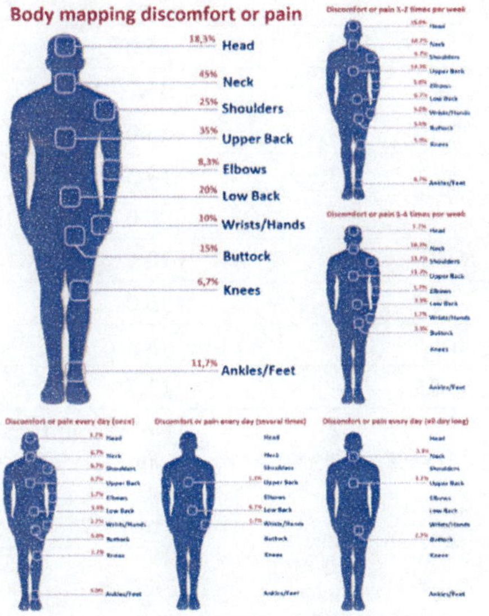

Fig. 2. Percentage of self-reporting discomfort or pain at the different body areas

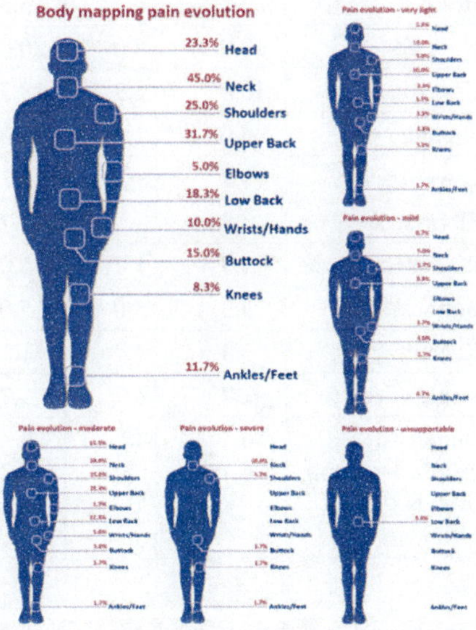

Fig. 3. Percentage of self-reporting evolution pain at different body areas

In a sitting position, the back of people (even those without any complaints) is not in a perfect upright state. Sitting prolonged hours in such a position could cause pains and become the reason for MSDs. Keeping good muscle tonus through physical exercises and sports activities should be a logical response to this issue, but in practice, those who do not currently suffer from any pain in most cases are reluctant to perform such activities. Among them, 40% do not exercise at all. Moreover, such people could have any existing discomforts without having the idea that they derive from prolonged sedentary behavior at PCs.

The degree of pain assessed by the participants is presented in Table 3 and Fig. 3. Most of the respondents (48.3%) have reported tolerable/moderate pain, and the three most painful areas are again the neck, upper back, and shoulders. They also have the highest overall pain rates –45% for the neck, 31.7% for the upper back, and 25% for the shoulders. It is interesting to note the discrepancy between the total values of the reported discomfort and pain and the assessment of the degree of pain in several areas of the body: head, upper back, elbow, low back, and knees. The largest discrepancy (5%) was reported in the headaches assessment, with 18.3% confirming discomfort and 23.3% assessing it as very mild to bearable. Headaches can have many different causes.

Working with a computer is associated with prolonged staring at the screen and a high concentration of attention, causing discomfort and headaches.

Table 3. Participant's evolution report for discomfort or pain from prolonged computer work

Body map area N = 60	Very light	Mild	Moderate	Severe	Unsupportable	Total pain
Head, %	3.3	6.7	13.3	–	–	**23.3**
Neck, %	10.0	5.0	20.0	10.0	–	**45.0**
Upper back, %	10.0	3.3	18.3	–	–	**31.7**
Shoulder, %	5.0	1.7	15.0	3.3	–	**25.0**
Lower back, %	1.7	–	13.3	–	3.3	**18.3**
Elbows, %	3.3	–	1.7	–	–	**5.0**
Wrists/Hands, %	3.3	1.7	5.0	–	–	**10.0**
Buttock, %	3.3	5.0	5.0	1.7	–	**15.0**
Knee, %	3.3	1.7	1.7	1.7	–	**8.3**
Ankles/Feet, %	1.7	6.7	1.7	1.7	–	**11.7**
Number of people according pain, N (%)	**16** (26.7%)	**15** (25%)	**29** (48.3%)	**9** (15%)	**2** (3.3%)	No discomfort or pain: **10** **(16.7%)**

4 Discussion

Our expectations for adequate background knowledge regarding musculoskeletal disorders among the respondents were not met. In 20% of the cases, they confess a lack of acquaintance with the MSDs-related risks and declare their willingness to obtain further information about the problem. In total, 60% of people judge themselves as aware to some extent, and only 20% feel entirely aware. The paradox is that even the most informed group does not comply with the recommendations for safety work. In total, 41.7% of them reported work uninterruptedly for more than two hours without changing the body's posture, and 50% of the same segment spend more than six hours in front of computers. The same *informed* respondents reported discomfort and pains related to computer work (66,7%), spread throughout the whole body once or twice per week-Fig. 4. Even this group follows the tendency for prevailing overall pain affecting the neck, shoulders, head, and upper back. Among them, 33,3% complained of chronic everyday pain – ones at the head, neck, upper back, and buttock, several times at lower back and all day long at neck and buttock. Even though the self-assessment tends towards "light to moderate pain", our conclusions refer to a strong necessity for risk reduction measures offsetting prolonged computer usage.

For most of the informed respondents, the pain is assessed as very light –50% and moderate –25% and is present in eight of the body's ten areas (Fig. 5). The highest total percentage is of people with neck pain (41.7%), followed by headaches (33.3%), and 25% feeling pain at shoulder, lower back, wrists/hands, and buttocks, caused according to them by the prolonged usage of a computer. Only 33.3% of the respondents reported that they are not affected by any discomfort or pain.

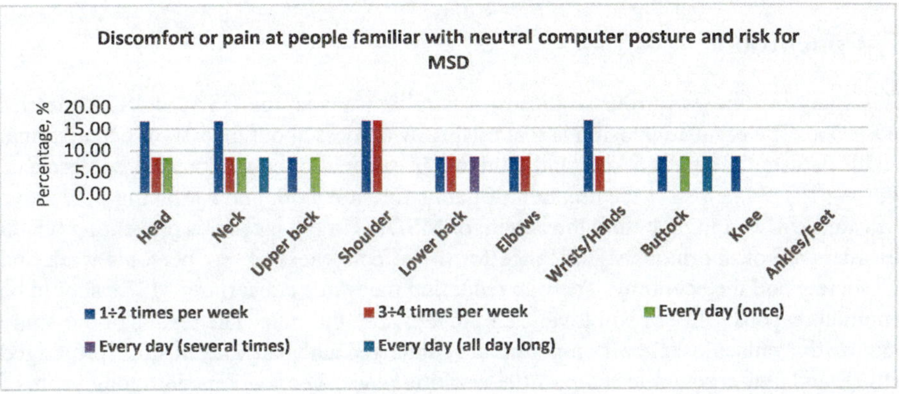

Fig. 4. Percentage of self-reporting familiar with neutral posture but feeling discomfort or pain at different body areas

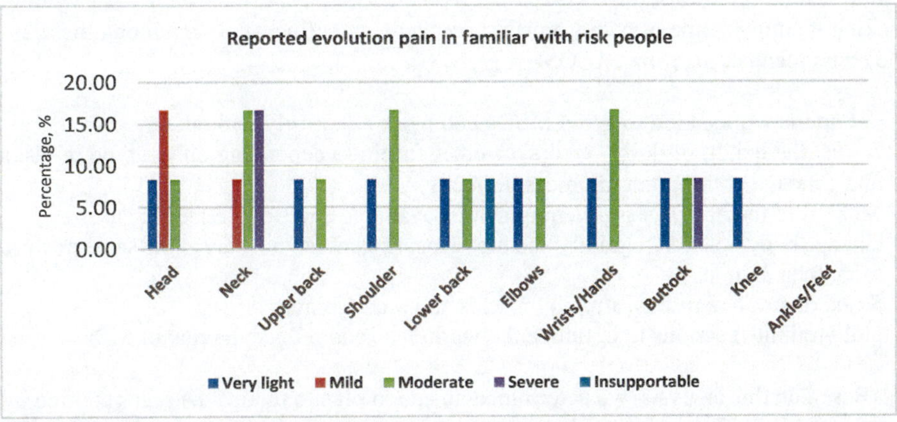

Fig. 5. Percentage of self-reporting familiar with neutral posture but feeling evolution pain at different body areas

Analyzing all data, we concluded that people are not aware of the existing risk for their health during prolonged work with a computer despite their background. Most people have accepted that discomfort and tolerable pain are an integral part of their daily work. The second conclusion is related to a possible overestimation of their knowledge due to the lack of specialized training on the risk of MSDs and the stereotype imposed in our country on the inevitability of health problems related to work. It should not be overlooked that caring for "tolerable" pain is a problem that workers deal with individually and often without consulting a specialist.

In our study, a gender-based comparison is not feasible due to the smaller number of male participants. Nevertheless, based on a study by The Swedish Work Environment Authority (SWEA) [12], we suppose that women might experience more discomfort than men due to the gender-specific differences in muscle tonus.

5 Conclusion

This study overviews various studies in Australia, Canada, the USA, and 11 countries in Europe. The main conclusion is that raising awareness about the risk of occupational health damage among workers and students in secondary education and adolescence [4, 6–9, 18, 21] helps for the creation of habits for safe work and for taking preventive measures early enough, before the advent of MSDs. Timely measures preventing health disorders are of enormous significance for the effectiveness of any person's work and the society and the economy. The risk reduction measures concerning MSDs should be a mutual responsibility of employers, employees, and the state. The results of our study confirm that musculoskeletal complaints are connected with bad body posture, prolonged sitting, and non-ergonomic setup of the working place. The workers with long years of service suffer more frequently than young employees. Seated computer work results in prolonged static loading associated with the development of MSDs confirmed throughout the current study by the reported pain discomfort and non-ergonomic postures. The simplest solution against MSDs is to train workers to safely perform their work without making it cumbersome or compromising precision and effectiveness. Among these are the recommendations from EU-OSHA [1, 5, 10, 12, 13, 22, 24]:

- Set up the workstation to avoid MSDs and work effectively without any pain;
- Adjust the height of desks, chairs, or other furniture depending on the type of tasks and personal anthropometric measurements;
- Make sure that the necessary equipment is available and trouble-free for use;
- Change body position regularly and move insufficient (do not stay at the same posture more than an hour);
- Stand or sit up regularly, stretch muscles and move joints;
- Hold training sessions to optimize the workplace and reduce the risk of MSDs.

Based on this analysis, we determined an action plan to reduce the pain reported by computer users. The general risk assessment procedure for MSD prevention is similar to other workplace hazards and have four steps:

S1 - Recognize and identify an MSD hazard (checklist questionnaire, workers conversation, observation, etc.);
S2 - Assess its risk to workers and try to eliminate the hazard (teach worker, change workplace hazard and organization to be safe for people);
S3 - Controls the risk of workers - put in place to reduce it (engineering and administrative control, periodical risk assessment);

S4 - Evaluate the effectiveness of the control (improve worker's health and work productivity).

Finally, we need to point out that the establishment of habits for proper and safe sitting and using a computer or laptop is recommended to start at school age because since then the initial skills to maintain a neutral posture are formed and since then start the first symptoms of health problems [21, 25, 28]. MSDs are the dominant cause of disability among working adults in the developed economies and a reason for both direct and indirect economic costs. The complexity and multifactorial nature of MSDs mean that they cannot easily be solved without an integrative approach - government, employer, or employee. They all should combine resources and efforts to improve the knowledge of work-related health problems and enhance their well-being.

Acknowledgments. This research was funded by Bulgarian National Science Fund (BNSF), under grant contract FNI № KP-06-N37/1, entitled "Ergonomic research on work-related health problems by innovative computer models with a focus on the prevention of Musculoskeletal Disorders."

References

1. European agency for safety and health at work, Musculoskeletal disorders. https://osha.europa.eu/en/themes/musculoskeletal-disorders. Accessed 19 Sep 2020
2. Bhattacharya, A.: Costs of occupational musculoskeletal disorders (MSDs) in the United States. Int. J. Ind. Ergon. **44**(3), 448–454 (2014). https://doi.org/10.1016/j.ergon.2014.01.008
3. Mustard, C.A., et al.: Time trends in musculoskeletal disorders attributed to work exposures in Ontario using three independent data sources, 2004–2011. Occup. Environ. Med. **72**, 252–257 (2015). https://doi.org/10.1136/oemed-2014-102442
4. Toh, S.H., et al.: A prospective longitudinal study of mobile touch screen device use and musculoskeletal symptoms and visual health in adolescents. Appl. Ergon. **85**, 103028 (2020)
5. Graveling, R., Giagloglou, E.: Prevention policy and practice: approaches to tackling work-related musculoskeletal disorders, European Risk Observatory (2020)
6. Barbieria, D.F., Divya Srinivasanb, D., Mathiassenc, S.F., Oliveira, A.B.: Variation in upper extremity, neck and trunk postures when performing computer work at a sit-stand station. Appl. Ergon. **75**, 120–128 (2019)
7. Waongenngarm, P., Rajaratnam, B.S., Janwantanakul, P.: Internal oblique and transversus abdominis muscle fatigue induced by slumped sitting posture after 1 hour of sitting in office workers. Saf. Health Work **7**, 49–54 (2016)
8. Kingma, I., Van Dieën, J.H.: Static and dynamic postural loadings during computer work in females: sitting on an office chair versus sitting on an exercise ball. Appl. Ergon. **40**, 199–205 (2009)
9. Antle, D.M., et al.: Lower limb blood flow and mean arterial pressure during standing and seated work: implications for workplace posture recommendations. Prev. Med. Rep. **10**, 117–122 (2018)
10. EU-OSHA homepage, EU directives. https://osha.europa.eu/en/legislation/directives/the-osh-framework-directive/the-osh-framework-directive-introduction. Accessed 25 Sep 2020
11. Caputo, F., Greco, A., Fera, M., Macchiaroli, R.: Digital twins to enhance the integration of ergonomics in the workplace design. Int. J. Ind. Ergon. **71**, 20–31 (2019)

12. Graveling, R., Giagloglou, E.: European Risk Observatory report, Prevention policy and practice: approaches to tackling work-related musculoskeletal disorders: EU-OSHA (2020). https://doi.org/10.2802/581582
13. De Kok, J., et al.: Work-related musculoskeletal disorders: prevalence, costs and demographics in the EU, European risk observatory report: EU-OSHA (2019). https://doi.org/10.2802/66947
14. Crawford, J.O.: The Nordic Musculoskeletal Questionnaire. Occup. Med. **57**, 300–301 (2007). https://doi.org/10.1093/occmed/kqm036
15. ISO 11226: 2006 Ergonomics — Evaluation of static working postures
16. Rafeemanesh, E., Yazdi, D.K., Ahmadifar, M., Bidel, H., Balochkhaneh, F.A.: Study of musculoskeletal disorders prevalence and ergonomic conditions by ROSA in administrative staff of Mashhad University of medical sciences. Int. J. Bus. Manag. **4**(2), 82–89 (2019)
17. Widiyawati, S., Lukodono, R.P., Lustyana, A.T., Pradana, I.A.: Investigation of the risk of daily officer work posture based on rapid upper limb assessment (Rula) method. Int. J. Hum. Mov. Sports Sci. **8**(1), 24–31 (2020). https://doi.org/10.13189/saj.2020.080103
18. Lima, T.M., Coelho, D.A.: Prevention of musculoskeletal disorders (MSDs) in office work: a case study. Pub. Med. **39**(4), 397–408 (2011). https://doi.org/10.3233/WOR-2011-1190
19. Crawford, J.O.R., et al.: European risk observatory-report, work-related musculoskeletal disorders: from research to practice. What can be learnt? EU-OSHA (2020). ISBN: 978-92-9479-357-7. https://doi.org/10.2802/118327
20. Schaub, K., Caragnano, G., Britzke, B., Bruger, R.: The European assembly worksheet. Theor. Issues Ergon. Sci. **14**(6), 1–23 (2012). https://doi.org/10.1080/1463922X.2012.678283
21. Filkova, S., Michaleva, V.: Main aspects of the consequences of poor body posture in children. Varna Med. Forum. **7**(Suppl. 3), 246–250 (2018)
22. Nielsen, K., Jørgensen, M.B., Milczarek, M., Munar, L.: Healthy workers, thriving companies - a practical guide to well-being at work. Tackling psychosocial risks and musculoskeletal disorders in small businesses. EU-OSHA. (2018)
23. ISO 9241-6: Ergonomic requirements for office work with visual display terminals (VDTs) — Part 6: Guidance on the work environment
24. Colombini, D., Occipinti, E., Grieco, A.: Risk Assessment and Management of Repetitive Movements and Exertions of Upper Limbs. Elsevier Ergonomics Book Series – vol. 2. Elsevier, Amsterdam (2002)
25. Mihajlova, M., Filkova, S., Michaleva, V.: Some aspects of the conservative treatment and prevention of lumbar disc disease. Scripta Scientifica Salutis Publicae **2**(Supplement 1), 95–100 (2016)
26. Vachinska, S.: Assessment study for occupational risk among computer operators in engineering higher education. In: Book of International Conference of Biomedical Innovation and Applications, Varna, pp. 85–89 (2019). ISBN: 978-1-7281-4754-3
27. Sanchez-Lite, A., Garcia, M., Domingo, R., Angel, S.M.: Novel ergonomic postural assessment method (NERPA) using product- process computer aided engineering for ergonomic workplace design. PLoS ONE **8**(8), e72703 (2013). https://doi.org/10.1371/journal.pone.0072703
28. Filkova, S., Michaleva. V.: The impact of the contemporary way of life on the posture of the body in children. Varna Med. Forum. **7**(Suppl. 4), 268–273 (2018)
29. Modified NMQ available on-line at. https://forms.gle/u2QXFhELtn73s8tKA. Accessed Dec 2020
30. Modified NMQ at Ergonomic study of the working environment through innovative computer models in order to prevent musculoskeletal disorders. http://isr.tu-varna.bg/ergo/

Technological Support to Musculoskeletal Disorders Assessment and Management

Valentina Markova(✉) and Todor Ganchev

Technical University of Varna, Studentska str.1, 9010 Varna, Bulgaria
{via,tganchev}@tu-varna.bg

Abstract. Work efficiency and quality of life largely depend on health. Daily computer use is an integral part of work, educational activities, and entertainment, and as it is already well known, it increases the risk of developing various musculoskeletal disorders (MSDs). MSDs are the main reason for chronic body pain, most often in the upper limbs, shoulders, or neck. In this regard, we present the ERGO-project's overall concept, which aims to develop appropriate resources and advanced technological tools supporting the early diagnosis and management of MSD among computer users who use PCs in their work routine. Our concept is based on the assumption that contemporary technology can support the standard therapeutic procedures used by medical staff to identify MSDs and the periodic assessment of MSDs progress. Specifically, we aim to develop a technological platform and appropriate tools for the automated estimation of the current body position based on video or images, questionnaire-based distress assessment, and concentration assessment based on physiological signals. Our project's main objectives are to develop a set of indicators and mobile apps that help manage MSDs in work-related computer use. The project also has educational aspects as we aim to make computer users aware of the good practices and the relevant regulations for long-term preservation of health and work efficiency.

Keywords: Ergonomics · Safety of work · Health and productivity preservation

1 Introduction

MSDs are disorders that negatively affect the musculoskeletal system, and negative emotions, stress, and higher cognitive could lead to serious health problems. The modern workplace is strongly associated with electronic devices (computers, smart-phones, tablets, etc.). Their intensive use leads to an increased risk of MSDs, which are common in school or university students (up to 65% of them have MSDs). Recent studies estimate the most common factor for MSDs is lack of rest (85%), passive load (70%), poor body posture (60%), overwork (30%), and other factors [1].

Safety and a healthy work environment are crucial for enhancing the potential and engagement of the workforce, as well as the competitiveness of businesses. Therefore, it is essential to pay attention to the quality of work to provide better, comfortable, and non-damaging work processes, guaranteeing aging in good health. The application of

S. S. Sotirov et al. (Eds.): BioInfoMed 2020, LNNS 374, pp. 445–454, 2022.
https://doi.org/10.1007/978-3-030-96638-6_46

ergonomic approaches and innovative solutions in academia and businesses is directly linked to the provision of healthy working conditions from the beginning to the end of the worker's career. EU-OSHA (European Agency for Safety and Health) campaigns related to this have been running since 2000. They respond to the growing need to raise awareness about health preservation at different levels by providing accessible information and science-based data. Campaigns, such as: "Healthy Workplaces for All Ages" during years 2017–2018 [2]; "Healthy Workplaces – Manage Stress" in the period 2014–2015 [3]; "Working Together to Prevent Risk" – 2012–2013, and the following awards for good practices [4], indicate the need for systematic work and scientific research in the field of human health issues prevention at work. The current EU-OSHA 2020–2022 Healthy Workplace campaign is focused on MSDs prevention activities and the management of chronic work-related MSDs.

Considering that people spend a significant amount of time at their workplace, it is essential to investigate the relationship between work environment and work-related health issues. One particularly important aspect is to study the influence of working posture on health and efficiency of work. Recent advances in stress detection methods through physiological signals and motion capture equipment helped establish models and technology to support biomechanics, human posture studies, and good life quality. This interdisciplinary approach provided new opportunities for stress reduction and lower rates of fatigue and muscle-skeletal disorders, increasing the worker's efficiency and improving their overall health.

In the following, we outline our approach used in the on-going ERGO-project, which is focused on the ergonomic assessment of the work or educational environment and identifying the risks due to prolonged work hours. Specifically, we adhere to the directives and methodology of the European Agency for Safety and Health at Work and the current 2020–2022 campaign, which focuses on preventing work-related MSDs. For that purpose, a system of indicators has been developed to study body posture, work environment, and behavior and assess the risks of MSDs. Specific actions focused on improving knowledge about the causes of MSDs have been developed to help people who use computers as part of their work duties, such as IT specialists, students, educators, etc. Besides, the influence of work-related stress, concentration, and work performance are investigated. As a result, the ERGO-project intends to implement adequate preventive measures against MSDs for people working on a computer. The overall project concept is briefly presented in Sect. 2. In Sect. 3, we outline the MSDs management methodology. The resources acquired in support of technology development are described in Sect. 4, and in Sect. 5, we conclude this work.

2 Overall Concept

In Bulgaria, each employer is obliged to comply with the National and EC legal framework for guaranteeing the safety of workers with respect to the corresponding regulations. However, in general, people are not aware of the following aspects of work-related risks (Fig. 1):

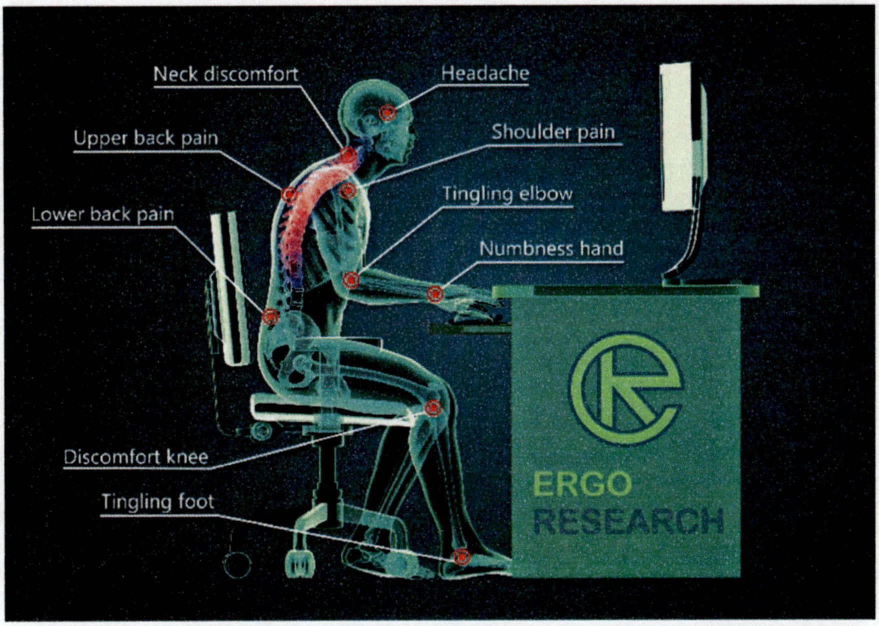

Fig. 1. Work-related risks of musculoskeletal disorders in people using computers

1) safety of working posture,
2) risk factors causing pain in back, neck, shoulders, arm, hands or legs,
3) the accumulation of damage in certain muscle groups over the years,
4) good practices for avoiding spinal distortions,
5) the risks associated with the absence of adequate prevention measures.

The abovementioned factors are the primary motivators for the present study on work-related MSDs. In the present work, we present the conceptual design and the methodology used in our modular platform that takes advantage of recent advances in information and communication technology in order to support the person-specific assessment and management of MSDs. With the establishment of such a technological platform, which provides additional functionalities, such as computer-supported analysis at some steps, we aim to facilitate the standard therapeutic procedures typically implemented by medical staff to identify MSDs and the periodic assessment of MSDs progress.

In the following sections, we focus on the methodology behind the design of our technology-supported MSDs monitoring and management platform and the resources required for the development of appropriate technological tools. These technological tools aim to facilitate the implementation of MSDs prevention, monitoring, and management by providing an automated estimation of the body position based on video or photographs and supporting self-assessments via questionnaires, assessing concentration, mental stress, concentration-related work-efficiency, based on physiological signals. In the rest of this section, we briefly outline the ERGO-project objectives in order to set the context for the discussion on the methodology for MSDs assessment and management.

In brief, in the ERGO-project, we focus on developing good practices and preventive measures to reduce the abovementioned risks associated with prolonged work with computers. In this regard, our project's primary goal is to identify the significant risk factors for students and educators. The project also implements specific activities aiming to increase awareness and knowledge about the significant risks and propose effective management of computer use habits. We intend to develop an MSDs prevention program, which would allow us to identify the person-specific risks, specify individual corrective recommendations, and evaluate their effectiveness. For that purpose, a set of indicators for automated detection of MSDs was developed. The ERGO-project also investigates the link between MSDs and stress and their influence on mental states, work effectiveness, and overall health and well-being.

The development of models and technological tools depends on the availability of resources such as datasets of physiological signals, photographs, videos, questionnaires, and posture assessments by medical experts — the photographs in various application scenarios, capturing fixed instances of the postural set. The videos are needed to capture the overall working behavior and register habits causing musculoskeletal changes. Self-assessment questionnaires are used to capture the individual perception of the computer-based working environment's influence on health and well-being. Medical experts' assessments of postures provide the ground-true references required to create statistical body models and automated tools for the detection of deviations from the correct posture. The expected outcome is new ergonomic models of the human-work environment system to prevent health issues and extend and enhance the workforce's productivity.

3 MSDs Management Methodology

Based on the concept presented in Sect. 2, we developed a flexible MSDs management methodology, which accounts for the person-specific context. In order to implement this methodology, we rely on novel signal and image processing algorithms, advanced statistical modeling, and machine-learning methods for the automated assessment of mental states and important body indicators. These technological tools, combined with other resources, such as recommendations, national and EC directives, standards, questionnaires, datasets, purposely-designed sets of stretching exercises, constitute the core of our methodology and the specialized technological platform that implements it. Specifically, the successful management strategy for MSDs prevention requires an iterative approach, which consists of five major steps:

(i) Identification of risks and bad habits, which do not comply with ergonomic principles;
(ii) Evaluation of risks and prioritization of risks exposure;

(iii) Selection of MSDs preventive actions;
(iv) Implementation of preventive actions and protective measures;
(v) Monitoring and regular reassessment of risks.

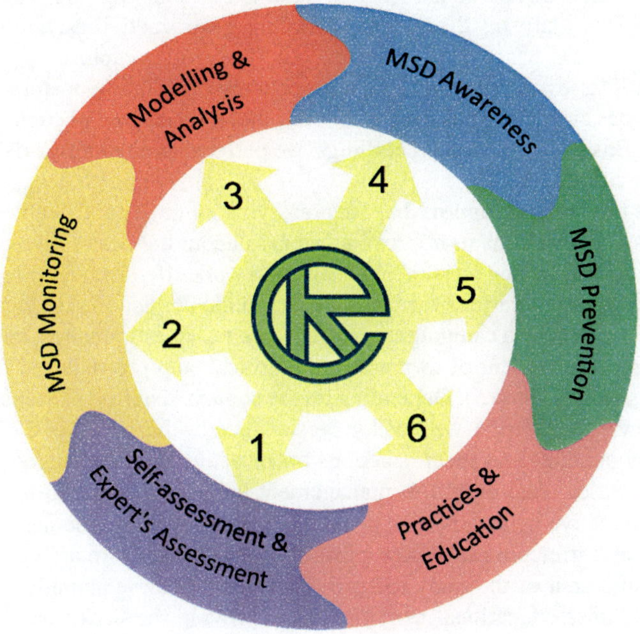

Fig. 2. The methodology of MSDs management

The methodology behind the technological platform developed in support of the MSDs monitoring and management is shown in Fig. 2. Specifically, at Step 1, we first start with an MSDs health-risk assessment procedure that includes several self-reported questionnaires on computer work-related risks, body pain, trunk deformity, and other related health issues. Among these are specialized questionnaires for acquiring general personal information regarding habits, behaviors, and levels of physical activity, such as the WHO-5 Well-being index test [5, 6], World Health Organization Quality of Life-BREF (WHOGOL-BREF 26) [7] and physical activity test. The WHO-5 test provides a five-point scale to assess subjective psychological well-being based on mood, vitality, and general interest. WHOGOL-BREF 26 is a broad, comprehensive instrument for the assessment of the quality of life. Other groups of questionnaires are related to trunk deformity, pain, self-perception of the physical appearance of the body, and indicator of Scoliosis, such as The Trunk Appearance Perception Scale (TAPS) [8, 9] and Scoliosis Research Society-22 SRS-22 [10–12]. The SRS-22 is a widely used tool to assess health-related quality of life (HRQL) in Adolescent Idiopathic Scoliosis (AIS) patients. In addition, an emotion assessment test, such as Positive and Negative Affect Schedule (PANAS) [13], and work-related stress [14] evaluate the overall condition of the person.

An important part of Step 1 is the medical expert-driven MSDs assessment of a person's current health status. In order to identify the most common risks and unhealthy habits from the ergonomic perspective, we collect videos, photos, and physiological signals needed for MSDs status assessment and ergonomic risk monitoring (Step 2). Besides, motion capture data are collected for a posture assessment.

In Step 3, based on the acquired data, automated modeling and analysis are performed. It involves analyzing the body posture during work activities, aiming to identify the presence of MSDs, based on position markers. The motion capture equipment is used as a reference in assessing the precision and accuracy of the body posture models. This step also includes evaluating negative emotional states, the degree of cognitive load, and mental stress. Based on the video recordings, we perform an ergonomic risk assessment of the work environment.

One of the fundamental approaches for preserving longevity and health is to plan and implement proactive actions to improve awareness about the destructive consequences of MSDs and their influence on personal health and work effectiveness (Step 4). The last is implemented through various means, such as mobile applications, educational seminars, awareness promotion campaigns, project websites, questionnaires, and advertising materials. The improvement of awareness is a critical step pre-required for managing and preventing MSDs (Step 5). Depending on the context acquired in Steps 1–3, and the improved knowledge (Step 4), we develop person-specific measures for the management of proper body postures and specific actions for musculoskeletal changes prevention.

The most critical step of MSDs management is the implementation of preventive actions and protective measures. The protective measures are implemented via sets of strengthening and stretching exercises, purposely developed for reducing pain in a specific problematic area of the body (Step 6). Besides the implementation of exercises, Step 6 also includes educational activities for improving the background on the risks associated with MSDs and their negative consequences on health and well-being. The educational activities include filling in questionnaires, participation in seminars, and practical training sessions. A good practice guide is provided for everyday reference to proper body postures and ergonomic guidelines and rules. In addition, a mobile application, referred to as Ergo-assistant, provides the necessary reminders about the recommended behaviors. Long-term risk reduction of MSDs requires continuous monitoring, iterative reassessment of health risks (Step 1), and the implementation of all procedures described at the subsequent Steps 1–6.

4 Development of Resources

The first step of implementing the strategy mentioned above (Fig. 2) is the consolidation of resources needed at the analysis and model establishment steps. In Fig. 3, we present the workflow, which we have used to develop the required resources. As shown in the figure, in order to implement the data collection, we adhered to a four-stage process, consisting of:

(i) Acquainting the participants with the experimental protocol, wearing and adjusting the equipment, as well as attaching markers at specific body parts;

(ii) Setting the participant in a specific body position setup (either sitting with or without a stability disk; full-face or side-view standing position, etc.)
(iii) Performing the actual data collection (capturing videos, taking photographs, acquiring physiological signals, logging the accelerometers data) corresponding to the specific body position setup;
(iv) Storing the person-specific dataset in the purposely-developed file repository and implementing the required data processing and annotation.

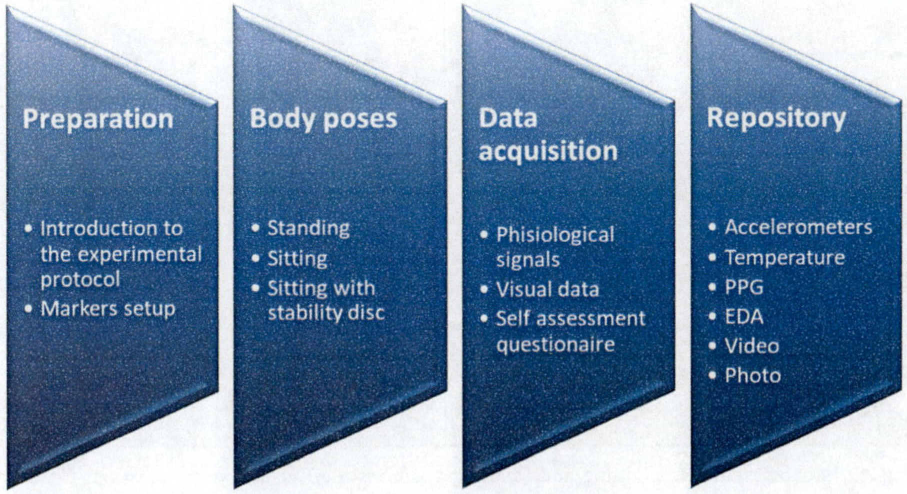

Fig. 3. Workflow for the creation of resources

All videos were captured with a web-camera with high resolution. The photographs were taken with a high-resolution professional camera. A motion capture system (Perception Neuron suit [15]) was used for the accelerometer data acquisition. The physiological signals, such as Photoplethysmography (PPG), Electrodermal activity of skin (EDA), skin temperature, were acquired with an Empatica E4 wristband device [16].

The resources collected in the file repository include:

(i) responses to purposely-designed self-reported questionnaires,
(ii) video recordings, images, and synchronized physiological signals,
(iii) the Stroop test responses.

We use these data for implementing the assessment, monitoring, and analysis specified in Step 1, Step 2, and Step 3, respectively (Fig. 2). In brief, the self-reported questionnaires were implemented via Google Forms in order to collect person-specific responses, as these provide essential information about the personal perspective and self-assessment (Step1). Next, the experts used the photos and videos in the assessment in Step 1. The video recordings, captured during work in the project-specific experimental setup, provide the evidence required for implementing the ergonomic assessment during work in front of a computer. Precisely, the photos were collected in various setups encompassing standing and sitting postures (Fig. 4).

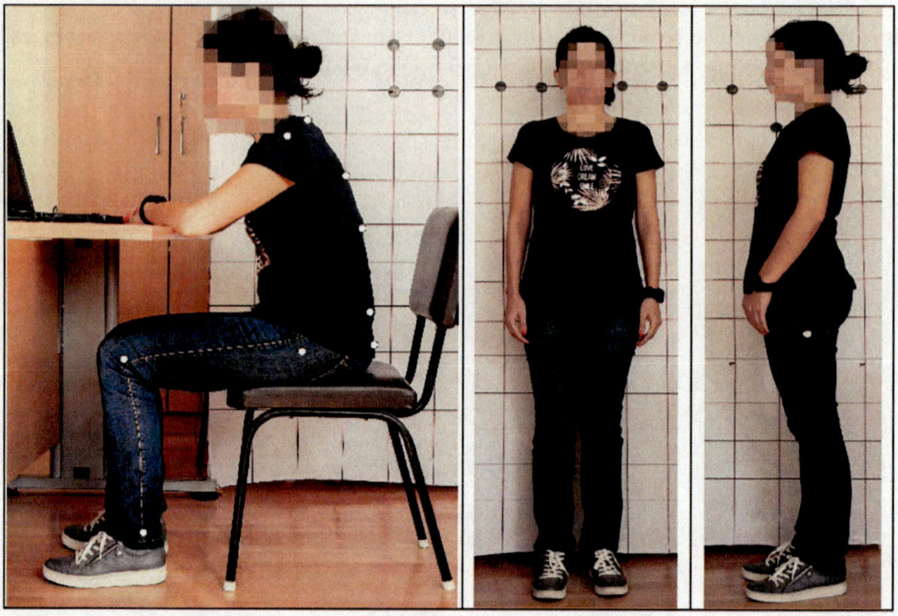

Fig. 4. Experimental setup: sitting and standing positions. *Left*: usual sitting position; *Center*: full-face photograph of standing position; *Right*: side-view photograph of standing position.

Each of these setups incorporates a set of pre-specified body postures. Specifically, in the sitting position, we simulated a typical computer workplace to evaluate person-specific behaviors and habits during work. In the sitting position setup, we collected photos, videos, and physiological signals, while in the standing position, we only took photos. Various physiological signals, such as the electrodermal response of the skin (EDA), Photoplethysmogram (PPG), skin temperature (T), as well as accelerometer data were collected. For that purpose, we used two different body positions: sitting with and without stability disk. Both were used in an experimental protocol based on the Stroop test, which we implemented in concentration and work efficiency models.

We collected various full-face and side view photographs (Fig. 4, *Center*, *Right*) of standing positions, which are used to assess whether there are any deviations from the gravity line and evaluate common postural tendencies. The full-face photographs are used to assess vertical and horizontal body posture alignment, as any observable misalignments might be an indicator for musculoskeletal changes.

Furthermore, the resources collected in Steps 1 and 2 serve as the basis for the development of tools, models, and applications required for the implementation of the methodology of MSDs management (Step 3). Specifically, the photographs are used in the computation of various indicators for postural misalignments, which are then incorporated in statistical models. The availability of such models would permit the automated recognition of the correct body postures in a computer-based working environment. Likewise, the video recordings are used in the ergonomic risk assessment. In Step 3, the aforementioned physiological signals are used to model mental states and develop an automated technology capable of recognizing the degree of concentration and evaluating concentration-related work efficiency.

In the long run, the synchronized recordings of heterogeneous data are necessary in order to evaluate our hypothesis that ergonomic working environment and the correct body posture are related to the physiological states such as work-related stress and the capacity for preserving efficiency, concentration, and high mental workload capacity. The prevention of MSDs and the proper monitoring and management of MSDs are preconditions for the long-term preservation of health and well-being.

5 Conclusion

In the current paper, we address the management of work-related health issues specific to computer users. Specifically, we present the conceptual design of MSDs management methodology, involving the complete cycle of information acquisition, risk assessment, and purposely-designed person-specific measures for reverting the negative consequences to health. The ERGO-project developed a modular technological platform, resources, and tools to implement this methodology. The proposed management methodology can help monitor MSDs-related indicators, and subsequently, cope with work-related stress, improve body posture, and enhance the ergonomic work environment. Summarizing the results of this study, we provide technical support to the long-term efforts to prevent, monitor, and manage MSDs, improving work-efficiency, preserving health, and enhancing the overall quality of life.

Acknowledgments. This research was funded by Bulgarian National Science Fund (BNSF), under grant contract FNI № KP-06-N37/1, entitled "Ergonomic research on work-related health problems by innovative computer models with a focus on the prevention of Musculoskeletal Disorders."

References

1. Sharan, D.: Musculoskeletal disorders in 115 students due to overuse of electronic devices: risk factors and clinical features. In: Bagnara, S., Tartaglia, R., Albolino, S., Alexander, T., Fujita, Y. (eds.) Proceedings of the 20th Congress of the International Ergonomics Association (IEA 2018). IEA 2018. Advances in Intelligent Systems and Computing, vol. 826, pp. 828–829. Springer, Cham (2019). https://doi.org/10.1007/978-3-319-96065-4_87
2. EU-Osha Homepage. http://www.osha.europa.eu/bg/healthy-workplaces-campaigns/2016-17-campaign-healthyworkplaces-all-ages. Accessed 29 Oct 2020
3. EU-Osha Homepage. http://www.osha.europa.eu/en/tools-and-publications/publications/reports/healthy-workplaces-goodpractice-awards-2014-2015. Accessed 29 Oct 2020
4. EU-Osha Homepage. http://www.osha.europa.eu/en/healthy-workplaces-campaigns/awards. Accessed 29 Oct 2020

5. Topp, C.W., Østergaard, S.D., Søndergaard, S., Bech, P.: The WHO-5 well-being index: a systematic review of the literature. Psychother. Psychosom. **84**, 167–176 (2015).https://doi.org/10.1159/000376585
6. Bech, P.: Measuring the dimensions of psychological general well-being by the WHO-5. QoL Newsl. **32**, 15–16 (2004)
7. WHO Website. http://www.who.int/substance_abuse/research_tools/whoqolbre. Accessed 21 Nov 2016
8. Bago, J., Sanchez-Raya, J., Perez-Grueso, F.J.S., Climent, J.M.: The trunk appearance perception scale (TAPS): a new tool to evaluate subjective impression of trunk deformity in patients with idiopathic Scoliosis. Scoliosis (2010)
9. Rigo, M., D'agata, E., Jelacic, M.: Trunk appearance perception scale (TAPS) discrepancy between scoliosis children and their parents influence the SRS-22 score. Scoliosis **7**(Suppl1), 03 (2021)
10. Asher, M., Lai, S.M., Burton, D., Manna, B.: The influence of spinal and trunk deformity on preoperative idiopathic scoliosis patient's health-related quality of life questionnaire responses. Spine **29**, 861–888 (2004)
11. Asher, M.A., Lai, S.M., Glattes, R.C., Burton, D.C., Alanay, A., Bago, J.: Refinement of the SRS-22 health-related quality of life questionnaire function domain. Spine **31**(5), 593–597 (2006)
12. Asher, M., Min Lai, S., Burton, D., Manna, B.: Scoliosis research society–22 patient questionnaire: responsiveness to change associated with surgical treatment. Spine **28**(1), 70–73 (2003)
13. Díaz-García, A., González-Robles, A., Mor, S., et al.: Positive and negative affect schedule (PANAS): psychometric properties of the online Spanish version in a clinical sample with emotional disorders. BMC Psychiatry **20**, 56 (2020)
14. Vachinska, S., Usheva, N.: Tools for analysis and assessment of health behavior and stress in higher education. In: Proceeding of International Conference UNITEX 2012, vol. 3, pp. 288–291 (2012)
15. Perception Neuron Series. https://www.noitom.com/perception-neuron-series. Accessed 18 Dec 2020
16. Empatica E4 wristband. https://empatica.com/research/e4/. Accessed 18 Dec 2020

Author Index

S. S. Sotirov et al. (Eds.): BioInfoMed 2020, LNNS 374, pp. 455–456, 2022.
https://doi.org/10.1007/978-3-030-96638-6